Applied Mathematics II

Applied Mathematics II

L. Bostock, B.Sc.

Senior Mathematics Lecturer,
Southgate Technical College.

S. Chandler, B.Sc.

Lecturer in Mathematics,
Southgate Technical College.

Stanley Thornes (Publishers) Ltd.

First published in 1976 by Stanley Thornes (Publishers) Ltd., Educa House, 32 Malmesbury Road, Kingsditch Estate, CHELTENHAM GL51 9PL

Reprinted with minor corrections 1977

ISBN 0 85950 024-1

Composed at the Alden Press
Oxford London and Northampton
Printed in Great Britain by offset lithography by
Billing & Sons Ltd, Guildford, London and Worcester

PREFACE

Volume II of APPLIED MATHEMATICS is a continuation of the work covered in Volume I and is intended to complete a full two year course in Mechanics. It caters for the mechanics section of such advanced level schemes as the University of London Syllabus 'D' in Further Mathematics as well as completing the work necessary for the Applied Mathematics syllabuses of most examining boards.

The text includes a very full analysis of vector algebra, and vector methods are used for solving a wide range of problems. The properties of straight lines and planes are dealt with using both vector methods and three dimensional coordinate geometry, the analogy between these forms of analysis being drawn at each stage of the work.

Certain topics introduced in Volume I are developed further and the solution of the problems in these more difficult sections requires a working knowledge of the solution of differential equations and a reasonable degree of competence in integration techniques. These sections contain much material which is particularly suitable for mathematical models.

Many worked examples are incorporated in the text to illustrate each main development of a topic and a set of straightforward problems follows each section. A selection of more demanding questions is given in the miscellaneous exercise at the end of each chapter while, as in Volume I, multiple choice exercises are included when appropriate.

We wish to express our sincere thanks to those friends and colleagues whose suggestions, criticisms and calculations have been most helpful. We are grateful to the following Examination Boards for their permission to reproduce questions from past examination papers:

The Associated Examining Board
Joint Matriculation Board
University of London University Entrance and School Examination Council
University of Cambridge Local Examinations Syndicate
Oxford Delegacy of Local Examinations
Southern Universities' Joint Board
Welsh Joint Education Committee

L. Bostock
S. Chandler

CONTENTS

NOTES ON USE OF THE BOOK

1. Notation Used in Diagrams

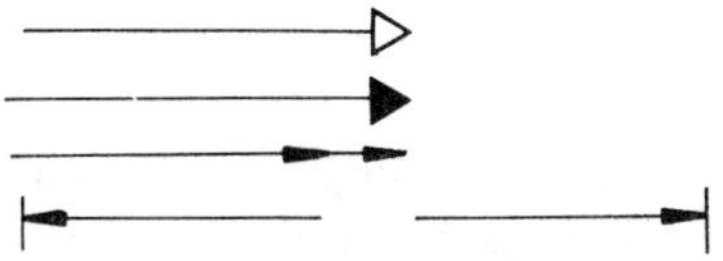

Force
Velocity
Acceleration
Dimensions

Where components and resultant are shown in one diagram the resultant is denoted by the larger arrow-head thus: 

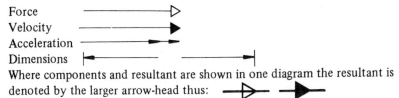

2. Instructions for Answering Multiple Choice Exercises

These exercises are at the end of most chapters. The questions are set in groups, each group representing one of the variations that may arise in examination papers. The answering techniques are different for each type of question and are classified as follows:

TYPE I

These questions consist of a problem followed by several alternative answers, only *one* of which is correct.
Write down the letter corresponding to the correct answer.

TYPE II

In this type of question some information is given and is followed by a number of possible responses. *One or more* of the suggested responses follow(s) directly from the information given.
Write down the letter(s) corresponding to the correct response(s).
e.g.: PQR is a triangle.
(a) $\hat{P} + \hat{Q} + \hat{R} = 180^\circ$.
(b) PQ + QR is less than PR.
(c) If $\hat{P}$ is obtuse, $\hat{Q}$ and $\hat{R}$ must both be acute.
(d) $\hat{P} = 90^\circ$, $\hat{Q} = 45^\circ$, $\hat{R} = 45^\circ$.
The correct responses are (a) and (c).
(b) is definitely incorrect and (d) may or may not be true of the triangle PQR.

There is not sufficient information given to allot a particular value to each angle. Responses of this kind, which require more information than is given, should not be regarded as correct.

TYPE III

Each problem contains two independent statements (a) and (b).

1) If (a) implies (b) but (b) does not imply (a)	write *A*.
2) If (b) implies (a) but (a) does not imply (b)	write *B*.
3) If (a) implies (b) *and* (b) implies (a)	write *C*.
4) If (a) denies (b) *and* (b) denies (a)	write *D*.
5) If none of the first four relationships apply	write *E*.

TYPE IV

A problem is introduced and followed by a number of pieces of information. You are not required to solve the problem but to decide whether:

1) the total amount of information given is insufficient to solve the problem. If so write *I*,

2) the given information is *all* needed to solve the problem. In this case write *A*,

3) the problem can be solved without using one or more of the given pieces of information. In this case write down the letter(s) corresponding to the items not needed.

TYPE V

A single statement is made. Write *T* if the statement is true and *F* if the statement is false.

CHAPTER 1

PROPERTIES OF VECTORS. STRAIGHT LINES IN THREE DIMENSIONS

BASIC PROPERTIES OF VECTORS

A quantity whose definition includes both magnitude and direction is a vector. Such quantities occur frequently in mechanics (e.g. force, velocity, acceleration) so the analysis of vectors is of fundamental importance in the study of mechanics. The basic properties of vectors were discussed in Volume One and these are now summarised.

1. A vector can be represented in magnitude and direction by a segment of a line.
2. A *free* vector has no specific location and can be represented by any one of a set of equal and parallel line segments. A *tied* vector has a particular location in space.
3. If $\mathbf{a}$ and $\mathbf{b}$ are two vectors such that $\mathbf{a} = \lambda\mathbf{b}$, then $\mathbf{a}$ and $\mathbf{b}$ are parallel.
   When λ is positive $\mathbf{a}$ and $\mathbf{b}$ have the same direction.
   When λ is negative $\mathbf{a}$ and $\mathbf{b}$ are opposite in direction.
   When $\lambda = 1$, $\mathbf{a}$ and $\mathbf{b}$ are *equal*.
   When $\lambda = -1$, $\mathbf{a}$ and $\mathbf{b}$ are *equal and opposite*.
4. If lines representing vectors in magnitude and direction are drawn consecutively, the line which completes the polygon represents the resultant vector. This property can be expressed as a vector equation.

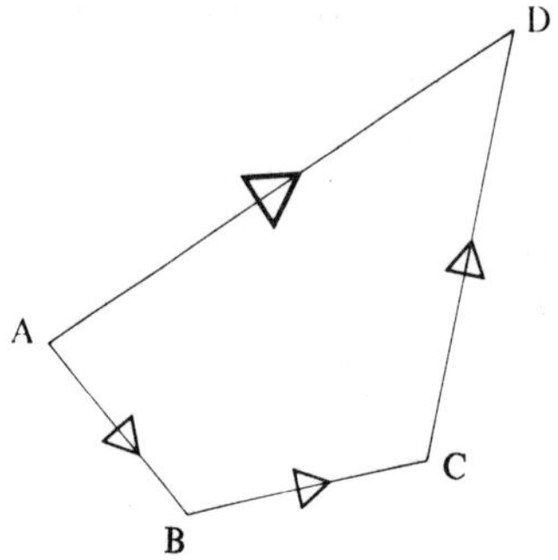

$$\overrightarrow{AD} = \overrightarrow{AB} + \overrightarrow{BC} + \overrightarrow{CD}$$

(A, B, C and D are not necessarily co-planar.)

$\overrightarrow{AD}$ represents the resultant vector.
$\overrightarrow{AB}$, $\overrightarrow{BC}$ and $\overrightarrow{CD}$ represent the components of $\overrightarrow{AD}$.
Thus: *A vector is equal to the vector sum of its components.*

5. If two vectors are represented by $\lambda\overrightarrow{OA}$ and $\mu\overrightarrow{OB}$, their resultant can be represented by $(\lambda + \mu)\overrightarrow{OC}$ where C is a point on AB such that $AC:CB = \mu:\lambda$.

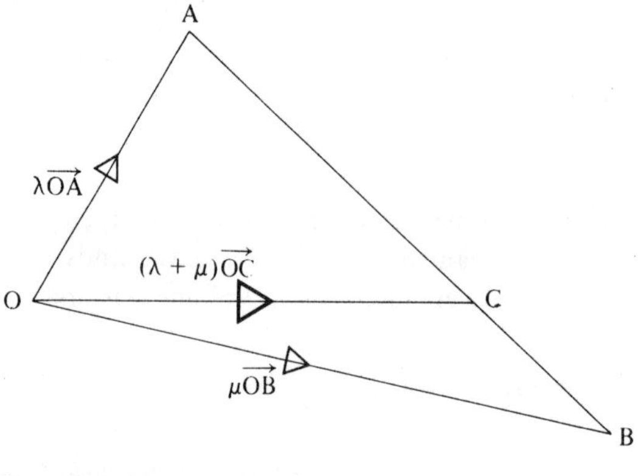

Cartesian Components

In three dimensional space any vector can be expressed in terms of its components in each of three perpendicular directions.

Suppose that Ox, Oy and Oz are three mutually perpendicular axes and:

i represents a vector of magnitude 1 unit in the direction Ox
j represents a vector of magnitude 1 unit in the direction Oy
k represents a vector of magnitude 1 unit in the direction Oz

Now consider a vector **V** whose components parallel to Ox, Oy and Oz are of magnitudes a, b and c respectively.
These components can be defined both in magnitude and direction by $a\mathbf{i}$, $b\mathbf{j}$ and $c\mathbf{k}$ respectively.

But a vector is equivalent in magnitude and direction to the vector sum of its components.

Therefore $$\mathbf{V} = a\mathbf{i} + b\mathbf{j} + c\mathbf{k}$$

The vector **V** can be represented by a line OP where P has co-ordinates (a, b, c) relative to O.

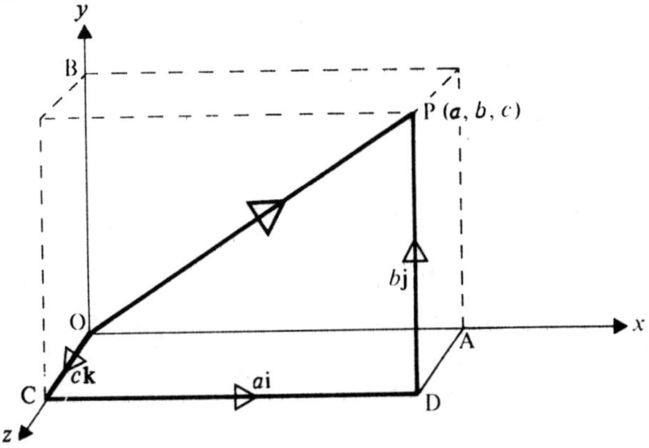

Note: OCDP is a vector polygon for which the corresponding vector equation is

$$\overrightarrow{OP} = \overrightarrow{OC} + \overrightarrow{CD} + \overrightarrow{DP}$$

Modulus

The magnitude or *modulus* of **V** is written as $|\mathbf{V}|$ or V and is represented by the length d of OP.

Hence $$|\mathbf{V}| = d = \sqrt{a^2 + b^2 + c^2}$$

Therefore, by its definition, $|\mathbf{V}|$ is always positive.

Parallel Vectors

If two vectors $\mathbf{V}_1$ and $\mathbf{V}_2$ are parallel then

$$\mathbf{V}_1 = \lambda \mathbf{V}_2$$

If λ is positive:

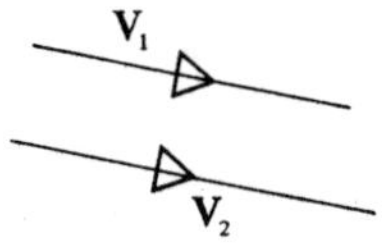

$\mathbf{V}_1$ and $\mathbf{V}_2$ are in the same sense and are said to have the *same direction.* $\mathbf{V}_1$ and $\mathbf{V}_2$ are called *like parallel vectors.*

If λ is negative

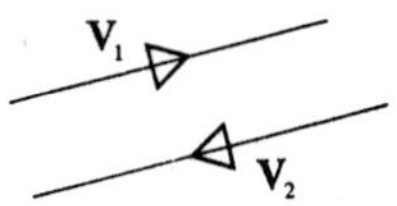

$\mathbf{V}_1$ and $\mathbf{V}_2$ are in opposite senses and are said to have *opposite directions.* $\mathbf{V}_1$ and $\mathbf{V}_2$ are called *unlike parallel vectors.*

Thus the statement that two vectors are parallel is ambiguous.
The statement that two vectors have the same direction, however, has only one meaning.

Direction Ratios and Direction Cosines

Now $a:b:c$ is the ratio of the components of **V** parallel to Ox, Oy and Oz respectively. As this ratio determines the direction of **V** relative to the axes, a, b and c are called the *direction ratios* of **V**.

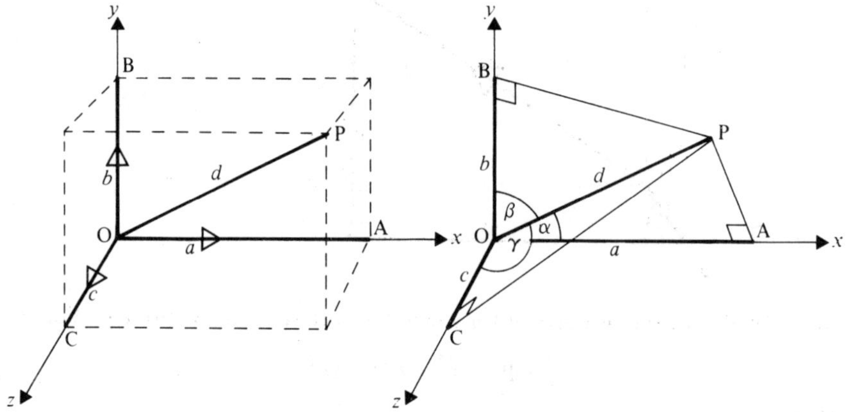

In the diagram, lines PA, PB and PC are drawn from P perpendicular to Ox, Oy and Oz respectively, so that angles PAO, PBO and PCO are each 90°.
But OA = a, OB = b, OC = c and OP = d.
So if OP makes angles α, β and γ with Ox, Oy and Oz respectively, then

$$\cos\alpha = \frac{a}{d}, \quad \cos\beta = \frac{b}{d}, \quad \cos\gamma = \frac{c}{d}$$

But the angles α, β and γ determine the direction of OP, so their cosines are called the *direction cosines* of **V** and are often given the symbols l, m and n, so that:

$$\cos\alpha = \frac{a}{d} = l$$

$$\cos\beta = \frac{b}{d} = m$$

$$\cos\gamma = \frac{c}{d} = n$$

Now $$l^2 + m^2 + n^2 = \left(\frac{a}{d}\right)^2 + \left(\frac{b}{d}\right)^2 + \left(\frac{c}{d}\right)^2$$

$$= \frac{a^2 + b^2 + c^2}{d^2}$$

But $$a^2 + b^2 + c^2 = d^2$$

Hence $$l^2 + m^2 + n^2 = 1$$

Thus *the sum of the squares of the direction cosines of any vector is unity*.

Direction Ratios and Direction Cosines of Parallel Vectors

Consider two parallel vectors $\mathbf{V}_1$ and $\mathbf{V}_2$

$$\mathbf{V}_2 = \lambda\mathbf{V}_1$$

Hence if $$\mathbf{V}_1 = a\mathbf{i} + b\mathbf{j} + c\mathbf{k}$$

then $$\mathbf{V}_2 = \lambda a\mathbf{i} + \lambda b\mathbf{j} + \lambda c\mathbf{k}$$

Now the direction ratios of $\mathbf{V}_1$ are $a:b:c$

and the direction ratios of $\mathbf{V}_2$ are $\lambda a:\lambda b:\lambda c$

But $\lambda a:\lambda b:\lambda c = a:b:c$ whatever the value of λ.

Hence parallel vectors have equal direction ratios.

Now considering the direction cosines of $\mathbf{V}_1$ and $\mathbf{V}_2$ we see that:

$\mathbf{V}_1$ has direction cosines $$\frac{a}{|\mathbf{V}_1|}, \frac{b}{|\mathbf{V}_1|}, \frac{c}{|\mathbf{V}_1|}$$

$\mathbf{V}_2$ has direction cosines $$\frac{\lambda a}{|\lambda\mathbf{V}_1|}, \frac{\lambda b}{|\lambda\mathbf{V}_1|}, \frac{\lambda c}{|\lambda\mathbf{V}_1|}$$

As $|\lambda\mathbf{V}_1|$ is always positive, $\mathbf{V}_1$ and $\mathbf{V}_2$ have the same direction cosines when λ is positive. But if λ is negative the direction cosines of $\mathbf{V}_2$ are opposite in sign to those of $\mathbf{V}_1$.

Hence *parallel vectors have equal direction ratios.*
Whereas *like parallel vectors have equal direction cosines* but *unlike parallel vectors have direction cosines equal in magnitude but opposite in sign*.

It follows that the direction ratios of a vector are not unique but its direction cosines are unique.

Unit Vector

A vector of magnitude 1 unit is a unit vector ($\mathbf{i}$, $\mathbf{j}$ and $\mathbf{k}$ are all unit vectors).

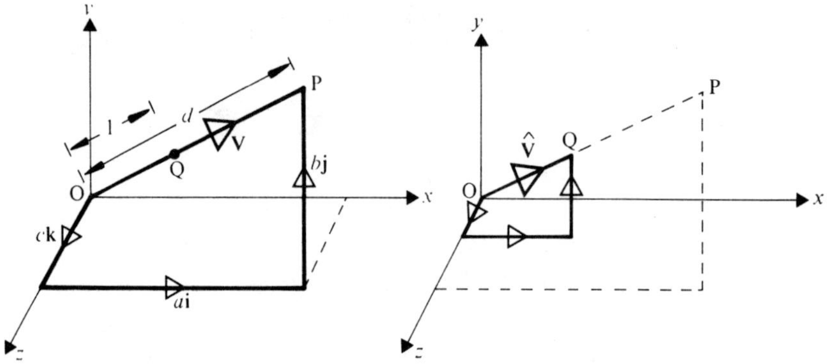

Consider a line OQP where $\overrightarrow{OP}$ represents a vector **V** and Q is distant 1 unit from O.
Then OQ represents a unit vector in the direction of **V**.
Such a unit vector is written $\hat{\mathbf{V}}$.

Now if	$OP = d$ units
and	$OQ = 1$ unit
then	$\overrightarrow{OP} = d\,\overrightarrow{OQ}$
But	$d = \lvert\mathbf{V}\rvert$
Hence	$\mathbf{V} = \lvert\mathbf{V}\rvert\,\hat{\mathbf{V}}$

This important relationship can be stated as follows:

Any vector is the product of its magnitude and a unit vector in the correct direction.

Now if P is the point (a, b, c) so that

$$\mathbf{V} = a\mathbf{i} + b\mathbf{j} + c\mathbf{k}$$

Then since $$\hat{\mathbf{V}} = \mathbf{V}/d$$

$$\hat{\mathbf{V}} = \frac{a}{d}\mathbf{i} + \frac{b}{d}\mathbf{j} + \frac{c}{d}\mathbf{k}$$

Hence $\hat{\mathbf{V}} = l\mathbf{i} + m\mathbf{j} + n\mathbf{k}$ showing that:

(i) the magnitude of the components of $\hat{\mathbf{V}}$ parallel to **i**, **j** and **k** are the direction cosines of **V**,

(ii) the co-ordinates of Q are (l, m, n).

Direction Vector

A vector which is used to specify the direction of another vector can be called a *direction vector*.

For example if we are told that a vector **V** of magnitude 14 units is in the direction of the vector 3**i** + 6**j** + 2**k** then 3**i** + 6**j** + 2**k** is the direction vector for **V**.

The unit direction vector $= \hat{\mathbf{V}}$

So $\quad \mathbf{V} = |\mathbf{V}| \times (\text{unit direction vector})$

In this case $\quad \mathbf{V} = 14\left(\dfrac{3\mathbf{i}+6\mathbf{j}+2\mathbf{k}}{7}\right) = 6\mathbf{i}+12\mathbf{j}+4\mathbf{k}$

Summarising, we have:

If $\quad \mathbf{V} = a\mathbf{i}+b\mathbf{j}+c\mathbf{k}$

$$|\mathbf{V}| = \sqrt{(a^2+b^2+c^2)} = d$$

V has direction ratios $a:b:c$

V has direction cosines $\dfrac{a}{d}, \dfrac{b}{d}, \dfrac{c}{d}$

or l, m, n

$$\hat{\mathbf{V}} = \mathbf{V}/d = l\mathbf{i}+m\mathbf{j}+n\mathbf{k}$$

$$l^2+m^2+n^2 = 1$$

EXAMPLES 1a

1) Find the modulus and direction cosines of the vector 2**i** + 3**j** − 6**k**.

$$\mathbf{V} = 2\mathbf{i}+3\mathbf{j}-6\mathbf{k}$$

Therefore $\quad |\mathbf{V}| = \sqrt{2^2+3^2+(-6)^2} = 7$

V has direction ratios $2:3:-6$

Hence $\quad$ **V** has direction cosines $\frac{2}{7}, \frac{3}{7}, -\frac{6}{7}$

2) A vector **V** is inclined to Ox at 45° and to Oy at 60°. Find its inclination to Oz. If the magnitude of **V** is 12 units, express **V** in the form a**i** + b**j** + c**k**.

The direction cosines of **V** are cos 45°, cos 60°, cos γ

So, $\quad l = \dfrac{1}{\sqrt{2}}, \quad m = \dfrac{1}{2}, \quad n = \cos\gamma$

But $\quad l^2+m^2+n^2 = 1$

Hence $\quad n^2 = 1-\frac{1}{2}-\frac{1}{4} = \frac{1}{4}$

$$n = \pm\tfrac{1}{2} = \cos\gamma$$

Therefore **V** is inclined to Oz either at 60° or at 120°.

Now $\quad l\mathbf{i}, m\mathbf{j}$ and $n\mathbf{k}$ are the components of $\hat{\mathbf{V}}$.

So
$$\hat{\mathbf{V}} = \frac{1}{\sqrt{2}}\mathbf{i} + \frac{1}{2}\mathbf{j} \pm \frac{1}{2}\mathbf{k}$$

But
$$\mathbf{V} = |\mathbf{V}|\hat{\mathbf{V}}$$

Hence
$$\mathbf{V} = 12\left(\frac{1}{\sqrt{2}}\mathbf{i} + \frac{1}{2}\mathbf{j} \pm \frac{1}{2}\mathbf{k}\right)$$
$$\mathbf{V} = 6\sqrt{2}\mathbf{i} + 6\mathbf{j} \pm 6\mathbf{k}$$

3) Express in the form $a\mathbf{i} + b\mathbf{j} + c\mathbf{k}$ a force vector **F** of magnitude 9 newtons which is parallel to the vector $2\mathbf{i} - \mathbf{j} + 2\mathbf{k}$.

The unit vector in the direction of $2\mathbf{i} - \mathbf{j} + 2\mathbf{k}$ is $\dfrac{2\mathbf{i} - \mathbf{j} + 2\mathbf{k}}{\sqrt{2^2 + (-1)^2 + 2^2}}$

Parallel vectors have either the same or opposite directions.

Therefore the unit vector in the direction of **F** is $\pm\dfrac{2\mathbf{i} - \mathbf{j} + 2\mathbf{k}}{3}$

But
$$\mathbf{F} = |\mathbf{F}|\hat{\mathbf{F}}$$

Hence
$$\mathbf{F} = \pm(9)\left(\frac{2\mathbf{i} - \mathbf{j} + 2\mathbf{k}}{3}\right)$$

Giving
$$\mathbf{F} = \pm(6\mathbf{i} - 3\mathbf{j} + 6\mathbf{k})$$

Note: In the example above $2\mathbf{i} - \mathbf{j} + 2\mathbf{k}$ is the direction vector of **F**.

EXERCISE 1a

1) Write down, in the form $a\mathbf{i} + b\mathbf{j} + c\mathbf{k}$, the vector represented by $\overrightarrow{OP}$ if P is a point with co-ordinates:
(i) (3, 6, 4), (ii) (1, −2, −7), (iii) (1, 0, −3).

2) $\overrightarrow{OP}$ represents a vector **V**. Write down the co-ordinates of P if:
(i) $\mathbf{V} = 5\mathbf{i} - 7\mathbf{j} + 2\mathbf{k}$,
(ii) $\mathbf{V} = \mathbf{i} + 4\mathbf{j}$,
(iii) $\mathbf{V} = \mathbf{j} - \mathbf{k}$.

3) Find the modulus and direction cosines of the vector **V** if:
(i) $\mathbf{V} = 2\mathbf{i} - 4\mathbf{j} + 4\mathbf{k}$,
(ii) $\mathbf{V} = 6\mathbf{i} + 2\mathbf{j} - 3\mathbf{k}$,
(iii) $\mathbf{V} = 11\mathbf{i} - 7\mathbf{j} - 6\mathbf{k}$.

4) Find the vector **V** if:
(i) $\mathbf{V} = \overrightarrow{OP}$ and P is the point (0, 4, 5),

(ii) $|\mathbf{V}| = 24$ units and $\hat{\mathbf{V}} = \frac{2}{3}\mathbf{i} - \frac{2}{3}\mathbf{j} - \frac{1}{3}\mathbf{k}$,
(iii) **V** is inclined at 60° to Oy and at 60° to Oz and is of magnitude 8 units,
(iv) **V** is parallel to the vector $8\mathbf{i} + \mathbf{j} + 4\mathbf{k}$ and is equal in magnitude to the vector $\mathbf{i} - 2\mathbf{j} + 2\mathbf{k}$.

5) Find the magnitude and the inclination to each of the co-ordinate axes of a vector **V** if:
(i) $\mathbf{V} = 3\mathbf{i} + 4\mathbf{j} + 5\mathbf{k}$,
(ii) $\mathbf{V} = -\mathbf{i} + \mathbf{j} - \mathbf{k}$,
(iii) **V** is represented by $\overrightarrow{OP}$ where P is the point (5, 1, 4).

6) Find $\hat{\mathbf{V}}$ in the form $a\mathbf{i} + b\mathbf{j} + c\mathbf{k}$ if:
(i) $\mathbf{V} = \mathbf{i} - \mathbf{j} + \mathbf{k}$,
(ii) $\mathbf{V} = \overrightarrow{OP}$ and P is the point (3, 2, −6).

7) A vector **V** is inclined at equal acute angles to Ox, Oy and Oz. If the magnitude of **V** is 6 units, find **V**.

8) Express in Cartesian vector form:
(i) a force vector of magnitude 14 N and with direction vector $6\mathbf{i} - 2\mathbf{j} + 3\mathbf{k}$,
(ii) the velocity of a particle moving with speed 27 ms^{-1} along a line with direction ratios 4 : 1 : 8.

Free Vector

A free vector is not located in any particular position. If a free vector **V** can be represented by $\overrightarrow{OP}$, it can equally well be represented by $\overrightarrow{AB}$ where OP and AB are equal in length and are in the same direction (i.e. $\overrightarrow{AB} = \overrightarrow{OP}$).

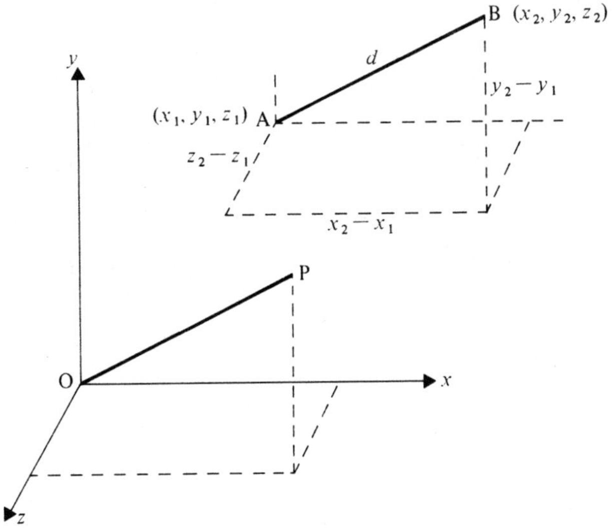

If A is the point (x_1, y_1, z_1) and B is the point (x_2, y_2, z_2) then the components of AB parallel to Ox, Oy and Oz are $(x_2 - x_1)$, $(y_2 - y_1)$ and $(z_2 - z_1)$ respectively.

Now AB = OP = |**V**| = d

Also AB is in the same direction as OP and so has the same direction cosines.

$$\text{Hence} \quad \cos\alpha = \frac{a}{d} = \frac{x_2 - x_1}{d}$$

$$\cos\beta = \frac{b}{d} = \frac{y_2 - y_1}{d}$$

$$\cos\gamma = \frac{c}{d} = \frac{z_2 - z_1}{d}$$

Position Vector

The position of any point P can be defined by the displacement vector $\overrightarrow{OP}$. The vector $\overrightarrow{OP}$ is the *position vector* of P relative to O and is unique.
Hence a point A, with co-ordinates (x_1, y_1, z_1), has a position vector $\overrightarrow{OA}$ where:

$$\overrightarrow{OA} = x_1\mathbf{i} + y_1\mathbf{j} + z_1\mathbf{k}$$

A position vector is a tied vector, i.e. it can be represented by one line segment only, unlike a free vector which can be represented by any one of a set of equal like parallel line segments.

For instance, any of the lines in diagram (i) is a true vector representation of a velocity vector 5**i** + 4**j** + 3**k**.

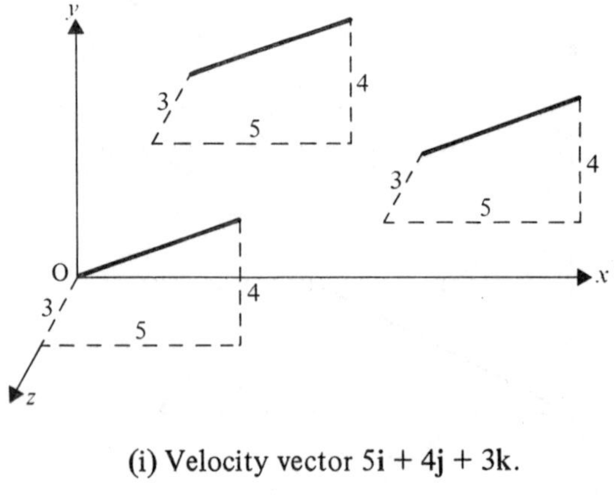

(i) Velocity vector 5**i** + 4**j** + 3**k**.

But a position vector $5\mathbf{i} + 4\mathbf{j} + 3\mathbf{k}$ can be represented only by $\overrightarrow{OA}$ in diagram (ii).

(ii) Position vector $5\mathbf{i} + 4\mathbf{j} + 3\mathbf{k}$

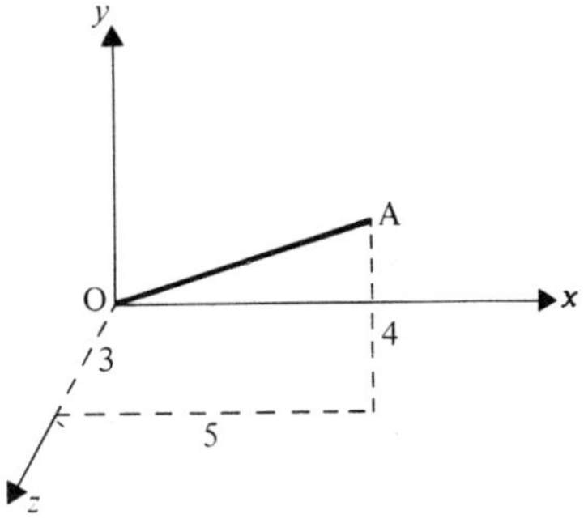

RESULTANT VECTORS

The resultant of a set of vectors is the vector sum of the set.
If each of the given vectors is expressed in the form $\mathbf{V}_r = x_r\mathbf{i} + y_r\mathbf{j} + z_r\mathbf{k}$, $(r = 1, 2, 3, \ldots, n)$ then the resultant vector $\mathbf{V}$ is found using

$$\mathbf{V} = \sum_1^n \left\{ x_r\mathbf{i} + y_r\mathbf{j} + z_r\mathbf{k} \right\} = \sum_1^n x_r\mathbf{i} + \sum_1^n y_r\mathbf{j} + \sum_1^n z_r\mathbf{k}$$

Suppose, for example, that three force vectors, $\mathbf{F}_1 = 2\mathbf{i} + 4\mathbf{j} - 5\mathbf{k}$, $\mathbf{F}_2 = 6\mathbf{i} + \mathbf{j} + 3\mathbf{k}$ and $\mathbf{F}_3 = \mathbf{i} - 3\mathbf{j} + \mathbf{k}$ act on a particle. The resultant force $\mathbf{F}$ acting on the particle is given by

$$\mathbf{F} = (2\mathbf{i} + 4\mathbf{j} - 5\mathbf{k}) + (6\mathbf{i} + \mathbf{j} + 3\mathbf{k}) + (\mathbf{i} - 3\mathbf{j} + \mathbf{k})$$

$$\mathbf{F} = 9\mathbf{i} + 2\mathbf{j} - \mathbf{k}$$

PROPERTIES OF A LINE JOINING TWO POINTS

Consider a line AB joining two points whose positions vectors are $\mathbf{a}$ and $\mathbf{b}$ respectively (i.e. $\overrightarrow{OA} = \mathbf{a}$ and $\overrightarrow{OB} = \mathbf{b}$).

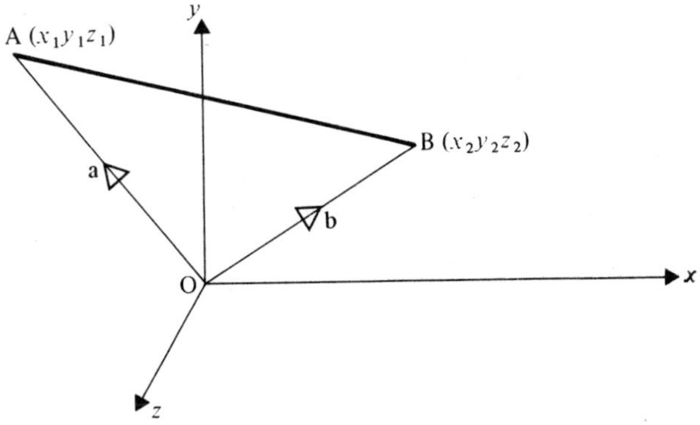

Modulus of $\overrightarrow{AB}$

$$\overrightarrow{AB} = \overrightarrow{AO} + \overrightarrow{OB}$$
$$= -\mathbf{a} + \mathbf{b}$$

Therefore
$$|\overrightarrow{AB}| = |\mathbf{b} - \mathbf{a}|$$

If A and B have co-ordinates (x_1, y_1, z_1) and (x_2, y_2, z_2)

then
$$\overrightarrow{AB} = (x_2\mathbf{i} + y_2\mathbf{j} + z_2\mathbf{k}) - (x_1\mathbf{i} + y_1\mathbf{j} + z_1\mathbf{k})$$
$$= (x_2 - x_1)\mathbf{i} + (y_2 - y_1)\mathbf{j} + (z_2 - z_1)\mathbf{k}$$

So
$$|\overrightarrow{AB}| = \sqrt{(x_2 - x_1)^2 + (y_2 - y_1)^2 + (z_2 - z_1)^2}$$

Direction Cosines of $\overrightarrow{AB}$

The direction cosines of $\overrightarrow{AB}$ are

$$\frac{x_2 - x_1}{|\overrightarrow{AB}|}, \frac{y_2 - y_1}{|\overrightarrow{AB}|}, \frac{z_2 - z_1}{|\overrightarrow{AB}|} \quad \text{(see p. 10)}$$

Note. Subtracting the coordinates of A and B in the order $x_1 - x_2, y_1 - y_2, z_1 - z_2$ gives the direction cosines of $\overrightarrow{BA}$.

Mid-Point of AB

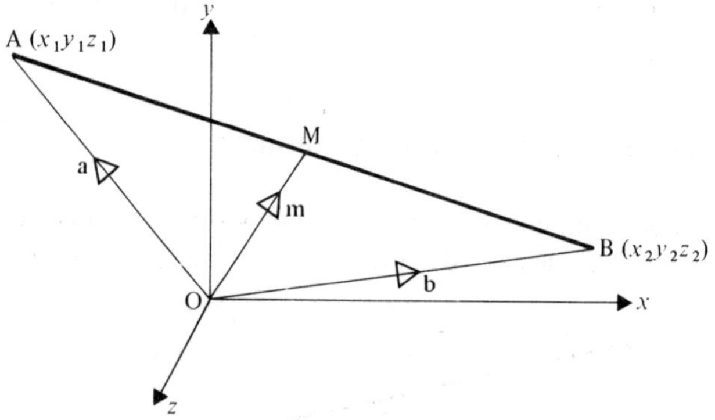

The position vector of M, the mid-point of AB, is $\overrightarrow{OM}$ where

$$\overrightarrow{OM} = \overrightarrow{OA} + \overrightarrow{AM}$$
$$= \overrightarrow{OA} + \tfrac{1}{2}\overrightarrow{AB}$$
$$= \mathbf{a} + \tfrac{1}{2}(\mathbf{b} - \mathbf{a})$$

$$= \tfrac{1}{2}(\mathbf{a} + \mathbf{b})$$

In Cartesian form,

$$\overrightarrow{OM} = \tfrac{1}{2}[(x_1 + x_2)\mathbf{i} + (y_1 + y_2)\mathbf{j} + (z_1 + z_2)\mathbf{k}]$$

The co-ordinates of M are $\left(\dfrac{x_1 + x_2}{2}, \dfrac{y_1 + y_2}{2}, \dfrac{z_1 + z_2}{2}\right)$

Point Dividing AB in a Given Ratio

Let R be the point on AB dividing AB internally in the ratio $\lambda : \mu$

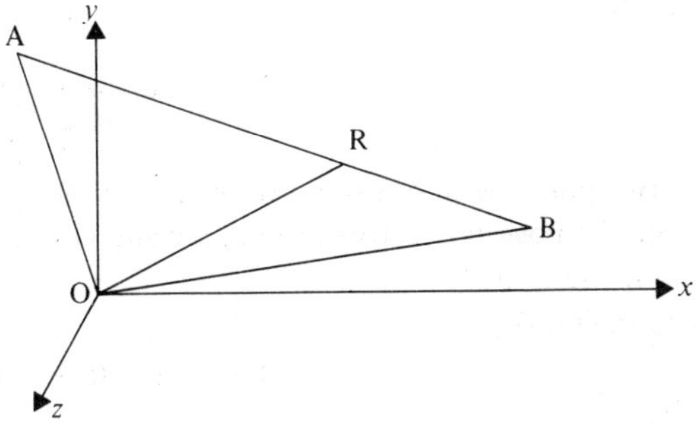

The position vector of R is $\overrightarrow{OR}$ where:

$$\overrightarrow{OR} = \overrightarrow{OA} + \overrightarrow{AR}$$

$$= \overrightarrow{OA} + \frac{AR}{AB}\,\overrightarrow{AB}$$

$$= \mathbf{a} + \frac{\lambda}{\lambda + \mu}(\mathbf{b} - \mathbf{a})$$

$$= \frac{\lambda\mathbf{b} + \mu\mathbf{a}}{\lambda + \mu}$$

In Cartesian form

$$\overrightarrow{OR} = \frac{\lambda(x_2\mathbf{i} + y_2\mathbf{j} + z_2\mathbf{k}) + \mu(x_1\mathbf{i} + y_1\mathbf{j} + z_1\mathbf{k})}{(\lambda + \mu)}$$

The co-ordinates of M are $\left(\dfrac{\lambda x_2 + \mu x_1}{\lambda + \mu}, \dfrac{\lambda y_2 + \mu y_1}{\lambda + \mu}, \dfrac{\lambda z_2 + \mu z_1}{\lambda + \mu}\right)$

Note: For external division the ratio is used in the form $\lambda : -\mu$
(See Examples 1b/2.)

EXAMPLES 1b

1) Find the modulus and direction cosines of the line joining the points A and B with position vectors $2\mathbf{i} + 7\mathbf{j} - \mathbf{k}$ and $4\mathbf{i} + \mathbf{j} + 2\mathbf{k}$ respectively.

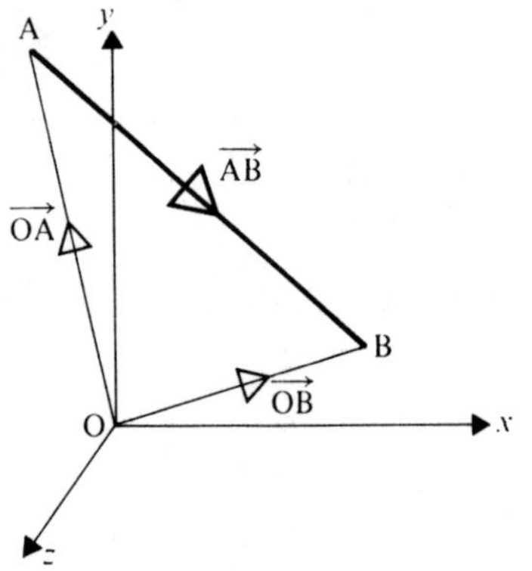

$$\overrightarrow{AB} = \overrightarrow{OB} - \overrightarrow{OA}$$
$$= (4\mathbf{i} + \mathbf{j} + 2\mathbf{k}) - (2\mathbf{i} + 7\mathbf{j} - \mathbf{k})$$
$$= 2\mathbf{i} - 6\mathbf{j} + 3\mathbf{k}$$

Hence

$$|\overrightarrow{AB}| = \sqrt{2^2 + (-6)^2 + 3^2} = 7$$

Also the direction ratios of $\overrightarrow{AB}$ are

$$2:-6:3.$$

So the direction cosines of $\overrightarrow{AB}$ are

$$\tfrac{2}{7}, -\tfrac{6}{7}, \tfrac{3}{7}.$$

2) A and B are two points with position vectors $\mathbf{i} - \mathbf{j} + 4\mathbf{k}$ and $7\mathbf{i} - \mathbf{j} - 2\mathbf{k}$ respectively. Find the position vectors of points P and Q which divide AB

(i) internally in the ratio 5:1,

(ii) externally in the ratio 3:2.

(i)

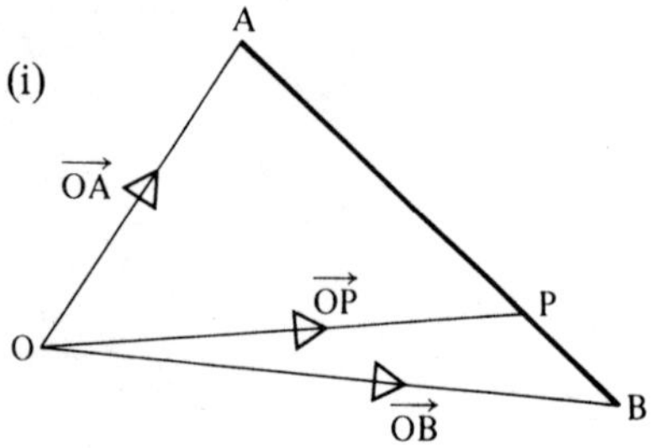

Since AP:PB = 5:1

$$\overrightarrow{OP} = \frac{5\overrightarrow{OB} + 1\overrightarrow{OA}}{5+1}$$
$$= \frac{5(7\mathbf{i} - \mathbf{j} - 2\mathbf{k}) + (\mathbf{i} - \mathbf{j} + 4\mathbf{k})}{6}$$
$$= 6\mathbf{i} - \mathbf{j} - \mathbf{k}$$

Hence P has position vector $6\mathbf{i} - \mathbf{j} - \mathbf{k}$.

(ii)

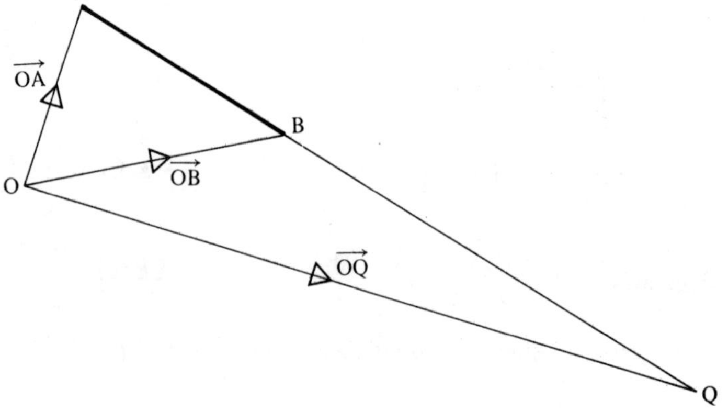

Since $\quad AQ:QB = 3:-2$

$$\overrightarrow{OQ} = \frac{3\overrightarrow{OB} - 2\overrightarrow{OA}}{3-2}$$

$$= \frac{3(7\mathbf{i} - \mathbf{j} - 2\mathbf{k}) - 2(\mathbf{i} - \mathbf{j} + 4\mathbf{k})}{3-2}$$

$$= 19\mathbf{i} - \mathbf{j} - 14\mathbf{k}$$

Hence Q has position vector $19\mathbf{i} - \mathbf{j} - 14\mathbf{k}$

3) Find the modulus and direction cosines of the line LM if L is the mid-point of AB, M is the mid-point of BC and A, B and C have position vectors $3\mathbf{i} - \mathbf{j} + 5\mathbf{k}$, $7\mathbf{i} + \mathbf{j} + 3\mathbf{k}$ and $-5\mathbf{i} + 9\mathbf{j} - \mathbf{k}$ respectively.

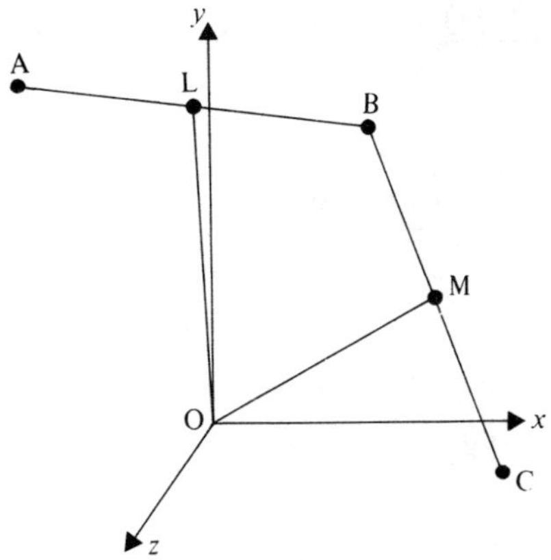

The co-ordinates of A, B and C are $(3, -1, 5)$, $(7, 1, 3)$ and $(-5, 9, -1)$ respectively.
Therefore at L

$$x = \frac{3+7}{2}, y = \frac{-1+1}{2}, z = \frac{5+3}{2}$$

And at M

$$x = \frac{7-5}{2}, y = \frac{1+9}{2}, z = \frac{3-1}{2}$$

Hence L is the point $(5, 0, 4)$ and M is the point $(1, 5, 1)$

Then $\quad |\overrightarrow{LM}| = \sqrt{(5-1)^2 + (0-5)^2 + (4-1)^2} = 5\sqrt{2}$

The direction ratios of $\overrightarrow{LM}$ are $(1-5):(5-0):(1-4)$

i.e. $-4:5:-3$

Hence the direction cosines of $\overrightarrow{LM}$ are $\dfrac{-4}{5\sqrt{2}}, \dfrac{5}{5\sqrt{2}}, \dfrac{-3}{5\sqrt{2}}$

i.e. $\dfrac{-2\sqrt{2}}{5}, \dfrac{\sqrt{2}}{2}, \dfrac{-3\sqrt{2}}{10}$

4) Find the position vector of the mid-point M of the line AB if A is a point with position vector $\mathbf{i} - 3\mathbf{j} - 4\mathbf{k}$ and B divides CD internally in the ratio 2:1. The position vectors of C and D are $\mathbf{i} + 2\mathbf{j} + 5\mathbf{k}$ and $4\mathbf{i} - \mathbf{j} + 2\mathbf{k}$ respectively. If E has position vector $-3\mathbf{i} + \mathbf{j}$ show that CE is parallel to MD.

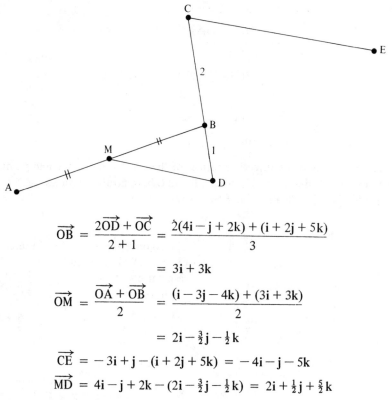

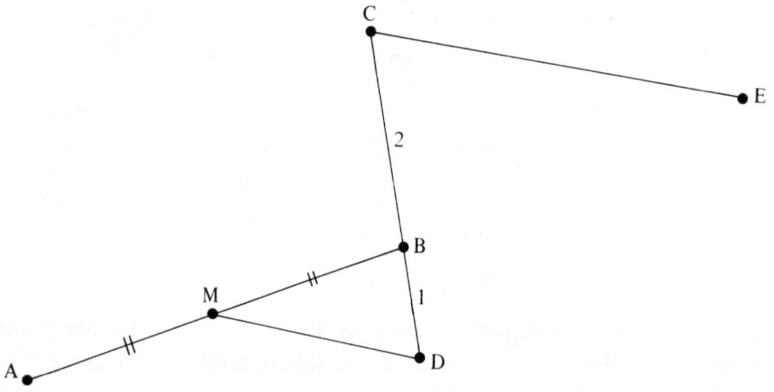

$$\overrightarrow{OB} = \frac{2\overrightarrow{OD} + \overrightarrow{OC}}{2+1} = \frac{2(4\mathbf{i} - \mathbf{j} + 2\mathbf{k}) + (\mathbf{i} + 2\mathbf{j} + 5\mathbf{k})}{3}$$

$$= 3\mathbf{i} + 3\mathbf{k}$$

$$\overrightarrow{OM} = \frac{\overrightarrow{OA} + \overrightarrow{OB}}{2} = \frac{(\mathbf{i} - 3\mathbf{j} - 4\mathbf{k}) + (3\mathbf{i} + 3\mathbf{k})}{2}$$

$$= 2\mathbf{i} - \tfrac{3}{2}\mathbf{j} - \tfrac{1}{2}\mathbf{k}$$

$$\overrightarrow{CE} = -3\mathbf{i} + \mathbf{j} - (\mathbf{i} + 2\mathbf{j} + 5\mathbf{k}) = -4\mathbf{i} - \mathbf{j} - 5\mathbf{k}$$

$$\overrightarrow{MD} = 4\mathbf{i} - \mathbf{j} + 2\mathbf{k} - (2\mathbf{i} - \tfrac{3}{2}\mathbf{j} - \tfrac{1}{2}\mathbf{k}) = 2\mathbf{i} + \tfrac{1}{2}\mathbf{j} + \tfrac{5}{2}\mathbf{k}$$

Direction ratios of CE are $-4:-1:-5 = 4:1:5$
Direction ratios of MD are $2:\frac{1}{2}:\frac{5}{2} = 4:1:5$
Therefore CE is parallel to MD.

5) Find the magnitude of the resultant of forces $\mathbf{F}_1$, $\mathbf{F}_2$ and $\mathbf{F}_3$ if $\mathbf{F}_1 = 3\mathbf{i} - 6\mathbf{j} - 4\mathbf{k}$, $\mathbf{F}_2$ is parallel to $\overrightarrow{AB}$ and is of magnitude 12 N, and $\mathbf{F}_3$ is parallel to $\overrightarrow{OM}$ and is of magnitude $3\sqrt{13}$ N.
The position vectors of A and B are $\mathbf{i} + 2\mathbf{j} + 3\mathbf{k}$ and $5\mathbf{i} - 2\mathbf{j} + \mathbf{k}$ respectively and M is the mid-point of AB.

For $\mathbf{F}_2$:

The direction vector is $(5\mathbf{i} - 2\mathbf{j} + \mathbf{k}) - (\mathbf{i} + 2\mathbf{j} + 3\mathbf{k})$

i.e. $4\mathbf{i} - 4\mathbf{j} - 2\mathbf{k}$

Therefore $\hat{\mathbf{F}}_2$ is $(4\mathbf{i} - 4\mathbf{j} - 2\mathbf{k})/\sqrt{4^2 + (-4)^2 + (-2)^2}$

i.e. $\frac{1}{6}(4\mathbf{i} - 4\mathbf{j} - 2\mathbf{k})$

$$\mathbf{F}_2 = |\mathbf{F}_2|\hat{\mathbf{F}}_2$$

$= 12 \times \frac{1}{6}(4\mathbf{i} - 4\mathbf{j} - 2\mathbf{k})$

Therefore $\mathbf{F}_2 = 8\mathbf{i} - 8\mathbf{j} - 4\mathbf{k}$

For $\mathbf{F}_3$: $\overrightarrow{OM} = \frac{1}{2}(\overrightarrow{OA} + \overrightarrow{OB})$

$= \frac{1}{2}(\mathbf{i} + 2\mathbf{j} + 3\mathbf{k} + 5\mathbf{i} - 2\mathbf{j} + \mathbf{k})$

$= 3\mathbf{i} + 2\mathbf{k}$

This is the direction vector of $\mathbf{F}_3$

Therefore $\hat{\mathbf{F}}_3 = (3\mathbf{i} + 2\mathbf{k})/\sqrt{3^2 + 2^2}$

$= \frac{1}{\sqrt{13}}(3\mathbf{i} + 2\mathbf{k})$

But $\mathbf{F}_3 = |\mathbf{F}_3|\hat{\mathbf{F}}_3$

$= 3\sqrt{13} \times \frac{1}{\sqrt{13}}(3\mathbf{i} + 2\mathbf{k})$

$= 9\mathbf{i} + 6\mathbf{k}$

Now the resultant force $\mathbf{F} = \mathbf{F}_1 + \mathbf{F}_2 + \mathbf{F}_3$

$= (3\mathbf{i} - 6\mathbf{j} - 4\mathbf{k}) + (8\mathbf{i} - 8\mathbf{j} - 4\mathbf{k}) + (9\mathbf{i} + 6\mathbf{k})$

$= 20\mathbf{i} - 14\mathbf{j} - 2\mathbf{k}$

Therefore $|\mathbf{F}| = \sqrt{(20)^2 + (-14)^2 + (-2)^2}$

$= 10\sqrt{6}$

The magnitude of the resultant force is $10\sqrt{6}$ N.

EXERCISE 1b

* In questions 1–5, A, B, C and D are points with position vectors $\mathbf{i} + \mathbf{j}$, $3\mathbf{i} - 2\mathbf{j} + \mathbf{k}$, $-3\mathbf{i} - 3\mathbf{j}$ and $7\mathbf{i} + 2\mathbf{j} + \mathbf{k}$ respectively.

1)* Find the modulus and direction cosines of $\overrightarrow{AB}$, $\overrightarrow{AC}$ and $\overrightarrow{BD}$.

2)* Find the position vector of a point which:
(i) divides AD internally in the ratio 1 : 2,
(ii) bisects AB,
(iii) divides BC externally in the ratio 3 : 1.

3)* Check whether the following pairs of lines are parallel:
(i) AB and CD,
(ii) AC and BD,
(iii) AD and BC.

4)* Find out whether:
(i) AC = BC,

(ii) AC = BD,
(iii) AD = BC,
(iv) AB = DC.

5)* If L and M are the mid-points of AD and BD respectively find the position vectors of L and M and show that LM is parallel to AB.

6) Find the force vector **F** if:
(i) **F** is of magnitude 16 N and is parallel to $\mathbf{i} + 2\mathbf{j} - 2\mathbf{k}$,
(ii) **F** acts along $\overrightarrow{AB}$ where the position vectors of A and B are $3\mathbf{i} + 4\mathbf{j} + \mathbf{k}$ and $\mathbf{i} - 2\mathbf{j} - 2\mathbf{k}$ respectively and **F** is equal in magnitude to the force vector $7\mathbf{i} - 14\mathbf{j} + 14\mathbf{k}$.

7) Find the magnitude and inclination to each of the axes Ox, Oy and Oz of the resultant of two vectors $\mathbf{V}_1$ and $\mathbf{V}_2$ if:
$\mathbf{V}_1$ is of magnitude 20 units and is inclined at acute angles arcsin $\sqrt{13}/5$, arctan $\sqrt{21}/2$ and arccos 3/5 to Ox, Oy and Oz respectively,
$\mathbf{V}_2$ is parallel to a vector with direction ratios $\sqrt{3} : -2 : -3$ and is of magnitude 16 units.

THE VECTOR EQUATION OF A STRAIGHT LINE

CASE 1. If a line has a specified direction and also passes through a fixed point, it is uniquely located in space.
Consider such a line which passes through a point A with position vector **a** and which is parallel to a vector **b**.

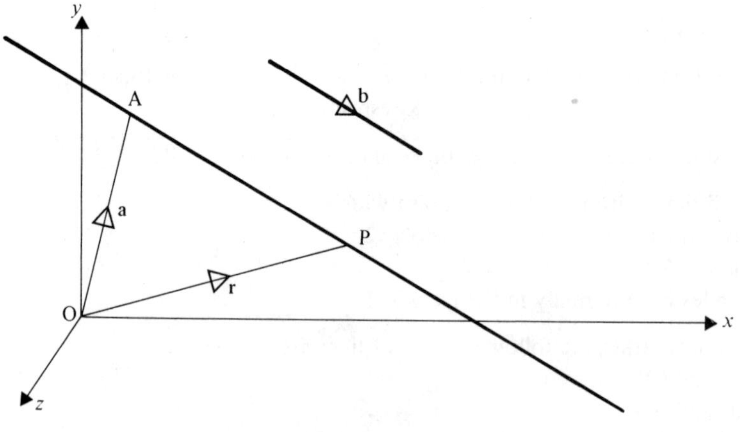

Let P be any point on this line and let **r** be the position vector of P.
$\overrightarrow{AP}$ is parallel to **b**.

Therefore $$\overrightarrow{AP} = \lambda\mathbf{b}$$

But $$\overrightarrow{OP} = \overrightarrow{OA} + \overrightarrow{AP}$$

Therefore $$\mathbf{r} = \mathbf{a} + \lambda\mathbf{b}$$

This equation gives the position vector of any point on the line and is called the *vector equation of the line.*

Each value of the parameter λ corresponds to one position of P.

For example, a line parallel to the vector $2\mathbf{i} + 3\mathbf{j} + \mathbf{k}$ and passing through the point with position vector $\mathbf{i} - \mathbf{j} + 4\mathbf{k}$ has a vector equation

$$\mathbf{r} = \mathbf{i} - \mathbf{j} + 4\mathbf{k} + \lambda(2\mathbf{i} + 3\mathbf{j} + \mathbf{k})$$

or $$\mathbf{r} = (1 + 2\lambda)\mathbf{i} + (-1 + 3\lambda)\mathbf{j} + (4 + \lambda)\mathbf{k}$$

From this second form we see that the co-ordinates of any point P on the line are

$$x = 1 + 2\lambda$$

$$y = -1 + 3\lambda$$

$$z = 4 + \lambda$$

These are the *parametric equations* of the specified line.

They can be rearranged to give

$$\lambda = \frac{x-1}{2}$$

$$\lambda = \frac{y+1}{3}$$

$$\lambda = \frac{z-4}{1}$$

so that $$\frac{x-1}{2} = \frac{y+1}{3} = \frac{z-4}{1} \quad (= \lambda)$$

These are the *Cartesian equations* of the specified line.

The equations of the line can thus be expressed in the following ways:

$\mathbf{r} = \mathbf{i} - \mathbf{j} + 4\mathbf{k} + \lambda(2\mathbf{i} + 3\mathbf{j} + \mathbf{k})$	Vector equation
$x = 1 + 2\lambda; \quad y = -1 + 3\lambda; \quad z = 4 + \lambda$	Parametric equations
$\dfrac{x-1}{2} = \dfrac{y+1}{3} = \dfrac{z-4}{1}$	Cartesian equations

Note: The direction ratios of the line, $2:3:1$, appear as coefficients of λ in the parametric equations and denominators in the Cartesian equations.

The line passes through the point $(1, -1, 4)$ by definition. This point corresponds to $\lambda = 0$. Other points on the line can be found by giving λ other values.

e.g. $\lambda = 1$

$$\mathbf{r} = 3\mathbf{i} + 2\mathbf{j} + 5\mathbf{k}$$

or $x = 3, \quad y = 2, \quad z = 5$

CASE 2. A line is uniquely defined if it passes through two fixed points. Consider such a line passing through two points A and B.

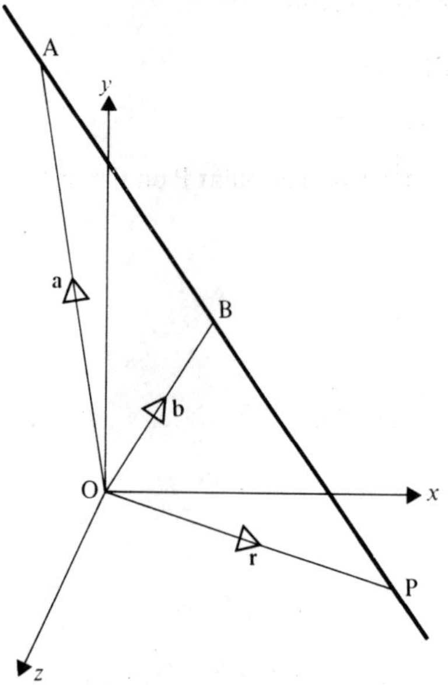

If $\overrightarrow{OA} = \mathbf{a}$

and $\overrightarrow{OB} = \mathbf{b}$

then $\overrightarrow{AB} = \mathbf{b} - \mathbf{a}$

If P is any point on the line through A and B, then

$$\overrightarrow{OP} = \overrightarrow{OA} + \overrightarrow{AP}$$

$$= \mathbf{a} + \lambda(\mathbf{b} - \mathbf{a})$$

(Since AP is parallel to AB)

Representing $\overrightarrow{OP}$ by **r** we have:

$$\mathbf{r} = \mathbf{a} + \lambda(\mathbf{b} - \mathbf{a})$$

This is the vector equation of the line through A and B and gives the position vector of any point P on that line (each value of λ gives one position of P). Alternative ways of expressing the equation are:

$$\mathbf{r} = \mathbf{a}(1 - \lambda) + \lambda\mathbf{b}$$

or

$$\mathbf{r} = \mu\mathbf{a} + (1 - \mu)\mathbf{b}$$

For example, if A and B are points with position vectors $2\mathbf{i} - 3\mathbf{j} + 4\mathbf{k}$ and $7\mathbf{i} + 5\mathbf{j} - \mathbf{k}$ respectively, a vector equation of the line through A and B is

$$\mathbf{r} = 2\mathbf{i} - 3\mathbf{j} + 4\mathbf{k} + \lambda[7\mathbf{i} + 5\mathbf{j} - \mathbf{k} - (2\mathbf{i} - 3\mathbf{j} + 4\mathbf{k})]$$

i.e.

$$\mathbf{r} = (2\mathbf{i} - 3\mathbf{j} + 4\mathbf{k}) + \lambda[5\mathbf{i} + 8\mathbf{j} - 5\mathbf{k}]$$

or

$$\mathbf{r} = (2 + 5\lambda)\mathbf{i} + (-3 + 8\lambda)\mathbf{j} + (4 - 5\lambda)\mathbf{k}$$

The above vector equation shows that, for any point P on the line through A and B,

$$x = 2 + 5\lambda$$

$$y = -3 + 8\lambda$$

$$z = 4 - 5\lambda$$

so that
$$\frac{x-2}{5} = \frac{y+3}{8} = \frac{z-4}{-5} \quad (=\lambda)$$

These are the parametric and Cartesian equations of the line.
Again the denominators in the Cartesian equations are seen to be the direction ratios of the line (i.e. direction ratios of $\overrightarrow{AB}$).
Summarising, we have:

$$\mathbf{r} = x_1\mathbf{i} + y_1\mathbf{j} + z_1\mathbf{k} + \lambda(a\mathbf{i} + b\mathbf{j} + c\mathbf{k})$$

represents the vector equation of a line passing through a point (x_1, y_1, z_1) and having direction ratios $a:b:c$.
The corresponding Cartesian equations are:

$$\frac{x-x_1}{a} = \frac{y-y_1}{b} = \frac{z-z_1}{c}$$

The parametric co-ordinates of any point on the line are

$$x = x_1 + \lambda a$$

$$y = y_1 + \lambda b$$

$$z = z_1 + \lambda c$$

The above equations are not the only ones which represent the specified line because (x_1, y_1, z_1) is only one of many fixed points through which the line passes.
(The co-ordinates of other suitable points can be obtained from the parametric equations above by giving λ various values.)

EXAMPLES 1c

1) (a) The Cartesian equations of a line are $\dfrac{x-5}{3} = \dfrac{y+4}{7} = \dfrac{z-6}{2}$.

Find a vector equation for the line.

(b) If the vector equation of a line is $\mathbf{r} = \mathbf{i} - 3\mathbf{j} + 2\mathbf{k} + \lambda(5\mathbf{i} + 2\mathbf{j} - \mathbf{k})$, express the equation of the line in parametric form and hence find the co-ordinates of the point where the line crosses the xy plane.

(a)
$$\frac{x-5}{3} = \frac{y+4}{7} = \frac{z-6}{2}$$

This line has direction ratios $3:7:2$. Hence its direction vector is $3\mathbf{i} + 7\mathbf{j} + 2\mathbf{k}$. One point on the line has co-ordinates $(5, -4, 6)$. The position vector of this point is $5\mathbf{i} - 4\mathbf{j} + 6\mathbf{k}$.

Hence a vector equation of the line is:

$$\mathbf{r} = 5\mathbf{i} - 4\mathbf{j} + 6\mathbf{k} + \lambda(3\mathbf{i} + 7\mathbf{j} + 2\mathbf{k})$$

(b)
$$\mathbf{r} = \mathbf{i} - 3\mathbf{j} + 2\mathbf{k} + \lambda(5\mathbf{i} + 2\mathbf{j} - \mathbf{k})$$

Therefore
$$\mathbf{r} = (1 + 5\lambda)\mathbf{i} + (-3 + 2\lambda)\mathbf{j} + (2 - \lambda)\mathbf{k}$$

A general point on the line has co-ordinates

$$x = 1 + 5\lambda; \quad y = -3 + 2\lambda; \quad z = 2 - \lambda$$

Hence $\lambda = \dfrac{x-1}{5} = \dfrac{y+3}{2} = \dfrac{z-2}{-1}$ are the parametric equations of the line.

At the point where the line crosses the xy plane, $z = 0$, so $2 - \lambda = 0$.
When $\lambda = 2, x = 11$ and $y = 1$.
Therefore the line crosses the xy plane at the point $(11, 1, 0)$.

2) A line passes through the point with position vector $2\mathbf{i} - \mathbf{j} + 4\mathbf{k}$ and is in the direction of $\mathbf{i} + \mathbf{j} - 2\mathbf{k}$. Find equations for the line in vector and in Cartesian form.

The equation of the line in vector form is

$$\mathbf{r} = 2\mathbf{i} - \mathbf{j} + 4\mathbf{k} + \lambda(\mathbf{i} + \mathbf{j} - 2\mathbf{k})$$

This shows that the co-ordinates of any point P on the line are:

$$[(2 + \lambda), (-1 + \lambda), (4 - 2\lambda)]$$

Hence the Cartesian equations are

$$x - 2 = y + 1 = \frac{z-4}{-2} \quad (= \lambda)$$

3) Find a vector equation for the line through the points $A(3, 4, -7)$ and $B(1, -1, 6)$.

$$\overrightarrow{OA} = \mathbf{a} = 3\mathbf{i} + 4\mathbf{j} - 7\mathbf{k}$$

$$\overrightarrow{OB} = \mathbf{b} = \mathbf{i} - \mathbf{j} + 6\mathbf{k}$$

For any point P on the line, $\overrightarrow{OP} = \mathbf{r}$, then

$$\mathbf{r} = \mathbf{a} + \lambda(\mathbf{b} - \mathbf{a})$$

$$= 3\mathbf{i} + 4\mathbf{j} - 7\mathbf{k} + \lambda(-2\mathbf{i} - 5\mathbf{j} + 13\mathbf{k})$$

or
$$\mathbf{r} = t(3\mathbf{i} + 4\mathbf{j} - 7\mathbf{k}) + (1 - t)(\mathbf{i} - \mathbf{j} + 6\mathbf{k})$$

4) Show that the line through the points $\mathbf{i} + \mathbf{j} - 3\mathbf{k}$ and $4\mathbf{i} + 7\mathbf{j} + \mathbf{k}$ is parallel to the line $\mathbf{r} = \mathbf{i} - \mathbf{k} + \lambda(\frac{3}{2}\mathbf{i} + 3\mathbf{j} + 2\mathbf{k})$.

The direction vector of the line joining the two given points is

$$4\mathbf{i}+7\mathbf{j}+\mathbf{k}-(\mathbf{i}+\mathbf{j}-3\mathbf{k})$$

i.e. $$3\mathbf{i}+6\mathbf{j}+4\mathbf{k}$$

The direction vector of the given line is $\frac{3}{2}\mathbf{i}+3\mathbf{j}+2\mathbf{k}$

$$=\tfrac{1}{2}(3\mathbf{i}+6\mathbf{j}+4\mathbf{k})$$

Hence the two lines are parallel.

5) Find the co-ordinates of the point where the line through A(3, 4, 1) and B(5, 1, 6) crosses the xy plane.

The vector equation of the line through A and B is

$$\mathbf{r} = 3\mathbf{i}+4\mathbf{j}+\mathbf{k}+\lambda(2\mathbf{i}-3\mathbf{j}+5\mathbf{k})$$

The line crosses the xy plane where $z = 0$. Any point on the line has coordinates $[(3+2\lambda), (4-3\lambda), (1+5\lambda)]$

If $z=0$, $\quad 1+5\lambda=0 \quad$ therefore $\lambda=-\frac{1}{5}$

Hence $$x = 3-\tfrac{2}{5} = \tfrac{13}{5};\quad y = 4+\tfrac{3}{5} = \tfrac{23}{5}$$

EXERCISE 1c

1) Convert the following vector equations to Cartesian form:

(i) $\mathbf{r} = 2\mathbf{i}+3\mathbf{j}-\mathbf{k}+\lambda(\mathbf{i}+\mathbf{j}+\mathbf{k})$,
(ii) $\mathbf{r} = 4\mathbf{j}+\lambda(3\mathbf{i}+5\mathbf{k})$,
(iii) $\mathbf{r} = \lambda(2\mathbf{i}+3\mathbf{j}+4\mathbf{k})$.

2) Convert to vector form, the following equations:

(i) $\dfrac{x-3}{4} = \dfrac{y-1}{2} = \dfrac{z-7}{6}$,

(ii) $x = 3\lambda+2,\quad y = \lambda-5,\quad z = 4\lambda+1$,

(iii) $\dfrac{1-x}{3} = \dfrac{y}{5} = z$.

3) Write down equations, in vector and in Cartesian form, for the line through a point A with position vector **a** and with a direction vector **b** if:

(i) $\mathbf{a} = \mathbf{i}-3\mathbf{j}+2\mathbf{k} \qquad \mathbf{b} = 5\mathbf{i}+4\mathbf{j}-\mathbf{k}$,
(ii) $\mathbf{a} = 2\mathbf{i}+\mathbf{j} \qquad \mathbf{b} = 3\mathbf{j}-\mathbf{k}$,
(iii) A is the origin $\qquad \mathbf{b} = \mathbf{i}-\mathbf{j}-\mathbf{k}$.

4) State whether or not the following pairs of lines are parallel:

(i) $\mathbf{r} = \mathbf{i}+\mathbf{j}-\mathbf{k}+\lambda(2\mathbf{i}-3\mathbf{j}+\mathbf{k})$
$\mathbf{r} = 2\mathbf{i}-4\mathbf{j}+5\mathbf{k}+\lambda(\mathbf{i}+\mathbf{j}-\mathbf{k})$,

(ii) $\dfrac{x-1}{2} = \dfrac{y-4}{3} = \dfrac{z+1}{-4}$

$$\frac{x}{4} = \frac{y+5}{6} = \frac{3-z}{8}.$$

(iii) $\mathbf{r} = 2\mathbf{i} - \mathbf{j} + 4\mathbf{k} + \lambda(\mathbf{i} + \mathbf{j} + 3\mathbf{k})$

$$x - 4 = y + 7 = \frac{z}{3},$$

(iv) $\mathbf{r} = \lambda(3\mathbf{i} - 3\mathbf{j} + 6\mathbf{k})$
$\mathbf{r} = 4\mathbf{j} + \lambda(-\mathbf{i} + \mathbf{j} - 2\mathbf{k}),$

(v) $\mathbf{r} = 3\mathbf{i} + \mathbf{k} + \lambda(\mathbf{i} - \mathbf{j} - 2\mathbf{k})$

$$\frac{x-3}{1} = \frac{y}{1} = \frac{z-1}{2}.$$

5) The points A(4, 5, 10), B(2, 3, 4) and C(1, 2, − 1) are three vertices of a parallelogram ABCD. Find vector and Cartesian equations for the sides AB and BC and find the co-ordinates of D.

6) Write down a vector equation for the line through A and B if:
(i) $\overrightarrow{OA}$ is $3\mathbf{i} + \mathbf{j} - 4\mathbf{k}$ and $\overrightarrow{OB}$ is $\mathbf{i} + 7\mathbf{j} + 8\mathbf{k}$,
(ii) A and B have co-ordinates (1, 1, 7) and (3, 4, 1).
Find, in each case, the co-ordinates of the points where the line crosses the xy plane, the yz plane and the zx plane.

7) A line has Cartesian equations $\dfrac{x-1}{3} = \dfrac{y+2}{4} = \dfrac{z-4}{5}$

Find a vector equation for a parallel line passing through the point with position vector $5\mathbf{i} - 2\mathbf{j} - 4\mathbf{k}$ and find the co-ordinates of the point on this line where $y = 0$.

PAIRS OF LINES

The location of two lines in space may be such that:
(a) the lines are parallel,
(b) the lines are not parallel and intersect,
(c) the lines are not parallel and do not intersect.

Parallel Lines

We have already seen that parallel lines have equal direction ratios. So if two lines are parallel, this property can be observed from their equations.

Non-parallel Lines

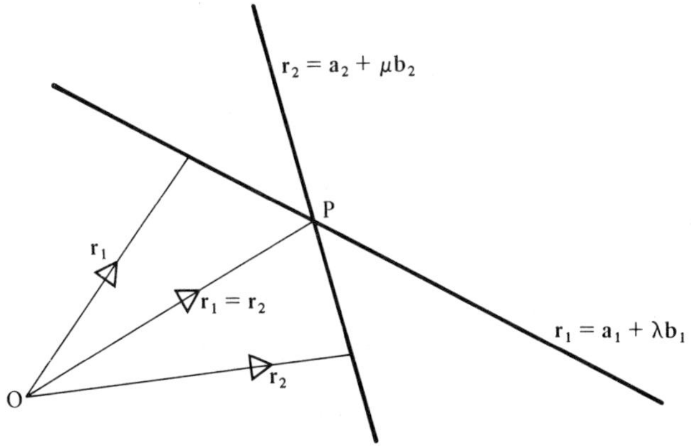

Consider two lines whose vector equations are $\mathbf{r}_1 = \mathbf{a}_1 + \lambda\mathbf{b}_1$ and $\mathbf{r}_2 = \mathbf{a}_2 + \mu\mathbf{b}_2$. In order that these lines shall intersect there must be unique values of λ and μ such that

$$\mathbf{a}_1 + \lambda\mathbf{b}_1 = \mathbf{a}_2 + \mu\mathbf{b}_2$$

If no such values can be found, the lines do not intersect.

EXAMPLE

Find out whether the following pairs of lines are parallel, non-parallel and intersecting, or non-parallel and non-intersecting.

(i) $\mathbf{r}_1 = \mathbf{i} + \mathbf{j} + 2\mathbf{k} + \lambda(3\mathbf{i} - 2\mathbf{j} + 4\mathbf{k})$
$\mathbf{r}_2 = 2\mathbf{i} - \mathbf{j} + 3\mathbf{k} + \mu(-6\mathbf{i} + 4\mathbf{j} - 8\mathbf{k})$,

(ii) $\mathbf{r}_1 = \mathbf{i} - \mathbf{j} + 3\mathbf{k} + \lambda(\mathbf{i} - \mathbf{j} + \mathbf{k})$
$\mathbf{r}_2 = 2\mathbf{i} + 4\mathbf{j} + 6\mathbf{k} + \mu(2\mathbf{i} + \mathbf{j} + 3\mathbf{k})$,

(iii) $\mathbf{r}_1 = \mathbf{i} + \mathbf{k} + \lambda(\mathbf{i} + 3\mathbf{j} + 4\mathbf{k})$
$\mathbf{r}_2 = 2\mathbf{i} + 3\mathbf{j} + \mu(4\mathbf{i} - \mathbf{j} + \mathbf{k})$.

(i) Checking first whether the lines are parallel we compare the direction ratios of two lines.
First line has direction ratios $3:-2:4$
Second line has direction ratios $-6:4:-8 = 3:-2:4$
Therefore these two lines are parallel.

(ii) In this case the two sets of direction ratios are $1:-1:1$ and $2:1:3$
These are not equal, so these two lines are not parallel.
Now if the lines intersect it will be at a point where $\mathbf{r}_1 = \mathbf{r}_2$

i.e. where

$$(1 + \lambda)\mathbf{i} - (1 + \lambda)\mathbf{j} + (3 + \lambda)\mathbf{k} = 2(1 + \mu)\mathbf{i} + (4 + \mu)\mathbf{j} + (6 + 3\mu)\mathbf{k}$$

Equating the coefficients of **i** and **j** we have

$$1+\lambda = 2(1+\mu)$$
$$-(1+\lambda) = 4+\mu$$

Hence $$\mu = -2, \quad \lambda = -3$$

Using these values for λ and μ, the coefficients of **k** become:

$$\left.\begin{array}{ll}\text{First line} & 3+\lambda = 0 \\ \text{Second line} & 6+3\mu = 0\end{array}\right\} \text{equal values}$$

So $\mathbf{r}_1 = \mathbf{r}_2$ when $\lambda = -3$ and $\mu = -2$

Therefore the lines *do* intersect at the point with position vector

$$(1-3)\mathbf{i} - (1-3)\mathbf{j} + (3-3)\mathbf{k} \qquad (\lambda = -3 \text{ in } \mathbf{r}_1)$$

i.e. $$-2\mathbf{i} + 2\mathbf{j}$$

(iii) The direction ratios of these two lines are not equal so the lines are not parallel.
If the lines intersect it will be where $\mathbf{r}_1 = \mathbf{r}_2$

i.e. where

$$(1+\lambda)\mathbf{i} + 3\lambda\mathbf{j} + (1+4\lambda)\mathbf{k} = (2+4\mu)\mathbf{i} + (3-\mu)\mathbf{j} + \mu\mathbf{k}$$

Equating the coefficients of **i** and **j** we have

$$1+\lambda = 2+4\mu$$
$$3\lambda = 3-\mu$$

Hence $$\mu = 0, \quad \lambda = 1$$

Using these values of λ and μ, the coefficients of **k** become:

$$\left.\begin{array}{ll}\text{First line} & 1+4\lambda = 5 \\ \text{Second line} & \mu = 0\end{array}\right\} \text{unequal values}$$

So there are no values of λ and μ for which $\mathbf{r}_1 = \mathbf{r}_2$ and these lines do not intersect.
Such lines are said to be *skew*.

Angle Between Two Lines

Consider two lines which have direction cosines l_1, m_1, n_1 and l_2, m_2, n_2 and which are inclined to each other at an angle θ. The angle between the lines depends only upon their directions and not upon their positions.
Hence we can find the required angle by considering two lines through O which are parallel to the given lines.

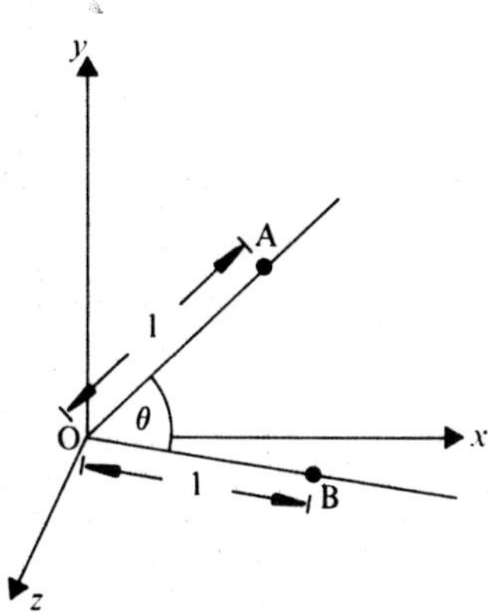

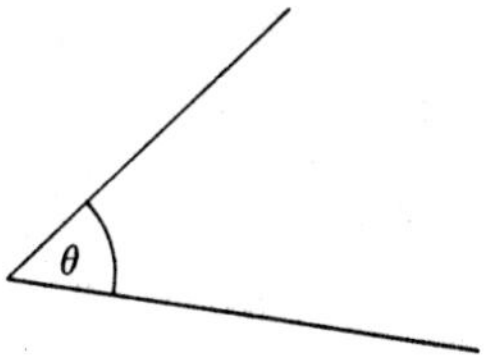

Let A and B be points, one on each line through O, each being one unit of distance from O.

For OA: $|\overrightarrow{OA}| = 1$

$\overrightarrow{OA}$ has direction cosines l_1, m_1, n_1

Therefore A has co-ordinates (l_1, m_1, n_1)

Similarly B has co-ordinates (l_2, m_2, n_2)

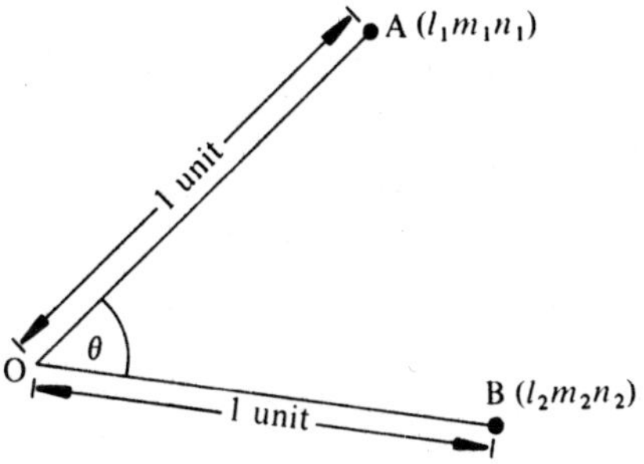

Now $$(AB)^2 = (l_2 - l_1)^2 + (m_2 - m_1)^2 + (n_2 - n_1)^2$$

$$= \sum l_1^2 + \sum l_2^2 - 2\sum l_1 l_2$$

But $\sum l_1^2 = 1$ and $\sum l_2^2 = 1$

Therefore $(AB)^2 = 1 + 1 - 2\sum l_1 l_2$

Using the cosine formula in triangle AOB gives:

$$\cos\theta = \frac{(OA)^2 + (OB)^2 - (AB)^2}{2(OA)(OB)}$$

$$= \frac{1+1-(1+1-2\,\Sigma\, l_1 l_2)}{2(1)(1)}$$

Hence $\qquad \cos\theta = \sum l_1 l_2$

Therefore *the angle θ between two lines with direction cosines l_1, m_1, n_1 and l_2, m_2, n_2 is given by*

$$\cos\theta = l_1 l_2 + m_1 m_2 + n_1 n_2$$

Perpendicular Lines

Since the angle θ between perpendicular lines is 90°, the above result becomes

$$l_1 l_2 + m_1 m_2 + n_1 n_2 = 0$$

EXAMPLES 1d

1) Find the angle between the lines:

$$\mathbf{r}_1 = \mathbf{i} - 2\mathbf{j} + 3\mathbf{k} + \lambda(2\mathbf{i} - 3\mathbf{j} + 6\mathbf{k}) \qquad (1)$$
$$\mathbf{r}_2 = 2\mathbf{i} - 7\mathbf{j} + 10\mathbf{k} + \mu(\mathbf{i} + 2\mathbf{j} + 2\mathbf{k}) \qquad (2)$$

The angle between the lines depends only upon their directions.
Line (1) has direction cosines $\frac{2}{7}, -\frac{3}{7}, \frac{6}{7}$
Line (2) has direction cosines $\frac{1}{3}, \frac{2}{3}, \frac{2}{3}$
The angle θ between the lines is given by

$$\cos\theta = (\tfrac{2}{7})(\tfrac{1}{3}) + (-\tfrac{3}{7})(\tfrac{2}{3}) + (\tfrac{6}{7})(\tfrac{2}{3})$$
$$\theta = \arccos \tfrac{8}{21}$$

2) Find the point of intersection of the lines:

$$\mathbf{r}_1 = \mathbf{i} + 4\mathbf{k} + \lambda(2\mathbf{j} - \mathbf{k})$$
$$\mathbf{r}_2 = 5\mathbf{i} + 2\mathbf{j} + 7\mathbf{k} + \mu(4\mathbf{i} + 4\mathbf{k})$$

If the lines intersect it will be where $\mathbf{r}_1 = \mathbf{r}_2$

i.e. $\qquad \mathbf{i} + 2\lambda\mathbf{j} + (4-\lambda)\mathbf{k} = (5+4\mu)\mathbf{i} + 2\mathbf{j} + (7+4\mu)\mathbf{k}$

Coefficients of **i** are equal if $\qquad 1 = 5 + 4\mu$

Coefficients of **j** are equal if $\qquad 2\lambda = 2$

Hence $\qquad \lambda = 1 \quad \text{and} \quad \mu = -1$

Using these values of λ and μ the coefficients of **k** become:

$$\left.\begin{array}{ll} \text{First line} & 4 - \lambda = 3 \\ \text{Second line} & 7 + 4\mu = 3 \end{array}\right\} \text{equal values}$$

Therefore the two lines intersect at a point with position vector

$$\mathbf{r} = \mathbf{i} + 4\mathbf{k} + 1(2\mathbf{j} - \mathbf{k})$$

i.e.
$$\mathbf{r} = \mathbf{i} + 2\mathbf{j} + 3\mathbf{k}$$

3) Two lines have equations

$$\frac{x-4}{2} = \frac{y-3}{1} = \frac{z-7}{2}$$

and
$$\frac{x+1}{3} = \frac{y}{2} = \frac{z+1}{6}$$

Show that the lines intersect and find the angle between them.

If we introduce a parameter t into the equations of the first line we get

$$\frac{x-4}{2} = \frac{y-3}{1} = \frac{z-7}{2} = t$$

so that
$$\begin{cases} x = 2t+4 \\ y = t+3 \\ z = 2t+7 \end{cases}$$

Similarly the equations of the second line become

$$\begin{cases} x = 3s-1 \\ y = 2s \\ z = 6s-1 \end{cases}$$

If the two lines intersect it will be at a point $P(x, y, z)$ whose co-ordinates satisfy both sets of equations.

Equating x co-ordinates $\quad 2t+4 = 3s-1$

Equating y co-ordinates $\quad t+3 = 2s$

Hence $\quad s = 1 \quad$ and $\quad t = -1$

Using these values for s and t the z co-ordinates become:

First line $\quad 2t+7 = 5$
Second line $\quad 6s-1 = 5$
} equal values

So the lines intersect at a point where $s = 1$

giving $\quad x = 2,\ y = 2,\ z = 5$

First line has direction ratios $\quad 2:1:2$

So its direction cosines are $\quad \frac{2}{3}, \frac{1}{3}, \frac{2}{3}$

Second line has direction ratios $\quad 3:2:6$

So its direction cosines are $\frac{3}{7}, \frac{2}{7}, \frac{6}{7}$

If θ is the angle between the lines then,

using $$\cos\theta = l_1 l_2 + m_1 m_2 + n_1 n_2$$

we have $$\cos\theta = (\tfrac{2}{3})(\tfrac{3}{7}) + (\tfrac{1}{3})(\tfrac{2}{7}) + (\tfrac{2}{3})(\tfrac{6}{7})$$

$$\cos\theta = \tfrac{20}{21}$$

The angle between the lines is arccos $\frac{20}{21}$.

4) Two forces, $\mathbf{F}_1 = 3\mathbf{i} + 4\mathbf{j} - 2\mathbf{k}$ and $\mathbf{F}_2 = \mathbf{i} - 5\mathbf{j} + 3\mathbf{k}$, act through points with position vectors $-5\mathbf{i} + 6\mathbf{j} - \mathbf{k}$ and $-\mathbf{i} + 5\mathbf{j}$ respectively. Find the equations of their lines of action; show that they intersect and hence find the equation of the line of action of their resultant.

The direction vector of the line of action of $\mathbf{F}_1$ is $3\mathbf{i} + 4\mathbf{j} - 2\mathbf{k}$
The equation of the line of action of $\mathbf{F}_1$ is therefore

$$\mathbf{r}_1 = -5\mathbf{i} + 6\mathbf{j} - \mathbf{k} + \lambda(3\mathbf{i} + 4\mathbf{j} - 2\mathbf{k})$$

The direction vector of the line of action of $\mathbf{F}_2$ is $\mathbf{i} - 5\mathbf{j} + 3\mathbf{k}$
The equation of the line of action of $\mathbf{F}_2$ is therefore

$$\mathbf{r}_2 = -\mathbf{i} + 5\mathbf{j} + \mu(\mathbf{i} - 5\mathbf{j} + 3\mathbf{k})$$

If the forces are concurrent it will be at a point where $\mathbf{r}_1 = \mathbf{r}_2$
i.e. where

$$(3\lambda - 5)\mathbf{i} + (4\lambda + 6)\mathbf{j} + (-2\lambda - 1)\mathbf{k} = (\mu - 1)\mathbf{i} + (5 - 5\mu)\mathbf{j} + 3\mu\mathbf{k}$$

Then $$3\lambda - 5 = \mu - 1 \quad \text{(coefficients of i)}$$

$$4\lambda + 6 = 5 - 5\mu \quad \text{(coefficients of j)}$$

These give $$\lambda = 1 \quad \text{and} \quad \mu = -1$$

With these values, the coefficients of k become:

$$\left.\begin{array}{ll} \text{From line of action of } \mathbf{F}_1 & -2\lambda - 1 = -3 \\ \text{From line of action of } \mathbf{F}_2 & 3\mu = -3 \end{array}\right\} \text{ equal values}$$

So the forces are concurrent at a point P with position vector

$$\mathbf{r} = -5\mathbf{i} + 6\mathbf{j} - \mathbf{k} + 1(3\mathbf{i} + 4\mathbf{j} - 2\mathbf{k})$$

i.e. $$\mathbf{r} = -2\mathbf{i} + 10\mathbf{j} - 3\mathbf{k}$$

Since $\mathbf{F}_1$ and $\mathbf{F}_2$ intersect, their resultant $\mathbf{F}$ will also pass through the point of intersection.

$$\begin{aligned} \mathbf{F} &= \mathbf{F}_1 + \mathbf{F}_2 \\ &= 3\mathbf{i} + 4\mathbf{j} - 2\mathbf{k} + \mathbf{i} - 5\mathbf{j} + 3\mathbf{k} \\ &= 4\mathbf{i} - \mathbf{j} + \mathbf{k} \end{aligned}$$

The equation of the line of action of $\mathbf{F}$ is therefore

$$\mathbf{r} = -2\mathbf{i} + 10\mathbf{j} - 3\mathbf{k} + \lambda(4\mathbf{i} - \mathbf{j} + \mathbf{k})$$

5) Three forces $\mathbf{F}_1$, $\mathbf{F}_2$ and $\mathbf{F}_3$ act in a plane.

$\mathbf{F}_1 = \mathbf{i} + 2\mathbf{j} + 3\mathbf{k}$ and acts at the point with position vector $3\mathbf{i} + \mathbf{j}$,

$\mathbf{F}_2 = 4\mathbf{i} - \mathbf{j} + \mathbf{k}$ and acts at the point with position vector $13\mathbf{i} - 6\mathbf{j} - 3\mathbf{k}$,

$\mathbf{F}_3 = -5\mathbf{i} - \mathbf{j} - 4\mathbf{k}$ and acts at the point with position vector $6\mathbf{i} - 2\mathbf{j} - 2\mathbf{k}$.

Show that the three forces are in equilibrium.

The resultant force $\mathbf{F} = \mathbf{F}_1 + \mathbf{F}_2 + \mathbf{F}_3$

Therefore $\mathbf{F} = (\mathbf{i} + 2\mathbf{j} + 3\mathbf{k}) + (4\mathbf{i} - \mathbf{j} - \mathbf{k}) + (-5\mathbf{i} - \mathbf{j} - 4\mathbf{k}) = \mathbf{0}$

Since the resultant force is zero, the system will be in equilibrium if the forces are concurrent.

The equation of the line of action of $\mathbf{F}_1$ is $\mathbf{r}_1 = 3\mathbf{i} + \mathbf{j} + \lambda(\mathbf{i} + 2\mathbf{j} + 3\mathbf{k})$

The equation of the line of action of $\mathbf{F}_2$ is $\mathbf{r}_2 = 13\mathbf{i} - 6\mathbf{j} - 3\mathbf{k} + \mu(4\mathbf{i} - \mathbf{j} + \mathbf{k})$

If these lines intersect it will be where

$$\left.\begin{aligned} 3 + \lambda &= 13 + 4\mu \\ 1 + 2\lambda &= -6 - \mu \end{aligned}\right\} \longrightarrow \mu = -3, \quad \lambda = -2$$

These values of λ and μ give $\mathbf{r}_1 = \mathbf{i} - 3\mathbf{j} - 6\mathbf{k}$

$$\mathbf{r}_2 = \mathbf{i} - 3\mathbf{j} - 6\mathbf{k}$$

Hence the lines of action of $\mathbf{F}_1$ and $\mathbf{F}_2$ meet at the point with position vector $\mathbf{i} - 3\mathbf{j} - 6\mathbf{k}$

Now the equation of the line of action of $\mathbf{F}_3$ is

$$\mathbf{r}_3 = 6\mathbf{i} - 2\mathbf{j} - 2\mathbf{k} + \eta(-5\mathbf{i} - \mathbf{j} - 4\mathbf{k})$$

This line also passes through $\mathbf{i} - 3\mathbf{j} - 6\mathbf{k}$ (putting $\eta = 1$).

So all three lines of action are concurrent and the three forces are in equilibrium.

EXERCISE 1d

1) Find the angle between each of the following pairs of lines

(a) $\mathbf{r}_1 = 3\mathbf{i} + 2\mathbf{j} - 4\mathbf{k} + \lambda(\mathbf{i} + 2\mathbf{j} + 2\mathbf{k})$

$\mathbf{r}_2 = 5\mathbf{j} - 2\mathbf{k} + \mu(3\mathbf{i} + 2\mathbf{j} + 6\mathbf{k})$,

(b) A line with direction ratios 4:4:2

A line joining (3, 1, 4) to (7, 2, 12),

(c) $\dfrac{x+4}{3} = \dfrac{y-1}{5} = \dfrac{z+3}{4}$

$\dfrac{x+1}{1} = \dfrac{y-4}{1} = \dfrac{z-5}{2}$

2) Find whether the following pairs of lines are parallel, intersecting or skew. In the case of intersection state the position vector of the common point.

(a) $\mathbf{r} = \mathbf{i}-\mathbf{j}+\mathbf{k}+\lambda(3\mathbf{i}-4\mathbf{j}+\mathbf{k})$
$\mathbf{r} = \mu(-9\mathbf{i}+12\mathbf{j}-3\mathbf{k})$,

(b) $\dfrac{x-4}{1} = \dfrac{y-8}{2} = \dfrac{z-3}{1}$

$\dfrac{x-7}{6} = \dfrac{y-6}{4} = \dfrac{z-5}{5}$,

(c) $\mathbf{r} = \mathbf{i}+3\mathbf{k}+\lambda(2\mathbf{i}+\mathbf{j}+\mathbf{k})$
$\mathbf{r} = 2\mathbf{i}-\mathbf{j}+\mathbf{k}+\mu(\mathbf{i}-2\mathbf{j})$.

3) Two forces $\mathbf{F}_1$ and $\mathbf{F}_2$ act on a particle at a point A with position vector $3\mathbf{i}+2\mathbf{j}+\mathbf{k}$.
$\mathbf{F}_1$ is of magnitude 14 N and is in the direction of the vector $6\mathbf{i}+3\mathbf{j}-2\mathbf{k}$,
$\mathbf{F}_2 = 4\mathbf{i}+7\mathbf{j}+6\mathbf{k}$.
Find a vector equation for the line of action of the resultant of $\mathbf{F}_1$ and $\mathbf{F}_2$.
A third force $\mathbf{F}_3$ has a line of action with equation

$$\mathbf{r} = \mathbf{i}-3\mathbf{j}+4\mathbf{k}+\lambda(2\mathbf{i}+5\mathbf{j}-3\mathbf{k})$$

Show that $\mathbf{F}_3$ also acts on the particle at A.

4) Two lines have equations:

$$\mathbf{r} = 2\mathbf{i}+9\mathbf{j}+13\mathbf{k}+\lambda(\mathbf{i}+2\mathbf{j}+3\mathbf{k})$$

$$\mathbf{r} = a\mathbf{i}+7\mathbf{j}-2\mathbf{k}+\mu(-\mathbf{i}+2\mathbf{j}-3\mathbf{k})$$

If they intersect, find the value of a and the position vector of the point of intersection.

5) Find the equation of the line of action of the resultant of forces $\mathbf{F}_1$ and $\mathbf{F}_2$ if the equations of the lines of action of $\mathbf{F}_1$ and $\mathbf{F}_2$ are respectively

$$\mathbf{r}_1 = 2\mathbf{i}-7\mathbf{j}+\mathbf{k}+\lambda(\mathbf{i}+4\mathbf{j}+2\mathbf{k}) \quad \text{and} \quad \mathbf{r}_2 = 3\mathbf{i}-4\mathbf{j}+\mu(\mathbf{i}+5\mathbf{j}+5\mathbf{k}).$$

and the magnitudes of $\mathbf{F}_1$ and $\mathbf{F}_2$ are $\sqrt{21}$ and $2\sqrt{51}$ respectively.

6) A(4, 7, 1), B(1, 2, 3) and C(−2, 0, 5) are the vertices of a triangle.
Find the equations of all three medians.
Use these equations to prove that the medians are concurrent and find the coordinates of the centroid of triangle ABC.

SUMMARY

1) If $\mathbf{V} = a\mathbf{i}+b\mathbf{j}+c\mathbf{k}$

$|\mathbf{V}| = \sqrt{a^2+b^2+c^2} = d$

$\mathbf{V}$ has direction ratios $a:b:c$

$\mathbf{V}$ has direction cosines $\dfrac{a}{d}, \dfrac{b}{d}, \dfrac{c}{d}$ or l, m, n

$$\hat{\mathbf{V}} = \mathbf{V}/d = l\mathbf{i} + m\mathbf{j} + n\mathbf{k}$$

$$l^2 + m^2 + n^2 = 1$$

2) If A and B are two points with position vectors

$$\mathbf{a} = x_1\mathbf{i} + y_1\mathbf{j} + z_1\mathbf{k} \quad \text{and} \quad \mathbf{b} = x_2\mathbf{i} + y_2\mathbf{j} + z_2\mathbf{k}$$

$$|\overrightarrow{AB}| = \sqrt{(x_2 - x_1)^2 + (y_2 - y_1)^2 + (z_2 - z_1)^2}$$

$\overrightarrow{AB}$ has direction ratios $(x_2 - x_1):(y_2 - y_1):(z_2 - z_1)$

The midpoint of AB has position vector $\frac{1}{2}(\mathbf{a} + \mathbf{b})$

A point which divides AB in the ratio $\lambda:\mu$ has position vector $\dfrac{\mu\mathbf{a} + \lambda\mathbf{b}}{\lambda + \mu}$

3) A line passing through a point (x_1, y_1, z_1) and with direction ratios $a:b:c$ has a vector equation $\mathbf{r} = x_1\mathbf{i} + y_1\mathbf{j} + z_1\mathbf{k} + \lambda(a\mathbf{i} + b\mathbf{j} + c\mathbf{k})$

and Cartesian equations $\dfrac{x - x_1}{a} = \dfrac{y - y_1}{b} = \dfrac{z - z_1}{c}$

4) Two lines with direction ratios $a_1 : b_1 : c_1$ and $a_2 : b_2 : c_2$ are parallel if $a_1/a_2 = b_1/b_2 = c_1/c_2$.

5) Two lines with equations:

$$\mathbf{r}_1 = x_1\mathbf{i} + y_1\mathbf{j} + z_1\mathbf{k} + \lambda(a_1\mathbf{i} + b_1\mathbf{j} + c_1\mathbf{k})$$

$$\mathbf{r}_2 = x_2\mathbf{i} + y_2\mathbf{j} + z_2\mathbf{k} + \mu(a_2\mathbf{i} + b_2\mathbf{j} + c_2\mathbf{k})$$

intersect if values of λ and μ can be found for which $\mathbf{r}_1 = \mathbf{r}_2$.

6) The angle θ between two lines with direction cosines l_1, m_1, n_1 and l_2, m_2, n_2 is given by

$$\cos\theta = l_1 l_2 + m_1 m_2 + n_1 n_2$$

Hence for perpendicular lines

$$l_1 l_2 + m_1 m_2 + n_1 n_2 = 0.$$

MULTIPLE CHOICE EXERCISE 1

Instructions for answering these questions are given on page (xi)

TYPE I

1) The modulus of a vector $6\mathbf{i} - 2\mathbf{j} - 3\mathbf{k}$ is:
(a) $\sqrt{23}$ (b) 7 (c) 1 (d) 49 (e) $\sqrt{11}$.

2) The direction cosines of a vector $\mathbf{i} + \mathbf{j} + \mathbf{k}$ are:

(a) 1, 1, 1 (b) $\frac{1}{3}, \frac{1}{3}, \frac{1}{3}$ (c) $\frac{1}{\sqrt{3}}, \frac{1}{\sqrt{3}}, \frac{1}{\sqrt{3}}$ (d) $\frac{1}{\sqrt{2}}, \frac{1}{\sqrt{2}}, \frac{1}{\sqrt{2}}$

(e) $\frac{-1}{\sqrt{3}}, \frac{-1}{\sqrt{3}}, \frac{-1}{\sqrt{3}}$.

3) Two vectors $4\mathbf{i} + 12\mathbf{j} - 6\mathbf{k}$ and $-2\mathbf{i} - 6\mathbf{j} + 3\mathbf{k}$ are:
(a) parallel (b) equal and opposite (c) equal (d) collinear
(e) none of these.

4) The equations:

$$\mathbf{r} = \mathbf{i} + a\mathbf{j} + \mathbf{k} + \lambda(\mathbf{i} + b\mathbf{j} + \mathbf{k})$$

and

$$\frac{x-1}{1} = \frac{y-2}{3} = \frac{1-z}{c}$$

represent the same line if:
(a) $b = 2, a = 3, c = 1$ (b) $b = 3, a = 2, c = 1$
(c) $b = -2, a = 3, c = -1$ (d) $b = 2, a = -3, c = -1$
(e) $b = 3, a = 2, c = -1$.

5) A force vector of magnitude 11 units with direction vector $2\mathbf{i} + 3\mathbf{j} + 6\mathbf{k}$ is:
(a) $11(2\mathbf{i} + 3\mathbf{j} + 6\mathbf{k})$ (b) $\frac{11}{7}(2\mathbf{i} + 3\mathbf{j} + 6\mathbf{k})$
(c) $\frac{1}{11}(2\mathbf{i} + 3\mathbf{j} + 6\mathbf{k})$ (d) $\frac{7}{11}(2\mathbf{i} + 3\mathbf{j} + 6\mathbf{k})$.

TYPE II

1) A line has vector equation $\mathbf{r} = \mathbf{i} + 2\mathbf{j} + 3\mathbf{k} + \lambda(4\mathbf{i} - \mathbf{j} + 7\mathbf{k})$.
(a) The line passes through the point (4, −1, 7).
(b) The line has a modulus $\sqrt{14}$.
(c) The line has direction cosines 4, −1, 7.
(d) The line passes through the point (1, 2, 3).

2) Two lines have equations

$$\mathbf{r} = \mathbf{i} + \mathbf{j} + \mathbf{k} + \lambda(2\mathbf{i} - 3\mathbf{j} + 5\mathbf{k})$$

$$\frac{x-3}{-2} = \frac{y+2}{3} = \frac{z-6}{-5}$$

(a) The lines are parallel.
(b) The first line passes through the point (1, 1, 1).
(c) The second line passes through the point (−2, 3, −5).
(d) The two equations represent the same line.

3) A vector $\mathbf{V} = 3\mathbf{i} + 3\mathbf{j} + 3\mathbf{k}$.
(a) $|\mathbf{V}| = 3\sqrt{3}$
(b) $\underset{\sim}{\mathbf{V}}$ is equally inclined to the three axes.
(c) $\hat{\mathbf{V}} = \mathbf{i} + \mathbf{j} + \mathbf{k}$.
(d) $\mathbf{V}$ is inclined to $\mathbf{i}$ at arccos 3.

4) The relationship between two vectors **a** and **b** is $\mathbf{a} = \lambda \mathbf{b}$.
(a) **a** and **b** are equal.
(b) **a** and **b** must be in the same sense.
(c) **a** and **b** must be collinear.
(d) **a** and **b** are parallel.

TYPE III

1) The position vectors of A, B and C are **a**, **b** and **c** respectively.
(a) $5\mathbf{c} = 3\mathbf{a} + 2\mathbf{b}$.
(b) AC:CB = 3:2.

2) The position vectors of points P, A and B are **r**, **a** and **b** respectively.
(a) $\mathbf{r} = \mathbf{a} + \lambda\mathbf{b}$.
(b) P is on the line through A and B.

3) AB and PQ have direction cosines l, m, n and l_1, m_1, n_1.
(a) AB is parallel to PQ.
(b) $ll_1 + mm_1 + nn_1 = 0$.

4) A, B and C are collinear points with position vectors **a**, **b** and **c** respectively.
(a) $2\mathbf{c} = \mathbf{a} - \mathbf{b}$.
(b) C is the mid-point of AB.

TYPE IV

1) Find the angle between two lines L_1 and L_2 if
(a) the direction of L_1 is given,
(b) the lines intersect,
(c) the direction of L_2 is given,
(d) L_1 passes through a given point A.

2) Find a vector equation for a line through A and B.
(a) A has position vector **a**,
(b) B has position vector **b**,
(c) the line has direction cosines l, m, n,
(d) A and B are collinear.

3) Find whether two lines L_1 and L_2 intersect, given:
(a) the direction of L_1,
(b) a point on L_1,
(c) L_1 and L_2 are not parallel,
(d) the direction of L_2.

4) Find the resultant of forces $\mathbf{F}_1$ and $\mathbf{F}_2$ if:
(a) $\mathbf{F}_1 = 2\mathbf{i} + 3\mathbf{j} + 4\mathbf{k}$,
(b) $|\mathbf{F}_2| = 6$,
(c) $\mathbf{F}_2$ passes through a point (4, 0, 7),

(d) $\mathbf{F}_2$ has direction cosines $\frac{1}{\sqrt{2}}, \frac{1}{2}, \frac{1}{2}$.

TYPE V

1) Two lines are parallel if their direction cosines are equal.

2) Two lines which do not intersect must be parallel.

3) If $a:b:c$ are the direction ratios of a vector, then $a^2 + b^2 + c^2 = 1$.

4) The equations $\frac{x-1}{2} = \frac{y-3}{1} = \frac{z-4}{3}$ and $\frac{x-3}{2} = \frac{y-4}{1} = \frac{z-7}{3}$

cannot represent the same line.

MISCELLANEOUS EXERCISE 1

1) The vertices of a tetrahedron are A(1, 0, 1), B(3, 1, 2), C(0, 3, 1) and D(5, 2, −7). Find:
(a) the lengths of AB and AC,
(b) the cosine of angle BCA,
(c) the area of triangle ABC.
Show that DA is perpendicular to BC.

2) If a point P has position vector $\mathbf{r} = 4\mathbf{i} - \mathbf{j} + 8\mathbf{k}$ find the magnitude of $\mathbf{r}$ and the angles between OP and the unit vectors $\mathbf{i}, \mathbf{j}, \mathbf{k}$.
Show that the line through P, with direction vector $2\mathbf{i} + 3\mathbf{j} + \mathbf{k}$, passes through the point Q with position vector $-2\mathbf{i} - 10\mathbf{j} + 5\mathbf{k}$.
Express the equation of the line OQ
(a) in vector form,
(b) in Cartesian form.

3) P and Q are two points with position vectors $5\mathbf{i} + 3\mathbf{j} - 13\mathbf{k}$ and $-3\mathbf{i} - \mathbf{j} + 7\mathbf{k}$, relative to the origin O. Find:
(a) a vector equation of the line PQ,
(b) a vector equation of the line through O that is perpendicular to PQ and which lies in the xy plane,
(c) the position vector of the point of intersection of these two lines.

4) Two lines have Cartesian equations

$$\frac{x-4}{1} = \frac{y-5}{2} = \frac{z-6}{2}$$

and

$$\frac{x+3}{3} = \frac{y-3}{2} = \frac{z+8}{6}$$

Show that these lines intersect and find the co-ordinates of the common point. Find also the cosine of the angle between the lines.

5) A, B and C are points with position vectors $3\mathbf{j} + 12\mathbf{k}$, $-3\mathbf{i} + 3\mathbf{j} + 6\mathbf{k}$ and $3\mathbf{i} - 6\mathbf{j} + 9\mathbf{k}$. If P, Q and R are points on BC, CA and AB respectively such that BP:PC = 2:1, CQ:QA = 1:2 and AR = RB, find:
(a) vector equations for the lines AP and BQ,
(b) the position vector of the point of intersection of AP and BQ.
Prove that CR, AP and BQ are concurrent.

6) A vector of magnitude XY acting from X towards Y is denoted by **XY**.
A straight line cuts the sides CA, AB of a triangle ABC internally at E, F respectively, and cuts BC produced at D. If BD/CD $= p$, CE/EA $= q$, AF/FB $= r$, prove that

$$\mathbf{EF} = \frac{1}{1+q}\mathbf{CA} + \frac{r}{1+r}\mathbf{AB}$$

and

$$\mathbf{DF} = \frac{p}{p-1}\mathbf{CA} + \frac{pr+1}{(p-1)(r+1)}\mathbf{AB}$$

Deduce that $pqr = 1$. (Cambridge)

7) If forces $\lambda\overrightarrow{AB}$ and $\mu\overrightarrow{AC}$ act along AB and AC, show that the resultant is $(\lambda + \mu)\overrightarrow{AD}$, where D is the point in BC such that BD:DC $= \mu:\lambda$.
The position vectors of the points A, B and C are $(\mathbf{i} + 2\mathbf{j} + 3\mathbf{k})$, $(4\mathbf{i} + 2\mathbf{j} - \mathbf{k})$ and $7\mathbf{i}$ respectively, the unit of distance being the metre.
Forces of magnitude 10 N and 7 N act along AB and AC respectively. If D is a point in BC, find the resultant of these forces in the form $n\overrightarrow{AD}$ giving the value of n and the position vector of D.
If these forces move a particle from A to D, find the work done by each force (stating the units). (AEB)

8) Show that each of the following systems of forces is in equilibrium:
(i) forces $3\overrightarrow{AB}$ and $4\overrightarrow{AC}$, where A, B and C have position vectors $(4\mathbf{i} - \mathbf{k})$, $(4\mathbf{j} + 3\mathbf{k})$ and $(7\mathbf{i} - 3\mathbf{j} - 4\mathbf{k})$ respectively;
(ii) force $\mathbf{F}_1 = 3\mathbf{i} + 4\mathbf{j} + 5\mathbf{k}$ acting at the point $L(7\mathbf{i} + 9\mathbf{j} + 11\mathbf{k})$
$\mathbf{F}_2 = \mathbf{i} + \mathbf{j} + \mathbf{k}$ acting at the point $M(4\mathbf{i} + 4\mathbf{j} + 4\mathbf{k})$
$\mathbf{F}_3 = -4\mathbf{i} - 5\mathbf{j} - 6\mathbf{k}$ acting at the point $N(5\mathbf{i} + 6\mathbf{j} + 7\mathbf{k})$.
Find the cosine of the angle between the lines of action of the forces $\mathbf{F}_1$ and $\mathbf{F}_2$.
(AEB)

9) Two forces $\mathbf{F}_1$ and $\mathbf{F}_2$ act through points with position vectors $\mathbf{a}_1$ and $\mathbf{a}_2$ respectively:
$\mathbf{F}_1 = 2\mathbf{i} - 2\mathbf{k}$, $\mathbf{F}_2 = \mathbf{i} - 2\mathbf{j} + \mathbf{k}$, $\mathbf{a}_1 = \mathbf{i} + 2\mathbf{j} - 3\mathbf{k}$ and $\mathbf{a}_2 = 4\mathbf{i} - p\mathbf{k}$.
Find the value of p if the lines of action of $\mathbf{F}_1$ and $\mathbf{F}_2$ intersect.
Find a vector equation of the line of action of the resultant of $\mathbf{F}_1$ and $\mathbf{F}_2$.

10) A force of magnitude XY acting along XY from X towards Y is denoted by **XY**.
The diagonals AC, BD of a quadrilateral ABCD meet at M, and AM/MC $= p$, BM/MD $= q$. Show that the four forces **AB**, p**CB**, q**AD**, pq**CD** produce equilibrium.
The side DA produced meets the side CB produced at N, and AN/DN $= \lambda$, BN/CN $= \mu$. Show that

$$\lambda[(1-\mu)\mathbf{ND} + \mu\mathbf{CD}] + (1-\lambda)\mathbf{BN} = \mathbf{BA}$$

and deduce that the forces

$$\mathbf{AB}, \quad \frac{\mu(1-\lambda)}{1-\mu}\mathbf{CB}, \quad \frac{\lambda(1-\mu)}{1-\lambda}\mathbf{AD}, \quad \mu\lambda\mathbf{CD}$$

produce equilibrium. (Cambridge)

11) Three forces $\mathbf{F}_1$, $\mathbf{F}_2$ and $\mathbf{F}_3$ act on a particle.
$\mathbf{F}_1$ has magnitude 14 N and acts in a direction parallel to the line with vector equation $\mathbf{r} = \mathbf{i} - 3\mathbf{j} + 8\mathbf{k} + \lambda(6\mathbf{i} - 2\mathbf{j} + 3\mathbf{k})$.
$\mathbf{F}_2 = a\mathbf{i} + b\mathbf{j} + c\mathbf{k}$. $\mathbf{F}_3$ is represented in magnitude and direction by the line AB where A and B have position vectors $7\mathbf{i} - 2\mathbf{j} + 5\mathbf{k}$ and $3\mathbf{i} + \mathbf{j} + \mathbf{k}$ respectively. Find the values of a, b and c if the particle does not accelerate.

12) Show that the vectors $\mathbf{v}_1 = \mathbf{i} + 2\mathbf{j} - \mathbf{k}$, $\mathbf{v}_2 = 4\mathbf{i} - \mathbf{j} + 2\mathbf{k}$ and $\mathbf{v}_3 = \mathbf{i} - 2\mathbf{j} - 3\mathbf{k}$ are mutually perpendicular. Find the magnitude and direction cosines of the resultant of $\mathbf{v}_1$, $\mathbf{v}_2$ and $\mathbf{v}_3$.

13) If the points A, B, C have position vectors **a**, **b**, **c** from an origin O, show that the equation $\mathbf{r} = t\mathbf{a} + (1-t)\mathbf{b}$, where t is a parameter, represents the straight line AB, and find the equation of BC.
Find the equation of the straight line joining L the mid-point of OA to M the mid-point of BC. Find also the position vector of the point in which the line LM meets the straight line joining the mid-point of OB to the mid-point of AC. (U of L)

14) A metal plate lies on a smooth horizontal table: **i** and **j** are unit vectors in two perpendicular lines Ox and Oy in the plane of the table. Three forces are applied to the plate as follows (the units of force and displacement are the newton and the metre).
The first force is $3\mathbf{i} + 4\mathbf{j}$ and it acts at a point with position vector $2\mathbf{i} - \mathbf{j}$.
The second force is $6\mathbf{i} - 2\mathbf{j}$ and it acts at a point with position vector $7\mathbf{i} + \mathbf{j}$. The third force is $3\mathbf{i} + 3\mathbf{j}$.
If the plate does not rotate, show that the third force passes through the point with position vector $\frac{21}{5}\mathbf{i} + \frac{29}{15}\mathbf{j}$.
Find the magnitude and direction of the resultant force acting on the plate and give the equation of its line of action in vector form.

15) Three points have position vectors $p\mathbf{x}$, $q\mathbf{y}$ and $r\mathbf{x} + s\mathbf{y}$, where $\mathbf{x}$ and $\mathbf{y}$ are non-parallel vectors and p, q, r, and s are scalars. If the points are collinear prove that $ps + qr = pq$.
Four points A, B, C, D have position vectors $\mathbf{a}$, $\mathbf{b}$, $5\mathbf{a}$, $3\mathbf{b}$ respectively, where $\mathbf{a}$ and $\mathbf{b}$ are non-parallel vectors. If the point of intersection of AB and CD has position vector $m\mathbf{a} + n\mathbf{b}$, find m and n. If, further, the point of intersection of AD and BC has position vector $m'\mathbf{a} + n'\mathbf{b}$, verify that $mn' + nm' = 0$.
(Cambridge)

16) The position vectors of the vertices B and C of a triangle ABC are respectively $8\mathbf{i} + 3\mathbf{j} + 5\mathbf{k}$ and $6\mathbf{i} + 4\mathbf{j} + 9\mathbf{k}$. Two forces $3\mathbf{i} + 2\mathbf{j} + \mathbf{k}$ and $4\mathbf{i} + 5\mathbf{j} + 6\mathbf{k}$ act along AB and AC respectively. A third force $\mathbf{F}$ acts through A.
If the system of forces is in equilibrium, find:
(a) the magnitude of the force $\mathbf{F}$,
(b) the position vector of A,
(c) the equation of the line of action of $\mathbf{F}$ in vector form. (U of L)

17) ABCD is a plane quadrilateral. The position vectors of the vertices are $\mathbf{a}$, $\mathbf{b}$, $\mathbf{c}$ and $\mathbf{d}$. If P is the mid-point of AB, Q is a point on BC such that BQ:QC = 2:1 and R divides CD externally in the ratio 2:1, find the position vectors of P, Q and R. Write down vector equations for the lines through (i) A and B (ii) P and Q. Find the position vector of the point where these lines intersect.

18) The position vectors of the vertices B and C of the triangle ABC are $-3\mathbf{i} + 3\mathbf{j} - 2\mathbf{k}$ and $-5\mathbf{i} + 14\mathbf{j} + \mathbf{k}$ respectively, relative to an origin O. The forces $4\mathbf{i} + 7\mathbf{j} + 5\mathbf{k}$ and $3\mathbf{i} - 2\mathbf{j} + \mathbf{k}$ act along the sides AB and AC respectively. The units of force are newtons. Calculate:
(i) the position vector of A,
(ii) the cosine of the angle BAC,
(iii) the magnitude of the resultant of the two forces. (AEB)

CHAPTER 2

CONSTANT VELOCITY VECTORS. SCALAR PRODUCT

When a body is moving with constant velocity, it is travelling in a straight line at constant speed.

If the constant velocity vector is $\mathbf{v}$, the constant speed is $|\mathbf{v}|$ and the displacement which the body undergoes in time t is of magnitude $t|\mathbf{v}|$ in the direction of $\mathbf{v}$.

Hence the displacement vector is $t\mathbf{v}$.

Then, if when $t = 0$ the body is at a point with position vector $\mathbf{a}$, its position vector after time t is given by

$$\mathbf{r} = \mathbf{a} + t\mathbf{v}$$

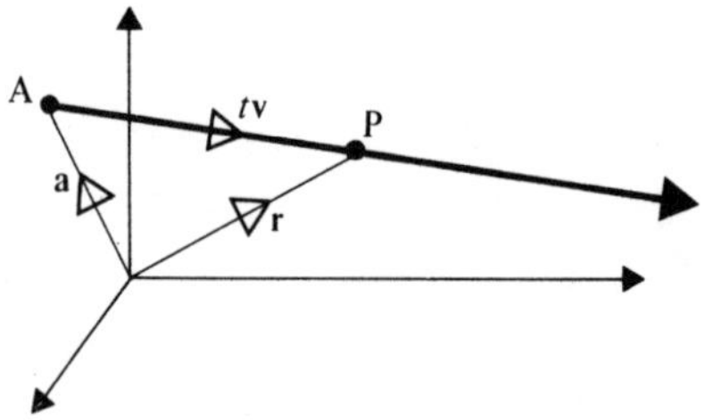

This is the vector equation of the line along which the body moves.

Consider two particles P_1 and P_2 moving with constant velocity vectors $\mathbf{v}_1$ and $\mathbf{v}_2$ respectively.

If P_1 and P_2 are initially at points with position vectors $\mathbf{a}_1$ and $\mathbf{a}_2$ and after time t they are at points with position vectors $\mathbf{r}_1$ and $\mathbf{r}_2$ respectively, then

$$\begin{cases} \mathbf{r}_1 = \mathbf{a}_1 + t\mathbf{v}_1 \\ \mathbf{r}_2 = \mathbf{a}_2 + t\mathbf{v}_2 \end{cases}$$

COLLISION

In order that P_1 and P_2 shall collide their paths must intersect, but also P_1 and P_2 must arrive simultaneously at the point of intersection.
Hence for collision there must be a value of t for which $\mathbf{r}_1 = \mathbf{r}_2$.

DISTANCE APART

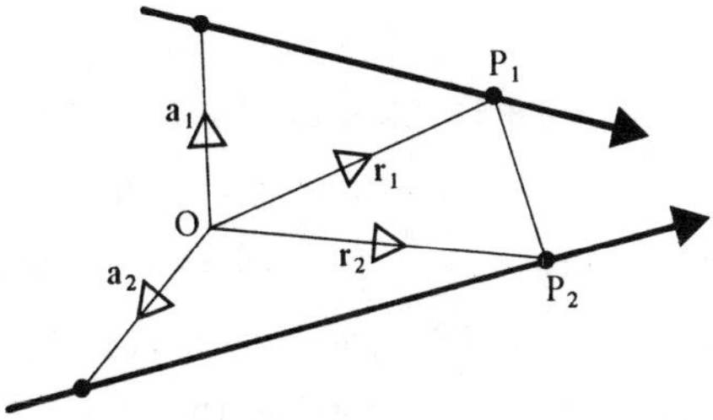

At any time t the line joining P_1 to P_2 represents the vector $\mathbf{r}_2 - \mathbf{r}_1$
The distance between the two particles is the length of the line P_1P_2, i.e. $|\mathbf{r}_2 - \mathbf{r}_1|$

EXAMPLES 2a

1) A particle P is moving with a constant speed of 6 units in a direction $2\mathbf{i} - \mathbf{j} - 2\mathbf{k}$. When $t = 0$, P is at a point with position vector $3\mathbf{i} + 4\mathbf{j} - 7\mathbf{k}$. Find the position vector of P after (a) t seconds, (b) 4 seconds.

If $\mathbf{v}$ is the velocity vector of P then the direction of $\mathbf{v}$ is $2\mathbf{i} - \mathbf{j} - 2\mathbf{k}$.

Hence $\qquad \hat{\mathbf{v}} = \frac{1}{3}(2\mathbf{i} - \mathbf{j} - 2\mathbf{k})$

But $\qquad |\mathbf{v}| = 6$

Therefore $\qquad \mathbf{v} = |\mathbf{v}|\hat{\mathbf{v}} = 4\mathbf{i} - 2\mathbf{j} - 4\mathbf{k}$

If $\mathbf{d}$ is the displacement vector of P in time t,

$$\mathbf{d} = t(4\mathbf{i} - 2\mathbf{j} - 4\mathbf{k})$$

But P is initially at $3\mathbf{i} + 4\mathbf{j} - 7\mathbf{k}$, so if $\mathbf{r}$ is the position vector of P after time t then:

(a) $$\mathbf{r} = 3\mathbf{i}+4\mathbf{j}-7\mathbf{k}+t(4\mathbf{i}-2\mathbf{j}-4\mathbf{k})$$

(b) When $t = 4$

$$\mathbf{r} = 3\mathbf{i}+4\mathbf{j}-7\mathbf{k}+4(4\mathbf{i}-2\mathbf{j}-4\mathbf{k})$$

Hence $$\mathbf{r} = 19\mathbf{i}-4\mathbf{j}-23\mathbf{k}$$

2) Two particles A and B pass simultaneously through two points A_0 and B_0 with position vectors $\mathbf{a}_0$ and $\mathbf{b}_0$. A and B are moving with constant velocities $\mathbf{v}_A$ and $\mathbf{v}_B$.
If $\mathbf{a}_0 = \mathbf{i}+4\mathbf{j}-26\mathbf{k}$, $\mathbf{b}_0 = 16\mathbf{i}+\mathbf{j}-2\mathbf{k}$, $\mathbf{v}_A = 4\mathbf{i}+\mathbf{j}+5\mathbf{k}$ and $\mathbf{v}_B = -\mathbf{i}+2\mathbf{j}-3\mathbf{k}$, show that A and B will collide.
Find the time at which collision occurs and the position vector of the point of collision.

If at time t after passing through A_0 and B_0, the position vectors of A and B are $\mathbf{r}_A$ and $\mathbf{r}_B$ then,

$$\begin{cases}\mathbf{r}_A = \mathbf{i}+4\mathbf{j}-26\mathbf{k}+t(4\mathbf{i}+\mathbf{j}+5\mathbf{k})\\ \mathbf{r}_B = 16\mathbf{i}+\mathbf{j}-2\mathbf{k}+t(-\mathbf{i}+2\mathbf{j}-3\mathbf{k})\end{cases}$$

or $$\begin{cases}\mathbf{r}_A = (1+4t)\mathbf{i}+(4+t)\mathbf{j}+(-26+5t)\mathbf{k}\\ \mathbf{r}_B = (16-t)\mathbf{i}+(1+2t)\mathbf{j}+(-2-3t)\mathbf{k}\end{cases}$$

Now A and B will collide if there is a particular value of t for which $\mathbf{r}_A = \mathbf{r}_B$.
Equating coefficients of $\mathbf{i}$, $1+4t = 16-t$ so $t = 3$
A collision will occur if, when $t = 3$, the coefficients of $\mathbf{j}$ and $\mathbf{k}$ are also equal for A and B.

Using $t = 3$, $$\left.\begin{aligned}\mathbf{r}_A &= 13\mathbf{i}+7\mathbf{j}-11\mathbf{k}\\ \mathbf{r}_B &= 13\mathbf{i}+7\mathbf{j}-11\mathbf{k}\end{aligned}\right\}\text{ equal coefficients of } \mathbf{i}, \mathbf{j} \text{ and } \mathbf{k}$$

Hence, when $t = 3$, A and B collide at a point with position vector
$$13\mathbf{i}+7\mathbf{j}-11\mathbf{k}.$$

3) Two ships P and Q are moving with constant velocity vectors $\mathbf{i}+4\mathbf{j}$ and $2\mathbf{i}+\mathbf{j}$ respectively. At a certain time P passes through a point P_0 with position vector $5\mathbf{i}-9\mathbf{j}$ and, two seconds later, Q passes through a point Q_0 with position vector $3\mathbf{i}-\mathbf{j}$. Show that the ships are not on collision course and find the bearing of Q from P four seconds after it passes through Q_0, using $\mathbf{i}$ and $\mathbf{j}$ as unit vectors to the East and North respectively.

At the time t seconds after Q passes through Q_0, P has been travelling for $(t+2)$ seconds after passing through P_0.

Therefore, for Q $$\mathbf{v}_Q = 2\mathbf{i}+\mathbf{j}$$

$$\begin{aligned}\mathbf{r}_Q &= 3\mathbf{i}-\mathbf{j}+t(2\mathbf{i}+\mathbf{j})\\ &= (3+2t)\mathbf{i}+(-1+t)\mathbf{j}\end{aligned}$$

and for P $\quad \mathbf{v_P} = \mathbf{i} + 4\mathbf{j}$

$$\mathbf{r_P} = 5\mathbf{i} - 9\mathbf{j} + (t+2)(\mathbf{i} + 4\mathbf{j})$$

$$= (7+t)\mathbf{i} + (-1+4t)\mathbf{j}$$

Now the ships will collide if there is a value of t which makes $\mathbf{r_P} = \mathbf{r_Q}$.

The coefficients of **i** are equal when $3 + 2t = 7 + t$

i.e. when $\quad t = 4$

At this time $\quad \mathbf{r_Q} = (3+8)\mathbf{i} + (-1+4)\mathbf{j} = 11\mathbf{i} + 3\mathbf{j}$

and $\quad \mathbf{r_P} = (7+4)\mathbf{i} + (-1+16)\mathbf{j} = 11\mathbf{i} + 15\mathbf{j}$

So when $t = 4$ the coefficients of **j** are not equal showing that the ships do not collide.

The displacement of Q from P when $t = 4$ is

$$\mathbf{r_Q} - \mathbf{r_P} = 11\mathbf{i} + 3\mathbf{j} - (11\mathbf{i} + 15\mathbf{j})$$

$$= -12\mathbf{j}$$

Now the direction of **j** is due North.

Therefore the bearing of Q from P when $t = 4$ is due South.

4) Two particles are moving with constant velocities. At a particular instant the particle with velocity vector $3\mathbf{i} - 4\mathbf{j} + 7\mathbf{k}$ passes through a point with position vector $\mathbf{i} + 2\mathbf{j} - 3\mathbf{k}$ and the particle with velocity vector $2\mathbf{i} - 6\mathbf{j} + 5\mathbf{k}$ passes through the origin. Find the shortest distance between the particles in the ensuing motion and the time when they are closest together.

Let the two particles be A and B.

Then $\quad \mathbf{v_A} = 3\mathbf{i} - 4\mathbf{j} + 7\mathbf{k}$

$$\mathbf{v_B} = 2\mathbf{i} - 6\mathbf{j} + 5\mathbf{k}$$

At time t $\quad \mathbf{r_A} = \mathbf{i} + 2\mathbf{j} - 3\mathbf{k} + t(3\mathbf{i} - 4\mathbf{j} + 7\mathbf{k})$

$$\mathbf{r_B} = \mathbf{0} + t(2\mathbf{i} - 6\mathbf{j} + 5\mathbf{k})$$

So that $\quad \mathbf{r_A} - \mathbf{r_B} = \mathbf{i} + 2\mathbf{j} - 3\mathbf{k} + t(\mathbf{i} + 2\mathbf{j} + 2\mathbf{k})$

The distance between A and B at time t is given by

$$l = |\mathbf{r_A} - \mathbf{r_B}|$$

$$l^2 = (t+1)^2 + (2t+2)^2 + (2t-3)^2$$

So that $\quad l^2 = 9t^2 - 2t + 14$

The least value of l can now be found using either of the following methods.

(i)
$$l^2 = 9t^2 - 2t + 14$$

Therefore
$$\frac{l^2}{9} = t^2 - \frac{2}{9}t + \frac{14}{9}$$
$$= \left(t - \frac{1}{9}\right)^2 + \frac{14}{9} - \left(\frac{1}{9}\right)^2$$

The expression on the right is least when $\left(t - \frac{1}{9}\right)^2 = 0$

Hence l is least when $t = \frac{1}{9}$, and then

$$l^2 = 9\left(\frac{14}{9} - \frac{1}{81}\right) = \frac{125}{9}$$

So the shortest distance between A and B is $\frac{5\sqrt{5}}{3}$ units

and occurs when $t = \frac{1}{9}$.

(ii)
$$l^2 = 9t^2 - 2t + 14$$
$$\frac{d(l^2)}{dt} = 18t - 2$$

and
$$\frac{d^2(l^2)}{dt^2} = 18$$

When l^2 is least, $\frac{d(l^2)}{dt} = 0$, therefore $t = \frac{1}{9}$

and since $\frac{d^2(l^2)}{dt^2}$ is positive, l^2 is minimum when $t = \frac{1}{9}$

Then
$$l^2 = 9\left(\frac{1}{9}\right)^2 - 2\left(\frac{1}{9}\right) + 14 = 14 - \frac{1}{9}$$

So the minimum value of l is $\frac{5\sqrt{5}}{3}$ units.

5) Two spheres A and B, with equal radii 3 units, are moving with constant velocity vectors $\mathbf{i} + 5\mathbf{j} - 3\mathbf{k}$ and $2\mathbf{i} + 9\mathbf{j} - 11\mathbf{k}$ respectively. Their centres pass simultaneously through two points with position vectors $6\mathbf{i} + 3\mathbf{j} + \mathbf{k}$ and $\mathbf{i} + 3\mathbf{j} + 11\mathbf{k}$ respectively. Show that they collide and find the time which elapses before impact.

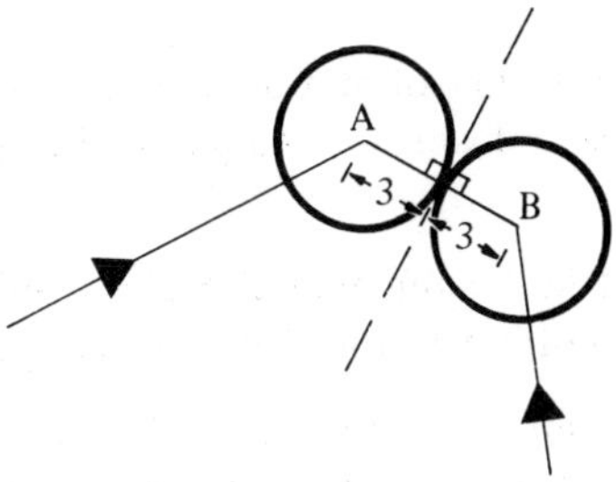

If the spheres collide their centres will be exactly 6 units apart at the instant of impact.
Considering the motion of the centre of each sphere, their position vectors at time t are respectively

$$\mathbf{r}_A = 6\mathbf{i} + 3\mathbf{j} + \mathbf{k} + t(\mathbf{i} + 5\mathbf{j} - 3\mathbf{k})$$
$$= (6 + t)\mathbf{i} + (3 + 5t)\mathbf{j} + (1 - 3t)\mathbf{k}$$
$$\mathbf{r}_B = \mathbf{i} + 3\mathbf{j} + 11\mathbf{k} + t(2\mathbf{i} + 9\mathbf{j} - 11\mathbf{k})$$
$$= (1 + 2t)\mathbf{i} + (3 + 9t)\mathbf{j} + (11 - 11t)\mathbf{k}$$

The distance l between the centres at any time t is given by

$$l = |\mathbf{r}_A - \mathbf{r}_B|$$

But $$\mathbf{r}_A - \mathbf{r}_B = (5 - t)\mathbf{i} + (-4t)\mathbf{j} + (-10 + 8t)\mathbf{k}$$

Therefore $$l^2 = (5 - t)^2 + (-4t)^2 + (-10 + 8t)^2$$
$$= 81t^2 - 170t + 125$$

Now for collision there must be a value of t for which $l = 6$

So that $$36 = 81t^2 - 170t + 125 \qquad (1)$$
$$0 = 81t^2 - 170t + 89$$
$$0 = (81t - 89)(t - 1)$$

Hence $$t = 1 \text{ or } t = \tfrac{89}{81}$$

But the spheres will collide only once and that will be after the shorter time interval.
So the spheres *do* collide after 1 unit of time.
Note: Had the roots of equation (1) been complex the conclusion would have been that A and B never collide.

EXERCISE 2a

1) Find a vector equation for the line of motion of a particle which has a constant velocity $2\mathbf{i} + 7\mathbf{j} - 6\mathbf{k}$ and which is at a point with position vector $\mathbf{i} - 5\mathbf{j} + 2\mathbf{k}$ when $t = 0$.

2) A particle P has a speed $6v$ in the direction $2\mathbf{i} + 2\mathbf{j} - \mathbf{k}$ and passes through the origin when $t = 0$. Find the position vector of P at time t.

3) A and B are points with position vectors $-3\mathbf{i} + 8\mathbf{j} + 11\mathbf{k}$ and $\mathbf{i} - 4\mathbf{j} + 3\mathbf{k}$ respectively. A particle P goes from A to B in 4 seconds with constant speed. Find the position vector of P seven seconds after passing through A.

4) P and Q are particles moving with constant velocities $\mathbf{v}_P$ and $\mathbf{v}_Q$. Determine whether they collide and if they do find the position vector of the point where collision occurs, given that:

(a) $\mathbf{v}_P = 3\mathbf{i} - \mathbf{j} - 2\mathbf{k}$ and $\mathbf{v}_Q = -7\mathbf{i} + 2\mathbf{j} + \mathbf{k}$. When $t = 0$, P is at the point $(0, 7, 0)$ and Q is at the point $(18, 1, 2)$.

(b) $\mathbf{v}_P = 3\mathbf{i} - 2\mathbf{j} - \mathbf{k}$ and $\mathbf{v}_Q = 3\mathbf{i} + 4\mathbf{j} + 5\mathbf{k}$. P is at a point with position vector $-8\mathbf{i} + 4\mathbf{j} + 4\mathbf{k}$ when $t = 0$ and one second later Q passes through the point with position vector $-5\mathbf{i} - 10\mathbf{j} - 9\mathbf{k}$.

(c) $\mathbf{v}_P = \mathbf{i} - \mathbf{j} - \mathbf{k}$ and $\mathbf{v}_Q = 2\mathbf{i} + 5\mathbf{j} - 3\mathbf{k}$. P and Q are at points $(1, 4, 3)$ and $(7, 6, 2)$ when $t = 0$.

5) Two particles P and Q are observed simultaneously when they are at points with position vectors $\mathbf{p}$ and $\mathbf{q}$. If they are travelling with constant velocity vectors $\mathbf{v}_P$ and $\mathbf{v}_Q$ respectively, find their distance apart after time t and hence find the least distance between them, given that:

(a) $\mathbf{p} = -\mathbf{i} + 8\mathbf{j} - 4\mathbf{k}$ $\quad$ $\mathbf{q} = 4\mathbf{i} + 6\mathbf{j} - 15\mathbf{k}$
$\mathbf{v}_P = 5\mathbf{i} + \mathbf{j} - 6\mathbf{k}$ $\quad$ $\mathbf{v}_Q = -3\mathbf{i} + \mathbf{j} - \mathbf{k}$

(b) $\mathbf{p} = -3\mathbf{i} - 3\mathbf{j} - 3\mathbf{k}$ $\quad$ $\mathbf{q} = -7\mathbf{i} - 4\mathbf{j} - 2\mathbf{k}$
$\mathbf{v}_P = \mathbf{i} + \mathbf{j} + \mathbf{k}$ $\quad$ $\mathbf{v}_Q = 3\mathbf{i} + 2\mathbf{j} + \mathbf{k}$

6) A ship A travelling at a speed of 15 knots in a direction $-3\mathbf{i} + 4\mathbf{j}$ sights a second ship B which is $20\sqrt{2}$ nautical miles from A in a direction $-7\mathbf{i} + \mathbf{j}$. B is moving in the direction $\mathbf{i} + 2\mathbf{j}$. Show that if B has a speed of $5\sqrt{5}$ knots the ships are on collision course. If, however, B's speed is $3\sqrt{5}$ knots, find the least distance between the ships.

7) Three particles A, B and C, moving with constant velocities $\mathbf{v}_A$, $\mathbf{v}_B$ and $\mathbf{v}_C$ are observed at position vectors $\mathbf{a}$, $\mathbf{b}$ and $\mathbf{c}$ at the times indicated below. Two of the three will collide. Find which two are on collision courses and determine the time and position of the collision.

$\mathbf{v}_A = 2\mathbf{i} + 4\mathbf{j} - 7\mathbf{k}$;	$\mathbf{a} = -5\mathbf{i} - 10\mathbf{j} + 24\mathbf{k}$	at a certain time
$\mathbf{v}_B = \mathbf{i} - \mathbf{j} + 2\mathbf{k}$;	$\mathbf{b} = 3\mathbf{i} + 4\mathbf{j} - \mathbf{k}$	one second later
$\mathbf{v}_C = -5\mathbf{i} + \mathbf{j} + 6\mathbf{k}$;	$\mathbf{c} = 6\mathbf{i} + \mathbf{j} - 3\mathbf{k}$	two seconds later

8) A sphere of radius 3 units is moving so that its centre, initially at the origin, has a constant speed of 5 units along Ox. Another sphere of radius 2 units has a constant velocity vector $4\mathbf{i} - \mathbf{j}$ and its centre is initially at the point with position vector $4\mathbf{i} + 11\mathbf{j}$. Show that the spheres collide when $t = 7$ and find the position vectors of the centres of both spheres at that time. (**i** and **j** are unit vectors parallel to Ox and Oy respectively.)

RELATIVE VELOCITY

The velocity of one object A relative to another object B is the velocity which A appears to have when viewed from B, i.e. it is the vector difference between the velocity of A and the velocity of B.
So, if $\mathbf{v}_A$ and $\mathbf{v}_B$ are the velocities of A and B,
the velocity of A relative to B is $\mathbf{v}_A - \mathbf{v}_B$.
Relative velocity does not depend upon the positions of the two objects.
Relative velocity problems can be solved using trigonometry, or by drawing and measuring a suitable scale diagram. These methods are dealt with in Volume One.
Certain problems, however, are more appropriately solved using vector analysis, as is illustrated in the following examples.

EXAMPLES 2b

1) Two bodies P and Q are moving with constant velocity vectors $4\mathbf{i} - 3\mathbf{j} + 7\mathbf{k}$ and $2\mathbf{i} + 3\mathbf{j} + 4\mathbf{k}$ respectively. Find their relative speed and a unit vector in the direction in which P seems to move when viewed from Q.

$$\mathbf{v}_P = 4\mathbf{i} - 3\mathbf{j} + 7\mathbf{k}$$
$$\mathbf{v}_Q = 2\mathbf{i} + 3\mathbf{j} + 4\mathbf{k}$$

The velocity of P relative is Q given by

$$\mathbf{v}_P - \mathbf{v}_Q = 2\mathbf{i} - 6\mathbf{j} + 3\mathbf{k}$$

The relative speed is $|\mathbf{v}_P - \mathbf{v}_Q| = \sqrt{2^2 + (-6)^2 + 3^2} = 7$

The direction of motion of P relative to Q is the direction of $\mathbf{v}_P - \mathbf{v}_Q$

Unit vector in this direction is $\frac{1}{7}(2\mathbf{i} - 6\mathbf{j} + 3\mathbf{k})$

i.e. $$\tfrac{2}{7}\mathbf{i} - \tfrac{6}{7}\mathbf{j} + \tfrac{3}{7}\mathbf{k}$$

[*Note*: The *velocity* of P relative to Q, $\mathbf{v}_P - \mathbf{v}_Q$, is *not* equal to the *velocity* of Q relative to P, $\mathbf{v}_Q - \mathbf{v}_P$, since these two *vectors* have opposite directions.
The *speed* of P relative to Q, $|\mathbf{v}_P - \mathbf{v}_Q|$, *is* however equal to the *speed* of Q relative to P, $|\mathbf{v}_Q - \mathbf{v}_P|$, since the relative speeds are *scalar* quantities.]

2) A boat P moves with constant speed $10\,\mathrm{ms}^{-1}$ from A to B where A and B have position vectors $2\mathbf{i} - \mathbf{j}$ and $5\mathbf{i} + 3\mathbf{j}$ respectively. A second boat Q appears

to P to have a velocity vector $-\mathbf{i}+4\mathbf{j}$. Using metres and seconds as units of distance and time, find the speed and the direction of motion of Q.

Let $\mathbf{v_P}$ and $\mathbf{v_Q}$ be the velocity vectors of P and Q.

For P: $|\mathbf{v_P}| = 10$

The direction vector of $\mathbf{v_P} = 5\mathbf{i}+3\mathbf{j}-(2\mathbf{i}-\mathbf{j})$

$= 3\mathbf{i}+4\mathbf{j}$

so $\hat{\mathbf{v}}_\mathbf{P} = \frac{1}{5}(3\mathbf{i}+4\mathbf{j})$

Hence $\mathbf{v_P} = |\mathbf{v_P}|\hat{\mathbf{v}}_\mathbf{P}$

$= 6\mathbf{i}+8\mathbf{j}$

Now $\mathbf{v_Q}-\mathbf{v_P} = -\mathbf{i}+4\mathbf{j}$

Therefore $\mathbf{v_Q} = -\mathbf{i}+4\mathbf{j}+6\mathbf{i}+8\mathbf{j}$

$= 5\mathbf{i}+12\mathbf{j}$

So Q's speed $= |\mathbf{v_Q}| = 13 \text{ ms}^{-1}$

And Q is moving in a direction $5\mathbf{i}+12\mathbf{j}$

3) A man is walking at constant speed u in the direction $\mathbf{j}$ and the wind appears to have a velocity $u_1(\sqrt{3}\mathbf{i}-3\mathbf{j})$. Without changing speed the man alters course so that he is walking in a direction $-\frac{\sqrt{3}}{2}\mathbf{i}+\frac{1}{2}\mathbf{j}$ and the velocity of the wind now appears to be $u_2\mathbf{i}$.
Find u_1 and u_2 in terms of u and express the true wind velocity in terms of u.

When the man's velocity is $u\mathbf{j}$, and the true wind velocity is $\mathbf{V}$

$$\mathbf{V}-u\mathbf{j} = u_1(\sqrt{3}\mathbf{i}-3\mathbf{j}) \qquad (1)$$

When the man is walking in the direction $-\frac{\sqrt{3}}{2}\mathbf{i}+\frac{1}{2}\mathbf{j}$, which is a unit vector, his velocity vector is $u\left(-\frac{\sqrt{3}}{2}\mathbf{i}+\frac{1}{2}\mathbf{j}\right)$.

So
$$\mathbf{V}-u\left(-\frac{\sqrt{3}}{2}\mathbf{i}+\frac{1}{2}\mathbf{j}\right) = u_2\mathbf{i} \qquad (2)$$

Eliminating $\mathbf{V}$ from (1) and (2) we have

$$\sqrt{3}u_1\mathbf{i}+(u-3u_1)\mathbf{j} = \left(u_2-\frac{\sqrt{3}}{2}u\right)\mathbf{i}+\frac{u}{2}\mathbf{j}$$

But two vectors are equal only if their components in specified directions are equal.

Therefore $$\sqrt{3}u_1 = u_2 - \frac{\sqrt{3}}{2}u$$

and $$u - 3u_1 = \frac{u}{2}$$

These equations give: $$u_1 = \frac{u}{6}; \quad u_2 = \frac{2\sqrt{3}}{3}u$$

But $$\mathbf{V} = u\mathbf{j} + u_1(\sqrt{3}\mathbf{i} - 3\mathbf{j})$$

$$\mathbf{V} = \frac{u}{6}(\sqrt{3}\mathbf{i} + 3\mathbf{j}).$$

EXERCISE 2b

1) Two bodies A and B have velocities $\mathbf{v}_A$ and $\mathbf{v}_B$ respectively. Find the velocity of A relative to B and hence the relative speed if:

(i) $\mathbf{v}_A = 3\mathbf{i} + 2\mathbf{j} - 4\mathbf{k}$ $\quad \mathbf{v}_B = \mathbf{i} - \mathbf{j} + 2\mathbf{k}$,

(ii) $\mathbf{v}_A = 5\mathbf{i} - 2\mathbf{j}$ $\quad \mathbf{v}_B = \mathbf{i} + \mathbf{j}$,

(iii) $\mathbf{v}_A = \mathbf{i} + \mathbf{j} + \mathbf{k}$ $\quad \mathbf{v}_B = \mathbf{i} - \mathbf{j} - \mathbf{k}$.

2) A boat is moving due East at 18 knots. The wind is blowing South West at 8 knots. What velocity does the wind appear to have to an observer on the boat? (Take **i** and **j** as unit vectors due South and due East respectively.)

3) A, B and C are three particles each moving with constant velocity. A's velocity is $3\mathbf{i} - 4\mathbf{j} + 5\mathbf{k}$. Relative to A, B's velocity is $\mathbf{i} - \mathbf{j} + 2\mathbf{k}$ and relative to C, B's velocity is $6\mathbf{i} - 2\mathbf{j} + 4\mathbf{k}$. Find the velocities of B and C.

4) To a cyclist travelling due S along a straight road at $10\,\text{kmh}^{-1}$ the wind appears to be blowing S 60°E. When he returns along the same road (travelling due N) at the same speed, the wind appears to be blowing N 30°E. Find the true velocity vector of the wind (**i** and **j** are unit vectors East and North respectively).

5) A, B and C have velocities $\mathbf{v}_A$, $\mathbf{v}_B$ and $\mathbf{v}_C$. If the velocity of C relative to B is equal to the velocity of B relative to A, show that the direction of motion of B is parallel to the resultant of the velocities of A and C.

6) A particle A moving with a constant velocity $\mathbf{i} + \mathbf{j} + \mathbf{k}$ passes through a point with position vector $3\mathbf{i} - 7\mathbf{j} - 4\mathbf{k}$ at the same instant as a particle B passes through a point with position vector $-\mathbf{i} + \mathbf{j} + p\mathbf{k}$. B has a constant velocity vector $2\mathbf{i} - \mathbf{j} - 5\mathbf{k}$.

(i) Find the velocity of B relative to A.

(ii) Find the value of p if A and B collide.

(iii) If $p = -\frac{1}{2}$, find the shortest distance between A and B in the subsequent motion.

7) Three particles A, B and C are moving with constant velocity vectors $\mathbf{v}_A$, $\mathbf{v}_B$ and $\mathbf{v}_C$. If $\mathbf{v}_C = -\mathbf{i}+\mathbf{j}+3\mathbf{k}$ and the velocities of A and B relative to C are $-3\mathbf{i}-8\mathbf{k}$ and $3\mathbf{i}-3\mathbf{j}-10\mathbf{k}$ respectively, find $\mathbf{v}_A$ and $\mathbf{v}_B$. At the same instant A and B are at points with position vectors $11\mathbf{i}-2\mathbf{j}+16\mathbf{k}$ and $-7\mathbf{i}+7\mathbf{j}+22\mathbf{k}$ respectively. Show that A and B collide.

THE SCALAR PRODUCT OF TWO VECTORS

Students should by now be aware that the addition of two vectors is a very different operation from the addition of two real numbers.
It is convenient to use the same terms (addition and subtraction) and the same symbols (+ and −) to represent both types of operation, but their interpretation depends on the form of the analysis being performed.
In the same way the multiplication of two vectors has very little similarity to the multiplication of two real numbers.
In fact there are two operations applied to vectors, both of which are called products.
The first of these results in a scalar quantity and it is therefore known as the *scalar product* of two vectors.
The second operation, which is referred to as the *vector product* of two vectors, produces a vector quantity.
Distinction is drawn between the two processes by using the multiplication "dot" symbol ($\mathbf{a}\cdot\mathbf{b}$) exclusively for the scalar product and the multiplication "cross" symbol ($\mathbf{a}\times\mathbf{b}$) exclusively for the vector product.
(This can cause some difficulty in vector work since the "dot" and "cross" can cause confusion if used to represent multiplication of numbers in the context of vector analysis. This problem can be avoided by the use of brackets.)

Consider two vectors **a** and **b** which are inclined to each other at an angle θ.

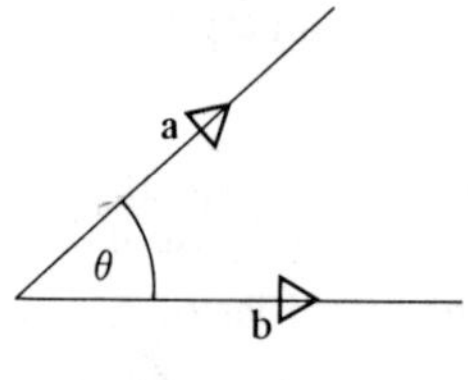

There are many occasions both in Mechanics and in Geometry when the use of vector methods involves $\cos\theta$ or $\sin\theta$ (e.g. resolving **a** in the direction **b**). For these reasons the scalar product and the vector product of **a** and **b** are defined so that the resulting expressions include $\cos\theta$ and $\sin\theta$ respectively.
Analysis of the scalar product now follows, while a study of the vector product appears in Chapter 4.

SCALAR PRODUCT

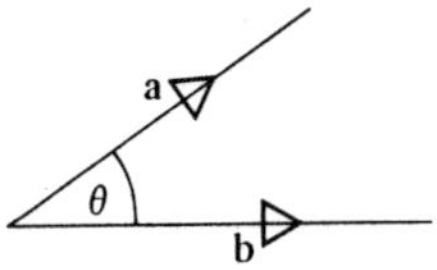

The scalar product of **a** and **b** is defined as

$$|\mathbf{a}||\mathbf{b}| \cos \theta$$

It is written as **a . b** and is spoken of as "**a** dot **b**" (in fact the scalar product is often referred to as the *dot product*).

Hence $$\mathbf{a}.\mathbf{b} = |\mathbf{a}||\mathbf{b}| \cos \theta$$

PROPERTIES OF THE SCALAR PRODUCT

1. The scalar product obeys the commutative and distributive laws since:

(i)
$$\begin{aligned} \mathbf{a}.\mathbf{b} &= |\mathbf{a}||\mathbf{b}| \cos \theta \\ &= |\mathbf{b}||\mathbf{a}| \cos \theta \\ &= \mathbf{b}.\mathbf{a} \end{aligned}$$

(ii)

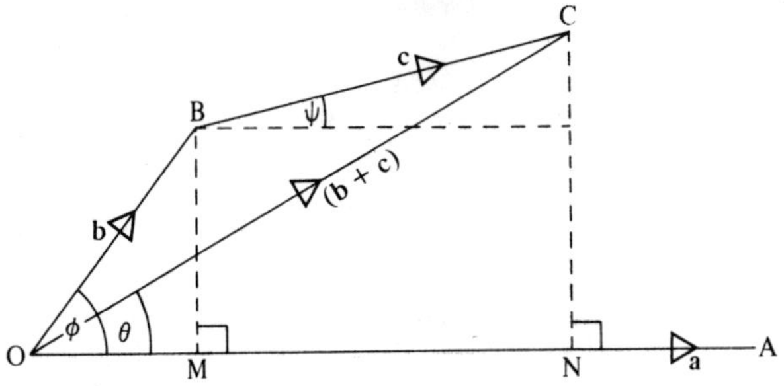

$$\begin{aligned} \mathbf{a}.(\mathbf{b}+\mathbf{c}) &= |\mathbf{a}||\mathbf{b}+\mathbf{c}| \cos \theta \\ &= (OA)(OC) \cos \theta \\ &= (OA)(ON) \\ &= (OA)(OM + MN) \\ &= (OA)(OB \cos \phi) + (OA)(BC \cos \psi) \\ &= \mathbf{a}.\mathbf{b} + \mathbf{a}.\mathbf{c} \end{aligned}$$

2. If two vectors **a** and **b** are perpendicular their scalar product is zero, since

$$\mathbf{a}.\mathbf{b} = |\mathbf{a}||\mathbf{b}| \cos 90^\circ = 0$$

In particular $\mathbf{i}.\mathbf{j} = \mathbf{j}.\mathbf{k} = \mathbf{k}.\mathbf{i} = 0$

3. If two vectors **a** and **b** are parallel their scalar product is $|\mathbf{a}||\mathbf{b}|$ since

$$\mathbf{a}.\mathbf{b} = |\mathbf{a}||\mathbf{b}| \cos 0^\circ = |\mathbf{a}||\mathbf{b}|$$

4. The scalar product of **a** with **a** is $|\mathbf{a}|^2$ since

$$\mathbf{a}.\mathbf{a} = |\mathbf{a}||\mathbf{a}| \cos 0^\circ = |\mathbf{a}|^2$$

In particular $\mathbf{i}.\mathbf{i} = \mathbf{j}.\mathbf{j} = \mathbf{k}.\mathbf{k} = 1$

CALCULATION OF THE SCALAR PRODUCT

Suppose that $\mathbf{a} = a_1\mathbf{i} + a_2\mathbf{j} + a_3\mathbf{k}$

and $\mathbf{b} = b_1\mathbf{i} + b_2\mathbf{j} + b_3\mathbf{k}$

Then

$$\begin{aligned}\mathbf{a}.\mathbf{b} &= (a_1\mathbf{i} + a_2\mathbf{j} + a_3\mathbf{k}).(b_1\mathbf{i} + b_2\mathbf{j} + b_3\mathbf{k}) \\ &= (a_1b_1\mathbf{i}.\mathbf{i} + a_2b_2\mathbf{j}.\mathbf{j} + a_3b_3\mathbf{k}.\mathbf{k}) \\ &\quad + (a_1b_2\mathbf{i}.\mathbf{j} + a_2b_3\mathbf{j}.\mathbf{k} + a_3b_1\mathbf{k}.\mathbf{i}) \\ &\quad + (a_2b_1\mathbf{j}.\mathbf{i} + a_3b_2\mathbf{k}.\mathbf{j} + a_1b_3\mathbf{i}.\mathbf{k}) \\ &= (a_1b_1 + a_2b_2 + a_3b_3) + (0) + (0)\end{aligned}$$

Hence

$$\mathbf{a}.\mathbf{b} = |\mathbf{a}||\mathbf{b}| \cos\theta = a_1b_1 + a_2b_2 + a_3b_3$$

Now

$$\cos\theta = \frac{\mathbf{a}.\mathbf{b}}{|\mathbf{a}||\mathbf{b}|} = \frac{a_1b_1 + a_2b_2 + a_3b_3}{|\mathbf{a}||\mathbf{b}|}$$

$$= \frac{a_1b_1}{|\mathbf{a}||\mathbf{b}|} + \frac{a_2b_2}{|\mathbf{a}||\mathbf{b}|} + \frac{a_3b_3}{|\mathbf{a}||\mathbf{b}|}$$

But $\frac{a_1}{|\mathbf{a}|}, \frac{a_2}{|\mathbf{a}|}, \frac{a_3}{|\mathbf{a}|}$ are the direction cosines of **a**;

i.e. l, m, n

Similarly $\frac{b_1}{|\mathbf{b}|}, \frac{b_2}{|\mathbf{b}|}, \frac{b_3}{|\mathbf{b}|}$ are the direction cosines of **b**;

i.e. l', m', n'

So $\cos\theta = ll' + mm' + nn'$

[This relationship was derived in Chapter 1 using a different approach.]

EXAMPLES 2c

1) Find the scalar product of $\mathbf{a} = 2\mathbf{i} - 3\mathbf{j} + 5\mathbf{k}$ and $\mathbf{b} = \mathbf{i} - 3\mathbf{j} + \mathbf{k}$ and hence find the cosine of the angle between **a** and **b**.

$$\mathbf{a}.\mathbf{b} = (2)(1) + (-3)(-3) + (5)(1) = 16$$

But $$\mathbf{a}.\mathbf{b} = |\mathbf{a}||\mathbf{b}|\cos\theta$$

$$|\mathbf{a}| = \sqrt{4+9+25} = \sqrt{38}$$

$$|\mathbf{b}| = \sqrt{1+9+1} = \sqrt{11}$$

Hence $$\cos\theta = \frac{\mathbf{a}.\mathbf{b}}{|\mathbf{a}||\mathbf{b}|} = \frac{16}{\sqrt{11}\sqrt{38}} = \frac{16}{\sqrt{418}}$$

2) If $\mathbf{a} = 10\mathbf{i} - 3\mathbf{j} + 5\mathbf{k}$, $\mathbf{b} = 2\mathbf{i} + 6\mathbf{j} - 3\mathbf{k}$ and $\mathbf{c} = \mathbf{i} + 10\mathbf{j} - 2\mathbf{k}$, verify that $\mathbf{a}.\mathbf{b} + \mathbf{a}.\mathbf{c} = \mathbf{a}.(\mathbf{b} + \mathbf{c})$.

$$\mathbf{a}.\mathbf{b} = (10)(2) + (-3)(6) + (5)(-3) = -13$$

$$\mathbf{a}.\mathbf{c} = (10)(1) + (-3)(10) + (5)(-2) = -30$$

$$\mathbf{b} + \mathbf{c} = 3\mathbf{i} + 16\mathbf{j} - 5\mathbf{k}$$

Hence $$\mathbf{a}.(\mathbf{b} + \mathbf{c}) = (10)(3) + (-3)(16) + (5)(-5) = -43$$

But $$\mathbf{a}.\mathbf{b} + \mathbf{a}.\mathbf{c} = -13 - 30 = -43$$

Therefore $$\mathbf{a}.\mathbf{b} + \mathbf{a}.\mathbf{c} = \mathbf{a}.(\mathbf{b} + \mathbf{c})$$

3) The resultant of two vectors **a** and **b** is perpendicular to **a**. If $|\mathbf{b}| = \sqrt{2}|\mathbf{a}|$, show that the resultant of 2**a** and **b** is perpendicular to **b**.

The resultant of **a** and **b** is $\mathbf{a} + \mathbf{b}$

Since $\mathbf{a} + \mathbf{b}$ is perpendicular to **a**,

$$(\mathbf{a} + \mathbf{b}).\mathbf{a} = 0$$

Hence $$\mathbf{a}.\mathbf{a} + \mathbf{b}.\mathbf{a} = 0$$

But $$\mathbf{a}.\mathbf{a} = |\mathbf{a}|^2$$

Hence $$\mathbf{b}.\mathbf{a} = -|\mathbf{a}|^2$$

Now the resultant of 2**a** and **b** is $2\mathbf{a} + \mathbf{b}$

The angle θ between $2\mathbf{a} + \mathbf{b}$ and **b** is given by

$$\cos\theta = \frac{(2\mathbf{a} + \mathbf{b}).\mathbf{b}}{|2\mathbf{a} + \mathbf{b}||\mathbf{b}|}$$

$$= \frac{2\mathbf{a}.\mathbf{b} + \mathbf{b}.\mathbf{b}}{|2\mathbf{a} + \mathbf{b}||\mathbf{b}|}$$

But $$\mathbf{b}\cdot\mathbf{b} = |\mathbf{b}|^2$$

And $$2\mathbf{a}\cdot\mathbf{b} = -2|\mathbf{a}|^2$$

Hence $$\cos\theta = \frac{-2|\mathbf{a}|^2 + |\mathbf{b}|^2}{|2\mathbf{a}+\mathbf{b}||\mathbf{b}|}$$

But $$|\mathbf{b}| = \sqrt{2}|\mathbf{a}| \qquad \text{(given)}$$

So $$\cos\theta = \frac{-2|\mathbf{a}|^2 + 2|\mathbf{a}|^2}{|2\mathbf{a}+\mathbf{b}||\mathbf{b}|} = 0$$

And $$\theta = \frac{\pi}{2}$$

Hence $2\mathbf{a}+\mathbf{b}$ is perpendicular to **b**.

EXERCISE 2c

1) Calculate $\mathbf{a}\cdot\mathbf{b}$ if:
(i) $\mathbf{a} = 2\mathbf{i}-4\mathbf{j}+5\mathbf{k}$, $\mathbf{b} = \mathbf{i}+3\mathbf{j}+8\mathbf{k}$,
(ii) $\mathbf{a} = 3\mathbf{i}-7\mathbf{j}+2\mathbf{k}$, $\mathbf{b} = 5\mathbf{i}+\mathbf{j}-4\mathbf{k}$,
(iii) $\mathbf{a} = 2\mathbf{i}-3\mathbf{j}+6\mathbf{k}$, $\mathbf{b} = \mathbf{i}+\mathbf{j}$.
What conclusion can you draw in (ii)?

2) Find $\mathbf{p}\cdot\mathbf{q}$ and the cosine of the angle between **p** and **q** if:
(i) $\mathbf{p} = 2\mathbf{i}+4\mathbf{j}+\mathbf{k}$, $\mathbf{q} = \mathbf{i}+\mathbf{j}+\mathbf{k}$,
(ii) $\mathbf{p} = -\mathbf{i}+3\mathbf{j}-2\mathbf{k}$, $\mathbf{q} = \mathbf{i}+\mathbf{j}-6\mathbf{k}$.

3) The angle between two vectors $\mathbf{v}_1$ and $\mathbf{v}_2$ is $\arccos\frac{4}{21}$.
If $\mathbf{v}_1 = 6\mathbf{i}+3\mathbf{j}-2\mathbf{k}$ and $\mathbf{v}_2 = -2\mathbf{i}+\lambda\mathbf{j}-4\mathbf{k}$, find the positive value of λ.

4) If $\mathbf{a} = 3\mathbf{i}+4\mathbf{j}-\mathbf{k}$, $\mathbf{b} = \mathbf{i}-\mathbf{j}+3\mathbf{k}$ and $\mathbf{c} = 2\mathbf{i}+\mathbf{j}-5\mathbf{k}$, find:
(i) $\mathbf{a}\cdot\mathbf{b}$ (ii) $\mathbf{a}\cdot\mathbf{c}$ (iii) $\mathbf{a}\cdot(\mathbf{b}+\mathbf{c})$ (iv) $(2\mathbf{a}+3\mathbf{b})\cdot\mathbf{c}$ (v) $(\mathbf{a}-\mathbf{b})\cdot\mathbf{c}$.

5) In a triangle ABC, $\overrightarrow{AB} = \mathbf{i}+2\mathbf{j}+3\mathbf{k}$ and $\overrightarrow{BC} = -\mathbf{i}+4\mathbf{j}$. Find the cosine of angle ABC.
Find the vector $\overrightarrow{AC}$ and use it to calculate the angle BAC.

6) A, B and C are points with position vectors **a**, **b** and **c** respectively, relative to the origin O. AB is perpendicular to OC and BC is perpendicular to OA. Show that AC is perpendicular to OB.

7) Given two vectors **a** and **b** ($\mathbf{a}\neq 0$, $\mathbf{b}\neq 0$), show that:
(i) if $\mathbf{a}+\mathbf{b}$ and $\mathbf{a}-\mathbf{b}$ are perpendicular then $|\mathbf{a}| = |\mathbf{b}|$,
(ii) if $|\mathbf{a}+\mathbf{b}| = |\mathbf{a}-\mathbf{b}|$ then **a** and **b** are perpendicular.

8) Three vectors **a**, **b** and **c** are such that $\mathbf{a}\neq\mathbf{b}\neq\mathbf{c}\neq 0$.
(i) If $\mathbf{a}\cdot(\mathbf{b}+\mathbf{c}) = \mathbf{b}\cdot(\mathbf{a}-\mathbf{c})$ prove that $\mathbf{c}\cdot(\mathbf{a}+\mathbf{b}) = 0$.
(ii) If $(\mathbf{a}\cdot\mathbf{b})\mathbf{c} = (\mathbf{b}\cdot\mathbf{c})\mathbf{a}$ show that **c** and **a** are parallel.

APPLICATIONS OF THE SCALAR PRODUCT

Work Done by a Constant Force

Consider a force vector **F** acting on a particle P which moves through a displacement vector **d**.

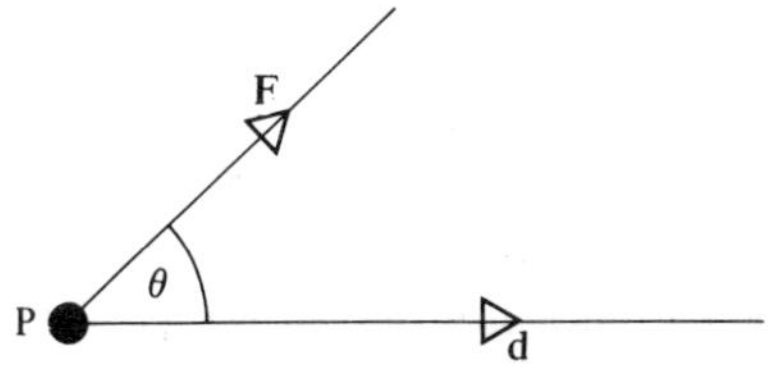

If **F** and **d** are inclined at an angle θ, the work done by **F** is given by multiplying the magnitude of the force component parallel to **d** by the distance through which P moves.

Hence $\quad$ Work done by $\mathbf{F} = |\mathbf{F}| \cos\theta |\mathbf{d}|$

But $\quad |\mathbf{F}||\mathbf{d}| \cos\theta = \mathbf{F}.\mathbf{d}$

Therefore the work done by a force **F** when its point of application undergoes a displacement **d** is $\mathbf{F}.\mathbf{d}$

Resolving a Vector in a Specified Direction

If a vector **v** is inclined at an angle θ to a direction vector **d**, the component of **v** in the direction **d** is of magnitude $|\mathbf{v}| \cos\theta$.

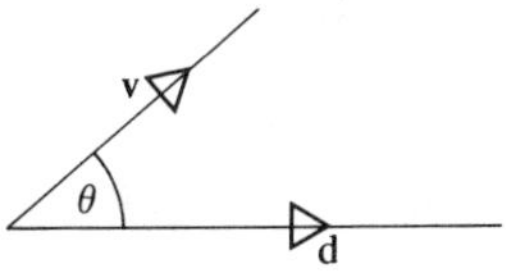

But $\quad |\mathbf{v}||\mathbf{d}| \cos\theta = \mathbf{v}.\mathbf{d}$

Hence $\quad |\mathbf{v}| \cos\theta = \dfrac{\mathbf{v}.\mathbf{d}}{|\mathbf{d}|} = \mathbf{v}.\hat{\mathbf{d}}$

In Mechanics problems it often happens that a vector has to be resolved in two perpendicular directions in a plane.
In such cases it is useful to notice that the vectors $a\mathbf{i} + b\mathbf{j}$ and $b\mathbf{i} - a\mathbf{j}$ are always perpendicular (since their scalar product is zero). For instance, if a vector **v** is to be resolved parallel and perpendicular to the direction vector $3\mathbf{i} + 2\mathbf{j}$, we will calculate the components parallel to $3\mathbf{i} + 2\mathbf{j}$ and to $2\mathbf{i} - 3\mathbf{j}$.

Now if $\mathbf{d}_1 = 3\mathbf{i} + 2\mathbf{j}$ and $\mathbf{d}_2 = 2\mathbf{i} - 3\mathbf{j}$

Then $\hat{\mathbf{d}}_1 = (3\mathbf{i} + 2\mathbf{j})/\sqrt{13}$ and $\hat{\mathbf{d}}_2 = (2\mathbf{i} - 3\mathbf{j})/\sqrt{13}$

The magnitudes of the components of $\mathbf{v}$ in the required directions are therefore:

$$\mathbf{v}.(3\mathbf{i} + 2\mathbf{j})/\sqrt{13} \quad \text{and} \quad \mathbf{v}.(2\mathbf{i} - 3\mathbf{j})/\sqrt{13}$$

[*Note*: An alternative expression for $\mathbf{d}_2$ is the vector $-2\mathbf{i} + 3\mathbf{j}$ since this vector also is perpendicular to $\mathbf{d}_1$.]

To Determine the Shortest Distance Between Two Skew Lines

The line of shortest length between two non-intersecting lines, is perpendicular to both of them. This property can be verified as follows:

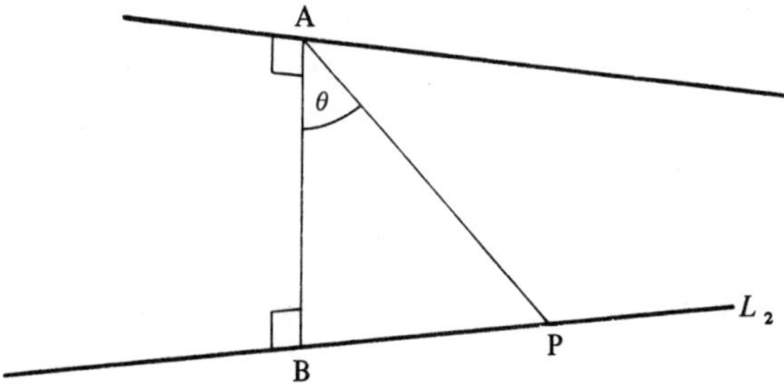

L_1 and L_2 are two skew lines,
A is the point on L_1 which is closest to L_2,
B is the point on L_2 such that AB is perpendicular to both L_1 and L_2,
P is any other point on L_2 and θ is the angle between AB and AP.
If AB is not the shortest distance between L_1 and L_2 there must be a point P such that AP < AB.
But in triangle ABP,

$$AB = AP \cos\theta$$

Therefore $$AB \leqslant AP$$

Hence AB is the shortest distance between L_1 and L_2.

So to determine the shortest distance between two lines we must find the length of their common perpendicular.

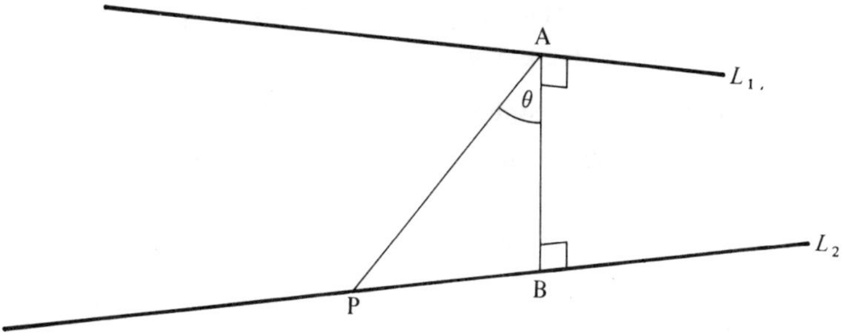

If L_1 has a direction vector $\mathbf{d}_1$ and passes through a point with position vector $\mathbf{a}_1$, its equation can be written

$$\mathbf{r}_1 = \mathbf{a}_1 + \lambda\mathbf{d}_1$$

Similarly the equation of L_2 can be written

$$\mathbf{r}_2 = \mathbf{a}_2 + \mu\mathbf{d}_2$$

Suppose that AB is the common normal where $\overrightarrow{AB} = \mathbf{n}$, and that A and P are the points where $\lambda = \lambda_1$ and $\mu = \mu_2$.

Then $\qquad \overrightarrow{AP} = \mathbf{a}_2 + \mu_2\mathbf{d}_2 - (\mathbf{a}_1 + \lambda_1\mathbf{d}_1)$

Now $\qquad |\overrightarrow{AB}| = |\overrightarrow{AP}|\cos\theta = \dfrac{\overrightarrow{AP}\,.\,\overrightarrow{AB}}{|\overrightarrow{AB}|}$

So $\qquad |\overrightarrow{AB}| = \dfrac{(\mathbf{a}_2 + \mu_2\mathbf{d}_2 - \mathbf{a}_1 - \lambda_1\mathbf{d}_1)\,.\,\mathbf{n}}{|\mathbf{n}|}$

$$= (\mathbf{a}_2 - \mathbf{a}_1)\,.\,\frac{\mathbf{n}}{|\mathbf{n}|} + \mu_2\mathbf{d}_2\,.\,\frac{\mathbf{n}}{|\mathbf{n}|} - \lambda_1\mathbf{d}_1\,.\,\frac{\mathbf{n}}{|\mathbf{n}|}.$$

But $\qquad \mathbf{d}_1\,.\,\mathbf{n} = 0 \qquad$ (AB is perpendicular to L_1)

And $\qquad \mathbf{d}_2\,.\,\mathbf{n} = 0 \qquad$ (AB is perpendicular to L_2)

Hence $\qquad |\overrightarrow{AB}| = (\mathbf{a}_2 - \mathbf{a}_1)\,.\,\hat{\mathbf{n}} \qquad \left(\dfrac{\mathbf{n}}{|\mathbf{n}|} = \hat{\mathbf{n}}\right)$

Thus the shortest distance between two skew lines is $(\mathbf{a}_2 - \mathbf{a}_1)\,.\,\hat{\mathbf{n}}$ where $\mathbf{a}_1$ and $\mathbf{a}_2$ are any two points, one on each line, and $\hat{\mathbf{n}}$ is a unit vector perpendicular to both lines.

EXAMPLE

The vector equations of two lines are

$$\mathbf{r}_1 = \mathbf{i} + \mathbf{j} + \lambda(2\mathbf{i} - \mathbf{j} + \mathbf{k})$$

$$\mathbf{r}_2 = 2\mathbf{i} + \mathbf{j} - \mathbf{k} + \mu(3\mathbf{i} - 5\mathbf{j} + 2\mathbf{k})$$

Show that the vector $-3\mathbf{i}+\mathbf{j}+7\mathbf{k}$ is perpendicular to both lines. Hence find the shortest distance between the lines.

The direction vector of the first line (L_1) is $2\mathbf{i}-\mathbf{j}+\mathbf{k}$

The direction vector of the second line (L_2) is $3\mathbf{i}-5\mathbf{j}+2\mathbf{k}$

Now $\quad (-3\mathbf{i}+\mathbf{j}+7\mathbf{k})\cdot(2\mathbf{i}-\mathbf{j}+\mathbf{k}) = -6-1+7 = 0$

Therefore $\quad -3\mathbf{i}+\mathbf{j}+7\mathbf{k}$ is perpendicular to L_1

Also $\quad (-3\mathbf{i}+\mathbf{j}+7\mathbf{k})\cdot(3\mathbf{i}-5\mathbf{j}+2\mathbf{k}) = -9-5+14 = 0$

Therefore $\quad -3\mathbf{i}+\mathbf{j}+7\mathbf{k}$ is perpendicular to L_2

So $\quad -3\mathbf{i}+\mathbf{j}+7\mathbf{k}$ is the common normal $\mathbf{n}$

And $\quad \hat{\mathbf{n}} = (-3\mathbf{i}+\mathbf{j}+7\mathbf{k})/\sqrt{59}$

One point on L_1 has position vector $\mathbf{i}+\mathbf{j}$

One point on L_2 has position vector $2\mathbf{i}+\mathbf{j}-\mathbf{k}$

Therefore the shortest distance between L_1 and L_2 is given by

$$\begin{aligned}&[\mathbf{i}+\mathbf{j}-(2\mathbf{i}+\mathbf{j}-\mathbf{k})]\cdot\hat{\mathbf{n}}\\ &= (-\mathbf{i}+\mathbf{k})\cdot(-3\mathbf{i}+\mathbf{j}+7\mathbf{k})/\sqrt{59}\\ &= (3+0+7)/\sqrt{59}\\ &= 10/\sqrt{59}\end{aligned}$$

(*Note*: An alternative solution to this problem, using the vector product, is given in Chapter 4.)

EXAMPLES 2d

1) Three forces $\mathbf{F}_1 = 2\mathbf{i}+4\mathbf{j}-\mathbf{k}$, $\mathbf{F}_2 = 3\mathbf{i}-7\mathbf{j}+6\mathbf{k}$ and $\mathbf{F}_3 = \mathbf{i}+5\mathbf{j}-4\mathbf{k}$ are acting on a particle which is given a displacement $12\mathbf{i}+4\mathbf{j}+2\mathbf{k}$.
Find the work done by each of the forces.
Find also the work done by the resultant force.

Let $\quad \mathbf{d} = 12\mathbf{i}+4\mathbf{j}+2\mathbf{k}$

Then the work done by $\mathbf{F}_1 = \mathbf{F}_1\cdot\mathbf{d}$

$$\begin{aligned}&= (2\mathbf{i}+4\mathbf{j}-\mathbf{k})\cdot(12\mathbf{i}+4\mathbf{j}+2\mathbf{k})\\ &= 24+16-2\\ &= 38 \text{ units}\end{aligned}$$

The work done by $\quad \mathbf{F}_2 = \mathbf{F}_2\cdot\mathbf{d}$

$$= (3\mathbf{i}-7\mathbf{j}+6\mathbf{k})\cdot(12\mathbf{i}+4\mathbf{j}+2\mathbf{k})$$

$= 20$ units

The work done by $\mathbf{F}_3 = (\mathbf{i} + 5\mathbf{j} - 4\mathbf{k}) . (12\mathbf{i} + 4\mathbf{j} + 2\mathbf{k})$

$= 24$ units

The resultant force $\mathbf{F} = \mathbf{F}_1 + \mathbf{F}_2 + \mathbf{F}_3$

$= 6\mathbf{i} + 2\mathbf{j} + \mathbf{k}$

So the work done by $\mathbf{F} = (6\mathbf{i} + 2\mathbf{j} + \mathbf{k}) . (12\mathbf{i} + 4\mathbf{j} + 2\mathbf{k})$

$= 72 + 8 + 2$

$= 82$ units

Note: It can be seen that, in the example above, the sum of the individual work done by $\mathbf{F}_1$, $\mathbf{F}_2$ and $\mathbf{F}_3$ is equal to the work done by the resultant force $\mathbf{F}$. This relationship applies to any set of forces acting on a particle and its validity can be proved as follows:

Suppose that a set of forces $\mathbf{F}_1, \mathbf{F}_2, \mathbf{F}_3, \ldots, \mathbf{F}_n$, act on a particle, causing it to be moved through a displacement $\mathbf{d}$.
The total work done by all the forces in the set is given by

$$\mathbf{F}_1 . \mathbf{d} + \mathbf{F}_2 . \mathbf{d} + \mathbf{F}_3 . \mathbf{d} + \ldots + \mathbf{F}_n . \mathbf{d}$$

$$= \mathbf{d} . (\mathbf{F}_1 + \mathbf{F}_2 + \mathbf{F}_3 + \ldots + \mathbf{F}_n)$$

But the resultant force $\mathbf{F} = \mathbf{F}_1 + \mathbf{F}_2 + \mathbf{F}_3 + \ldots + \mathbf{F}_n$

Hence the work done by $\mathbf{F}$ is given by

$$\mathbf{F} . \mathbf{d} = (\mathbf{F}_1 + \mathbf{F}_2 + \mathbf{F}_3 + \ldots + \mathbf{F}_n) . \mathbf{d}$$

Thus, when a set of forces act on a particle and displace it, the total work done by the individual forces is equal to the work done by the resultant force.

2) Find the magnitude of the component of a vector $\mathbf{v} = 4\mathbf{i} - 3\mathbf{j} + \mathbf{k}$ in the direction (i) $\mathbf{i} - 2\mathbf{j}$ (ii) $2\mathbf{i} - \mathbf{j} - 12\mathbf{k}$ (iii) $\mathbf{i} + 2\mathbf{j} + 2\mathbf{k}$.
What conclusions can you draw in parts (ii) and (iii)?

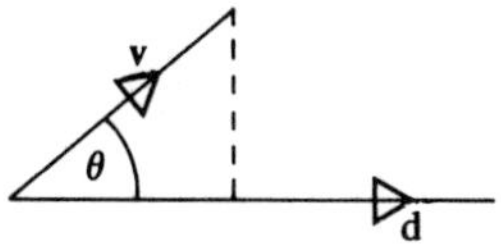

The magnitude of the component of $\mathbf{v}$ in the direction $\mathbf{d}$ is $|\mathbf{v}| \cos\theta = \mathbf{v} . \hat{\mathbf{d}}$

(i) When $\mathbf{d} = \mathbf{i} - 2\mathbf{j}$, $\hat{\mathbf{d}} = (\mathbf{i} - 2\mathbf{j})/\sqrt{5}$

Hence the required component of $\mathbf{v}$ is of magnitude

$$(4\mathbf{i} - 3\mathbf{j} + \mathbf{k}) . (\mathbf{i} - 2\mathbf{j})/\sqrt{5} = (4 + 6 + 0)/\sqrt{5} = 2\sqrt{5}$$

(ii) When $\mathbf{d} = 2\mathbf{i} - \mathbf{j} - 12\mathbf{k}$, $\hat{\mathbf{d}} = (2\mathbf{i} - \mathbf{j} - 12\mathbf{k})/\sqrt{149}$

The required component of **v** is therefore of magnitude

$$(4\mathbf{i}-3\mathbf{j}+\mathbf{k})\cdot(2\mathbf{i}-\mathbf{j}-12\mathbf{k})/\sqrt{149} = -1/\sqrt{149}$$

(iii) When $\mathbf{d}=\mathbf{i}+2\mathbf{j}+2\mathbf{k}$ the magnitude of the required component of **v** is

$$(4\mathbf{i}-3\mathbf{j}+\mathbf{k})\cdot(\mathbf{i}+2\mathbf{j}+2\mathbf{k})/3 = (4-6+2)/3 = 0$$

In part (ii) the sign of the component is negative showing that the actual component of **v** is in the direction $-(2\mathbf{i}-\mathbf{j}-12\mathbf{k})$
In part (iii), since the component of **v** in the direction of **d** is zero we conclude that **v** is perpendicular to **d**.

3) Resolve the vector $3\mathbf{i}+4\mathbf{j}$ parallel and perpendicular to the direction vector $\mathbf{i}-2\mathbf{j}$.

Let $\quad \mathbf{d}_1 = \mathbf{i}-2\mathbf{j}$ so that $\hat{\mathbf{d}}_1 = (\mathbf{i}-2\mathbf{j})/\sqrt{5}$

Then $\quad \mathbf{d}_2 = 2\mathbf{i}+\mathbf{j}$ is perpendicular to $\mathbf{d}_1$

and $\quad \hat{\mathbf{d}}_2 = (2\mathbf{i}+\mathbf{j})/\sqrt{5}$

Then if $\mathbf{v}=3\mathbf{i}+4\mathbf{j}$, the components of **v** in the directions $\mathbf{d}_1$ and $\mathbf{d}_2$ are of magnitudes $\mathbf{v}\cdot\hat{\mathbf{d}}_1$ and $\mathbf{v}\cdot\hat{\mathbf{d}}_2$ respectively.

i.e. $\quad (3\mathbf{i}+4\mathbf{j})\cdot(\mathbf{i}-2\mathbf{j})/\sqrt{5}$ and $(3\mathbf{i}+4\mathbf{j})\cdot(2\mathbf{i}+\mathbf{j})/\sqrt{5}$

i.e. $\quad -5/\sqrt{5}$ and $10/\sqrt{5}$

Hence the resolved parts of **v** are:

$-\sqrt{5}$ units in the direction $\mathbf{i}-2\mathbf{j} = -\sqrt{5}\hat{\mathbf{d}}_1 = -\mathbf{i}+2\mathbf{j}$

$2\sqrt{5}$ units in the direction $2\mathbf{i}+\mathbf{j} = 2\sqrt{5}\hat{\mathbf{d}}_2 = 2(2\mathbf{i}+\mathbf{j})$

(Note that the first of these components is equivalent to $+\sqrt{5}$ units in the direction $-\mathbf{i}+2\mathbf{j}$)

4) Find a unit vector which is perpendicular to AB and AC if $\overrightarrow{AB}=\mathbf{i}+2\mathbf{j}+3\mathbf{k}$ and $\overrightarrow{AC}=4\mathbf{i}-\mathbf{j}+2\mathbf{k}$.

Let $a\mathbf{i}+b\mathbf{j}+c\mathbf{k}$ be a vector perpendicular both to AB and to AC.
It is perpendicular to AB so $\quad (a\mathbf{i}+b\mathbf{j}+c\mathbf{k})\cdot(\mathbf{i}+2\mathbf{j}+3\mathbf{k}) = 0$

It is perpendicular to AC so $\quad (a\mathbf{i}+b\mathbf{j}+c\mathbf{k})\cdot(4\mathbf{i}-\mathbf{j}+2\mathbf{k}) = 0$

Therefore
$$\begin{aligned} a+2b+3c &= 0\\ 4a-b+2c &= 0 \end{aligned}$$

Eliminating b gives $\quad a = -\dfrac{7}{9}c$

Eliminating a gives $\quad b = -\dfrac{10}{9}c$

Hence
$$a\mathbf{i} + b\mathbf{j} + c\mathbf{k} = -\frac{7}{9}c\mathbf{i} - \frac{10}{9}c\mathbf{j} + c\mathbf{k}$$
$$= \frac{c}{9}(-7\mathbf{i} - 10\mathbf{j} + 9\mathbf{k})$$

Then if $a\mathbf{i} + b\mathbf{j} + c\mathbf{k}$ is perpendicular to both AB and AC, so also is the vector $-7\mathbf{i} - 10\mathbf{j} + 9\mathbf{k}$.
A unit vector perpendicular to AB and AC is therefore $(-7\mathbf{i} - 10\mathbf{j} + 9\mathbf{k})/\sqrt{230}$.

EXERCISE 2d

1) Find the work done when a force $\mathbf{F} = 3\mathbf{i} + 5\mathbf{j} + 6\mathbf{k}$ moves its point of application through a displacement $\mathbf{d} = 3\mathbf{i} - 2\mathbf{j} + 4\mathbf{k}$.

2) Forces of magnitudes 6, 7 and 9 N act in the directions $2\mathbf{i} + 2\mathbf{j} + \mathbf{k}$, $6\mathbf{i} - 3\mathbf{j} + 2\mathbf{k}$ and $7\mathbf{i} + 4\mathbf{j} - 4\mathbf{k}$ respectively. These forces act on a particle causing a displacement $34\mathbf{i} + 10\mathbf{j}$ (metres). Find the work done by each force and show that the total work done by the three forces is equal to the work done by the resultant force.

3) Resolve a vector $\mathbf{v}$ parallel and perpendicular to a direction vector $\mathbf{d}$ if:
(i) $\mathbf{v} = 2\mathbf{i} + 3\mathbf{j}$, $\mathbf{d} = \mathbf{i} - \mathbf{j}$,
(ii) $\mathbf{v} = 4\mathbf{i} - 3\mathbf{j}$, $\mathbf{d} = 2\mathbf{i} + \mathbf{j}$,
(iii) $\mathbf{v} = \mathbf{i} + \mathbf{j}$, $\mathbf{d} = 3\mathbf{i} + 4\mathbf{j}$.
In each case, sketch $\mathbf{v}$ and the two components in the appropriate directions.

4) Resolve a velocity $\mathbf{v} = 3\mathbf{i} + 2\mathbf{j}$ parallel and perpendicular to the line whose vector equation is $\mathbf{r} = \mathbf{i} - 4\mathbf{j} + \lambda(4\mathbf{i} - \mathbf{j})$.

5) Find a unit vector which is perpendicular to AB and to AC if:
(i) $\overrightarrow{AB} = 6\mathbf{i} + \mathbf{j} + 3\mathbf{k}$, $\overrightarrow{AC} = 5\mathbf{i} + \mathbf{k}$,
(ii) $\overrightarrow{AB} = \mathbf{i} - \mathbf{j} - \mathbf{k}$, $\overrightarrow{AC} = 7\mathbf{i} - 2\mathbf{j} + 3\mathbf{k}$.

6) Two skew lines have vector equations:
$$\mathbf{r}_1 = 3\mathbf{j} + \lambda(2\mathbf{i} - 5\mathbf{j} + 3\mathbf{k}) \quad \text{and} \quad \mathbf{r}_2 = 2\mathbf{i} + \mathbf{k} + \mu(\mathbf{j} + \mathbf{k})$$
Show that $4\mathbf{i} + \mathbf{j} - \mathbf{k}$ is a vector which is perpendicular to both lines and hence find the shortest distance between the lines.

7) Find the shortest distance between two lines L_1 and L_2 if the vector equations of L_1 and L_2 are:
(i) $\mathbf{r}_1 = 7\mathbf{i} + \mathbf{j} - 2\mathbf{k} + \lambda(3\mathbf{i} - 5\mathbf{j} + 2\mathbf{k})$,
(ii) $\mathbf{r}_2 = \mathbf{i} + 9\mathbf{j} + 5\mathbf{k} + \mu(7\mathbf{i} + \mathbf{j} - 8\mathbf{k})$.

8) Two constant forces $\mathbf{F}_1$ and $\mathbf{F}_2$ are the only forces acting on a particle P, of mass 4 kg, which is initially at rest at a point with position vector $\mathbf{i} + 2\mathbf{j} - \mathbf{k}$. Four seconds later P is at the point with position vector $9\mathbf{i} - 2\mathbf{j} + 11\mathbf{k}$.

If $\mathbf{F}_1 = 3\mathbf{i} - \mathbf{j} + 4\mathbf{k}$ find:

(i) the work done by $\mathbf{F}_1$ during the four seconds,
(ii) the force vector $\mathbf{F}_2$,
(iii) the total work done during the four seconds.

9) Forces $\mathbf{F}_1$, $\mathbf{F}_2$ and $\mathbf{F}_3$ act for a time t on a particle of mass 6 kg which undergoes a displacement from rest of $2\mathbf{i} + 7\mathbf{j} - \mathbf{k}$ (metres). If $\mathbf{F}_1 = \mathbf{i} + 3\mathbf{k}$, $\mathbf{F}_2 = 4\mathbf{j} - \mathbf{k}$, $\mathbf{F}_3 = 2\mathbf{i} - 3\mathbf{j} - \mathbf{k}$ (newtons) find the total work done by the three forces.
If no other work is done, use the principle of work and energy to find the speed of the particle at time t.

MISCELLANEOUS EXERCISE 2

1) The velocity vectors of the particles P_1, P_2 are

$$u_1\mathbf{i} + v_1\mathbf{j}, \quad u_2\mathbf{i} + v_2\mathbf{j}$$

respectively. Their relative velocity has the same magnitude as that of the velocity of P_1. If the velocity of one particle is reversed, the magnitude of the relative velocity is doubled. Find the ratio of the speeds of P_1 and P_2 and the sine of the angle between their directions. (U of L)

2) A man bicycling at a constant speed u finds that when his velocity is $u\mathbf{j}$ the velocity of the wind appears to be

$$\tfrac{1}{2}v_1(\mathbf{i} - \sqrt{3}\mathbf{j}),$$

where **i** and **j** are unit vectors in the east and north directions respectively: but when his velocity is $\tfrac{1}{2}u(-\sqrt{3}\mathbf{i} + \mathbf{j})$ the velocity of the wind appears to be $v_2\mathbf{i}$. Prove that the true velocity of the wind is

$$\tfrac{1}{6}\sqrt{3}u(\mathbf{i} + \sqrt{3}\mathbf{j}),$$

and find v_1 and v_2 in terms of u. (U of L)

3) Two particles A and B are moving with constant velocity vectors $\mathbf{v}_1 = 5\mathbf{i} + 3\mathbf{j} - \mathbf{k}$ and $\mathbf{v}_2 = 3\mathbf{i} + 4\mathbf{j} - 3\mathbf{k}$ respectively.
Find the velocity vector of A relative to B. At time $t = 0$ the particle A is at the point whose position vector is $-4\mathbf{i} + 7\mathbf{j} - 6\mathbf{k}$.
If A collides with B when $t = 5$, find the position vector of B at $t = 0$.
The velocity of A relative to a third moving particle C is in the direction of the vector $2\mathbf{i} + \mathbf{j} - 2\mathbf{k}$ and the velocity of B relative to C is in the direction of the vector $2\mathbf{i} + 3\mathbf{j} - 6\mathbf{k}$. Find the magnitude and direction of the velocity of C.
(U of L)

4) Three points A, B, C have position vectors

$$4\mathbf{i} + 3\mathbf{j}, \quad 3\mathbf{i} + \mathbf{j} + 2\mathbf{k}, \quad 7\mathbf{i} + 4\mathbf{j} + 2\mathbf{k},$$

respectively referred to an origin O, the unit of distance being the metre. A particle P leaves O and simultaneously a second particle Q leaves B. Each particle

moves with constant velocity, P moving along OA and Q along BA. The speed of Q is 6 m/s and the velocity of P relative to Q is parallel to OC.
Find the speed of P.
Find also, in the subsequent motion,
(a) the least distance between P and Q,
(b) the least distance between P and the line BC. (U of L)

5) At time $t = 0$ the position vectors of two particles P and Q are $\mathbf{i} + \mathbf{j} + 3\mathbf{k}$ and $4\mathbf{i} + 5\mathbf{j} + \mathbf{k}$ respectively. The particles have constant velocity vectors $2\mathbf{i} + \mathbf{j} + 2\mathbf{k}$ and $-4\mathbf{j} + 3\mathbf{k}$ respectively. Find the position vector of Q relative to P when $t = T$. Show that the distance between the two particles is a minimum when $t = \frac{14}{15}$ and find the minimum distance. Also find the position vector of Q relative to P at this instant. (U of L)

6) One particle starts from the point A whose position vector is $2\mathbf{i} + 3\mathbf{j}$ and moves towards B, whose position vector is $5\mathbf{i} + 7\mathbf{j}$, with constant speed of 10 units. A second particle starts simultaneously from C whose position vector is $-27\mathbf{i} + \mathbf{j}$ and moves, with a speed of $3\sqrt{37}$ units, towards D, whose position vector is $9\mathbf{i} + 7\mathbf{j}$. Find the shortest distance between the two particles during the subsequent motion and also the coordinates of the position of each particle at this moment of closest approach.
As the particles start to move a third particle leaves D and travels with a constant speed of $\sqrt{17}$ units to strike the first particle. Find the two possible values of the velocity vectors of the third particle. (AEB)

7) A particle A starts from the point P whose position vector is $(a\mathbf{i} + 3\mathbf{j} + 2\mathbf{k})$ with constant velocity vector $(\mathbf{i} - \mathbf{j} + \mathbf{k})$. Simultaneously a second particle B leaves the point Q whose position vector is $(7\mathbf{i} - 2\mathbf{j} + 4\mathbf{k})$ with constant velocity vector $(-2\mathbf{i} + \mathbf{j} + 4\mathbf{k})$. Find the velocity vector of A relative to B and the position vector of A relative to B at t units of time after the start. Find also the distance between A and B at this time. If the distance AB is least when $t = 2$, find the value of a. (U of L)

8) Particles A and B start simultaneously from points which have position vectors $-11\mathbf{i} + 17\mathbf{j} - 14\mathbf{k}$ and $-9\mathbf{i} + 9\mathbf{j} - 32\mathbf{k}$ respectively. The velocities of A and B are constant and represented by $6\mathbf{i} - 7\mathbf{j} + 8\mathbf{k}$ and $5\mathbf{i} - 3\mathbf{j} + 17\mathbf{k}$ respectively. Show that A and B will collide.
A third particle C moves so that its velocity relative to A is parallel to the vector $2\mathbf{i} + 3\mathbf{j} + 4\mathbf{k}$ and its velocity relative to B is parallel to the vector $\mathbf{i} + 2\mathbf{j} + 3\mathbf{k}$. Find the velocity of C and its initial position if all three particles collide simultaneously. (U of L)

9) Distances being measured in nautical miles and speeds in knots, a motor boat sets out at 11 a.m from a position $-6\mathbf{i} - 2\mathbf{j}$ relative to a marker buoy, and travels at a steady speed of magnitude $\sqrt{53}$ on a direct course to intercept a ship. The ship maintains a steady velocity vector $3\mathbf{i} + 4\mathbf{j}$ and at 12 noon is at a

position $3\mathbf{j} - \mathbf{j}$ from the buoy. Find the velocity vector of the motor boat, the time of interception, and the position vector of the point of interception from the buoy. (U of L)

10) At time $t = 0$ one particle leaves the point whose position vector is $\mathbf{i} + 2\mathbf{j} + 3\mathbf{k}$ with constant velocity vector $2\mathbf{i} - \mathbf{j} + 3\mathbf{k}$. Simultaneously a second particle leaves the point whose position vector is $-3\mathbf{j} - 2\mathbf{k}$ with constant velocity vector $3\mathbf{i} + \mathbf{j} + 2\mathbf{k}$. Find the shortest distance between the particles in the subsequent motion and find the value of t when they are at this distance apart.
At time $t = 1$ a third particle leaves the point whose position vector is $5\mathbf{i} - 6\mathbf{j} - 2\mathbf{k}$ with constant velocity of magnitude $\sqrt{22}$ and subsequently collides with the second particle. Find the velocity vector of the third particle and the value of t when the collision occurs. (AEB)

11) Two particles A and B move with constant velocity vectors $(4\mathbf{i} + \mathbf{j} - 2\mathbf{k})$ and $(6\mathbf{j} + 3\mathbf{k})$ respectively, the unit of speed being the metre per second. At time $t = 0$, A is at the point with position vector $(-\mathbf{i} + 20\mathbf{j} + 21\mathbf{k})$ and B is at the point with position vector $(\mathbf{i} + 3\mathbf{k})$, the unit of distance being the metre. Find the value of t for which the distance between A and B is least and find also this least distance. (U of L)

12) One particle A has velocity vector $4\mathbf{i} + 3\mathbf{j} + 2\mathbf{k}$ and another particle B has velocity vector $2\mathbf{i} - \mathbf{j} + 4\mathbf{k}$. The velocity of a third particle C relative to A is $4\mathbf{i} - 10\mathbf{j} + 4\mathbf{k}$. Find the velocity vector of C and the velocity of C relative to B. If A is at the origin when B is at the point with position vector $8\mathbf{i} + 16\mathbf{j} - 8\mathbf{k}$, show that A and B will collide. Determine the position vector of the point of collision. (U of L)

13) If forces $\lambda\overrightarrow{AB}$ and $\mu\overrightarrow{AC}$ act along AB and AC, show that the resultant is $(\lambda + \mu)\overrightarrow{AD}$, where D is the point in BC such that $BD:DC = \mu:\lambda$. The position vectors of the points A, B and C are $(\mathbf{i} + 2\mathbf{j} + 3\mathbf{k})$, $(4\mathbf{i} + 2\mathbf{j} - \mathbf{k})$ and $7\mathbf{i}$ respectively, the unit of distance being the metre.
Forces of magnitude 10 N and 7 N act along AB and AC respectively. If D is a point in BC, find the resultant of these forces in the form $n\overrightarrow{AD}$, giving the value of n and the position vector of D.
If these forces move a particle from A to D, find the work done by each force (stating the units). (AEB)

14) Show that the straight line with vector equation

$$\mathbf{r} = \mathbf{i} + t\mathbf{j} + t\mathbf{k}$$

passes through the points P and Q which have position vectors $(\mathbf{i} - \mathbf{j} - \mathbf{k})$ and $(\mathbf{i} + \mathbf{j} + \mathbf{k})$ respectively. Find the vector equation of the straight line through the points R and S which have position vectors $(\mathbf{i} + \mathbf{j})$ and $(2\mathbf{i} + \mathbf{j} + \mathbf{k})$ respectively. Show that PQ and RS are inclined at 60° to one another.

Show that the vector

$$\mathbf{a} = \mathbf{i}+\mathbf{j}-\mathbf{k}$$

is perpendicular to both PQ and RS. Find the scalar product of **a** with $\overrightarrow{PR}$, and hence find the shortest distance between PQ and RS. (U of L)

15) Define the resolute of a vector **a** in a given direction and show that if α, β, γ are the resolutes of **a**, **b**, **a** + **b** respectively in a given direction, where **a**, **b** are any two vectors, then $\gamma = \alpha + \beta$.
[Note: **a**, **b** and the given direction are not necessarily coplanar.]
Show that $(\mathbf{a}+\mathbf{b})\cdot\mathbf{c} = \mathbf{a}\cdot\mathbf{c}+\mathbf{b}\cdot\mathbf{c}$, where **a**, **b**, **c** are any three vectors.
O, A, B, C are any four points, and P, Q, R are the mid-points of BC, CA, AB respectively. Show that

$$\overrightarrow{OP}\cdot\overrightarrow{BC}+\overrightarrow{OQ}\cdot\overrightarrow{CA}+\overrightarrow{OR}\cdot\overrightarrow{AB} = 0.$$ (WJEC)

16) Show that the vectors $\mathbf{r}_1 = 4\mathbf{i}+3\mathbf{j}$ and $\mathbf{r}_2 = 3\mathbf{i}-4\mathbf{j}$ are perpendicular.
A billiard table PQRS has one edge PQ in the direction $4\mathbf{i}+3\mathbf{j}$. Two billiard balls A and B are moving across the table with constant velocity vectors $\mathbf{v}_A = 2\mathbf{i}+\mathbf{j}$ and $\mathbf{v}_B = \mathbf{i}-5\mathbf{j}$.
Find the magnitudes of the velocity components of A and B parallel to PQ and QR and find the velocity vector of A relative to B in the direction $\overrightarrow{PQ}$.

17) P and Q are the points on the line $\mathbf{r} = \mathbf{i}+\mathbf{j}-2\mathbf{k}+\lambda(\mathbf{i}-2\mathbf{j}+3\mathbf{k})$ at which $\lambda = 2$ and $\lambda = 5$. If the force $\mathbf{F} = 4\mathbf{i}-2\mathbf{j}+\mathbf{k}$ moves a particle from P to Q find the work done by **F**.
Find also the angle between PQ and the line $\mathbf{r} = 3\mathbf{j}+\mathbf{k}+\mu(4\mathbf{i}-\mathbf{j}-2\mathbf{k})$.

18) If **i**, **j** and **k** are mutually perpendicular unit vectors and $\mathbf{v} = a\mathbf{i}+b\mathbf{j}+c\mathbf{k}$, show that $a = \mathbf{v}\cdot\mathbf{i}$, $b = \mathbf{v}\cdot\mathbf{j}$, $c = \mathbf{v}\cdot\mathbf{k}$.
If $\sqrt{2}\mathbf{v}_1 = \mathbf{i}-\mathbf{j}$, $\sqrt{3}\mathbf{v}_2 = \mathbf{i}+\mathbf{j}+\mathbf{k}$ and $2\sqrt{3}\mathbf{v}_3 = \sqrt{2}(\mathbf{i}+\mathbf{j}-2\mathbf{k})$ show that $\mathbf{v}_1, \mathbf{v}_2$ and $\mathbf{v}_3$ are mutually perpendicular unit vectors and express **i** in the form $p\mathbf{v}_1+q\mathbf{v}_2+r\mathbf{v}_3$.

19) A bead of mass 2 kg is constrained to move along a smooth straight horizontal wire whose equation is $\mathbf{r} = \mathbf{i}+\lambda(3\mathbf{i}+2\mathbf{j}+\mathbf{k})$. The bead moves from rest at the point P(4, 2, 1) to the point Q(16, 10, 5) under the action of a force $\mathbf{F} = 7\mathbf{i}+2\mathbf{k}$. Using the newton and the metre as units for force and distance find:
(i) the work done by **F**,
(ii) the speed of the bead at Q.

20) If the points P_1, P_2 have position vectors $2\mathbf{i}-5\mathbf{j}+\mathbf{k}$, $-8\mathbf{i}-\mathbf{j}+4\mathbf{k}$ respectively, and the points Q_1, Q_2 have position vectors $-13\mathbf{j}+5\mathbf{k}$, $4\mathbf{i}+3\mathbf{j}-3\mathbf{k}$ respectively, prove that the lines P_1P_2 and Q_1Q_2 intersect at right angles. Find the position vector of their point of intersection.
A force of magnitude F acting in the direction P_1Q_2 moves a particle from P_2 to Q_2. Determine the work done by the force. (U of L)

CHAPTER 3

VECTOR EQUATIONS OF CURVES. MOTION OF A PARTICLE ON A CURVE

VECTOR ANALYSIS OF CURVES

Before investigating the general motion of a particle using vector methods, it is necessary to consider the vector equations of some familiar curves. With the exception of the helix these curves are plane curves, considered in the xy plane in the following discussion, although they can obviously exist in any plane. Throughout this chapter, **i**, **j** and **k** are unit vectors in the directions Ox, Oy, Oz respectively.

The Position Vector of a Point on a Circle

Consider the circle $x^2 + y^2 = a^2$ with **r** the position vector (relative to O) of any point P on this circle.

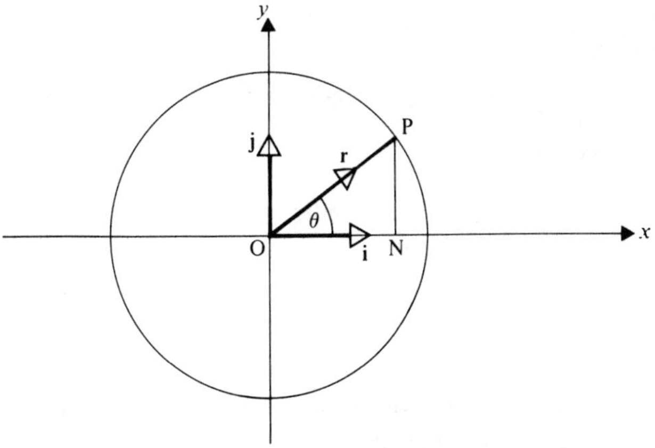

The radius of this circle is a, so the x coordinate of P is $a \cos \theta$,

therefore $\quad \text{ON} = a \cos \theta \quad$ and $\quad \overrightarrow{\text{ON}} = a \cos \theta \mathbf{i}$

Similarly the y coordinate of P is $a \sin \theta$,

therefore $\quad \overrightarrow{\text{NP}} = a \sin \theta \mathbf{j}$

From the diagram, by vector addition,

$$\overrightarrow{\text{OP}} = \overrightarrow{\text{ON}} + \overrightarrow{\text{NP}}$$

hence $\quad \mathbf{r} = a \cos \theta \mathbf{i} + a \sin \theta \mathbf{j}$

This equation gives the position vector of any point on the circle and is called the *vector equation of the circle.*

The Position Vector of a Point on a Parabola

Consider the parabola $y^2 = 4ax$ with $\mathbf{r}$ the position vector (relative to O) of any point P on this parabola.

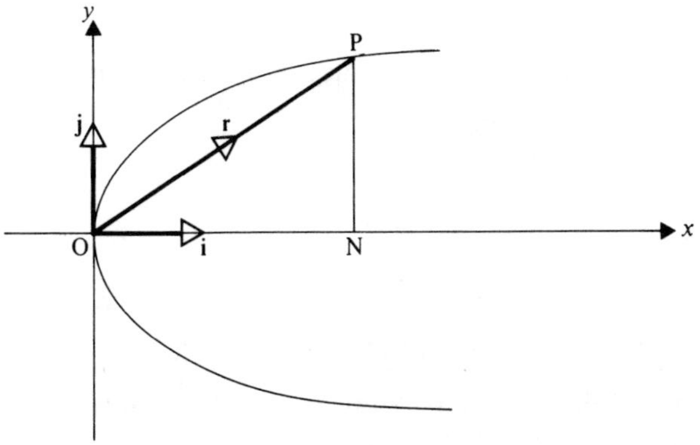

The parametric equations of $y^2 = 4ax$ are $x = at^2, y = 2at$,
so the x coordinate of P is at^2,

therefore $\quad ON = at^2 \quad$ and $\quad \overrightarrow{ON} = at^2\mathbf{i}$

Similarly the y coordinate of P is $2at$,

therefore $\quad \overrightarrow{NP} = 2at\mathbf{j}$

From the diagram, by vector addition,

$$\overrightarrow{OP} = \overrightarrow{ON} + \overrightarrow{NP}$$

hence $\quad \mathbf{r} = at^2\mathbf{i} + 2at\mathbf{j}$

This equation gives the position vector of any point P on this parabola and is called the *vector equation of the parabola.*

The Position Vector of a Point on an Ellipse

Consider the ellipse $\dfrac{x^2}{a^2} + \dfrac{y^2}{b^2} = 1$ with $\mathbf{r}$ the position vector (relative to O) of any point P on this ellipse.

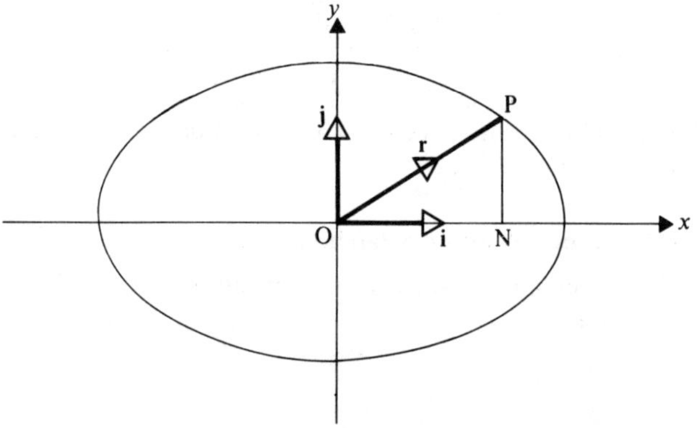

The parametric equations of the ellipse are $x = a\cos\theta, y = b\sin\theta$.

Therefore $\quad \overrightarrow{ON} = a\cos\theta\mathbf{i}$

and $\quad \overrightarrow{NP} = b\sin\theta\mathbf{j}$

From the diagram, by vector addition,

$$\overrightarrow{OP} = \overrightarrow{ON} + \overrightarrow{NP}$$

hence $\quad \mathbf{r} = a\cos\theta\mathbf{i} + b\sin\theta\mathbf{j}$

This equation gives the position vector of any point P on this ellipse and is called the *vector equation of the ellipse*.

The Position Vector of a Point on a Hyperbola

Consider the hyperbola $\frac{x^2}{a^2} - \frac{y^2}{b^2} = 1$ with **r** the position vector (relative to O) of any point on this hyperbola.

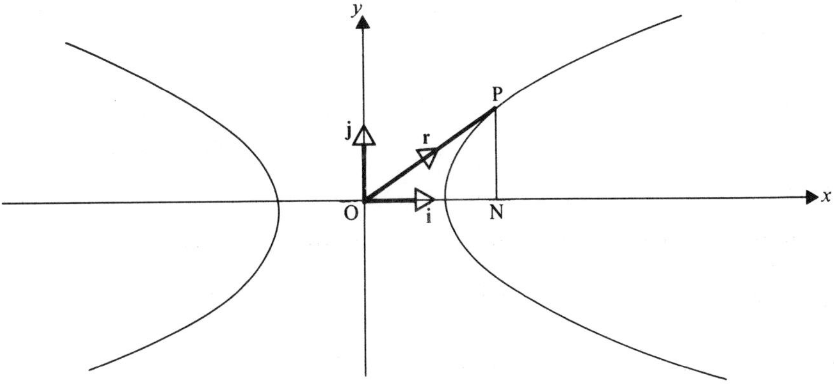

The parametric equations of the hyperbola are $x = a \sec \theta, y = b \tan \theta$.

Therefore $\qquad \overrightarrow{ON} = a \sec \theta \mathbf{i}$.

and $\qquad \overrightarrow{NP} = b \tan \theta \mathbf{j}$

From the diagram, by vector addition,

$$\overrightarrow{OP} = \overrightarrow{ON} + \overrightarrow{NP}$$

hence $\qquad \mathbf{r} = a \sec \theta \mathbf{i} + b \tan \theta \mathbf{j}$

This equation gives the position vector of any point on the hyperbola and is called the *vector equation of the hyperbola*.

The Position Vector of a Point on a Rectangular Hyperbola

Consider the hyperbola $xy = c^2$ with **r** the position vector (relative to O) of any point P on the hyperbola.

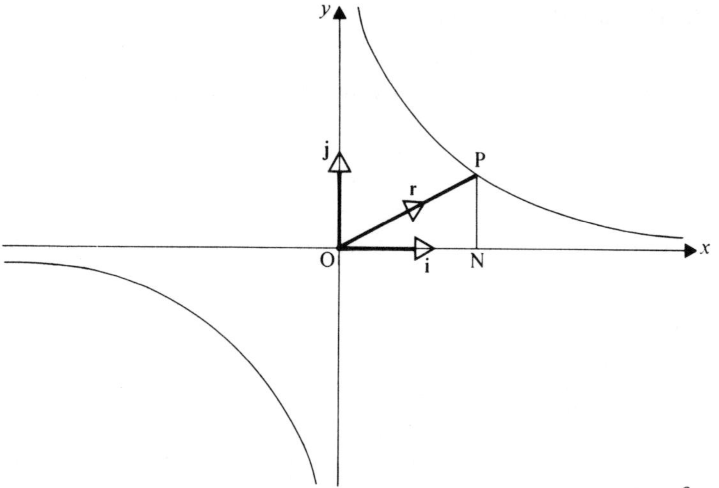

The parametric equations of the rectangular hyperbola are $x = ct, y = \dfrac{c}{t}$,

therefore $\qquad \overrightarrow{ON} = ct\mathbf{i}$

and $\qquad \overrightarrow{NP} = \dfrac{c}{t}\mathbf{j}$

From the diagram, by vector addition,

$$\overrightarrow{OP} = \overrightarrow{ON} + \overrightarrow{NP}$$

hence $\qquad \mathbf{r} = ct\mathbf{i} + \dfrac{c}{t}\mathbf{j}$

This equation gives the position vector of any point P on the rectangular hyperbola and is called the *vector equation of the rectangular hyperbola*.

The Position Vector of a Point on a Helix

A helix is the locus of a point that rotates about a fixed axis in such a way that its distance from that axis is constant and it either advances or withdraws a fixed distance along that axis for each complete rotation. A thread wound round a cylinder in such a way that the turns are equally spaced is an example of a helix. The distance between consecutive turns is called the *pitch* of the helix. Let **r** be the position vector of any point P on the helix of pitch p, advancing along the axis Oz, and which passes through the point with position vector $a\mathbf{i}$.

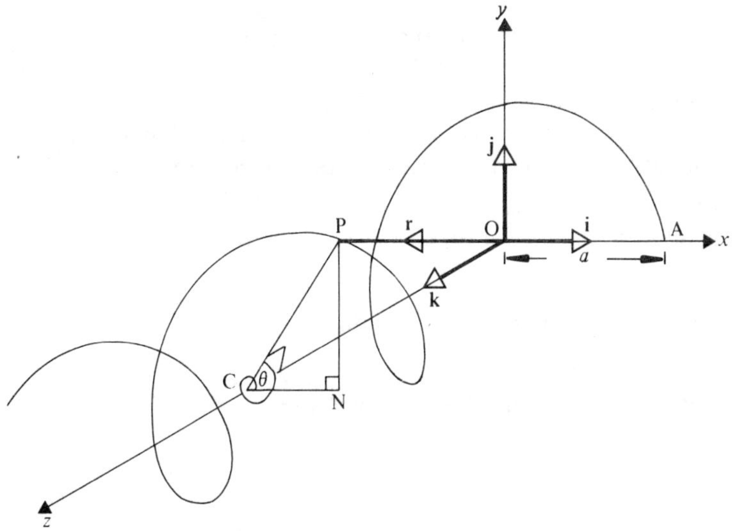

Suppose as a point moves from A to P on the curve it rotates through an angle θ about Oz. Since the helix passes through the point with position vector $a\mathbf{i}$, every other point on the curve is at a distance a from Oz.

Therefore $\quad CP = a, \quad CN = a\cos\theta, \quad NP = a\sin\theta$

For a turn of 2π the curve advances p along Oz,

so for a turn of θ it advances $\dfrac{p\theta}{2\pi}$, thus $OC = \dfrac{p\theta}{2\pi}$.

Therefore $\quad \overrightarrow{CN} = a\cos\theta\mathbf{i}, \quad \overrightarrow{NP} = a\sin\theta\mathbf{j}, \quad \overrightarrow{OC} = \dfrac{p\theta}{2\pi}\mathbf{k}$

From the diagram, by vector addition,

$$\overrightarrow{OP} = \overrightarrow{CN} + \overrightarrow{NP} + \overrightarrow{OC}$$

hence

$$\mathbf{r} = a\cos\theta\mathbf{i} + a\sin\theta\mathbf{j} + \frac{p\theta}{2\pi}\mathbf{k}$$

This equation gives the position vector of any point on this helix and is called the *vector equation of the helix*.

The curve we have just considered is a typical helix but slight variations in the behaviour of the moving point P have some effect on the vector equation:

(a) When the point describing the helix advances along the axis (progresses positively), the progression component is positive. If the point withdraws (progresses negatively) the progression component is negative.

Point advancing: $\quad \mathbf{r} = a\cos\theta\mathbf{i} + a\sin\theta\mathbf{j} + \dfrac{p\theta}{2\pi}\mathbf{k}$

Point withdrawing: $\mathbf{r} = a\cos\theta\mathbf{i} + a\sin\theta\mathbf{j} - \frac{p\theta}{2\pi}\mathbf{k}$

(b) If the point describing the helix rotates in the positive sense ($x \to y \to z \to x$) then θ is a positive angle whose cosine and sine are both positive. For negative rotation θ is a negative angle with positive cosine but negative sine.

Positive rotation: $\mathbf{r} = a\cos\theta\mathbf{i} + a\sin\theta\mathbf{j} + \frac{p\theta}{2\pi}\mathbf{k}$

Negative rotation: $\mathbf{r} = a\cos\theta\mathbf{i} - a\sin\theta\mathbf{j} + \frac{p\theta}{2\pi}\mathbf{k}$

General Position Vector of a Point on any Curve

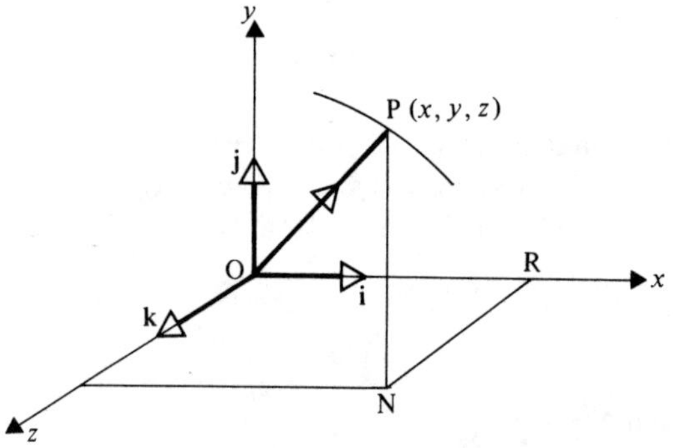

If $\mathbf{r}$ is the position vector of any point P(x, y, z) on a curve, then by vector addition

$$\mathbf{r} = \overrightarrow{OR} + \overrightarrow{NP} + \overrightarrow{RN} \qquad (1)$$

But $\overrightarrow{OR} = x\mathbf{i}, \quad \overrightarrow{NP} = y\mathbf{j}, \quad \overrightarrow{RN} = z\mathbf{k}$

If x, y and z can be expressed in parametric form:
i.e. $x = f(\lambda), y = g(\lambda), z = h(\lambda)$ where λ is a parameter then (1) can be written

$$\mathbf{r} = f(\lambda)\mathbf{i} + g(\lambda)\mathbf{j} + h(\lambda)\mathbf{k} \qquad (2)$$

and as λ varies $\mathbf{r}$ is the position vector of any point on the curve and is called the vector equation of that curve.
If the vector equation of any curve is given in the form of equation (2), then

$$x = f(\lambda), \quad y = g(\lambda), \quad z = h(\lambda)$$

and the Cartesian equations of the curve can be found by eliminating λ from these three equations.

SUMMARY

Curve	*Vector equation*
Circle $x^2 + y^2 = a^2$	$\mathbf{r} = a\cos\theta\mathbf{i} + a\sin\theta\mathbf{j}$
Parabola $y^2 = 4ax$	$\mathbf{r} = at^2\mathbf{i} + 2at\mathbf{j}$
Ellipse $\dfrac{x^2}{a^2} + \dfrac{y^2}{b^2} = 1$	$\mathbf{r} = a\cos\theta\mathbf{i} + b\sin\theta\mathbf{j}$
Hyperbola $\dfrac{x^2}{a^2} - \dfrac{y^2}{b^2} = 1$	$\mathbf{r} = a\sec\theta\mathbf{i} + b\tan\theta\mathbf{j}$
Rectangular hyperbola $xy = c^2$	$\mathbf{r} = ct\mathbf{i} + \dfrac{c}{t}\mathbf{j}$
Helix, axis Oz, pitch p, positive progression	$\mathbf{r} = a\cos\theta\mathbf{i} + a\sin\theta\mathbf{j} + \dfrac{p\theta}{2\pi}\mathbf{k}$

EXAMPLES 3a

1) Sketch the curve $\mathbf{r} = 5\cos\theta\mathbf{i} + 3\sin\theta\mathbf{j}$ and find its Cartesian equation.

The vector equation of the curve is given as

$$\mathbf{r} = 5\cos\theta\mathbf{i} + 3\sin\theta\mathbf{j}$$

Therefore $$x = 5\cos\theta \quad (1)$$

and $$y = 3\sin\theta \quad (2)$$

[Equations (1) and (2) are the parametric equations of an ellipse, of major axis 10 units and minor axis 6 units and centre O.]

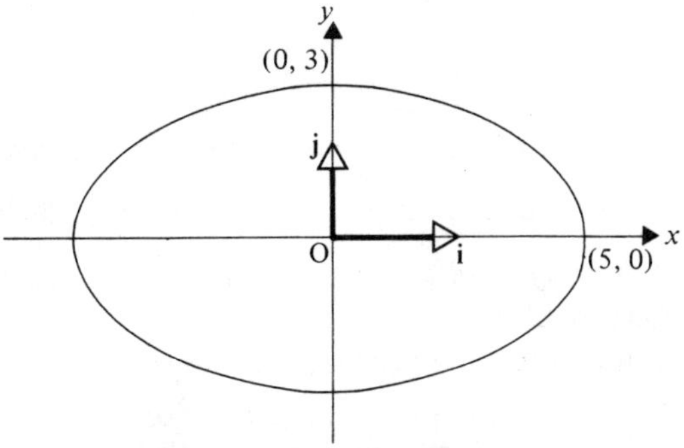

The Cartesian equation is found by eliminating θ from (1) and (2).

From (1) $$\cos\theta = x/5$$

From (2) $$\sin\theta = y/3$$

But $$\cos^2\theta + \sin^2\theta \equiv 1$$

Therefore $$\frac{x^2}{25} + \frac{y^2}{9} = 1$$

2) Sketch the curve $\mathbf{r} = (2\cos\theta + 3)\mathbf{i} + (2\sin\theta + 6)\mathbf{j}$ and find its Cartesian equation.

The vector equation of the curve can be rearranged in the form

$$\mathbf{r} - (3\mathbf{i} + 6\mathbf{j}) = 2\cos\theta\mathbf{i} + 2\sin\theta\mathbf{j} \qquad (1)$$

Comparing this with the vector equation of a circle of radius a and centre at $\mathbf{r} = \mathbf{0}$, which is

$$\mathbf{r} = a\cos\theta\mathbf{i} + a\sin\theta\mathbf{j}$$

we see that equation (1) represents a circle of radius 2 and centre at $3\mathbf{i} + 6\mathbf{j}$.

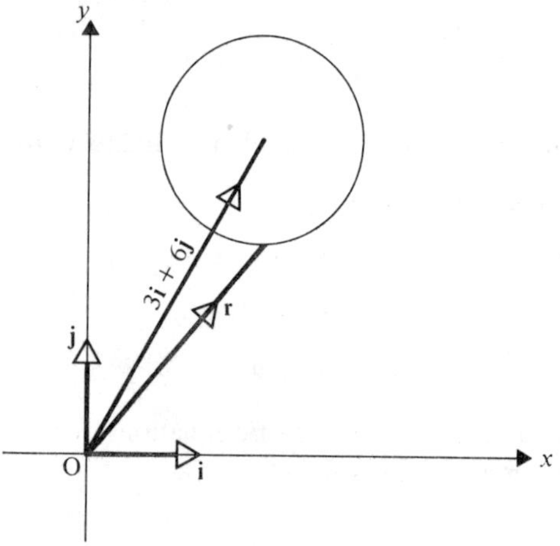

Alternatively since $\mathbf{r} = (2\cos\theta + 3)\mathbf{i} + (2\sin\theta + 6)\mathbf{j}$

then $x = 2\cos\theta + 3$ and $y = 2\sin\theta + 6$

or $x - 3 = 2\cos\theta$ and $y - 6 = 2\sin\theta$

This pair of equations may be recognised as the parametric equations of a circle with radius 2 and centre (3, 6)

The Cartesian equation is found by eliminating θ using $\cos^2\theta + \sin^2\theta = 1$.

Hence $$\left(\frac{x-3}{2}\right)^2 + \left(\frac{y-6}{2}\right)^2 = 1$$

Therefore the Cartesian equation of the curve is

$$x^2 + y^2 - 6x - 12y + 41 = 0$$

3) Find the vector equation of the circle in the **i**, **j** plane, of radius a and whose centre has position vector $a\mathbf{i} - 3a\mathbf{j}$.

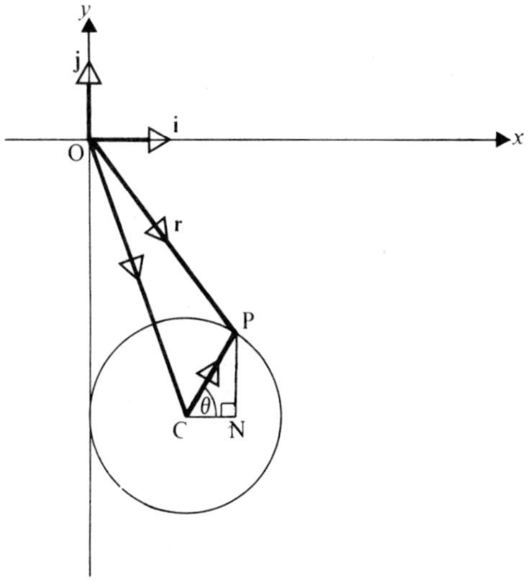

Let P be any point on this circle with position vector **r** relative to O.

From the diagram $\quad \mathbf{r} = \overrightarrow{OC} + \overrightarrow{CP}$

$$= \overrightarrow{OC} + \overrightarrow{CN} + \overrightarrow{NP}$$

$$= (a\mathbf{i} - 3a\mathbf{j}) + (a\cos\theta\,\mathbf{i} + a\sin\theta\,\mathbf{j})$$

Therefore $\quad \mathbf{r} = (a + a\cos\theta)\mathbf{i} + (a\sin\theta - 3a)\mathbf{j}$

4) Sketch the curve $\mathbf{r} = t^2\mathbf{i} + 2t\mathbf{j} + 3\mathbf{k}$ and find its Cartesian equations.

$$\mathbf{r} = t^2\mathbf{i} + 2t\mathbf{j} + 3\mathbf{k} \qquad (1)$$

The component of **r** in the direction of **k** is constant.

Therefore the curve must be in the plane $\mathbf{r} = 3\mathbf{k}$.

Rearranging (1) so that the variable components of **r** are isolated:

$$\mathbf{r} - 3\mathbf{k} = t^2\mathbf{i} + 2t\mathbf{j} \qquad (2)$$

[Comparing (2) with $\mathbf{r} = at^2\mathbf{i} + 2at\mathbf{j}$ it can be seen that (2) is the vector equation of a parabola in the plane $\mathbf{r} = 3\mathbf{k}$ as already stated.]

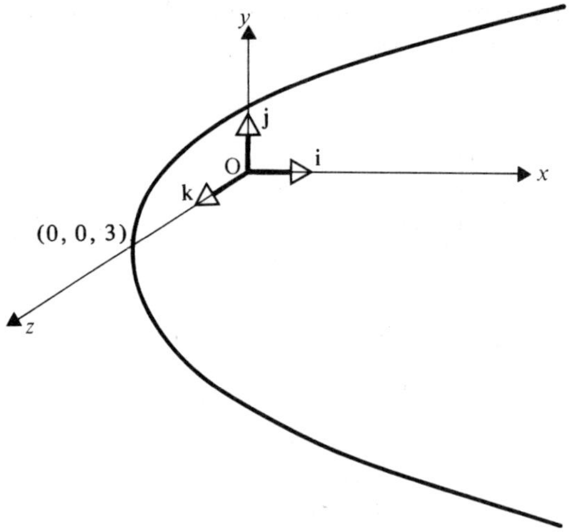

From (1) $x = t^2, \quad y = 2t, \quad z = 3$

Eliminating t from the first two equations gives $y^2 = 4x$

Therefore the Cartesian equations of this curve are

$$y^2 = 4x, \quad z = 3$$

5) Find the position vectors of the points of intersection of the line $\mathbf{r} = (2+s)\mathbf{i} - (2+3s)\mathbf{j}$ with the curve $\mathbf{r} = 4t^2\mathbf{i} - 8t^3\mathbf{j}$.

At a point of intersection

$$(2+s)\mathbf{i} - (2+3s)\mathbf{j} = 4t^2\mathbf{i} - 8t^3\mathbf{j}$$

Therefore $$2+s = 4t^2 \quad (1)$$

and $$2+3s = 8t^3 \quad (2)$$

Eliminating s from (1) and (2) gives

$$2t^3 - 3t^2 + 1 = 0$$

$$(t-1)(t-1)(2t+1) = 0$$

Hence $$t = 1 \quad \text{or} \quad -\tfrac{1}{2}.$$

Substituting these values of t into the vector equation of the curve gives

$$\mathbf{r} = 4\mathbf{i} - 8\mathbf{j}$$

and $$\mathbf{r} = \mathbf{i} + \mathbf{j}$$

These are the position vectors of the two points of intersection.
Check: when $t = 1$, $s = 2$ and when $t = -\frac{1}{2}$, $s = -1$.

Substituting these values of s into the vector equation of the *line* also gives

$$\mathbf{r} = 4\mathbf{i} - 8\mathbf{j} \quad \text{and} \quad \mathbf{r} = \mathbf{i} + \mathbf{j}$$

EXERCISE 3a

1) Sketch the following plane curves:
(a) $\mathbf{r} = \cos\theta\mathbf{i} + \sin\theta\mathbf{j}$,
(b) $\mathbf{r} = 4t^2\mathbf{i} + 8t\mathbf{j}$,
(c) $\mathbf{r} = 3\cos\theta\mathbf{i} + 4\sin\theta\mathbf{j}$,
(d) $\mathbf{r} = 4t\mathbf{i} + \dfrac{4}{t}\mathbf{j}$,
(e) $\mathbf{r} = (1 + \cos\theta)\mathbf{i} + (\sin\theta - 2)\mathbf{j}$,
(f) $\mathbf{r} = (3 + t^2)\mathbf{i} + (2t^2 - 4)\mathbf{j}$,
(g) $\mathbf{r} = 3\sec\theta\mathbf{i} + 2\tan\theta\mathbf{j}$,
(h) $\mathbf{r} = (1 + 2\cos\theta)\mathbf{i} - (2 + \sin\theta)\mathbf{j}$.

2) Sketch the following three dimensional curves:
(a) $\mathbf{r} = 2\cos\theta\mathbf{i} + 2\sin\theta\mathbf{j} + \dfrac{\theta}{2\pi}\mathbf{k}$,
(b) $\mathbf{r} = 3\cos\theta\mathbf{i} + 3\sin\theta\mathbf{j} + \mathbf{k}$,
(c) $\mathbf{r} = -\dfrac{\theta}{2\pi}\mathbf{i} + \cos\theta\mathbf{j} + \sin\theta\mathbf{k}$,
(d) $\mathbf{r} = at^2\mathbf{i} + 2at\mathbf{j} + 3\mathbf{k}$.

3) Find the Cartesian equations of the following plane curves:
(a) $\mathbf{r} = 2\cos\theta\mathbf{i} + 2\sin\theta\mathbf{j}$
(b) $\mathbf{r} = (2 - \sin\theta)\mathbf{i} + (4 - \cos\theta)\mathbf{j}$
(c) $\mathbf{r} = 2\sec\theta\mathbf{i} + 2\tan\theta\mathbf{j}$
(d) $\mathbf{r} = (6 - at^2)\mathbf{i} + (4 - 2at)\mathbf{j}$

4) Find the Cartesian equations of the following curves:
(a) $\mathbf{r} = a\cos\theta\mathbf{i} + a\sin\theta\mathbf{k}$,
(b) $\mathbf{r} = t\mathbf{j} + t^2\mathbf{k}$,
(c) $\mathbf{r} = 2\cos\phi\mathbf{i} + 3\sin\phi\mathbf{k}$,
(d) $\mathbf{r} = \cos\theta\mathbf{i} + \sin\theta\mathbf{j} + \mathbf{k}$.

5) Find the position vectors of the points of intersection of:
(a) $\mathbf{r} = \cos\theta\mathbf{i} + \sin\theta\mathbf{j}$
$\mathbf{r} = \lambda\mathbf{i} + 2\lambda\mathbf{j}$,
(b) $\mathbf{r} = 3\cos\theta\mathbf{i} + 3\sin\theta\mathbf{j}$
$\mathbf{r} = p^2\mathbf{i} + p\mathbf{j}$,
(c) $\mathbf{r} = 3\cos\theta\mathbf{i} + 3\sin\theta\mathbf{j} + \dfrac{2\theta}{\pi}\mathbf{k}$
$\mathbf{r} = (3 - s)\mathbf{i} + s\mathbf{j} + (s - 2)\mathbf{k}$.

6) Find the vector equation of the ellipse in the **i**, **j** plane, centre $\mathbf{i} - 2\mathbf{j}$, major axis of length $2a$ and minor axis of length $2b$.

7) Find the vector equation of the parabola in the **i**, **j** plane, with axis parallel to **i**, latus rectum of length $4a$ and whose vertex has position vector $\mathbf{i} - 2\mathbf{j}$.

8) Find the vector equation of the helix of pitch 4 units which advances along its axis Oz, which has positive rotation and passes through the point with position vector $3\mathbf{i}$.

9) Find the vector equation of the helix of pitch 3 units, which rotates in the sense from x to y, advances along its axis Oz and which passes through the point with position vector $3\mathbf{i} - 4\mathbf{j}$.

10) Find the position vector of any point on the circle in the plane parallel to $\mathbf{i}$ and $\mathbf{k}$ which has a radius of 3 units and whose centre has position vector $2\mathbf{i} + \mathbf{j}$.

11) Show that $\mathbf{r} = (t^2 - 3)\mathbf{i} + (5 + 3t)\mathbf{j}$ is the position vector of a point on a parabola and find the vector equation of its directrix.

MOTION OF A PARTICLE ON A CURVE USING VECTOR ANALYSIS

Velocity

The velocity of a particle is defined as the rate of increase of its displacement relative to a fixed point,
i.e. if, at time t, the displacement of a particle from a fixed point O is $\mathbf{r}$, from the definition the velocity of the particle $\mathbf{v}$ is given by $\mathbf{v} = \frac{d\mathbf{r}}{dt}$.

Let us now, from first principles, investigate the nature of $\frac{d\mathbf{r}}{dt}$ and verify that the magnitude of $\frac{d\mathbf{r}}{dt}$ is the speed of the particle and that the direction of $\frac{d\mathbf{r}}{dt}$ is the direction of motion.
Consider a particle moving along a curve such that
at time t it is at the point P on the curve
and at time $t + \delta t$ it is at the point Q on the curve

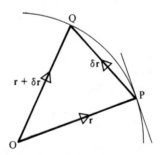

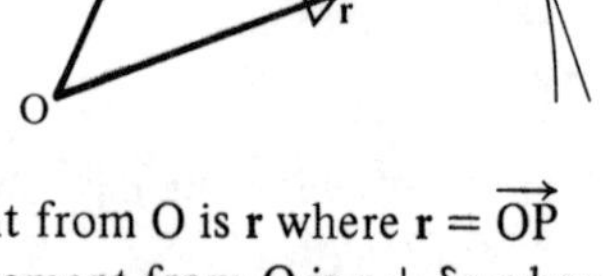

At time t the displacement from O is $\mathbf{r}$ where $\mathbf{r} = \overrightarrow{OP}$
At time $t + \delta t$ the displacement from O is $\mathbf{r} + \delta\mathbf{r}$ where $\mathbf{r} + \delta\mathbf{r} = \overrightarrow{OQ}$
Thus the increase in displacement ($\delta\mathbf{r}$) in the interval of time δt is

$$\overrightarrow{OQ} - \overrightarrow{OP}$$

But $$\overrightarrow{OQ} - \overrightarrow{OP} = \overrightarrow{PQ}$$

Therefore $$\delta\mathbf{r} = \overrightarrow{PQ}$$

Hence the average velocity during the interval of time δt is $\dfrac{\delta\mathbf{r}}{\delta t}$ where

$$\frac{\delta\mathbf{r}}{\delta t} = \frac{\overrightarrow{PQ}}{\delta t}$$

As $\delta t \to 0$, the direction of $\overrightarrow{PQ}$ $\left(\text{and hence of } \dfrac{\overrightarrow{PQ}}{\delta t}\right)$ tends to the direction of the tangent to the curve at P

and $\dfrac{PQ}{\delta t}$ tends to the speed of the particle at P.

Thus $\lim\limits_{\delta t \to 0} \left(\dfrac{\delta\mathbf{r}}{\delta t}\right)$ is a vector whose magnitude is the speed and whose direction is the direction of motion of a particle whose position vector is $\mathbf{r}$ at time t,

i.e. $\dfrac{d\mathbf{r}}{dt} = \mathbf{v}$, the velocity vector of the particle.

For example if $\mathbf{r} = 2\mathbf{i} + t^2\mathbf{j} + (2t - 1)\mathbf{k}$ is the position vector of a particle at time t, its velocity vector $\mathbf{v}$ is given by

$$\begin{aligned}\mathbf{v} &= \frac{d\mathbf{r}}{dt}\\ &= 2t\mathbf{j} + 2\mathbf{k}\end{aligned}$$

Conversely if $\mathbf{v} = 2\mathbf{i} - 3t^2\mathbf{j}$ is the velocity vector of a particle at time t, the position vector $\mathbf{r}$ of the particle at time t is given by

$$\begin{aligned}\mathbf{r} &= \int \mathbf{v}\,dt\\ &= 2t\mathbf{i} - t^3\mathbf{j} + \mathbf{c}\end{aligned}$$

where the vector $\mathbf{c}$ is a vector of integration. If the position vector of the particle is known at some specified time, $\mathbf{c}$ can be evaluated.

ACCELERATION

Acceleration is defined as the rate of increase of velocity.
If $\mathbf{v}$ is the velocity vector and $\mathbf{a}$ is the acceleration vector of a particle at time t, then

$$\mathbf{a} = \frac{d\mathbf{v}}{dt}$$

and conversely $$\mathbf{v} = \int \mathbf{a}\,dt$$

Note that the velocity vector is defined in terms of magnitude and direction only. i.e. a *velocity vector is a free vector.*
Similarly an *acceleration vector is a free vector*, but *displacement from a specified point is a tied vector.*

SUMMARY

$$\mathbf{v} = \frac{d}{dt}(\mathbf{r}) \qquad \mathbf{r} = \int \mathbf{v}\, dt$$

$$\mathbf{a} = \frac{d}{dt}(\mathbf{v}) \qquad \mathbf{v} = \int \mathbf{a}\, dt$$

where **r**, **v** and **a** are the position, velocity and acceleration vectors respectively of a particle.

WORK DONE BY A VARIABLE FORCE

Consider a particle P moving along a curve under the action of a variable force.

If at time t the force is **F** and the position vector of the particle is **r**, and at time $t + \delta t$ the position vector of the particle is $\mathbf{r} + \delta\mathbf{r}$

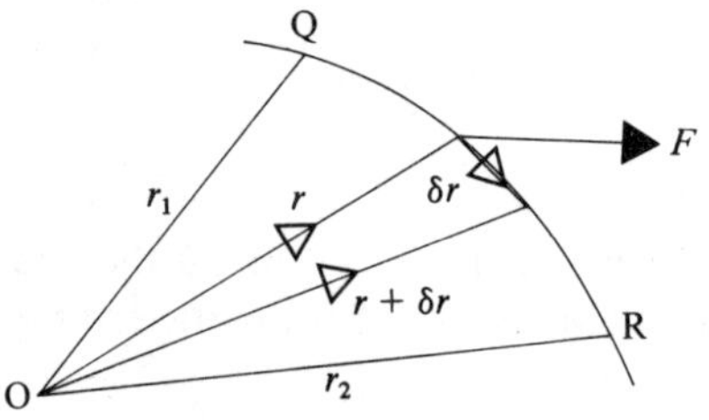

then the work done in the interval of time δt is approximately

$$\mathbf{F}.\delta\mathbf{r}$$

Therefore the work done in moving the particle along the arc QR is

$$\int_{r_1}^{r_2} \mathbf{F}.d\mathbf{r}$$

Note that if **F** *is constant this becomes* $\mathbf{F}.(\mathbf{r}_2 - \mathbf{r}_1)$.

EXAMPLES 3b

1) The position vector of a particle at time t is given by $\mathbf{r} = 3\cos 2t\mathbf{i} + 5\sin 2t\mathbf{j}$. Find the velocity and acceleration vectors of the particle when $t = \pi/4$ and show that the angle between them at this time is $\pi/2$.

The velocity vector **v** is given by

$$\mathbf{v} = \frac{d}{dt}(\mathbf{r}) = \frac{d}{dt}(3\cos 2t\mathbf{i} + 5\sin 2t\mathbf{j})$$

$$= -6\sin 2t\mathbf{i} + 10\cos 2t\mathbf{j}$$

hence when $t = \pi/4 \quad \mathbf{v} = -6\mathbf{i}$

The acceleration vector **a** is given by

$$\mathbf{a} = \frac{d}{dt}(\mathbf{v}) = \frac{d}{dt}(-6\sin 2t\mathbf{i} + 10\cos 2t\mathbf{j})$$

$$= -12\cos 2t\mathbf{i} - 20\sin 2t\mathbf{j}$$

hence when $t = \pi/4 \quad \mathbf{a} = -20\mathbf{j}$

The angle between **v** and **a** can be found using $\cos\theta = \dfrac{\mathbf{v}\cdot\mathbf{a}}{va}$

when $t = \pi/4 \quad \cos\theta = \dfrac{(-6\mathbf{i})\cdot(-20\mathbf{j})}{va} = 0$

Therefore $\theta = \pi/2$

so when $t = \pi/4$ the angle between **v** and **a** is $\pi/2$.

2) A particle of mass 4 units is acted upon by constant forces $\mathbf{F}_1$ and $\mathbf{F}_2$ where $\mathbf{F}_1 = 2\mathbf{i} - \mathbf{j}$ and $\mathbf{F}_2 = 3\mathbf{i} + 2\mathbf{j}$.
Initially the particle is at rest at the point with position vector $\mathbf{i} - 2\mathbf{k}$. Find the position vector of the particle at time t and also find the momentum and kinetic energy of the particle when $t = 2$.

The resultant force **F** acting on the particle is $\mathbf{F}_1 + \mathbf{F}_2$,

hence $\mathbf{F} = 5\mathbf{i} + \mathbf{j}$

If **a** is the acceleration vector,

$$\mathbf{F} = 4\mathbf{a} \quad \text{(Newton's Law)}$$

therefore $\mathbf{a} = \frac{1}{4}(5\mathbf{i} + \mathbf{j})$

The velocity vector $\mathbf{v} = \int \mathbf{a}\, dt$

$$= \tfrac{1}{4}(5t\mathbf{i} + t\mathbf{j}) + \mathbf{c}$$

$\mathbf{v} = \mathbf{0}$ when $t = 0$, therefore $\mathbf{c} = \mathbf{0}$

giving $\mathbf{v} = \frac{1}{4}(5t\mathbf{i} + t\mathbf{j})$

The position vector $\mathbf{r} = \int \mathbf{v}\, dt$

$$= \tfrac{1}{8}(5t^2\mathbf{i} + t^2\mathbf{j}) + \mathbf{d}$$

$\mathbf{r} = \mathbf{i} - 2\mathbf{k}$ when $t = 0$, therefore $\mathbf{d} = \mathbf{i} - 2\mathbf{k}$

giving $$\mathbf{r} = \tfrac{1}{8}[(5t^2 + 8)\mathbf{i} + t^2\mathbf{j} - 16\mathbf{k}]$$

The momentum of the particle is $m\mathbf{v}$.

When $t = 2$, $\mathbf{v} = \frac{1}{2}(5\mathbf{i} + \mathbf{j})$.

Therefore the momentum of the particle when $t = 2$ is $10\mathbf{i} + 2\mathbf{j}$.
The kinetic energy of the particle is $\frac{1}{2}mv^2$.
When $t = 2$, $v = \frac{1}{2}\sqrt{26}$.
Therefore the kinetic energy of the particle when $t = 2$ is 13 units.

3) The velocity vector of a particle P is $2\mathbf{i} - 3t^2\mathbf{j}$. The velocity vector of a second particle Q relative to P is $2t\mathbf{i} + 2\mathbf{j}$. If Q is initially at the point with position vector $\mathbf{i} - \mathbf{j}$ find the vector equation of the path of Q.

If $\mathbf{v}_P$ and $\mathbf{v}_Q$ are the velocity vectors of P and Q and ${}_Q\mathbf{v}_P$ is the velocity vector of Q relative to P, then

$$\mathbf{v}_P = 2\mathbf{i} - 3t^2\mathbf{j}$$

$${}_Q\mathbf{v}_P = 2t\mathbf{i} + 2\mathbf{j}$$

$$\mathbf{v}_Q = {}_Q\mathbf{v}_P + \mathbf{v}_P$$

hence $$\mathbf{v}_Q = (2 + 2t)\mathbf{i} + (2 - 3t^2)\mathbf{j}$$

therefore the position vector of Q is given by

$$\mathbf{r} = (2t + t^2)\mathbf{i} + (2t - t^3)\mathbf{j} + \mathbf{c}$$

When $t = 0$, $\mathbf{r} = \mathbf{i} - \mathbf{j}$, so $\mathbf{c} = \mathbf{i} - \mathbf{j}$

therefore $$\mathbf{r} = (2t + t^2 + 1)\mathbf{i} + (2t - t^3 - 1)\mathbf{j}$$

This is the vector equation of the path of Q.

4) A particle moves on the circle $\mathbf{r} = (2 + \cos\theta)\mathbf{i} + (\sin\theta - 1)\mathbf{j}$ with constant speed v in the sense $\mathbf{i} \rightarrow \mathbf{j}$. Find the position vector of the particle at time t if when $t = 0$ it is at the point $2\mathbf{i} - 2\mathbf{j}$, and show that it has an acceleration of constant magnitude.

$\mathbf{r} = (2 + \cos\theta)\mathbf{i} + (\sin\theta - 1)\mathbf{j}$ represents a circle of radius 1 unit and centre $2\mathbf{i} - \mathbf{j}$.

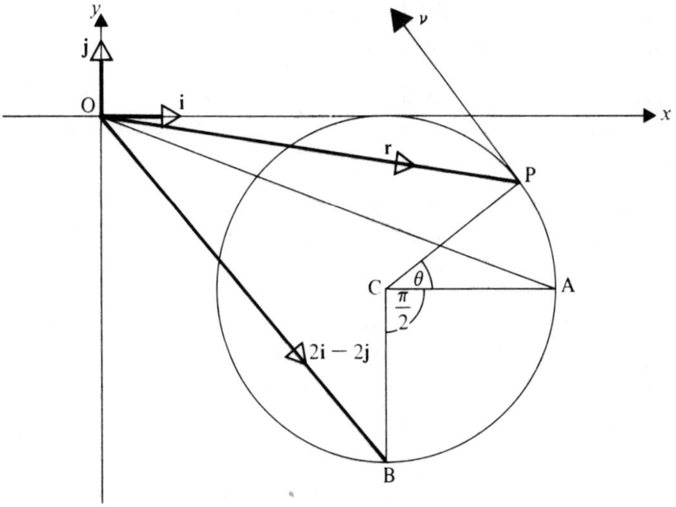

When $\theta = 0$, $\mathbf{r} = \overrightarrow{OA}$.

When $\mathbf{r} = 2\mathbf{i} - 2\mathbf{j}$, $\theta = -\pi/2$, so when $t = 0$, $\theta = -\pi/2$.

The angular velocity is $\dfrac{v}{r}$, but since $r = 1$

$$\dot{\theta} = v$$

Integrating with respect to t gives

$$\theta = vt - \pi/2$$

hence $\quad \mathbf{r} = [2 + \cos(vt - \pi/2)]\mathbf{i} + [\sin(vt - \pi/2) - 1]\mathbf{j}$

where $\mathbf{r}$ is the position vector of the particle at time t.

Now $\quad \mathbf{v} = \dfrac{d\mathbf{r}}{dt} = -v\sin(vt - \pi/2)\mathbf{i} + v\cos(vt - \pi/2)\mathbf{j}$

and $\quad \mathbf{a} = \dfrac{d\mathbf{v}}{dt} = -v^2\cos(vt - \pi/2)\mathbf{i} - v^2\sin(vt - \pi/2)\mathbf{j}$

therefore $\quad |\mathbf{a}| = [v^4\cos^2(vt - \pi/2) + v^4\sin^2(vt - \pi/2)]^{1/2} = v^2$

showing that the magnitude of the acceleration is constant.
[Note: for a particle rotating in the sense $\mathbf{j} \rightarrow \mathbf{i}$, $\dot{\theta} = -v$ and $\theta = -vt - \pi/2$]

EXERCISE 3b

1) The position vector of a particle at time t is $at^3\mathbf{i} + at\mathbf{j} + at^2\mathbf{k}$ where a is a constant. Find its velocity and acceleration vector at time t.

2) The position vector of a particle is $\mathbf{r} = \ln t\mathbf{i} + t\mathbf{j}$ where t is the time. Find its velocity and acceleration vector when $t = 4$.

3) The position vector of a particle is given by $\mathbf{r} = \sin pt\mathbf{i} + \cos pt\mathbf{j} + \mathbf{k}$ where t is the time, and p is a constant. Find its velocity and acceleration vectors when $t = \pi/3p$.

4) The position vector of a particle is $\mathbf{r} = e^t\mathbf{i} + 2t\mathbf{j}$ where t is the time. Show that the acceleration vector has no component in the direction of $\mathbf{j}$ and find the speed of the particle when $t = 4$.

5) The acceleration of a particle at time t is given by $\mathbf{a} = 2t\mathbf{i} + \mathbf{j}$. Initially it is at the origin with velocity vector $\mathbf{i} + 2\mathbf{j}$. Find the velocity and position vector of the particle at time t.

6) The velocity vector of a particle is $3t^2\mathbf{i} + (3t - 1)\mathbf{j}$ at time t. Initially the particle is at the point $\mathbf{i} + \mathbf{j}$.
Find the position vector of the particle when its speed is 13 units.

7) The acceleration of a particle is $\mathbf{a} = \sin t\mathbf{i} + \cos t\mathbf{j}$ where t is the time. It is at the origin with velocity vector $\frac{1}{2}(-\sqrt{3}\mathbf{i} + \mathbf{j})$ when $t = \pi/6$. Find its position vector at time t.

8) The velocity vector of a particle at time t is $-\sin wt\mathbf{i} - \cos wt\mathbf{j}$. The particle is at the point with position vector $\dfrac{1}{w}\mathbf{i}$ when $t = 0$. Find the position and acceleration vectors of the particle at time t. Show that the angle between the velocity and acceleration vectors at any time t is independent of t and describe the motion.

9) A particle of mass 3 kg is initially at rest at the point whose position vector is $\mathbf{i} - 2\mathbf{j}$ and is acted on by a constant force $2\mathbf{i} - \mathbf{j} + \mathbf{k}$. Write down vector expressions for its acceleration, velocity and position at time t. (The unit of force is the newton and the unit of distance is the metre.)

10) A particle of mass m is acted on by two forces $\mathbf{F}_1$ and $\mathbf{F}_2$ where $\mathbf{F}_1 = 3\mathbf{i} - \mathbf{k}$ and $\mathbf{F}_2 = \mathbf{i} + \mathbf{j}$.
Find the position vector of the particle at time t if it is initially at rest at the origin.

11) A variable force $\mathbf{F}$ acts on a particle of mass m such that the position vector of the particle at time t is $\mathbf{r} = 2\sin pt\mathbf{i} + t^2\mathbf{j}$. Find the greatest value of the magnitude of $\mathbf{F}$.

12) A particle moves on the helix $\mathbf{r} = \cos\theta\mathbf{i} - \sin\theta\mathbf{j} + \dfrac{\theta}{2\pi}\mathbf{k}$ with constant speed speed $v\sqrt{(1 + 1/4\pi^2)}$ such that it advances along the axis while rotating in a negative sense about that axis. Find vector expressions for its position, velocity and acceleration at time t if when $t = 0$ it is at the point $\mathbf{i}$.

13) A particle moves so that its position vector is $\mathbf{r} = t^2\mathbf{i} - \dfrac{2}{t}\mathbf{j}$ where t is the time. Find the velocity and position vector of the particle when it is moving in the direction of the vector $\mathbf{i} + \mathbf{j}$.

14) A particle is projected from the origin with initial velocity vector $v \cos \alpha\mathbf{i} + v \sin \alpha\mathbf{j}$. Find the vector equation of its path if $\mathbf{i}$ and $\mathbf{j}$ are in the horizontal and vertically upward directions respectively.

15) The position vector of a particle P is given by $\mathbf{r} = t\mathbf{i} + t^2\mathbf{j}$ where t is the time. The velocity of a second particle Q relative to P is $\mathbf{i} + \mathbf{j}$. If Q passes through the origin when $t = 0$, find the position vector of Q at time t and hence the Cartesian equation of its path.

DIFFERENTIATION OF A VECTOR WITH RESPECT TO A SCALAR VARIABLE

We have shown that if a vector is expressed in the form

$$\mathbf{v} = f(t)\mathbf{i} + g(t)\mathbf{j} + h(t)\mathbf{k}$$

the differential of $\mathbf{v}$ with respect to t is

$$\frac{d\mathbf{v}}{dt} = f'(t)\mathbf{i} + g'(t)\mathbf{j} + h'(t)\mathbf{k}$$

where t represents the scalar variable time. However as far as the process of differentiation is concerned, t can represent *any* scalar variable quantity.

Thus if $\qquad \mathbf{r} = f(\theta)\mathbf{i} + g(\theta)\mathbf{j} + h(\theta)\mathbf{k} \quad$ where θ is scalar

$$\frac{d\mathbf{r}}{d\theta} = f'(\theta)\mathbf{i} + g'(\theta)\mathbf{j} + h'(\theta)\mathbf{k}$$

If we wish to differentiate $\mathbf{r}$ with respect to another scalar variable t then

$$\frac{d\mathbf{r}}{dt} = \frac{d\mathbf{r}}{d\theta}\frac{d\theta}{dt} = f'(\theta)\frac{d\theta}{dt}\mathbf{i} + g'(\theta)\frac{d\theta}{dt}\mathbf{j} + h'(\theta)\frac{d\theta}{dt}\mathbf{k}$$

DIFFERENTIATION OF A UNIT VECTOR WITH RESPECT TO A SCALAR

Consider a unit vector $\hat{\mathbf{r}}$ which has a variable direction.

Now $\qquad \hat{\mathbf{r}} = l\mathbf{i} + m\mathbf{j} + n\mathbf{k}$

where l, m and n are the (variable) direction cosines of $\hat{\mathbf{r}}$ and therefore are subject to the condition $l^2 + m^2 + n^2 = 1$.
The differential of $\hat{\mathbf{r}}$ with respect to a scalar variable t is

$$\frac{d\hat{\mathbf{r}}}{dt} = \frac{dl}{dt}\mathbf{i} + \frac{dm}{dt}\mathbf{j} + \frac{dn}{dt}\mathbf{k}$$

Now $$\hat{\mathbf{r}} \cdot \frac{d\hat{\mathbf{r}}}{dt} = (l\mathbf{i} + m\mathbf{j} + n\mathbf{k}) \cdot \left(\frac{dl}{dt}\mathbf{i} + \frac{dm}{dt}\mathbf{j} + \frac{dn}{dt}\mathbf{k}\right)$$

$$= l\frac{dl}{dt} + m\frac{dm}{dt} + n\frac{dn}{dt}$$

$$= \frac{1}{2}\frac{d}{dt}(l^2 + m^2 + n^2)$$

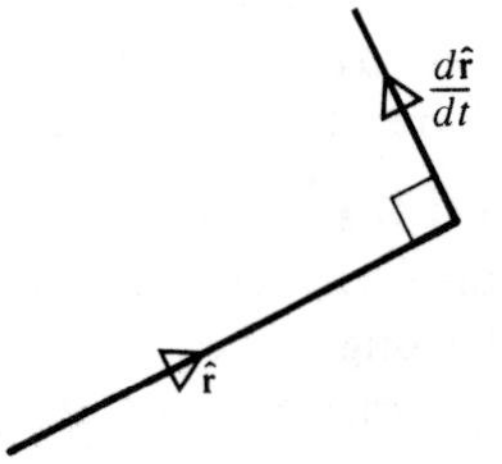

But $l^2 + m^2 + n^2 = 1$

Therefore $$\hat{\mathbf{r}} \cdot \frac{d\hat{\mathbf{r}}}{dt} = 0$$

i.e. *a unit vector and its differential with respect to a scalar are perpendicular* (But note that $\frac{d\hat{\mathbf{r}}}{dt}$ is *not* necessarily a unit vector).

In the special case in which $\hat{\mathbf{r}}$ is free to move in the xy plane only, $\hat{\mathbf{r}}$ may be expressed in terms of θ where

$$\hat{\mathbf{r}} = \cos\theta\mathbf{i} + \sin\theta\mathbf{j}$$

then $$\frac{d\hat{\mathbf{r}}}{dt} = \frac{d\hat{\mathbf{r}}}{d\theta}\frac{d\theta}{dt}$$

$$= (-\sin\theta\mathbf{i} + \cos\theta\mathbf{j})\frac{d\theta}{dt}$$

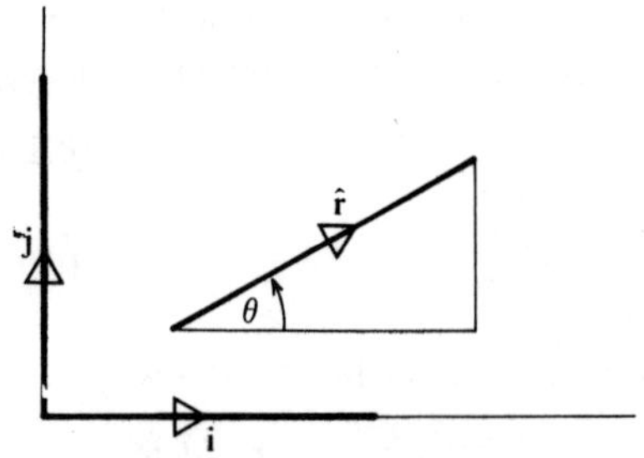

Now $-\sin\theta\mathbf{i} + \cos\theta\mathbf{j}$ is a unit vector in the direction given by rotating $\hat{\mathbf{r}}$ through a *positive* right angle.

Therefore $$\frac{d\hat{\mathbf{r}}}{dt} = \frac{d\theta}{dt}\hat{\mathbf{s}}$$

where $\hat{\mathbf{s}} = -\sin\theta\mathbf{i} + \cos\theta\mathbf{j}$

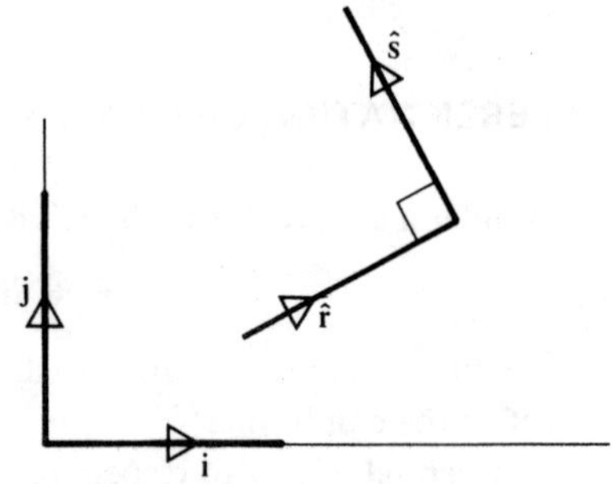

MOTION OF A PARTICLE IN A PLANE USING POLAR CO-ORDINATES

Consider a particle P moving along a curve in the xy plane such that, at time t, P is at the point with polar co-ordinates (r, θ).

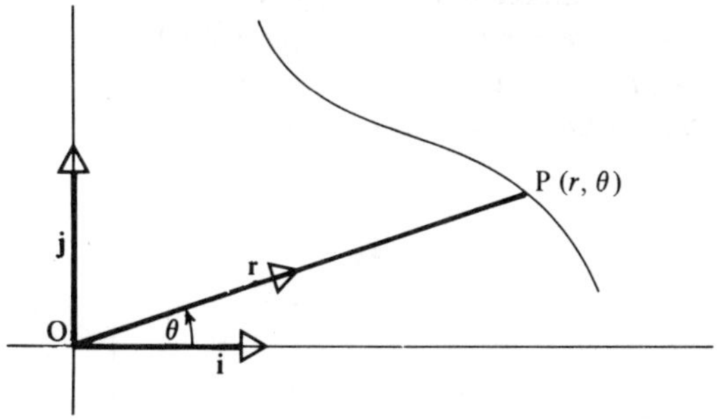

The position vector of P at time t is **r** where

$$\mathbf{r} = r(\cos\theta\mathbf{i} + \sin\theta\mathbf{j})$$

the velocity vector of P at time t is **v** where

$$\mathbf{v} = \frac{d\mathbf{r}}{dt} = \dot{r}(\cos\theta\mathbf{i} + \sin\theta\mathbf{j}) + r\dot{\theta}(-\sin\theta\mathbf{i} + \cos\theta\mathbf{j})$$

i.e.
$$\mathbf{v} = \dot{r}\hat{\mathbf{r}} + r\dot{\theta}\hat{\mathbf{s}}$$

where $\hat{\mathbf{s}}$ is inclined at $\theta + \pi/2$ to Ox.

The acceleration vector of P at time t is **a** where

$$\mathbf{a} = \frac{d\mathbf{v}}{dt} = \ddot{r}\hat{\mathbf{r}} + \dot{r}\dot{\theta}\hat{\mathbf{s}} + \dot{r}\dot{\theta}\hat{\mathbf{s}} + r\ddot{\theta}\hat{\mathbf{s}} + r\dot{\theta}\frac{d\hat{\mathbf{s}}}{dt}$$

Now $\dfrac{d\hat{\mathbf{s}}}{dt} = \dot{\theta}(-\hat{\mathbf{r}})$ as the direction of $\dfrac{d\hat{\mathbf{s}}}{dt}$ is given by rotating $\hat{\mathbf{s}}$ through one positive right angle, i.e. rotating $\hat{\mathbf{r}}$ through two right angles,

therefore
$$\mathbf{a} = (\ddot{r} - r\dot{\theta}^2)\hat{\mathbf{r}} + (r\ddot{\theta} + 2\dot{r}\dot{\theta})\hat{\mathbf{s}}$$

Now θ is the angle that OP (called the radius vector) makes with Ox at time t.
Therefore $\dot{\theta}$ is the angular velocity of OP (or simply of P)
and $\ddot{\theta}$ is the angular acceleration of OP.

RADIAL AND TRANSVERSE COMPONENTS OF VELOCITY AND ACCELERATION

From the results obtained above for **a** and **v** we see that the velocity and acceleration vectors of P can both be expressed as the sum of two component vectors, one in the direction OP and one in the direction given by rotating $\overrightarrow{OP}$ through $+\pi/2$.

The component along OP is called the radial component and the component perpendicular to OP is called the transverse component.

Therefore *the radial component of velocity is* $\dot{r}$

the transverse component of velocity is $r\dot{\theta}$

and *the radial component of acceleration is* $\ddot{r} - r\dot{\theta}^2$

the transverse component of acceleration is $r\ddot{\theta} + 2\dot{r}\dot{\theta}$

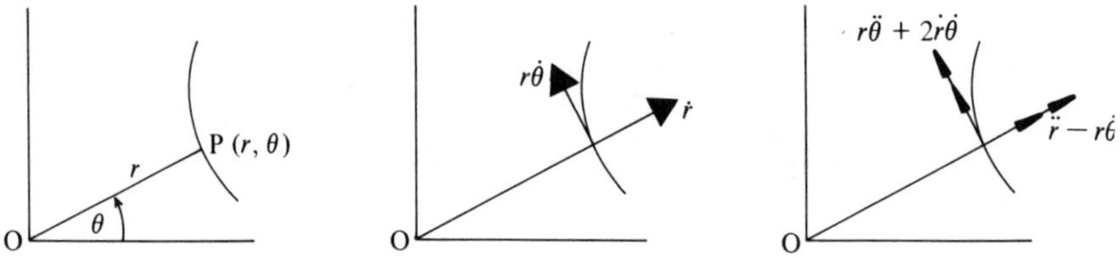

EXAMPLES 3c

1) A particle moves round the curve $r = a(1 + \cos\theta)$ with constant angular velocity ω. Find the radial and transverse components of velocity and acceleration at any time t in terms of ω and θ.

As the particle moves round the curve with constant angular velocity ω, $\dot{\theta} = \omega$ and $\ddot{\theta} = 0$.

Now $$r = a(1 + \cos\theta)$$

Hence $$\dot{r} = -a\sin\theta\dot{\theta} = -a\omega\sin\theta$$

and $$\ddot{r} = -a\omega\cos\theta\dot{\theta} = -a\omega^2\cos\theta$$

Therefore the radial component of velocity is $\dot{r} = -a\omega\sin\theta$

and the transverse component of velocity is $r\dot{\theta} = a(1 + \cos\theta)\omega$

The radial component of acceleration is

$$\ddot{r} - r\dot{\theta}^2 = (-a\omega^2\cos\theta) - \omega^2 a(1 + \cos\theta)$$
$$= -a\omega^2(1 + 2\cos\theta)$$

and the transverse component of acceleration is

$$r\ddot{\theta} + 2\dot{r}\dot{\theta} = 0 + 2\omega(-a\omega\sin\theta)$$
$$= -2a\omega^2\sin\theta$$

(*Note*: The radial and transverse components are perpendicular.)

2) Use the results obtained in example 1 to find the point on the curve $r = a(1 + \cos\theta)$ at which a particle moving on the curve with constant angular velocity has:

(a) minimum speed,
(b) minimum acceleration.

(a) The radial and transverse components of velocity at time t are

$$-a\omega \sin\theta \quad \text{and} \quad a\omega(1 + \cos\theta)$$

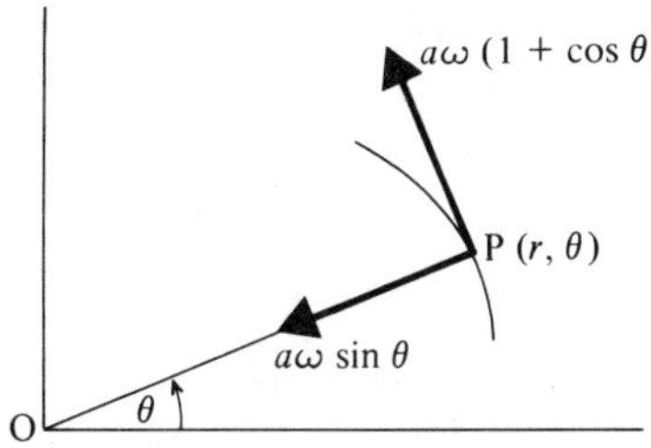

Therefore the speed at time t $= \sqrt{[(a\omega \sin\theta)^2 + (a\omega + a\omega \cos\theta)^2]}$
$= a\omega\sqrt{(2 + 2\cos\theta)}$

The speed is minimum when $2 + 2\cos\theta$ is minimum

i.e. when $\cos\theta = -1$

$$\theta = +\pi$$

From the equation of the curve, when $\theta = \pi, r = 0$.

Therefore the speed is minimum as the particle passes through O.
(b) The radial and transverse components of acceleration are:

$$-a\omega^2(1 + 2\cos\theta) \quad \text{and} \quad -2a\omega^2 \sin\theta$$

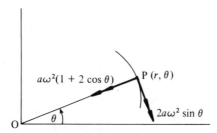

Therefore the magnitude of the acceleration is

$$a\omega^2\sqrt{[(1 + 2\cos\theta)^2 + 4\sin^2\theta]} = a\omega^2\sqrt{(5 + 4\cos\theta)}$$

The acceleration is minimum when $5 + 4\cos\theta$ is minimum

i.e. when $\cos\theta = -1$

Therefore the acceleration is minimum as the particle passes through O.

[These results show that the particle has minimum velocity and minimum acceleration at the same point on the curve (the origin). This is unexpected but a sketch of the curve shows that the result is reasonable.]

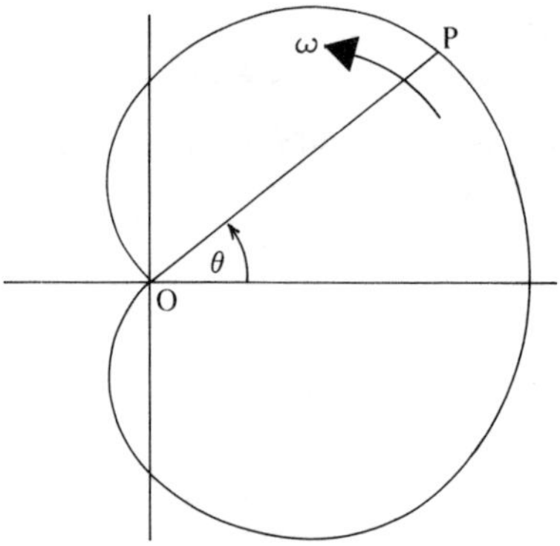

3) A point P moves round a curve with a constant angular velocity ω. At time t, its polar co-ordinates are (r, θ) and its radial component of acceleration is $-2r\omega^2$. Initially when $r = a$ and $\theta = 0$, the radial component of velocity is $a\omega$. Find the equation of the locus of P.

As the particle moves with constant angular velocity

$$\dot{\theta} = \omega \quad \text{and} \quad \ddot{\theta} = 0$$

The general radial component of acceleration at time t is $\ddot{r} - r\dot{\theta}^2$. Therefore, for the particle P,

$$\ddot{r} - r\omega^2 = -2r\omega^2$$

Hence

$$\ddot{r} = -r\omega^2$$

Writing $\ddot{r}$ as $\dot{r}\dfrac{d\dot{r}}{dr}$ gives

$$\dot{r}\frac{d\dot{r}}{dr} = -r\omega^2$$

Therefore

$$\int_{a\omega}^{\dot{r}} \dot{r}\, d\dot{r} = -\int_a^r r\omega^2\, dr$$

$$\dot{r}^2 - a^2\omega^2 = -r^2\omega^2 + a^2\omega^2$$

$$\dot{r} = \omega\sqrt{(2a^2 - r^2)}$$

Now

$$\dot{r} = \frac{dr}{d\theta}\frac{d\theta}{dt} = \omega\frac{dr}{d\theta}$$

Therefore

$$\frac{dr}{d\theta} = \sqrt{(2a^2 - r^2)}$$

Separating the variables gives

$$\int_a^r \frac{dr}{\sqrt{(2a^2 - r^2)}} = \int_0^\theta d\theta$$

i.e. $$\arcsin \frac{r}{a\sqrt{2}} - \arcsin \frac{1}{\sqrt{2}} = \theta$$

Hence $$r = a\sqrt{2} \sin (\theta + \pi/4)$$

EXERCISE 3c

1) A particle moves with constant angular velocity ω round the curve with polar equation:
(a) $r = a\theta$,
(b) $r = a(1 + \theta)$,
(c) $r = a(1 + \sin \theta)$,
(d) $r = a(\cos \theta + \sin \theta)$.
In each case find, in terms of θ and ω, the radial and transverse components of acceleration when the particle is at the point (r, θ).

2) Find the point or points on each of the curves given in question 1, at which the acceleration of the particle is maximum.

3) A particle moves with constant angular velocity ω round the curve $r = a \cos \theta$. Find expressions for the radial and transverse components of acceleration in terms of r and ω when the particle is at the point (r, θ).

4) A particle moves round a curve with constant angular velocity ω. When the particle is at the point (r, θ) it has a radial component of acceleration of $2\omega^2(1 - r)$. Initially $r = 3$, $\theta = 0$, and the radial component of velocity is zero. Find the polar equation of the curve.

5) A particle moves round the curve $r(1 + \cos \theta) = a$ with constant angular velocity ω. Find the radial and transverse components of velocity at any instant as functions of θ.

6) A particle P moves with constant angular velocity ω about an origin O. At time t, P is at the point with polar co-ordinates (r, θ) and its radial component of acceleration is zero. Initially $\theta = 0$, $r = 1$ and $\dot{r} = \omega$. Find the polar equation of the locus of P.

MULTIPLE CHOICE EXERCISE 3

The instructions for answering these questions are given on page (xi)

TYPE I

1) A circle of radius 2 units has its centre at the point (3, 2). The position vector of any point on the circumference of the circle is given by:

(a) $\mathbf{r} = 2\cos\theta\mathbf{i} + 2\sin\theta\mathbf{j}$, (b) $\mathbf{r} = 2(\cos\theta + 3)\mathbf{i} + 2(\sin\theta + 2)\mathbf{j}$,
(c) $\mathbf{r} = (2\cos\theta + 3)\mathbf{i} + (2\sin\theta + 2)\mathbf{j}$, (d) $\mathbf{r} = 3\cos\theta\mathbf{i} + 2\sin\theta\mathbf{j}$,
(e) $\mathbf{r} = (2\cos\theta - 3)\mathbf{i} + (2\sin\theta - 2)\mathbf{j}$.

2) The position vector of a particle at time t is $t^2\mathbf{i} - \ln t\mathbf{j} + \mathbf{k}$.
The acceleration vector of the particle is:

(a) $2\mathbf{i} - e^t\mathbf{j}$, (b) $2\mathbf{i} + \dfrac{1}{t^2}\mathbf{j}$, (c) $2\mathbf{i} - \dfrac{1}{t^2}\mathbf{j}$, (d) $2t\mathbf{i} - \dfrac{1}{t}\mathbf{j}$, (e) $-\dfrac{1}{2t^3}\mathbf{j}$.

3) A point P moves round the curve $r = a\theta$ such that OP rotates with constant angular velocity ω. The transverse component of acceleration when P is at the point (r, θ) is:
(a) $a\omega^2$ (b) $-r\omega^2$ (c) $-2a\omega^2$ (d) $2a\omega^2$ (e) $r\omega^2$.

4) The vector equation of a curve is $\mathbf{r} = 2t\mathbf{i} + t^2\mathbf{j}$. The Cartesian equation of this curve is:
(a) $y^2 = 4x$ (b) $4y + x^2 = 0$ (c) $4y = x^2$ (d) $y^2 = 4ax$ (e) $xy = 2$.

5) A particle of unit mass starts from rest at the origin and moves under the action of a force which t seconds after the particle leaves O is $6t\mathbf{i} - \mathbf{j} + t^2\mathbf{k}$. The position vector of the particle at time t is:
(a) $t^3\mathbf{i} - \frac{1}{2}t^2\mathbf{j} + \frac{1}{12}t^4\mathbf{k}$ (b) $2\mathbf{k}$ (c) $3t^2\mathbf{i} - t\mathbf{j} + \frac{1}{3}t^3\mathbf{k}$ (d) $6\mathbf{i} + 2t\mathbf{k}$
(e) $t^3\mathbf{i} - t\mathbf{j} + 2\mathbf{k}$.

TYPE II

6) The vector equation of a curve is $\mathbf{r} = 3\cos\theta\mathbf{i} + 2\sin\theta\mathbf{j}$.
(a) The curve is an ellipse.
(b) The Cartesian equation of the curve is $\dfrac{x^2}{9} - \dfrac{y^2}{4} = 1$.
(c) The centre of the curve is at the point (3, 2).

7) A particle moves with constant angular velocity ω round the curve with polar equation $r = a\cos\theta$.
(a) The radial component of acceleration is $-r\omega^2$.
(b) The radial component of velocity is $-a\omega\sin\theta$.
(c) The transverse component of acceleration is $2\omega\dot{r}$.

8) A particle moves with constant speed round the curve whose vector equation is $\mathbf{r} = (2 + \sin\theta)\mathbf{i} + (\cos\theta - 3)\mathbf{j}$.
(a) The acceleration of the particle is constant.
(b) The particle is always the same distance from the point $(2, -3)$.
(c) The speed of the particle is equal to 2 units.

9) A particle moves round a curve such that at time t it is at the point with polar coordinates (a, θ), has a radial component of acceleration of $-a\omega^2$ and zero transverse component of acceleration. where a and ω are constants.
(a) $\ddot{r} = 0$ (b) $\dot{\theta} \neq 0$ (c) the particle moves with constant speed.

TYPE III

10) (a) At time t, the position vector of a particle is $\mathbf{r} = \cos \omega t\mathbf{i} + \sin \omega t\mathbf{j}$ where ω is constant.

(b) A particle is moving round a circle with constant speed ω.

11) (a) $\hat{\mathbf{r}} \cdot \frac{d\hat{\mathbf{r}}}{dt} = 0$.

(b) $\mathbf{r} \cdot \frac{d\mathbf{r}}{dt} = 0$.

12) (a) $\mathbf{r}$ is a vector of constant magnitude.

(b) $\frac{d\mathbf{r}}{dt}$ is perpendicular to $\mathbf{r}$ for all values of t.

TYPE IV

13) A particle moves under the action of a force $\mathbf{F}$. Find the position vector of the particle at time t.
(a) The mass of the particle is m.
(b) $\mathbf{F} = \mathbf{i} + 3\mathbf{j}$.
(c) The initial velocity vector of the particle is $\mathbf{j} - 2\mathbf{k}$.

14) Find, in terms of a, θ and t, the transverse component of acceleration of a particle moving round a curve such that:
(a) the initial speed of the particle is u,
(b) the mass of the particle is m,
(c) the polar equation of the curve is $r(1 + \cos\theta) = a$.

15) A particle moves on a plane curve. Find its position vector when $t = 2$.
(a) The initial velocity vector is $\mathbf{i} + \mathbf{j}$.
(b) The acceleration vector at time t is $3\mathbf{i} - 2t\mathbf{j}$.
(c) When $t = 3$ the particle is at the point (2, 3).

16) Find the vector equation of the ellipse such that:
(a) the centre of the ellipse is at the point (3, 2),
(b) the major axis is of length $2a$,
(c) the minor axis is of length $2b$,
(d) the minor axis is parallel to the x axis.

17) At time t the position vector of a particle A is $\mathbf{r}$ and the velocity vector of a particle B relative to A is $\mathbf{v}$. Find the position vector of B at time t.
(a) $\mathbf{r} = (2\cos 3t)\mathbf{i} + (\sin t)\mathbf{j}$.
(b) $\mathbf{v} = (3\sin 3t)\mathbf{i} + (2\cos t)\mathbf{j}$.
(c) A is initially at the point $2\mathbf{i}$.
(d) B is initially at the origin.

TYPE V

18) The differential of a unit vector with respect to a scalar is also a unit vector.

19) $\hat{\mathbf{r}} = k\dfrac{d^2\hat{\mathbf{r}}}{dt^2}$ where k is a scalar.

20) The transverse component of acceleration for a particle moving round a curve is parallel to the tangent to the curve.

MISCELLANEOUS EXERCISE 3

1) A particle of unit mass moves in the $\mathbf{i}, \mathbf{j}$ plane under the action of a variable force $\mathbf{F}$, where $\mathbf{F} = \mathbf{i} + (\sin at)\mathbf{j}$. Initially the particle is at the origin with velocity vector $v\mathbf{i}$. If a and v are constants and t is the time, write down an expression for the particle's acceleration vector and hence find the vector equation of its path.

2) Sketch the circle $\mathbf{r} = \cos\theta\mathbf{i} + \sin\theta\mathbf{j}$ and the straight line $\mathbf{r} = (t+1)\mathbf{i} + (2t+a)\mathbf{j}$ where a is a constant. Find the range of values of a for which the line cuts the circle in two real distinct points.

3) A particle of mass 2 units is moving under the action of a force such that its position vector at time t is

$$\mathbf{r} = 5\cos t\mathbf{i} + 3\sin t\mathbf{j}$$

Sketch the path of the particle and find the force acting on it. Find also the kinetic energy of the particle when $t = \pi/4$.

4) A particle of mass 4 units starts from rest. If it is acted upon by forces $\mathbf{F}_1$ and $\mathbf{F}_2$ where $\mathbf{F}_1 = \mathbf{i} + \mathbf{j}$ and $\mathbf{F}_2 = 2\mathbf{i} - 4\mathbf{k}$, find the momentum of the particle after 5 seconds. Find also the work done on the particle in the third second of its motion.

5) A particle of mass m kg is acted upon by a constant force of $21\,m$ newtons in the direction of the vector $3\mathbf{i} + 2\mathbf{j} - 6\mathbf{k}$. Initially the particle is at the origin moving with speed 18 ms^{-1} in the direction of the vector $7\mathbf{i} - 4\mathbf{j} + 4\mathbf{k}$. Find the position vector of the particle 4 seconds later and find the work done by the force in this period of 4 seconds. (U of L)

6) A particle P moves such that its position vector at time t is $\mathbf{r} = \cos wt\mathbf{i} + \sin wt\mathbf{j}$. A second particle Q moves such that at time t its position vector is $\mathbf{r} = (t+1)\mathbf{i} + 2t\mathbf{j}$. Find an expression for the angle between the velocity vectors of P and Q at time t and hence find the values of t for which the paths of the two particles are perpendicular.

7) Two particles move so that their position vectors at time t are $\mathbf{r}_1 = (3\sin wt)\mathbf{i} + (4\cos wt)\mathbf{j} + \mathbf{k}$ and $\mathbf{r}_2 = t^2\mathbf{i} + 2t\mathbf{j} + \mathbf{k}$. Obtain the Cartesian

equations of each path and sketch these paths. Give, in both Cartesian and vector form, the equation of the plane in which the particles move. Show that the paths are perpendicular when $t = 0$ and find the values of t when the accelerations are perpendicular. (AEB)

8) A point moves such that its position vector at time t is $\mathbf{r} = ct\mathbf{i} + \frac{c}{t}\mathbf{j}$. Sketch its path and find the acceleration vector of the point. Show that this acceleration vector is equal to $k\mathbf{j}(\mathbf{r}.\mathbf{j})^3$ where k is a constant, and find k in terms of c.

9) If $\mathbf{u} = 3\mathbf{i} + 4\mathbf{j} + 5\mathbf{k}$ is the position vector of a point P, find the magnitude of $\mathbf{u}$ and calculate to the nearest minute the angles made by $\mathbf{u}$ with the unit vectors $\mathbf{i}, \mathbf{j}, \mathbf{k}$. Find also the unit vector in the same direction as $\mathbf{u}$. Show that the equation $\mathbf{r} = 4(\mathbf{i}\cos p + \mathbf{j}\sin p)$, where p is a parameter, represents a circle. Find the position vectors of the points on this circle which are nearest to and farthest from the point P. (U of L)

10) A particle starts from the origin O, and moves in a horizontal plane with constant acceleration $(-2\mathbf{i} + \mathbf{j})\ \text{cm s}^{-2}$, $\mathbf{i}$ and $\mathbf{j}$ being unit vectors in the directions of the coordinate axes Ox and Oy respectively, the unit of distance being 1 cm on each axis. The initial velocity of the particle is $(9\mathbf{i} - 4\mathbf{j})\ \text{cm s}^{-1}$. By considering the resolved parts of the motion of the particle in the directions Ox and Oy, or otherwise, show that the position vector of the particle after 5 seconds is $20\mathbf{i} - (15/2)\mathbf{j}$.
Find the position vector of the particle after t seconds and hence show that the particle moves along the curve whose equation is

$$x^2 + 4xy + 4y^2 - 8x - 18y = 0.$$ (AEB)

11) Find the position vectors of the points of intersection of the circle $\mathbf{r} = (1 - \cos\theta)\mathbf{i} + \sin\theta\mathbf{j}$ and the line $\mathbf{r} = a\sqrt{3}\mathbf{i} + a\mathbf{j}$. Two particles start simultaneously at one point of intersection and arrive simultaneously at the other. The first particle moves with constant speed u along the shorter arc of the circle and the second particle moves from rest with constant acceleration f along the straight line. Find f in terms of u.

12) Sketch the curves

$$\mathbf{r} = (2a\cos\theta)\mathbf{i} + (2a\sin\theta)\mathbf{j}$$
$$\mathbf{r} = 3ap^2\mathbf{i} + 3ap\mathbf{j}.$$

Find the position vectors of the points in which they meet.
A particle describes the first curve in such a way that $\theta = \omega t$, where ω is constant and t is the time. Another particle describes the second curve in such a way that $p = kt$, where k is a constant.
Show that the acceleration of each particle is constant in magnitude and find the times at which these accelerations are at right angles. (U of L)

13) The position vector of a particle of mass m at time t is

$$\mathbf{r} = \mathbf{i}a\sin t + \mathbf{j}a\cos t + \mathbf{k}at.$$

Find the position vector and the velocity vector of the particle at $t = 0$ and $t = \frac{1}{2}\pi$. Find also the angle between the paths of the particle at $t = 0$ and $t = \frac{1}{2}\pi$. Show that the kinetic energy of the particle and the magnitude of the force acting on it are constant and find their respective magnitudes. Find the magnitude of the moment of the force about the origin at $t = \pi/4$. (AEB)

14) A particle is moving such that its position vector at time t is $\mathbf{r} = \cos wt\mathbf{i} + \sin wt\mathbf{j}$. The velocity of a second particle relative to the first is $-w \cos wt\mathbf{j}$. Find the acceleration vector of the second particle at time t. If the second particle has position vector $\mathbf{i}$ when $t = 0$, find its position vector at time t, and show that its motion is simple harmonic.

15) A particle is describing simple harmonic motion with period $2\pi/w$ between the points $\pm a\mathbf{i}$ and has acceleration vector $-w^2a\mathbf{i}$ when $t = 0$. Find the acceleration of the particle at time t.

16) A particle of mass m which is free to move in the $\mathbf{i}, \mathbf{j}$ plane has position vector $\mathbf{r} = \mathbf{0}$ and velocity vector $\mathbf{v} = v\mathbf{i}$ when time $t = 0$. If it is acted upon by a force $mue^{-t}\mathbf{j}$ throughout its motion, show that when $t = \ln 2$, $\mathbf{v} = v\mathbf{i} + \dfrac{u}{2}\mathbf{j}$. Show also that the path of the particle is asymptotic to $\mathbf{r} = vt\mathbf{i} + u(t-1)\mathbf{j}$. (U of L)

17) Obtain expressions for the radial and transverse components of acceleration of a point (r, θ) moving along a plane curve. If the angular velocity has the constant value ω, find the equation of the curve, given that the radial component of acceleration is $(a - 2r)\omega^2$ and the transverse component is $4a\omega^2 \cos\theta$, where a is a constant. Sketch the curve and find the point on it at which the resultant acceleration is a maximum. (U of L)

18) Show that the radial and transverse components of acceleration of a point (r, θ) moving in a plane curve are $\ddot{r} - r\dot{\theta}^2$ and $2\dot{r}\dot{\theta} + r\ddot{\theta}$ respectively. The straight line OA rotates with constant angular velocity ω about the fixed point O. The radial acceleration of a point P(r, θ) on OA, such that initially $r = a, \theta = 0$, is always $kr\omega^2$, and the initial radial velocity of P is $a\omega$. Find the equations of the locus of P when $k = 0, k = -1, k = -2$, and sketch the locus in each case. (U of L)

19) A particle is describing a plane curve and at time t is at the point P whose polar coordinates are (r, θ) referred to pole O. Find expressions for the components of the acceleration of the particle along and perpendicular to the radius vector OP.
A particle P of mass m is travelling along the curve $r(1 + \cos\theta) = 2a$ in a horizontal plane so that the radius vector OP is rotating at constant angular speed ω. Prove that the speed of the particle at any instant is $\omega\sqrt{r^3/a}$. When $\theta = \frac{1}{2}\pi$, find the resultant horizontal force acting on the particle. (U of L)

CHAPTER 4

THE VECTOR PRODUCT AND ITS APPLICATIONS

The vector product of two vectors **a** and **b** which are inclined at an angle θ is written as **a** × **b** (or sometimes **a** ∧ **b**) and is defined as
a *vector of magnitude ab* sin θ in a *direction perpendicular to the plane containing* **a** *and* **b** in the sense of a right-handed screw turned from **a** to **b**.

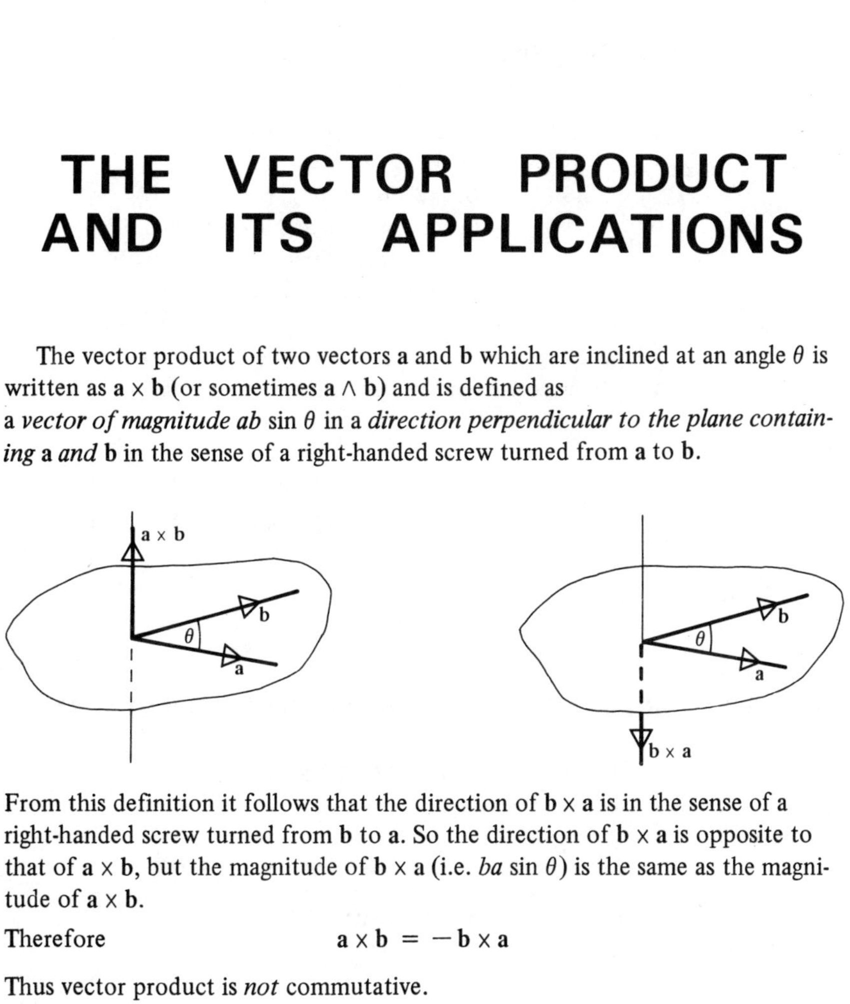

From this definition it follows that the direction of **b** × **a** is in the sense of a right-handed screw turned from **b** to **a**. So the direction of **b** × **a** is opposite to that of **a** × **b**, but the magnitude of **b** × **a** (i.e. *ba* sin θ) is the same as the magnitude of **a** × **b**.

Therefore $$\mathbf{a} \times \mathbf{b} = -\mathbf{b} \times \mathbf{a}$$

Thus vector product is *not* commutative.

VECTOR PRODUCT OF PARALLEL VECTORS

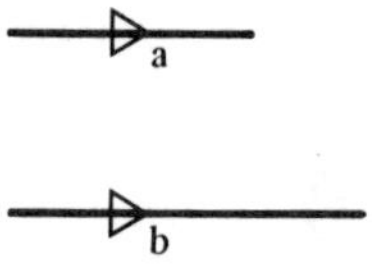

If **a** and **b** are parallel vectors

$$|\mathbf{a} \times \mathbf{b}| = ab \sin\theta$$

but $\quad \sin\theta = 0$

therefore $\quad \mathbf{a} \times \mathbf{b} = \mathbf{0}$

VECTOR PRODUCT OF PERPENDICULAR VECTORS

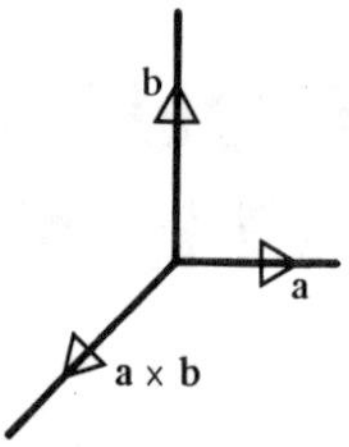

If **a** and **b** are perpendicular vectors, $\sin\theta = 1$ and $|\mathbf{a} \times \mathbf{b}| = ab$
In this case **a**, **b** and **a** × **b** form a right-handed set of three mutually perpendicular vectors as shown in the diagram.

This result is particularly important in the case of the unit vectors **i**, **j** and **k**:

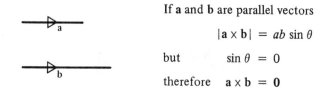

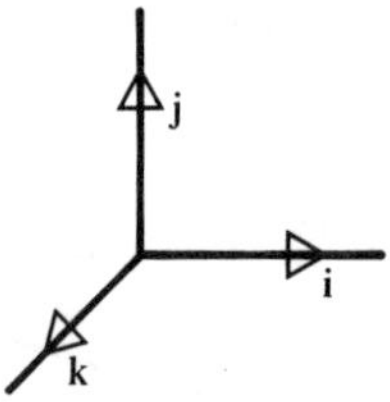

thus $\mathbf{i} \times \mathbf{j} = \mathbf{k}$ and $\mathbf{j} \times \mathbf{i} = -\mathbf{k}$

$\mathbf{j} \times \mathbf{k} = \mathbf{i}$ and $\mathbf{k} \times \mathbf{j} = -\mathbf{i}$

$\mathbf{k} \times \mathbf{i} = \mathbf{j}$ and $\mathbf{i} \times \mathbf{k} = -\mathbf{j}$

also $\mathbf{i} \times \mathbf{i} = \mathbf{j} \times \mathbf{j} = \mathbf{k} \times \mathbf{k} = \mathbf{0}$

VECTOR PRODUCT OF VECTORS IN CARTESIAN COMPONENT FORM

Vector product is distributive: i.e. $\mathbf{a} \times (\mathbf{b} + \mathbf{c}) = \mathbf{a} \times \mathbf{b} + \mathbf{a} \times \mathbf{c}$
(This property is proved on page 104.)
Therefore if

$$\mathbf{a} = x_1\mathbf{i} + y_1\mathbf{j} + z_1\mathbf{k} \quad \text{and} \quad \mathbf{b} = x_2\mathbf{i} + y_2\mathbf{j} + z_2\mathbf{k}$$

$$\begin{aligned}
\mathbf{a} \times \mathbf{b} &= (x_1\mathbf{i} + y_1\mathbf{j} + z_1\mathbf{k}) \times (x_2\mathbf{i} + y_2\mathbf{j} + z_2\mathbf{k}) \\
&= x_1x_2(\mathbf{i} \times \mathbf{i}) + x_1y_2(\mathbf{i} \times \mathbf{j}) + x_1z_2(\mathbf{i} \times \mathbf{k}) + y_1x_2(\mathbf{j} \times \mathbf{i}) \\
&\quad + y_1y_2(\mathbf{j} \times \mathbf{j}) + y_1z_2(\mathbf{j} \times \mathbf{k}) + z_1x_2(\mathbf{k} \times \mathbf{i}) + z_1y_2(\mathbf{k} \times \mathbf{j}) + z_1z_2(\mathbf{k} \times \mathbf{k}) \\
&= x_1y_2\mathbf{k} - x_1z_2\mathbf{j} - y_1x_2\mathbf{k} + y_1z_2\mathbf{i} + z_1x_2\mathbf{j} - z_1y_2\mathbf{i}
\end{aligned}$$

(using the results above)

$$= (y_1z_2 - z_1y_2)\mathbf{i} - (x_1z_2 - z_1x_2)\mathbf{j} + (x_1y_2 - y_1x_2)\mathbf{k}$$

This expression is the expansion of the determinant $\begin{vmatrix} \mathbf{i} & \mathbf{j} & \mathbf{k} \\ x_1 & y_1 & z_1 \\ x_2 & y_2 & z_2 \end{vmatrix}$

Therefore $$(x_1\mathbf{i}+y_1\mathbf{j}+z_1\mathbf{k}) \times (x_2\mathbf{i}+y_2\mathbf{j}+z_2\mathbf{k}) = \begin{vmatrix} \mathbf{i} & \mathbf{j} & \mathbf{k} \\ x_1 & y_1 & z_1 \\ x_2 & y_2 & z_2 \end{vmatrix}$$

Thus $$(2\mathbf{i}+\mathbf{j}-2\mathbf{k}) \times (\mathbf{j}+3\mathbf{k}) = \begin{vmatrix} \mathbf{i} & \mathbf{j} & \mathbf{k} \\ 2 & 1 & -2 \\ 0 & 1 & 3 \end{vmatrix}$$
$$= 5\mathbf{i}-6\mathbf{j}+2\mathbf{k}$$

Some calculations involve a mixture of vector and scalar product, e.g. $\mathbf{a} \times \mathbf{b} \cdot \mathbf{c}$ Brackets are unnecessary in expressions of this type as the cross product *must* be calculated first. If $\mathbf{b} \cdot \mathbf{c}$ were worked first this would lead to the vector product of **a** and a scalar quantity, which is meaningless.

SUMMARY

$\mathbf{a} \times \mathbf{b} = ab \sin\theta\,\hat{\mathbf{n}}$ where $\hat{\mathbf{n}}$ is a unit vector perpendicular to **a** and **b** in the direction of a right-handed screw turned from **a** to **b**.

$\mathbf{a} \times \mathbf{b} = -\mathbf{b} \times \mathbf{a}$ so the order of the vectors in the product matters.

If **a** and **b** are parallel $\mathbf{a} \times \mathbf{b} = \mathbf{0}$.

If **a** and **b** are perpendicular **a**, **b** and $\mathbf{a} \times \mathbf{b}$ form a right-handed set of mutually perpendicular vectors.

EXAMPLES 4a

1) Simplify (i) $\mathbf{a} \times (\mathbf{a}-\mathbf{b})$, (ii) $(\mathbf{a} \times \mathbf{b}) \cdot \mathbf{a}$

(i) $$\mathbf{a} \times (\mathbf{a}-\mathbf{b}) = \mathbf{a} \times \mathbf{a} - \mathbf{a} \times \mathbf{b}$$
$$= \mathbf{0} - \mathbf{a} \times \mathbf{b}$$
$$= \mathbf{b} \times \mathbf{a}$$

(ii) By definition $\mathbf{a} \times \mathbf{b}$ is perpendicular to **a** and the scalar product of perpendicular vectors is zero.
Therefore $(\mathbf{a} \times \mathbf{b}) \cdot \mathbf{a} = 0$.

2) If $\mathbf{a} = (2\mathbf{i}+\mathbf{j}-\mathbf{k})$, $\mathbf{b} = (3\mathbf{i}-\mathbf{j}+\mathbf{k})$ and $\mathbf{c} = (\mathbf{i}+2\mathbf{j})$ find $\mathbf{a} \times (\mathbf{b} \times \mathbf{c})$.

The brackets indicate the order in which the vector product is to be evaluated.

$$\mathbf{b} \times \mathbf{c} = \begin{vmatrix} \mathbf{i} & \mathbf{j} & \mathbf{k} \\ 3 & -1 & 1 \\ 1 & 2 & 0 \end{vmatrix}$$

$$= -2\mathbf{i} + \mathbf{j} + 7\mathbf{k}$$

Therefore
$$\mathbf{a} \times (\mathbf{b} \times \mathbf{c}) = \begin{vmatrix} \mathbf{i} & \mathbf{j} & \mathbf{k} \\ 2 & 1 & -1 \\ -2 & 1 & 7 \end{vmatrix}$$

$$= 8\mathbf{i} - 12\mathbf{j} + 4\mathbf{k}$$

$$= 4(2\mathbf{i} - 3\mathbf{j} + \mathbf{k})$$

3) Find the sine of the angle between AB and BC where A, B and C are the points (0, 1, 3), (− 1, 0, 1) and (1, − 1, − 2) respectively. Find also a unit vector which is perpendicular to the plane containing A, B and C.

$$\left.\begin{aligned} \overrightarrow{OA} &= \mathbf{j} + 3\mathbf{k} \\ \overrightarrow{OB} &= -\mathbf{i} + \mathbf{k} \\ \overrightarrow{OC} &= \mathbf{i} - \mathbf{j} - 2\mathbf{k} \end{aligned}\right\} \qquad \begin{aligned} \overrightarrow{BA} &= \overrightarrow{OA} - \overrightarrow{OB} = \mathbf{i} + \mathbf{j} + 2\mathbf{k} \\ \overrightarrow{BC} &= \overrightarrow{OC} - \overrightarrow{OB} = 2\mathbf{i} - \mathbf{j} - 3\mathbf{k} \end{aligned}$$

$$|\overrightarrow{BA} \times \overrightarrow{BC}| = (BA)(BC) \sin\theta$$

Therefore
$$\sin\theta = \frac{|\overrightarrow{BA} \times \overrightarrow{BC}|}{(BA)(BC)}$$

$$\overrightarrow{BA} \times \overrightarrow{BC} = \begin{vmatrix} \mathbf{i} & \mathbf{j} & \mathbf{k} \\ 1 & 1 & 2 \\ 2 & -1 & -3 \end{vmatrix} = -\mathbf{i} + 7\mathbf{j} - 3\mathbf{k}$$

Therefore
$$|\overrightarrow{BA} \times \overrightarrow{BC}| = \sqrt{59}$$

$$BA = |\mathbf{i} + \mathbf{j} + 2\mathbf{k}| = \sqrt{6}$$

$$BC = |2\mathbf{i} - \mathbf{j} - 3\mathbf{k}| = \sqrt{14}$$

Therefore
$$\sin\theta = \sqrt{59}/(\sqrt{6}\sqrt{14}) = \sqrt{59}/\sqrt{84}$$

$\overrightarrow{BA} \times \overrightarrow{BC}$ is a vector which is perpendicular to both BA and BC and is therefore perpendicular to the plane ABC.

Therefore the unit vector perpendicular to the plane ABC is $\dfrac{\overrightarrow{BA} \times \overrightarrow{BC}}{|\overrightarrow{BA} \times \overrightarrow{BC}|}$

$$= \frac{1}{\sqrt{59}}(-\mathbf{i} + 7\mathbf{j} - 3\mathbf{k})$$

4) **a**, **b** and **c** are three vectors such that $\mathbf{a} \times \mathbf{b} = \mathbf{c} \times \mathbf{a}$, $\mathbf{a} \neq \mathbf{0}$.
Find a linear relationship between **a**, **b** and **c**.

As $\quad \mathbf{a} \times \mathbf{b} = \mathbf{c} \times \mathbf{a}$

$$\mathbf{a} \times \mathbf{b} = -\mathbf{a} \times \mathbf{c}$$

Therefore $\quad \mathbf{a} \times \mathbf{b} + \mathbf{a} \times \mathbf{c} = \mathbf{0}$

Therefore $\quad \mathbf{a} \times (\mathbf{b} + \mathbf{c}) = \mathbf{0} \quad$ (distributive law)

As the vector product of $\mathbf{a}$ and $\mathbf{b} + \mathbf{c}$ is zero, $\mathbf{a}$ and $\mathbf{b} + \mathbf{c}$ are parallel vectors, or $\mathbf{b} + \mathbf{c} = \mathbf{0}$,

i.e. $\quad \mathbf{a} = k(\mathbf{b} + \mathbf{c}) \quad$ where k is a scalar quantity

or $\quad \mathbf{b} = -\mathbf{c}$

EXERCISE 4a

1) Simplify the following:

(i) $(\mathbf{a} + \mathbf{b}) \times \mathbf{b}$ (iv) $\mathbf{a} \times (\mathbf{b} + \mathbf{c}) \cdot \mathbf{b}$
(ii) $(\mathbf{a} + \mathbf{b}) \times (\mathbf{a} + \mathbf{b})$ (v) $\mathbf{a} \cdot (\mathbf{b} + \mathbf{c}) \times \mathbf{a}$
(iii) $(\mathbf{a} - \mathbf{b}) \times (\mathbf{a} + \mathbf{b})$ (vi) $(\mathbf{a} \times \mathbf{b}) \cdot \mathbf{a} + \mathbf{b} \cdot (\mathbf{a} \times \mathbf{b})$

2) If $\mathbf{a} = \mathbf{i} + \mathbf{j} - \mathbf{k}$ and $\mathbf{b} = 2\mathbf{i} - \mathbf{j} + \mathbf{k}$ find:
(i) $\mathbf{a} \times \mathbf{b}$ (ii) $\mathbf{a} \times (\mathbf{a} + \mathbf{b})$
and verify that $\mathbf{a} \cdot (\mathbf{a} \times \mathbf{b}) = 0$.

3) If $\mathbf{a} = \mathbf{i} + 2\mathbf{j} - \mathbf{k}$ and $\mathbf{b} = \mathbf{j} + \mathbf{k}$ find the unit vector perpendicular to both $\mathbf{a}$ and $\mathbf{b}$. Calculate also the sine of the angle between $\mathbf{a}$ and $\mathbf{b}$.

4) If $\mathbf{a} = \mathbf{i} + \mathbf{j} - \mathbf{k}$, $\mathbf{b} = \mathbf{i} - \mathbf{j}$ and $\mathbf{c} = 2\mathbf{i} + \mathbf{k}$, find $(\mathbf{a} \times \mathbf{b}) \times \mathbf{c}$ and $\mathbf{a} \times (\mathbf{b} \times \mathbf{c})$. Verify that $\mathbf{a} \cdot \mathbf{b} \times \mathbf{c} = \mathbf{a} \times \mathbf{b} \cdot \mathbf{c}$.

5) If $\mathbf{a} = \mathbf{i} + \mathbf{j}$ and $\mathbf{b} = 2\mathbf{i} + \mathbf{k}$ find the sine of the angle between $\mathbf{a}$ and $\mathbf{b}$, and the unit vector perpendicular to both $\mathbf{a}$ and $\mathbf{b}$.

6) A, B and C are the points $(0, 1, 2)$, $(3, 2, 1)$ and $(1, -1, 0)$ respectively. Find the unit vector which is perpendicular to the plane ABC.

7) Three vectors $\mathbf{a}$, $\mathbf{b}$ and $\mathbf{c}$ are such that $\mathbf{a} \times \mathbf{b} = \mathbf{a} \times \mathbf{c} \quad (\mathbf{a} \neq \mathbf{0})$.
Show that $\mathbf{b} - \mathbf{c} = k\mathbf{a}$ where k is a scalar.

8) Three vectors $\mathbf{a}$, $\mathbf{b}$ and $\mathbf{c}$ are such that $\mathbf{a} \times 3\mathbf{b} = 2\mathbf{a} \times \mathbf{c}$. Find a linear relationship between $\mathbf{a}$, $\mathbf{b}$ and $\mathbf{c}$.

9) Show that $\mathbf{u} = (\mathbf{i} + \mathbf{j})$ is a solution of the equation

$$\mathbf{u} \times (\mathbf{i} + 4\mathbf{j}) = 3\mathbf{k}$$

Show also that the general solution to this equation is

$$\mathbf{u} = -3\mathbf{j} + t(\mathbf{i} + 4\mathbf{j})$$

10) Find a general solution to the equation

$$\mathbf{u} \times (\mathbf{i} - 3\mathbf{k}) = 2\mathbf{j}$$

GEOMETRIC APPLICATIONS OF THE VECTOR PRODUCT

1. Area of a Parallelogram

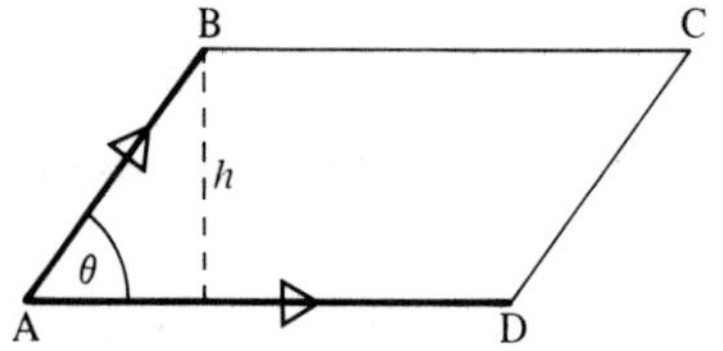

The area of a parallelogram

= base × height

= (AD)(h)

= (AD)(AB sin θ)

= $|\overrightarrow{AD} \times \overrightarrow{AB}|$

Therefore the area of a parallelogram is the magnitude of the vector product of two adjacent sides.

2. Area of a Triangle

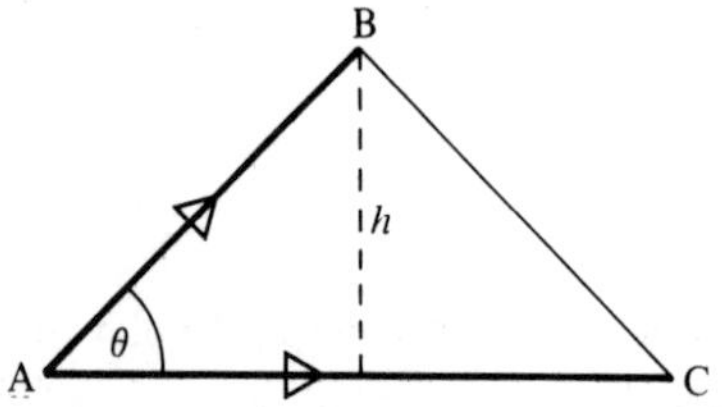

The area of ΔABC

= $\frac{1}{2}$ base × height

= $\frac{1}{2}$(AC)(h)

= $\frac{1}{2}$(AC)(AB sin θ)

= $\frac{1}{2}|\overrightarrow{AC} \times \overrightarrow{AB}|$

Therefore the area of a triangle is half the magnitude of the vector product of two sides.

EXAMPLE

A triangle ABC has its vertices at the points A(1, 2, 1), B(1, 0, 3), C(− 1, 2, − 1). Find the area of ΔABC.

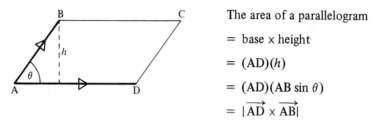

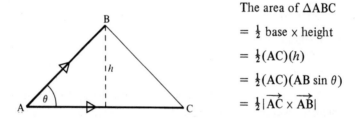

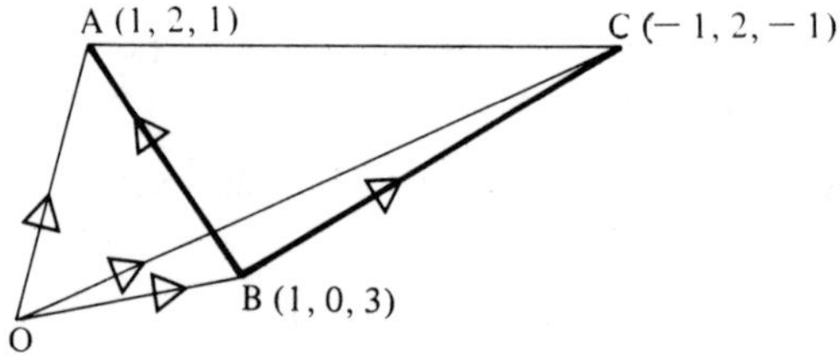

If O is the origin then $\overrightarrow{OA} = \mathbf{i} + 2\mathbf{j} + \mathbf{k}$

$\overrightarrow{OB} = \mathbf{i} + 3\mathbf{k}$

$\overrightarrow{OC} = -\mathbf{i} + 2\mathbf{j} - \mathbf{k}$

Therefore $\overrightarrow{BA} = \overrightarrow{OA} - \overrightarrow{OB} = 2\mathbf{j} - 2\mathbf{k}$

$$\overrightarrow{BC} = \overrightarrow{OC} - \overrightarrow{OB} = -2i + 2j - 4k$$

$$\text{Area of } \Delta ABC = \tfrac{1}{2}|\overrightarrow{BA} \times \overrightarrow{BC}|$$

$$\overrightarrow{BA} \times \overrightarrow{BC} = \begin{vmatrix} i & j & k \\ 0 & 2 & -2 \\ -2 & 2 & -4 \end{vmatrix} = -4i + 4j + 4k$$

Therefore area of $\Delta ABC = \frac{1}{2}|-4\mathbf{i} + 4\mathbf{j} + 4\mathbf{k}| = 2\sqrt{3}$

3. Volume of a Parallelepiped

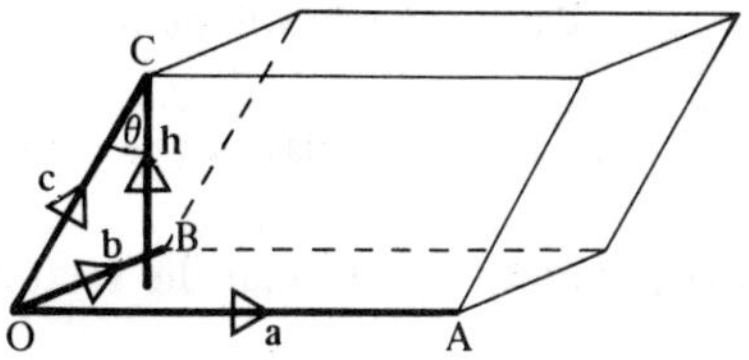

The volume of a parallelepiped

$= $ area of base $\times$ height

$= |\mathbf{a} \times \mathbf{b}|h$

$= |\mathbf{a} \times \mathbf{b}||\mathbf{c}| \cos \theta$

θ is the angle between **c** and **h** and as **h** is perpendicular to **a** and **b**, θ is also the angle between $\mathbf{a} \times \mathbf{b}$ and **c**.
$\therefore |\mathbf{a} \times \mathbf{b}||\mathbf{c}| \cos \theta$ is the scalar product of the vectors $\mathbf{a} \times \mathbf{b}$ and **c**.

Therefore the volume of the parallelepiped is the numerical value of $\mathbf{a} \times \mathbf{b}.\mathbf{c}$

4. Volume of a Tetrahedron

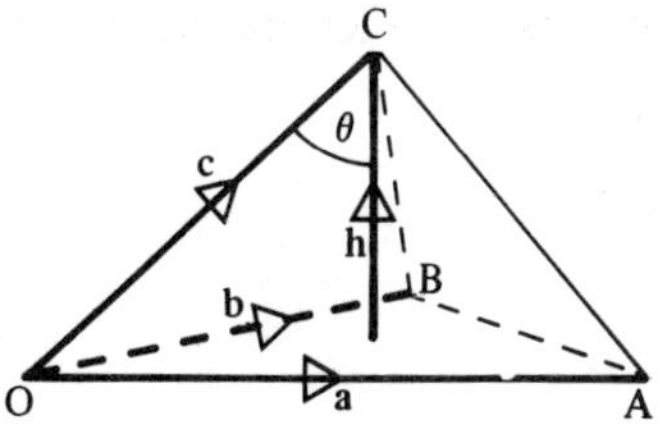

The volume of a tetrahedron

$= \frac{1}{3}$ area of base $\times$ height

$= \frac{1}{3}(\frac{1}{2}|\mathbf{a} \times \mathbf{b}|)|\mathbf{c}| \cos \theta$

$= |\frac{1}{6}\, \mathbf{a} \times \mathbf{b}.\mathbf{c}|$

EXAMPLE

Find the volume of the tetrahedron OABC where O is the origin and A, B, C are the points (2, 1, 1), (0, − 1, 1), (− 1, 3, 0).

The volume of OABC is $|\frac{1}{6}\mathbf{OA} \times \mathbf{OB}.\mathbf{OC}|$

$$= |\tfrac{1}{6}(2i + j + k) \times (-j + k).(-i + 3j)|$$

$$= |\tfrac{1}{6}(2i - 2j - 2k).(-i + 3j)|$$

$$= 1\tfrac{1}{3}$$

Triple Scalar Product and the Proof of the Distributive Law for Vector Products

Expressions of the form $\mathbf{a} \times \mathbf{b} \cdot \mathbf{c}$ are referred to as a triple scalar product. The volume of the parallelepiped above could have been obtained by considering the side defined by **b** and **c** as the base, in which case the result would have been in the form $\mathbf{b} \times \mathbf{c} \cdot \mathbf{a}$.

i.e. $$\mathbf{a} \times \mathbf{b} \cdot \mathbf{c} = \mathbf{b} \times \mathbf{c} \cdot \mathbf{a}$$

The order in which a scalar product is performed does not matter.

So $$\mathbf{b} \times \mathbf{c} \cdot \mathbf{a} = \mathbf{a} \cdot \mathbf{b} \times \mathbf{c}$$

Therefore $$\mathbf{a} \times \mathbf{b} \cdot \mathbf{c} = \mathbf{a} \cdot \mathbf{b} \times \mathbf{c}$$

i.e. in a triple scalar product, the "cross" and "dot" may be interchanged without altering the value of the expression. This property will be referred to as the *triple scalar product property*.
(But if the order of the vectors is altered, the expression may not remain the same: e.g. $\mathbf{a} \times \mathbf{b} \cdot \mathbf{c} = -\mathbf{b} \times \mathbf{a} \cdot \mathbf{c}$)

We will now use this property of a triple scalar product to prove that the vector product is distributive:

Consider: $\mathbf{d} \cdot (\mathbf{a} \times \mathbf{b} + \mathbf{a} \times \mathbf{c})$

$= \mathbf{d} \cdot \mathbf{a} \times \mathbf{b} + \mathbf{d} \cdot \mathbf{a} \times \mathbf{c}$ (scalar product is distributive)

$= \mathbf{d} \times \mathbf{a} \cdot \mathbf{b} + \mathbf{d} \times \mathbf{a} \cdot \mathbf{c}$ (triple scalar product property)

$= \mathbf{d} \times \mathbf{a} \cdot (\mathbf{b} + \mathbf{c})$ (scalar product is distributive)

$= \mathbf{d} \cdot \mathbf{a} \times (\mathbf{b} + \mathbf{c})$ (triple scalar product property)

i.e. $$\mathbf{d} \cdot (\mathbf{a} \times \mathbf{b} + \mathbf{a} \times \mathbf{c}) = \mathbf{d} \cdot \mathbf{a} \times (\mathbf{b} + \mathbf{c})$$

therefore $$\mathbf{a} \times \mathbf{b} + \mathbf{a} \times \mathbf{c} = \mathbf{a} \times (\mathbf{b} + \mathbf{c})$$

i.e. the vector product is distributive.

5. Distance of a Point from a Line

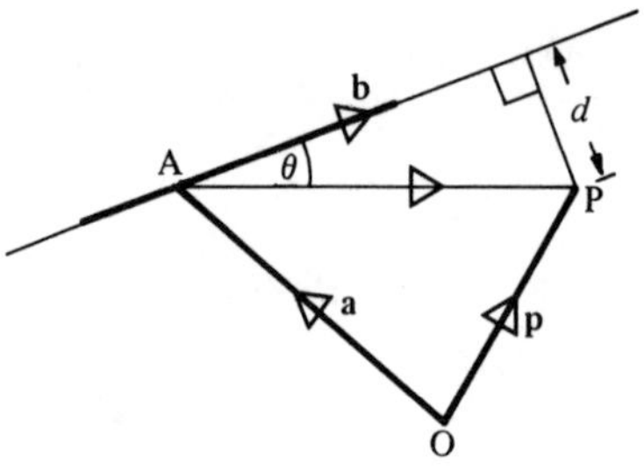

Consider the line whose vector equation is $\mathbf{r} = \mathbf{a} + \lambda\mathbf{b}$ and the point P whose position vector is **p**. If A is the point on the line with position vector **a** then $\overrightarrow{AP} = \mathbf{p} - \mathbf{a}$.

If θ is the angle between $\overrightarrow{AP}$ and the line (which is parallel to **b**)

then $$d = AP \sin\theta$$
$$= \frac{|\mathbf{b} \times \overrightarrow{AP}|}{|\mathbf{b}|} = \frac{|\mathbf{b} \times (\mathbf{p} - \mathbf{a})|}{|\mathbf{b}|}$$

Therefore the perpendicular distance of a point P whose position vector is **p** from the line whose vector equation is $\mathbf{r} = \mathbf{a} + \lambda\mathbf{b}$ is $\dfrac{|\mathbf{b} \times (\mathbf{p} - \mathbf{a})|}{|\mathbf{b}|}$.

Thus the distance of the point P(1, 3, 2) from the line $\mathbf{r} = (1 - t)\mathbf{i} + (2 - 3t)\mathbf{j} + 2\mathbf{k}$ can be found as follows:

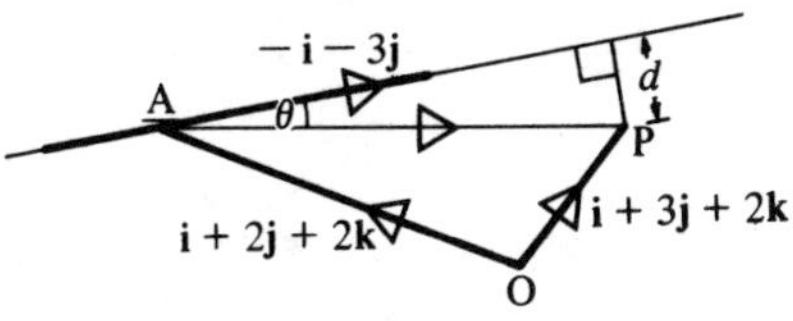

Rewriting the equation of the line in the form

$$\mathbf{r} = \mathbf{i} + 2\mathbf{j} + 2\mathbf{k} + t(-\mathbf{i} - 3\mathbf{j})$$

we see that $\mathbf{i} + 2\mathbf{j} + 2\mathbf{k}$ is the position vector of a point on the line (say A)

therefore $$\overrightarrow{AP} = (\mathbf{i} + 3\mathbf{j} + 2\mathbf{k}) - (\mathbf{i} + 2\mathbf{j} + 2\mathbf{k})$$
$$= \mathbf{j}$$

If d is the distance of P from the line

$$\mathbf{d} = \frac{\mathbf{j} \times (-\mathbf{i} - 3\mathbf{j})}{|-\mathbf{i} - 3\mathbf{j}|} = \frac{1}{\sqrt{10}}\begin{vmatrix} \mathbf{i} & \mathbf{j} & \mathbf{k} \\ 0 & 1 & 0 \\ -1 & -3 & 0 \end{vmatrix} = \frac{1}{\sqrt{10}}\mathbf{k}$$

$$d = \frac{1}{\sqrt{10}}$$

6. Shortest Distance between Two Skew Lines

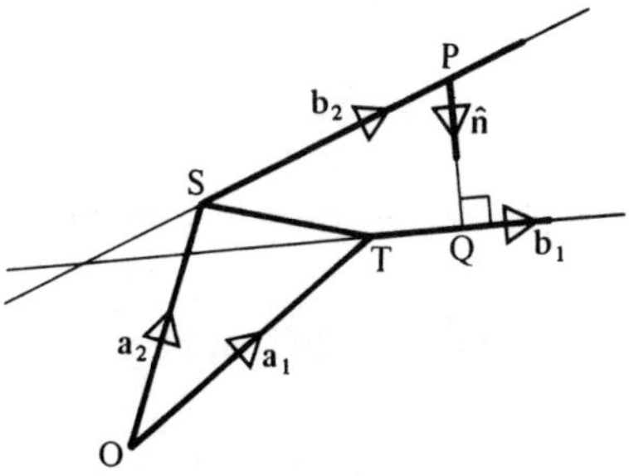

Consider two lines whose vector equations are

$$\mathbf{l}_1 = \mathbf{a}_1 + \lambda\mathbf{b}_1$$
$$\mathbf{l}_2 = \mathbf{a}_2 + \mu\mathbf{b}_2$$

If PQ is the shortest distance between the lines then PQ is perpendicular to both lines. i.e. $\overrightarrow{PQ}$ is parallel to $\mathbf{b}_1 \times \mathbf{b}_2$

Therefore the unit vector $\hat{\mathbf{n}}$ in the direction of $\overrightarrow{PQ}$ is $\dfrac{\mathbf{b}_1 \times \mathbf{b}_2}{|\mathbf{b}_1 \times \mathbf{b}_2|}$

$\mathbf{a}_1 - \mathbf{a}_2$ represents the displacement from a point S on the first line to a point T on the second line.
If θ is the angle between $\overrightarrow{ST}$ and $\overrightarrow{PQ}$

then
$$\begin{aligned} PQ &= ST \cos\theta \\ &= |\overrightarrow{ST}\cdot\hat{\mathbf{n}}| \\ &= \left|\frac{(\mathbf{a}_1 - \mathbf{a}_2)\cdot\mathbf{b}_1 \times \mathbf{b}_2}{|\mathbf{b}_1 \times \mathbf{b}_2|}\right| \end{aligned}$$

EXAMPLE

Find the shortest distance between the lines whose vector equations are $\mathbf{r}_1 = \mathbf{i} + \mathbf{j} + \lambda(2\mathbf{i} - \mathbf{j} + \mathbf{k})$ and $\mathbf{r}_2 = 2\mathbf{i} + \mathbf{j} - \mathbf{k} + \mu(3\mathbf{i} - 5\mathbf{j} + 2\mathbf{k})$.

A vector which is perpendicular to both lines is

$$(2\mathbf{i} - \mathbf{j} + \mathbf{k}) \times (3\mathbf{i} - 5\mathbf{j} + 2\mathbf{k}) = 3\mathbf{i} - \mathbf{j} - 7\mathbf{k}$$

Therefore a unit vector perpendicular to both lines is $(3\mathbf{i} - \mathbf{j} - 7\mathbf{k})/\sqrt{59}$.
A displacement vector from a point on one line to a point on the other is $(\mathbf{i} + \mathbf{j}) - (2\mathbf{i} + \mathbf{j} - \mathbf{k}) = -\mathbf{i} + \mathbf{k}$
Thus the least distance between the two lines is given by

$$|(-\mathbf{i} + \mathbf{k})\cdot(3\mathbf{i} - \mathbf{j} - 7\mathbf{k})/\sqrt{59}| = 10/\sqrt{59}$$

Compare this with the solution given on p. 57.

Condition That Two Lines Intersect

Two lines with equations $\mathbf{r}_1 = \mathbf{a}_1 + \lambda\mathbf{b}_1$, $\mathbf{r}_2 = \mathbf{a}_2 + \mu\mathbf{b}_2$ intersect if the least distance between them is zero. i.e. if $(\mathbf{a}_1 - \mathbf{a}_2)\cdot\mathbf{b}_1 \times \mathbf{b}_2 = 0$.
(This is the condition that two lines shall intersect. It does not give the position vector of the point of intersection. If this is required use the method given in Chapter 1.)

EXERCISE 4b

1) The vectors $(2\mathbf{i} + 3\mathbf{j} - \mathbf{k})$ and $(\mathbf{i} + 2\mathbf{j} + \mathbf{k})$ represent two sides of a triangle. Find the area of the triangle.

2) The triangle ABC has its vertices at the points A(0, 0, 1), B(1, 0, 1), C(2, 1, 3). Find the area of the triangle ABC.

3) The vertices of a triangle are at the points with position vectors **a**, **b** and **c**. Prove that the area of the triangle is $\frac{1}{2}|\mathbf{a}\times\mathbf{b}+\mathbf{b}\times\mathbf{c}+\mathbf{c}\times\mathbf{a}|$.

4) A parallelogram OABC has one vertex O at the origin and the vertices A and B at the points (0, 1, 3), (0, 2, 5). Find the area of OABC.

5) The parallelogram ABCD has three of its vertices A, B and C at the points with position vectors A(1, 2, − 1), B(1, 3, 2), C(− 1, − 3, − 1). Find the area of the parallelogram.

6) The vectors $\overrightarrow{OA}$, $\overrightarrow{OB}$, $\overrightarrow{OC}$ are three edges of a parallelepiped where O is the origin and A, B and C are the points (2, 1, 0), (− 1, − 1, 1), (0, 2, − 1). Find the volume of the parallelepiped.

7) Find the volume of the tetrahedron OABC where O is the origin and A, B and C are the points (2, 0, 1), (3, 1, 2) and (− 1, 3, 0).

8) ABCD is a tetrahedron and A, B, C and D are the points (0, 1, 0), (0, 0, 4), (1, 1, 1) and (− 1, 3, 2). Find the volume of the tetrahedron.

9) The four vertices of a tetrahedron are at the points with position vectors **a**, **b**, **c**, **d**. Find the volume of the tetrahedron.

10) Find the perpendicular distance of the given point from the given line in the following cases:
(a) (1, 0, 2) $\quad \mathbf{l}=\mathbf{i}+\mathbf{j}+\lambda(\mathbf{i}-\mathbf{j}+\mathbf{k})$,
(b) (0, 0, 0) $\quad \mathbf{l}=(1-\lambda)\mathbf{i}+(2\lambda-1)\mathbf{j}+\lambda\mathbf{k}$,
(c) point with position vector **a**, $\quad \mathbf{l}=\mathbf{a}-\mathbf{b}+\lambda\mathbf{c}$,
(d) point with position vector **r**, $\quad \mathbf{l}=(\lambda-1)\mathbf{r}+(2\lambda+1)\mathbf{p}$.

11) The vector equations of two lines are:

$$\mathbf{l}_1 = \mathbf{i}+2\mathbf{j}+\mathbf{k}+\lambda(\mathbf{i}-\mathbf{j}+\mathbf{k})$$

$$\mathbf{l}_2 = 2\mathbf{i}-\mathbf{j}-\mathbf{k}+\mu(2\mathbf{i}+\mathbf{j}+2\mathbf{k})$$

Find the shortest distance between these lines.

12) The vector equations of two lines are:

$$\mathbf{r}_1 = (1-t)\mathbf{i}+(t-2)\mathbf{j}+(3-2t)\mathbf{k}$$

$$\mathbf{r}_2 = (s+1)\mathbf{i}+(2s-1)\mathbf{j}-(2s+1)\mathbf{k}$$

Find the shortest distance between these lines.

13) Determine whether the following pairs of lines intersect.
(a) $\mathbf{r}=\mathbf{i}-\mathbf{j}+\lambda(2\mathbf{i}+\mathbf{k})$, $\quad \mathbf{r}=2\mathbf{i}-\mathbf{j}+\mu(\mathbf{i}+\mathbf{j}-\mathbf{k})$.
(b) $\mathbf{r}=\mathbf{i}+\mathbf{j}-\mathbf{k}+\lambda(3\mathbf{i}-\mathbf{j})$, $\quad \mathbf{r}=4\mathbf{i}-\mathbf{k}+\mu(2\mathbf{i}+3\mathbf{k})$.

THE VECTOR MOMENT OF A FORCE ABOUT A POINT

The term "moment of a force about a point" is meaningless in itself but is a common abbreviation for the turning effect of a force about an axis passing through the point and which is perpendicular to the plane containing the force and the point.

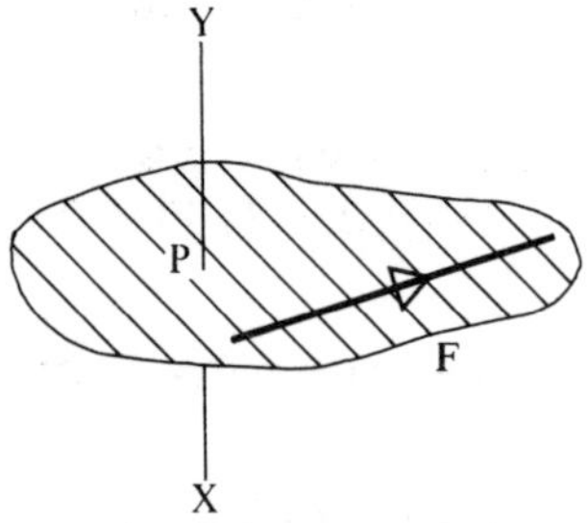

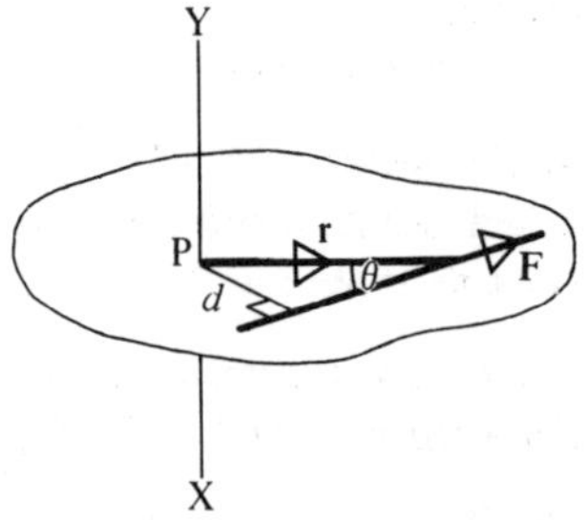

i.e. in the diagram the moment of **F** about P is the turning effect of **F** about XY. Let **r** be the position vector, relative to P, of *any* point on the line of action of **F**. The magnitude of the moment of **F** about P is

$$dF = (r \sin\theta)F = |\mathbf{r} \times \mathbf{F}|$$

The axis of rotation XY is perpendicular to **r** and **F**, so $\overrightarrow{XY}$ is parallel to **r** × **F**. Therefore the vector **r** × **F** has magnitude equal to the magnitude of the moment of **F** about the axis XY, and direction along the axis of rotation in the sense of a right-handed screw turned from **r** to **F**.

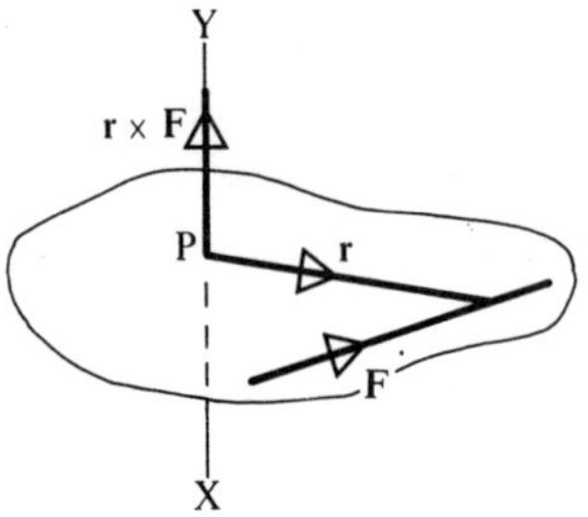

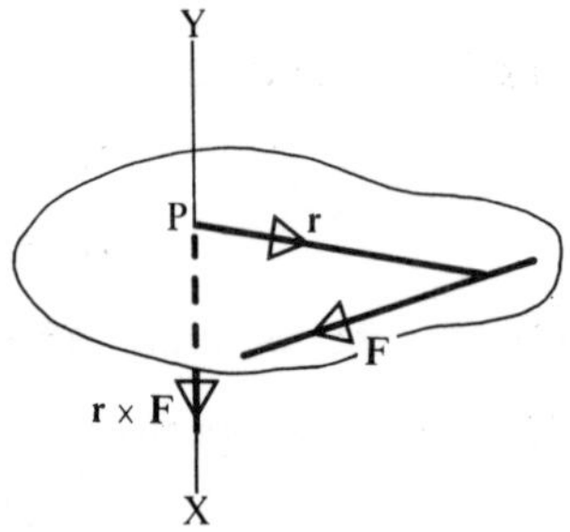

r × **F** is called the vector moment of **F** about P.

Note. Given that the moment of **F** about O is **a** we have **r** × **F** = **a**. As **r** is the position vector of *any* point on the line of action of **F**, the equation **r** × **F** = **a** is the vector equation of the line of action of **F**.

EXAMPLES 4c

1) A force **F** acts through the point with position vector **i** + 2**j** − **k**. If **F** = **i** + 2**k** find the vector moment of **F** about (a) the origin (b) the point with position vector **i** − **j**.

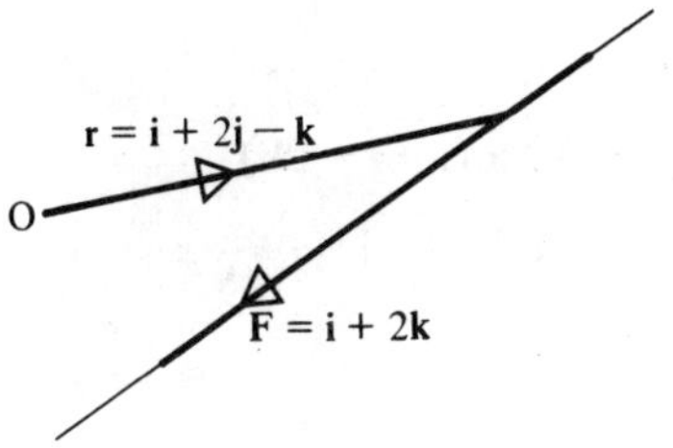

The moment of **F** about O

$= \mathbf{r} \times \mathbf{F}$

$= (\mathbf{i} + 2\mathbf{j} - \mathbf{k}) \times (\mathbf{i} + 2\mathbf{k})$

$= \begin{vmatrix} \mathbf{i} & \mathbf{j} & \mathbf{k} \\ 1 & 2 & -1 \\ 1 & 0 & 2 \end{vmatrix}$

$= 4\mathbf{i} - 3\mathbf{j} - 2\mathbf{k}$

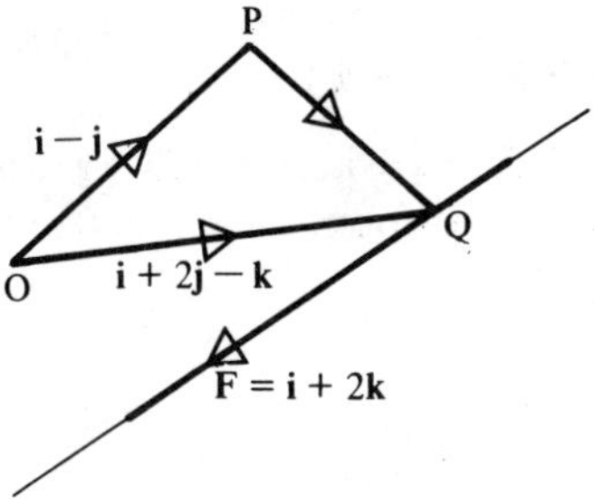

The moment of **F** about P

$= \overrightarrow{PQ} \times \mathbf{F}$

$= (\mathbf{r} - \overrightarrow{OP}) \times \mathbf{F}$

$= (3\mathbf{j} - \mathbf{k}) \times (\mathbf{i} + 2\mathbf{k})$

$= \begin{vmatrix} \mathbf{i} & \mathbf{j} & \mathbf{k} \\ 0 & 3 & -1 \\ 1 & 0 & 2 \end{vmatrix}$

$= 6\mathbf{i} - \mathbf{j} - 3\mathbf{k}$

2) A force **F**, where $\mathbf{F} = \mathbf{i} - \mathbf{j} + 2\mathbf{k}$, acts through the point with position vector $2\mathbf{i} + \mathbf{k}$. Find the magnitude of the moment of **F** about the point P where the position vector of P is $\mathbf{j} - 2\mathbf{k}$. Find also the vector equation of the axis through P.

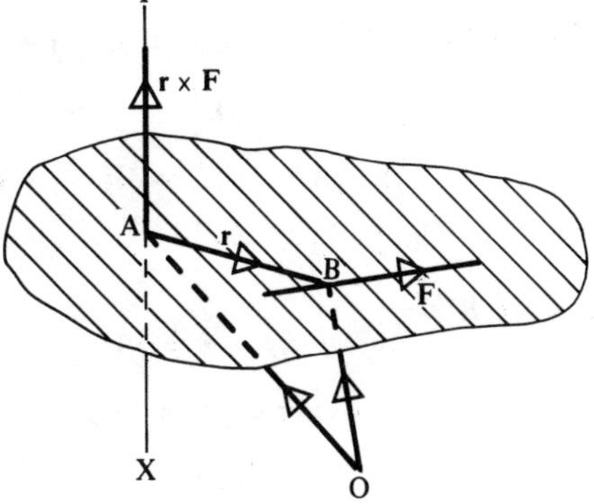

$\left.\begin{aligned} \overrightarrow{OB} &= 2\mathbf{i} + \mathbf{k} \\ \overrightarrow{OA} &= \mathbf{j} - 2\mathbf{k} \end{aligned}\right\}$ Therefore $\overrightarrow{AB} = \overrightarrow{OB} - \overrightarrow{OA} = 2\mathbf{i} - \mathbf{j} + 3\mathbf{k}$

$$
\begin{aligned}
\text{The vector moment of } \mathbf{F} \text{ about A} &= \mathbf{r} \times \mathbf{F} \\
&= \overrightarrow{AB} \times \mathbf{F} \\
&= (2\mathbf{i} - \mathbf{j} + 3\mathbf{k}) \times (\mathbf{i} - \mathbf{j} + 2\mathbf{k}) \\
&= \begin{vmatrix} \mathbf{i} & \mathbf{j} & \mathbf{k} \\ 2 & -1 & 3 \\ 1 & -1 & 2 \end{vmatrix} \\
&= \mathbf{i} - \mathbf{j} - \mathbf{k}
\end{aligned}
$$

Therefore the magnitude of the moment of **F** about A $= \sqrt{3}$.
The direction of $\mathbf{r} \times \mathbf{F}$ is parallel to XY and as A is a point on XY, the vector equation of XY is

$$
\begin{aligned}
\mathbf{r} &= \overrightarrow{OA} + \lambda(\mathbf{r} \times \mathbf{F}) \\
&= \mathbf{j} - 2\mathbf{k} + \lambda(\mathbf{i} - \mathbf{j} - \mathbf{k})
\end{aligned}
$$

3) The moment of a force **F** about the origin is $3\mathbf{i} - \mathbf{j} + \mathbf{k}$. If $\mathbf{F} = 2\mathbf{i} + \mathbf{j} - 5\mathbf{k}$ find an equation for the line of action of **F** in vector form.

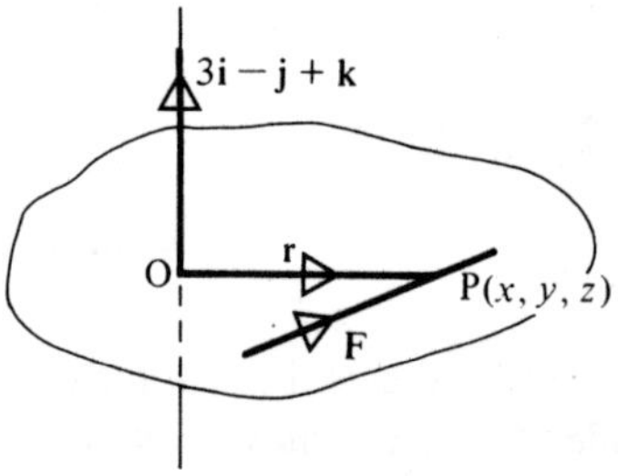

If **r** is the position vector of any point $P(x, y, z)$ on the line of action of **F** then

$$\mathbf{r} \times \mathbf{F} = 3\mathbf{i} - \mathbf{j} + \mathbf{k}$$

i.e.
$$(x\mathbf{i} + y\mathbf{j} + z\mathbf{k}) \times (2\mathbf{i} + \mathbf{j} - 5\mathbf{k}) = 3\mathbf{i} - \mathbf{j} + \mathbf{k}$$

$$\begin{vmatrix} \mathbf{i} & \mathbf{j} & \mathbf{k} \\ x & y & z \\ 2 & 1 & -5 \end{vmatrix} = 3\mathbf{i} - \mathbf{j} + \mathbf{k}$$

equating coefficients of **i**, **j**, **k**:

$$-5y - z = 3 \qquad (1)$$

$$5x + 2z = -1 \qquad (2)$$

$$x - 2y = 1 \qquad (3)$$

(These three equations are not independent. If they were, unique values for x, y, z would be found whereas we know that **r** is *any* position vector on the line of action of **F**.)

From (1) and (3): $y = \dfrac{z+3}{-5} = \dfrac{x-1}{2} = \lambda$ (say)

Therefore $\mathbf{r} = \mathbf{i} - 3\mathbf{k} + \lambda(2\mathbf{i} + \mathbf{j} - 5\mathbf{k})$ is the equation of the line of action of **F** in vector form.

VECTOR MOMENT OF A COUPLE

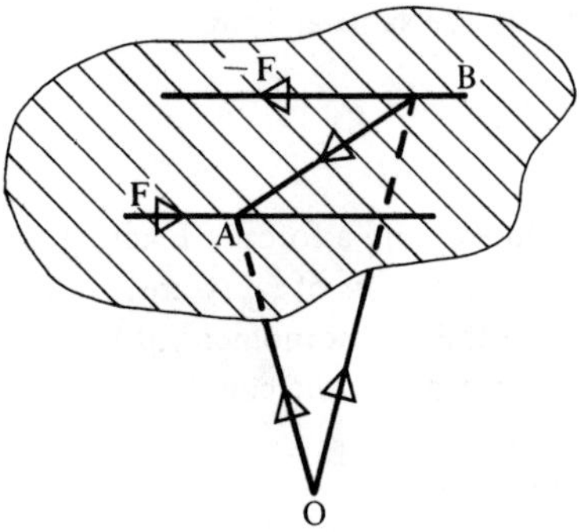

Consider the forces **F** and − **F** acting along parallel lines. Such a pair of forces is a couple and we know that the moment of a couple is the same about any point in the plane of the couple. Now consider the sum of the moments of these forces about a point O, not in the plane of the couple.

If A is a point on the line of action of **F** and B is a point on the line of action of − **F**, the sum of the moments of the forces about O is

$$\begin{aligned} & \overrightarrow{OA} \times \mathbf{F} + \overrightarrow{OB} \times (-\mathbf{F}) \\ = \; & (\overrightarrow{OA} - \overrightarrow{OB}) \times \mathbf{F} \\ = \; & \overrightarrow{BA} \times \mathbf{F} \end{aligned}$$

The expression $\overrightarrow{BA} \times \mathbf{F}$ is called the vector moment of the couple. Investigating its nature we see that:

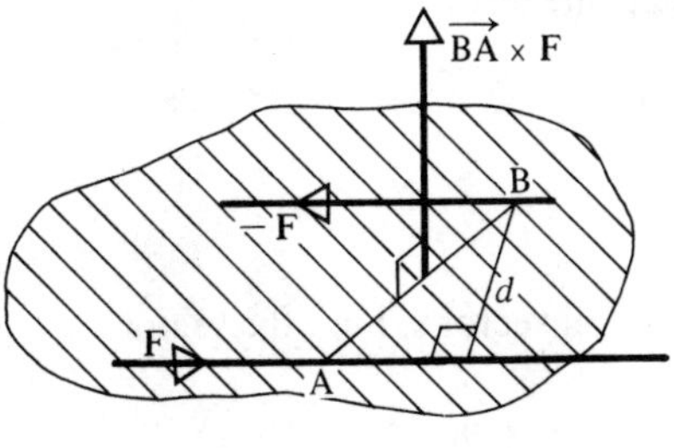

(a) $\overrightarrow{BA}$ is independent of O.

(b) $\overrightarrow{BA} \times \mathbf{F}$ is perpendicular to the plane of the couple.

(c) $|\overrightarrow{BA} \times \mathbf{F}| = Fd$ which is the magnitude of the moment of the couple.

Therefore the vector moment of a couple is the same about any point whether or not that point is in the plane of the couple.
When a problem refers to a single vector **G** as representing a couple it must be remembered that **G** is the vector moment of that couple and so is a vector perpendicular to the plane containing the couple.

Equivalent couples

Two couples are equivalent if their vector moments are equal, i.e. the magnitude and direction of their vector moments must be the same and this can only be so if the couples act in parallel planes.

EXAMPLES 4c (continued)

4) A couple comprises a force **F** acting along the line l_1 and a force $-$ **F** acting along the line l_2. If $\mathbf{F} = 3(\mathbf{i}-\mathbf{j}+\mathbf{k})$ and the vector equations of l_1, l_2 are $\mathbf{r} = \mathbf{i}+\mathbf{j}+t(\mathbf{i}-\mathbf{j}+\mathbf{k})$, $\mathbf{r} = \mathbf{i}+\mathbf{k}+s(\mathbf{i}-\mathbf{j}+\mathbf{k})$ find the vector moment of the couple and hence a unit vector perpendicular to the plane of the couple.

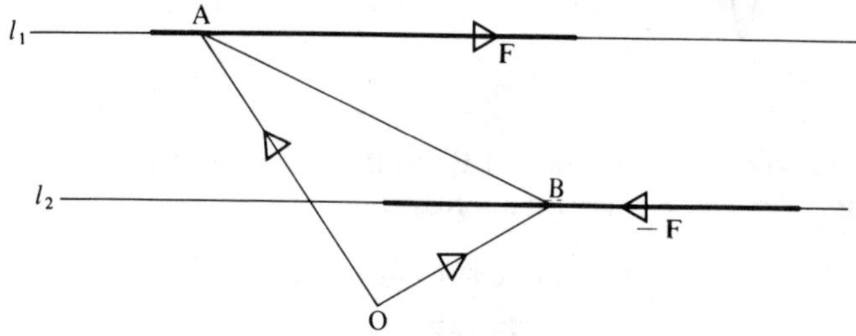

As **F** acts along the line $\mathbf{r} = \mathbf{i}+\mathbf{j}+t(\mathbf{i}-\mathbf{j}+\mathbf{k})$ then $\mathbf{i}+\mathbf{j}$ is the position vector of a point A on the line l_1. Similarly $\mathbf{i}+\mathbf{k}$ is the position vector of a point B on the line l_2.
The vector moment of the couple is $\overrightarrow{BA} \times \mathbf{F}$

$$= (\mathbf{j}-\mathbf{k}) \times 3(\mathbf{i}-\mathbf{j}+\mathbf{k})$$

$$= 3(-\mathbf{j}-\mathbf{k})$$

Therefore a unit vector perpendicular to the plane of the couple is

$$\frac{1}{\sqrt{2}}(-\mathbf{j}-\mathbf{k})$$

EXERCISE 4c

1) A force **F** acts through the point with position vector **r**. Find the vector moment of **F** about the origin where:

(a) $\mathbf{F} = \mathbf{i}+2\mathbf{j}$, $\mathbf{r} = \mathbf{i}-\mathbf{j}$ (c) $\mathbf{F} = \mathbf{i}-\mathbf{j}$, $\mathbf{r} = 3\mathbf{k}$

(b) $\mathbf{F} = \mathbf{i}+2\mathbf{k}$, $\mathbf{r} = 2\mathbf{i}-\mathbf{k}$ (d) $\mathbf{F} = \mathbf{i}+\mathbf{j}-2\mathbf{k}$, $\mathbf{r} = 3\mathbf{i}-\mathbf{j}+\mathbf{k}$

2) A force $\mathbf{F}$ acts through the point with position vector $\mathbf{r}$. Find the vector moment of $\mathbf{F}$ about the point P where:
(a) $\mathbf{F} = 2\mathbf{i} - \mathbf{j}$, $\mathbf{r} = \mathbf{i} + 2\mathbf{j}$, P(0, 1),
(b) $\mathbf{F} = \mathbf{i} + \mathbf{j} - \mathbf{k}$, $\mathbf{r} = 3\mathbf{i} - 2\mathbf{j}$, P(0, 1, 0),
(c) $\mathbf{F} = \mathbf{i} + 2\mathbf{k}$, $\mathbf{r} = \mathbf{i} + \mathbf{j} - \mathbf{k}$, P(2, 0, 2).

3) A force $\mathbf{F}$ acts through the point with position vector $\mathbf{a}$. Find, in terms of $\mathbf{F}, \mathbf{a}, \mathbf{b}$, the vector moment of $\mathbf{F}$ about the point with position vector $\mathbf{b}$.

4) Two forces $\mathbf{F}_1$ and $\mathbf{F}_2$ act through the point with position vector $\mathbf{i} + 2\mathbf{j}$. Find the vector moment of $\mathbf{F}_1$ and $\mathbf{F}_2$ about the origin and also find the vector moment of the resultant of $\mathbf{F}_1$ and $\mathbf{F}_2$ about the origin, where

$$\mathbf{F}_1 = \mathbf{i} - \mathbf{j} - \mathbf{k} \quad \text{and} \quad \mathbf{F}_2 = 2\mathbf{i} + \mathbf{k}$$

5) Find the magnitude of the moment of the force $\mathbf{F}$ about the origin where $\mathbf{r}$ is the position vector of a point on the line of action of $\mathbf{F}$, given:
(a) $\mathbf{F} = \mathbf{i} + 2\mathbf{j}$, $\mathbf{r} = 2\mathbf{i} - \mathbf{j}$, (b) $\mathbf{F} = \mathbf{i} + 2\mathbf{j} - \mathbf{k}$, $\mathbf{r} = \mathbf{i} - \mathbf{j} + 2\mathbf{k}$.

6) A force $\mathbf{F}$ acts through the point A. Find the vector equation of the axis through the point B about which the moment of $\mathbf{F}$ is calculated, given:
(a) $\mathbf{F} = \mathbf{i} - 2\mathbf{j}$, A(0, 1, 0), B(0, 0, 0),
(b) $\mathbf{F} = \mathbf{j} + 2\mathbf{k}$, A($\mathbf{i} + \mathbf{k}$), B($\mathbf{i} + 2\mathbf{j} - \mathbf{k}$),
(c) $\mathbf{F} = 2\mathbf{i} + \mathbf{j} - \mathbf{k}$, A(1, 1, 2), B(−1, 2, −1).

7) A force $\mathbf{F}$ has a vector moment $\mathbf{a}$ about the origin. Find the vector equation of the line of action of $\mathbf{F}$ if:
(a) $\mathbf{F} = (\mathbf{i} + \mathbf{j})$, $\mathbf{a} = 4\mathbf{k}$,
(b) $\mathbf{F} = (2\mathbf{i} - \mathbf{j})$, $\mathbf{a} = (\mathbf{i} + 2\mathbf{j})$,
(c) $\mathbf{F} = (\mathbf{i} + \mathbf{j} - \mathbf{k})$, $\mathbf{a} = (3\mathbf{i} - 2\mathbf{j} + \mathbf{k})$.

8) Two forces $\mathbf{F}$ and $-\mathbf{F}$ act through the point with position vectors $\mathbf{i} + \mathbf{j}$ and $\mathbf{i} - \mathbf{k}$ respectively. Find the moment of this couple and hence find the distance between the lines of action of the forces comprising the couple given $\mathbf{F} = \mathbf{i} - 2\mathbf{j} + 2\mathbf{k}$.

9) A couple is formed from two forces of magnitude $\sqrt{3}$ units acting along the lines whose vector equations are $\mathbf{r} = \mathbf{i} + 2\mathbf{j} + t(\mathbf{i} - \mathbf{j} + \mathbf{k})$ and $\mathbf{r} = 2\mathbf{i} - \mathbf{k} + s(\mathbf{i} - \mathbf{j} + \mathbf{k})$. Find the magnitude of the moment of this couple. Find also a vector which is perpendicular to the plane containing the couple.

SYSTEMS OF FORCES IN THREE DIMENSIONS

Two systems of forces are *equivalent* if their effect on a body is the same in all respects.
The *resultant* of a set of forces is the simplest possible equivalent system and the original set of forces is said to *reduce* to its resultant system.

We know that a system of coplanar forces is equivalent either to a single force or to a couple or that the system is in equilibrium. But if we consider a set of non-coplanar forces a fourth possibility exists: consider, for example, two forces whose lines of action are skew. Such a pair of forces cannot be in equilibrium and they cannot reduce either to a single force or to a couple but are equivalent to a combination of a force and couple where the plane of the couple does not contain the force.

Thus a set of *non-coplanar* forces is equivalent to:

either (a) a single force,

or (b) a couple,

or (c) a combination of a couple and non-coplanar force,

or (d) the system is in equilibrium.

RESULTANT VECTOR MOMENT OF A SET OF FORCES

The resultant vector moment about a point of a set of forces is the vector sum of the moments of all the forces about the same point.

Consider first a set of forces S which is equivalent to a single force $\mathbf{R}$ acting through a point a.

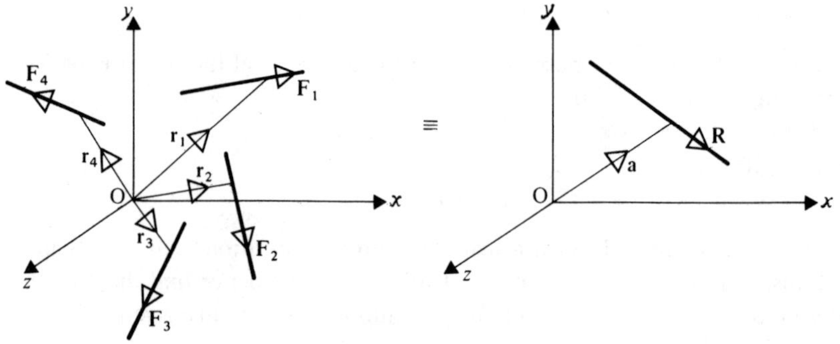

To investigate the vector sum of the moments of the forces in S about O, consider first a simple system of two non parallel forces $\mathbf{F}_1$ and $\mathbf{F}_2$

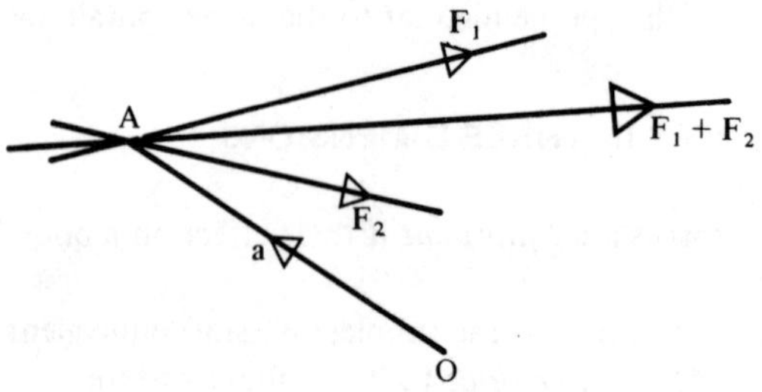

If $\mathbf{F}_1$ and $\mathbf{F}_2$ have a single resultant their lines of action must be concurrent. If their lines of action intersect at A, the resultant of $\mathbf{F}_1$ and $\mathbf{F}_2$ (i.e. $\mathbf{F}_1 + \mathbf{F}_2$) also acts through A.
The moment of $\mathbf{F}_1$ about O plus the moment of $\mathbf{F}_2$ about O is

$$\begin{aligned} &\mathbf{a} \times \mathbf{F}_1 + \mathbf{a} \times \mathbf{F}_2 \\ = \ &\mathbf{a} \times (\mathbf{F}_1 + \mathbf{F}_2) \end{aligned}$$

which is the moment of the resultant force about O. This argument can be applied successively to include further forces thus establishing that *if a set of forces reduces to a single force the vector sum of the moments of all the forces about a point is equal to the moment of the resultant force about the same point.*

i.e. $$\sum \mathbf{r} \times \mathbf{F} = \mathbf{a} \times \mathbf{R}$$

Consider next a system of forces S which is equivalent to a couple of moment $\mathbf{G}$.

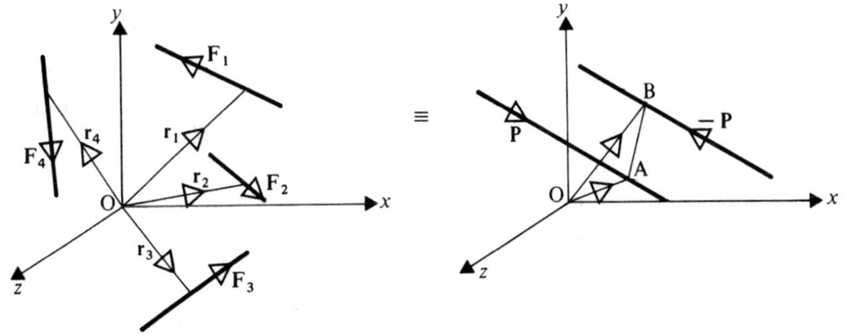

If S is divided into two subsets of forces S_1 and S_2 such that S_1 contains those forces whose resultant is $\mathbf{P}$ acting through the point A and S_2 contains the forces whose resultant is $-\mathbf{P}$ acting through the point B where $\mathbf{P}$ and $-\mathbf{P}$ form a couple whose vector moment is $\mathbf{G}$ then the vector sum of the moments of the forces in S_1 about O is $\overrightarrow{OA} \times \mathbf{P}$. The vector sum of the moments of the forces in S_2 about O is $\overrightarrow{OB} \times -\mathbf{P}$. Adding these two results gives the vector sum of the moments of all the forces in S.

i.e. $$\begin{aligned} \sum \mathbf{r} \times \mathbf{F} &= \overrightarrow{OA} \times \mathbf{P} + \overrightarrow{OB} \times -\mathbf{P} \\ &= (\overrightarrow{OA} - \overrightarrow{OB}) \times \mathbf{P} \\ &= \overrightarrow{BA} \times \mathbf{P} \\ &= \mathbf{G} \end{aligned}$$

Thus *if a system of forces reduces to a couple the vector sum of the moments of all the forces about a point is equal to the vector moment of the couple.*

Now consider a system of forces which is in equilibrium

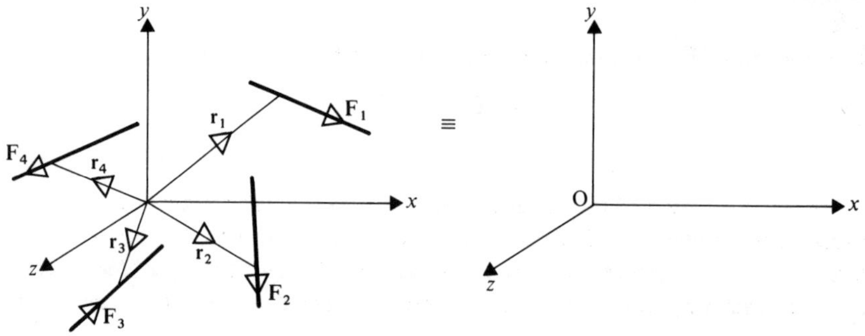

There is zero resultant force and zero turning effect. Thus *if a system of forces is in equilibrium the vector sum of the moments of all the forces about any point is zero.*

PROPERTIES OF A SYSTEM OF FORCES

S is a system of forces $\mathbf{F}_1, \mathbf{F}_2, \mathbf{F}_3, \ldots$ acting through points with position vectors $\mathbf{r}_1, \mathbf{r}_2, \mathbf{r}_3, \ldots$

If S reduces to a single force **R** acting through the point **a**

$$\text{then} \quad \sum \mathbf{F} = \mathbf{R}$$

$$\text{and} \quad \sum \mathbf{r} \times \mathbf{F} = \mathbf{a} \times \mathbf{R}$$

If S reduces to a couple of moment **G**

$$\text{then} \quad \sum \mathbf{F} = \mathbf{0}$$

$$\text{and} \quad \sum \mathbf{r} \times \mathbf{F} = \mathbf{G}$$

If S is in equilibrium

$$\text{then} \quad \sum \mathbf{F} = \mathbf{0}$$

$$\text{and} \quad \sum \mathbf{r} \times \mathbf{F} = \mathbf{0}$$

Conversely *to prove that a system of forces is in equilibrium* it is necessary to show both that $\Sigma\, \mathbf{F} = \mathbf{0}$ and $\Sigma\, \mathbf{r} \times \mathbf{F} = \mathbf{0}$, either condition on its own not being sufficient. ($\Sigma\,\mathbf{F} = \mathbf{0}$ is true for a system reducing to a couple and $\Sigma\,\mathbf{r}\times\mathbf{F} = 0$ is true for a system reducing to a single force whose line of action passes through O.)

To prove that a system of forces reduces to a couple it is necessary to show both that $\Sigma\,\mathbf{F} = \mathbf{0}$ and that $\Sigma\,\mathbf{r}\times\mathbf{F}$ is not zero, either condition on its own again being insufficient.

To prove that a system of forces reduces to a single force is not so straight forward. If for a given system of forces $\Sigma\,\mathbf{F}$ is not zero, the system could reduce to either a single force or a combination of a couple and non-coplanar force. To distinguish between these two possibilities, in general, requires work beyond the scope of this book.

However there are some cases when we can rule out the possibility of a resultant force and a non-coplanar couple. If the lines of action of the forces are concurrent, all passing through the point A say, the vector sum of the moments of the forces about A is zero so there is no resultant turning effect about A and the system must reduce to a single force. Alternatively, if the forces are coplanar, the resultant system will act in the same plane thus eliminating the possibility of a non-coplanar couple.

EXAMPLES 4d

1) Prove that the following system of forces is in equilibrium:

$\mathbf{F}_1 = 3\mathbf{i}-\mathbf{j}+\mathbf{k}$ acting at the point $\mathbf{r}_1 = \mathbf{i}+\mathbf{j}$,

$\mathbf{F}_2 = -2\mathbf{i}+4\mathbf{j}-5\mathbf{k}$ acting at the point $\mathbf{r}_2 = 3\mathbf{i}-3\mathbf{j}+5\mathbf{k}$,

$\mathbf{F}_3 = -\mathbf{i}-3\mathbf{j}+4\mathbf{k}$ acting at the point $\mathbf{r}_3 = -2\mathbf{j}+4\mathbf{k}$.

To show that this system is in equilibrium we must prove that $\Sigma\,\mathbf{F} = \mathbf{0}$ and that $\Sigma\,\mathbf{r}\times\mathbf{F} = \mathbf{0}$.

$$\sum \mathbf{F} = (3\mathbf{i}-\mathbf{j}+\mathbf{k})+(-2\mathbf{i}+4\mathbf{j}-5\mathbf{k})+(-\mathbf{i}-3\mathbf{j}+4\mathbf{k}) = \mathbf{0}$$

$$\begin{aligned}\sum \mathbf{r}\times\mathbf{F} &= (\mathbf{i}+\mathbf{j})\times(3\mathbf{i}-\mathbf{j}+\mathbf{k})+(3\mathbf{i}-3\mathbf{j}+5\mathbf{k})\times(-2\mathbf{i}+4\mathbf{j}-5\mathbf{k})\\ &\quad+(-2\mathbf{j}+4\mathbf{k})\times(-\mathbf{i}-3\mathbf{j}+4\mathbf{k})\\ &= \begin{vmatrix}\mathbf{i} & \mathbf{j} & \mathbf{k}\\ 1 & 1 & 0\\ 3 & -1 & 1\end{vmatrix} + \begin{vmatrix}\mathbf{i} & \mathbf{j} & \mathbf{k}\\ 3 & -3 & 5\\ -2 & 4 & -5\end{vmatrix} + \begin{vmatrix}\mathbf{i} & \mathbf{j} & \mathbf{k}\\ 0 & -2 & 4\\ -1 & -3 & 4\end{vmatrix}\\ &= (\mathbf{i}-\mathbf{j}-4\mathbf{k})+(-5\mathbf{i}+5\mathbf{j}+6\mathbf{k})+(4\mathbf{i}-4\mathbf{j}-2\mathbf{k}) = \mathbf{0}\end{aligned}$$

Therefore the system is in equilibrium.

(Alternatively having shown that $\Sigma\,\mathbf{F} = \mathbf{0}$ we could prove that the lines of action of the three forces are concurrent thus establishing that the three forces are in equilibrium).

2) Forces $\mathbf{F}_1$, $\mathbf{F}_2$ and $\mathbf{F}_3$ act through the points with position vectors $\mathbf{r}_1$, $\mathbf{r}_2$ and $\mathbf{r}_3$ where:

$$\begin{aligned} \mathbf{F}_1 &= \mathbf{i}+\mathbf{j}-\mathbf{k}, & \mathbf{r}_1 &= \mathbf{i}+\mathbf{j}+\mathbf{k} \\ \mathbf{F}_2 &= 2\mathbf{i}-\mathbf{j}+3\mathbf{k}, & \mathbf{r}_2 &= -\mathbf{i}-2\mathbf{j}+\mathbf{k} \\ \mathbf{F}_3 &= -3\mathbf{i}-4\mathbf{j}+\mathbf{k}, & \mathbf{r}_3 &= 4\mathbf{i}+5\mathbf{j} \end{aligned}$$

When a fourth force $\mathbf{F}_4$ is added it reduces the system to equilibrium. Find $\mathbf{F}_4$ and a vector equation of its line of action.

The system is in equilibrium

Therefore $$\mathbf{F}_1+\mathbf{F}_2+\mathbf{F}_3+\mathbf{F}_4 = \mathbf{0}$$

$$(\mathbf{i}+\mathbf{j}-\mathbf{k})+(2\mathbf{i}-\mathbf{j}+3\mathbf{k})+(-3\mathbf{i}-4\mathbf{j}+\mathbf{k})+\mathbf{F}_4 = \mathbf{0}$$

Therefore $$\mathbf{F}_4 = 4\mathbf{j}-3\mathbf{k}$$

Let $\mathbf{r}_4$ be the position vector of any point on the line of action of $\mathbf{F}_4$.
As the system is in equilibrium $\Sigma\, \mathbf{r}\times\mathbf{F} = \mathbf{0}$

Therefore $(\mathbf{i}+\mathbf{j}+\mathbf{k})\times(\mathbf{i}+\mathbf{j}-\mathbf{k})+(-\mathbf{i}-2\mathbf{j}+\mathbf{k})\times(2\mathbf{i}-\mathbf{j}+3\mathbf{k})$
$+(4\mathbf{i}+5\mathbf{j})\times(-3\mathbf{i}-4\mathbf{j}+\mathbf{k})+\mathbf{r}_4\times(4\mathbf{j}-3\mathbf{k}) = \mathbf{0}$

If $\mathbf{r}_4 = x\mathbf{i}+y\mathbf{j}+z\mathbf{k}$

$$\begin{vmatrix} \mathbf{i} & \mathbf{j} & \mathbf{k} \\ 1 & 1 & 1 \\ 1 & 1 & -1 \end{vmatrix} + \begin{vmatrix} \mathbf{i} & \mathbf{j} & \mathbf{k} \\ -1 & -2 & 1 \\ 2 & -1 & 3 \end{vmatrix} + \begin{vmatrix} \mathbf{i} & \mathbf{j} & \mathbf{k} \\ 4 & 5 & 0 \\ -3 & -4 & 1 \end{vmatrix} + \begin{vmatrix} \mathbf{i} & \mathbf{j} & \mathbf{k} \\ x & y & z \\ 0 & 4 & -3 \end{vmatrix} = \mathbf{0}$$

$$(-2\mathbf{i}+2\mathbf{j})+(-5\mathbf{i}+5\mathbf{j}+5\mathbf{k})+(5\mathbf{i}-4\mathbf{j}-\mathbf{k})+(-3y-4z)\mathbf{i}+3x\mathbf{j}+4x\mathbf{k} = \mathbf{0}$$

Therefore $$-2\mathbf{i}+3\mathbf{j}+4\mathbf{k}+(-3y-4z)\mathbf{i}+3x\mathbf{j}+4x\mathbf{k} = \mathbf{0}$$

or $$(-2-3y-4z)\mathbf{i}+(3+3x)\mathbf{j}+(4+4x)\mathbf{k} = \mathbf{0}$$

giving $$x = -1 \quad \text{and} \quad 3y+4z = -2$$

$$y = \frac{-2-4z}{3} = \lambda \quad \text{(say)}$$

Therefore $$\mathbf{r} = -\mathbf{i}+\lambda\mathbf{j}+\frac{(-2-3\lambda)}{4}\mathbf{k}$$

which is a vector equation of the line of action of $\mathbf{F}_4$.

3) The points A(0, 1, 1), B(3, − 1, 0), C(2, 0, 4), D(1, 3, − 1) form a skew quadrilateral ABCD. Forces $\overrightarrow{AB}$, $\overrightarrow{BC}$, $\overrightarrow{CD}$, $\overrightarrow{DA}$ act along the sides AB, BC, CD, DA of this quadrilateral. Prove that this system of forces is equivalent to a couple and find its vector moment.

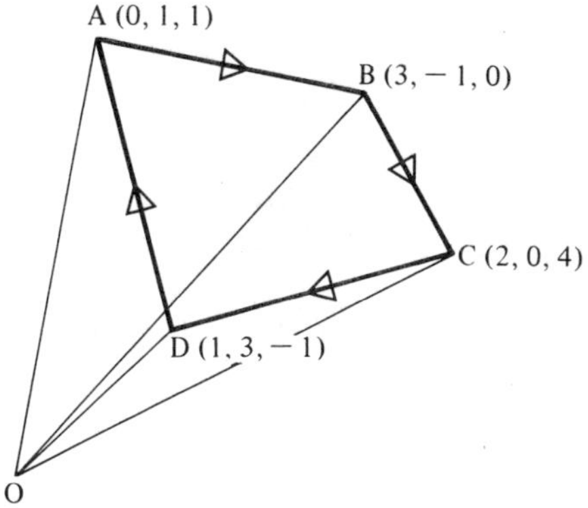

$$\left.\begin{aligned}\overrightarrow{AB} &= \overrightarrow{OB} - \overrightarrow{OA} = 3\mathbf{i} - 2\mathbf{j} - \mathbf{k}\\ \overrightarrow{BC} &= \overrightarrow{OC} - \overrightarrow{OB} = -\mathbf{i} + \mathbf{j} + 4\mathbf{k}\\ \overrightarrow{CD} &= \overrightarrow{OD} - \overrightarrow{OC} = -\mathbf{i} + 3\mathbf{j} - 5\mathbf{k}\\ \overrightarrow{DA} &= \overrightarrow{OA} - \overrightarrow{OD} = -\mathbf{i} - 2\mathbf{j} + 2\mathbf{k}\end{aligned}\right\} \quad \text{Therefore } \overrightarrow{AB} + \overrightarrow{BC} + \overrightarrow{CD} + \overrightarrow{DA} = \mathbf{0}$$

Therefore the resultant force is zero.
The sum of the moments of the forces about O is

$$\overrightarrow{OA} \times \overrightarrow{AB} + \overrightarrow{OB} \times \overrightarrow{BC} + \overrightarrow{OC} \times \overrightarrow{CD} + \overrightarrow{OD} \times \overrightarrow{DA}$$

$$= \begin{vmatrix} \mathbf{i} & \mathbf{j} & \mathbf{k} \\ 0 & 1 & 1 \\ 3 & -2 & -1 \end{vmatrix} + \begin{vmatrix} \mathbf{i} & \mathbf{j} & \mathbf{k} \\ 3 & -1 & 0 \\ -1 & 1 & 4 \end{vmatrix} + \begin{vmatrix} \mathbf{i} & \mathbf{j} & \mathbf{k} \\ 2 & 0 & 4 \\ -1 & 3 & -5 \end{vmatrix} + \begin{vmatrix} \mathbf{i} & \mathbf{j} & \mathbf{k} \\ 1 & 3 & -1 \\ -1 & -2 & 2 \end{vmatrix}$$

$$= (\mathbf{i} + 3\mathbf{j} - 3\mathbf{k}) + (-4\mathbf{i} - 12\mathbf{j} + 2\mathbf{k}) + (-12\mathbf{i} + 6\mathbf{j} + 6\mathbf{k}) + (4\mathbf{i} - \mathbf{j} + \mathbf{k})$$

$$= -11\mathbf{i} - 4\mathbf{j} + 6\mathbf{k}$$

As this result is not zero, the forces reduce to a couple of vector moment $-11\mathbf{i} - 4\mathbf{j} + 6\mathbf{k}$.

4) The vertices of a triangle ABC are at the points A(0, 1, 2), B(3, − 1, 1), C(1, 0, 1). Forces $\mathbf{F}_1, \mathbf{F}_2, \mathbf{F}_3$ of magnitudes $\sqrt{14}$ N, $2\sqrt{5}$ N, $3\sqrt{3}$ N act along the sides AB, BC, CA of the triangle in the sense indicated by the letters. Prove that this system of forces is equivalent to a single force and find the cartesian equation of its line of action.

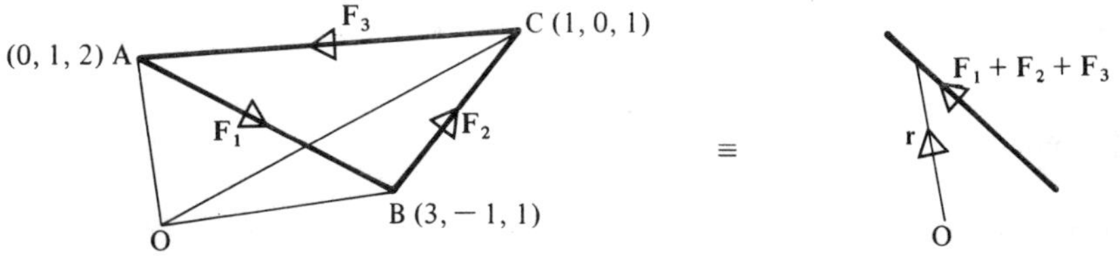

$$\left.\begin{aligned}\overrightarrow{AB} &= \overrightarrow{OB}-\overrightarrow{OA} = 3\mathbf{i}-2\mathbf{j}-\mathbf{k}\\ \overrightarrow{BC} &= \overrightarrow{OC}-\overrightarrow{OB} = -2\mathbf{i}+\mathbf{j}\\ \overrightarrow{CA} &= \overrightarrow{OA}-\overrightarrow{OC} = -\mathbf{i}+\mathbf{j}+\mathbf{k}\end{aligned}\right\}\quad \text{Therefore}\quad \begin{aligned}\mathbf{F}_1 &= \sqrt{14}\,\widehat{\mathbf{AB}} = 3\mathbf{i}-2\mathbf{j}-\mathbf{k}\\ \mathbf{F}_2 &= 2\sqrt{5}\,\widehat{\mathbf{BC}} = -4\mathbf{i}+2\mathbf{j}\\ \mathbf{F}_3 &= 3\sqrt{3}\,\widehat{\mathbf{CA}} = -3\mathbf{i}+3\mathbf{j}+3\mathbf{k}\end{aligned}$$

Therefore $\sum \mathbf{F} = -4\mathbf{i}+3\mathbf{j}+2\mathbf{k}$

Therefore the forces cannot be in equilibrium, nor can they be equivalent to a couple.

Also the points A, B and C are coplanar (any three points are coplanar) so $\mathbf{F}_1$, $\mathbf{F}_2$ and $\mathbf{F}_3$ are coplanar.

Therefore the forces are equivalent to the single force $-4\mathbf{i}+3\mathbf{j}+2\mathbf{k}$.

If $\mathbf{r} = x\mathbf{i}+y\mathbf{j}+z\mathbf{k}$ is the position vector of any point on the line of action of the resultant force then:

$$\overrightarrow{OA}\times\mathbf{F}_1+\overrightarrow{OB}\times\mathbf{F}_2+\overrightarrow{OC}\times\mathbf{F}_3 = \mathbf{r}\times(\mathbf{F}_1+\mathbf{F}_2+\mathbf{F}_3)$$

$$\begin{vmatrix}\mathbf{i} & \mathbf{j} & \mathbf{k}\\ 0 & 1 & 2\\ 3 & -2 & -1\end{vmatrix} + \begin{vmatrix}\mathbf{i} & \mathbf{j} & \mathbf{k}\\ 3 & -1 & 1\\ -4 & 2 & 0\end{vmatrix} + \begin{vmatrix}\mathbf{i} & \mathbf{j} & \mathbf{k}\\ 1 & 0 & 1\\ -3 & 3 & 3\end{vmatrix} = \begin{vmatrix}\mathbf{i} & \mathbf{j} & \mathbf{k}\\ x & y & z\\ -4 & 3 & 2\end{vmatrix}$$

$$-2\mathbf{i}-4\mathbf{j}+2\mathbf{k} = (2y-3z)\mathbf{i}+(-2x-4z)\mathbf{j}+(3x+4y)\mathbf{k}$$

Therefore
$$2y-3z = -2 \qquad (1)$$
$$x+2z = 2 \qquad (2)$$
$$3x+4y = 2 \qquad (3)$$

Because $\mathbf{r}$ is not a unique position vector we expect to find that these equations are not independent. [(3) − 2(1) = 3(2).] (This fact is a useful check on the calculations.)

Any pair of equations from (1), (2) and (3) will serve as the Cartesian equations of the line of action of the resultant force. Using equations (1) and (3) and writing them in standard form:

From (1) $$y = \frac{3z-2}{2}$$

From (2) $$y = \frac{2-3x}{4} = \frac{3x-2}{-4}$$

Therefore $$\frac{x-\frac{2}{3}}{-4} = \frac{y}{3} = \frac{z-\frac{2}{3}}{2}$$

are the Cartesian equations of the line of action of the resultant force.

5) A force **F** acts at the point **r** where $\mathbf{F} = \mathbf{i} - \mathbf{j} + 2\mathbf{k}$ and $\mathbf{r} = \mathbf{i} - \mathbf{k}$. Prove that this force is equivalent to a force **F** acting at the point $\mathbf{i} + 2\mathbf{j} + \mathbf{k}$ together with a couple and find the vector moment of the couple.

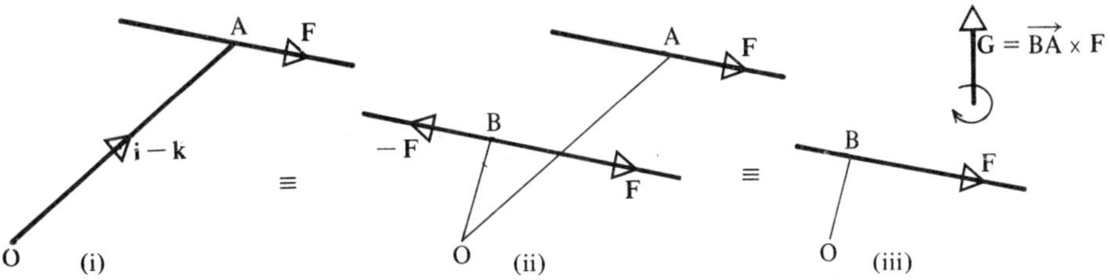

If we introduce forces **F** and $-\mathbf{F}$ acting at B ($\overrightarrow{OB} = \mathbf{i} + 2\mathbf{j} + \mathbf{k}$), these two forces, being in equilibrium, do not alter the effect of the original force.
Therefore the force in (i) is equivalent to the forces in (ii). The force **F** acting at A together with the force $-\mathbf{F}$ acting at B form a couple of moment

$$\overrightarrow{BA} \times \mathbf{F} = (-2\mathbf{j} - 2\mathbf{k}) \times (\mathbf{i} - \mathbf{j} + 2\mathbf{k})$$

$$= \begin{vmatrix} \mathbf{i} & \mathbf{j} & \mathbf{k} \\ 0 & -2 & -2 \\ 1 & -1 & 2 \end{vmatrix} = -6\mathbf{i} - 2\mathbf{j} + 2\mathbf{k}$$

Therefore the force **F** acting at $(\mathbf{i} - \mathbf{k})$ is equivalent to the force **F** acting at $(\mathbf{i} + 2\mathbf{j} + \mathbf{k})$ together with a couple of moment $-6\mathbf{i} - 2\mathbf{j} + 2\mathbf{k}$.
Any force acting through a given point can be replaced by an equal force acting through another point together with a coplanar couple. This is shown in general terms in Volume One.

6) The points A, B, C have position vectors **a**, **b**, **c** respectively. Forces AB, BC and CA act round the sides of the triangle ABC in the sense indicated by the order of the letters. Verify that these forces are equivalent to a couple.

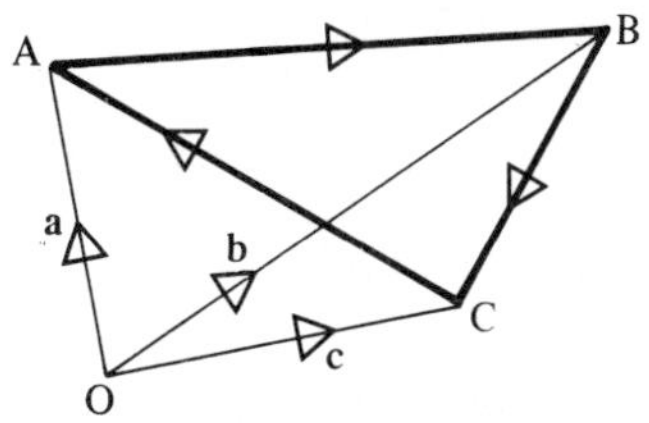

$$\left.\begin{aligned} \overrightarrow{AB} &= \mathbf{b} - \mathbf{a} \\ \overrightarrow{BC} &= \mathbf{c} - \mathbf{b} \\ \overrightarrow{CA} &= \mathbf{a} - \mathbf{c} \end{aligned}\right\} \quad \text{Therefore } \sum \mathbf{F} = \mathbf{0}$$

(Therefore the forces are either in equilibrium or reduce to a couple.)

Taking moments about A:

$$\mathbf{0} \times \overrightarrow{AB} + \overrightarrow{AB} \times \overrightarrow{BC} + \mathbf{0} \times \overrightarrow{AC}$$

$$= \overrightarrow{AB} \times \overrightarrow{BC}$$

But $\frac{1}{2}|\overrightarrow{AB} \times \overrightarrow{BC}| =$ area of ΔABC.
Therefore $\overrightarrow{AB} \times \overrightarrow{BC}$ is not zero.
Therefore the forces are equivalent to a couple of magnitude equal to twice the area of Δ ABC. (This property is proved in Volume One.)

EXERCISE 4d

1) Prove that the following system of forces is in equilibrium:
$\mathbf{F}_1 = \mathbf{i} - \mathbf{j}$ acting at the point $\mathbf{r}_1 = \mathbf{i} + \mathbf{k}$,
$\mathbf{F}_2 = \mathbf{i} - \mathbf{k}$ acting at the point $\mathbf{r}_2 = 2\mathbf{i}$,
$\mathbf{F}_3 = 2\mathbf{j} + \mathbf{k}$ acting at the point $\mathbf{r}_3 = \mathbf{i} - 2\mathbf{j}$,
$\mathbf{F}_4 = -2\mathbf{i} - \mathbf{j}$ acting at the point $\mathbf{r}_4 = 3\mathbf{i} + \mathbf{j} + \mathbf{k}$.

2) Prove that the following system of forces reduces to a couple and find the vector moment of the couple.
$\mathbf{F}_1 = \mathbf{i} + 2\mathbf{j} - 3\mathbf{k}$ acting at the point $\mathbf{r}_1 = \mathbf{i} - \mathbf{j} + 2\mathbf{k}$,
$\mathbf{F}_2 = \mathbf{i} - \mathbf{j} + 2\mathbf{k}$ acting at the point $\mathbf{r}_2 = 2\mathbf{i} + \mathbf{j} - \mathbf{k}$,
$\mathbf{F}_3 = -3\mathbf{i} + \mathbf{j} - 3\mathbf{k}$ acting at the point $\mathbf{r}_3 = 3\mathbf{i} + \mathbf{k}$,
$\mathbf{F}_4 = \mathbf{i} - 2\mathbf{j} + 4\mathbf{k}$ acting at the point $\mathbf{r}_4 = \mathbf{j} - 2\mathbf{k}$.

3) The forces $\mathbf{F}_1, \mathbf{F}_2, \mathbf{F}_3$ act through the points with position vectors $\mathbf{r}_1, \mathbf{r}_2, \mathbf{r}_3$ where:
$\mathbf{F}_1 = \mathbf{i} - 2\mathbf{j}$ $\quad \mathbf{r}_1 = 2\mathbf{i} - \mathbf{j}$,
$\mathbf{F}_2 = \mathbf{j} + \mathbf{k}$ $\quad \mathbf{r}_2 = 2\mathbf{i} + \mathbf{k}$,
$\mathbf{F}_3 = 3\mathbf{i} + \mathbf{j} - \mathbf{k}$ $\quad \mathbf{r}_3 = 8\mathbf{i} + \mathbf{j} - 2\mathbf{k}$.
Find the vector equations of the lines of action of $\mathbf{F}_1, \mathbf{F}_2, \mathbf{F}_3$ and show that these lines are concurrent. Prove that the system of forces $\mathbf{F}_1, \mathbf{F}_2$ and $\mathbf{F}_3$ is equivalent to a single force and find a vector equation for its line of action.

4) The points A(0, 1, 1), B(1, −1, 0), C(1, 1, 0), D(0, 0, 1), E(−1, 0, 1) from a skew pentagon ABCDE. Forces **AB**, **BC**, **CD**, **DE**, **EA** act round the sides AB, BC, CD, DE, EA of the pentagon. Prove that this system of forces is equivalent to a couple and find the magnitude of the moment of this couple.

5) In each of the following cases find the simplest system of forces which is equivalent to the system given:
(a) $\mathbf{F}_1 = 2\mathbf{i} - \mathbf{j}$ acting at the point $\mathbf{r}_1 = \mathbf{i} + 2\mathbf{j}$,
$\mathbf{F}_2 = 3\mathbf{i} + \mathbf{j}$ acting at the point $\mathbf{r}_2 = 2\mathbf{i} + 3\mathbf{j}$,
$\mathbf{F}_3 = -\mathbf{i} + \mathbf{j}$ acting at the point $\mathbf{r}_3 = -2\mathbf{j}$.
(b) $\overrightarrow{AB}$, $\overrightarrow{BC}$, $\overrightarrow{CA}$ where the points A, B, C have position vectors **a**, **b**, and **c** respectively.

(c) $\mathbf{F} = \mathbf{i} + 3\mathbf{j}$ acting at the origin and a couple of vector moment $3\mathbf{k}$.
(d) $\mathbf{F}_1 = \mathbf{i} - \mathbf{j} + \mathbf{k}$ acting at the point $\mathbf{r}_1 = \mathbf{i} - 2\mathbf{k}$,
$\mathbf{F}_2 = -2\mathbf{i} + 3\mathbf{j} + \mathbf{k}$ acting at the point $\mathbf{r}_2 = -\mathbf{i} + 2\mathbf{j} - \mathbf{k}$,
$\mathbf{F}_3 = -2\mathbf{j} - \mathbf{k}$ acting at the point $\mathbf{r}_3 = 3\mathbf{j}$.

6) Forces $\mathbf{F}_1, \mathbf{F}_2, \mathbf{F}_3$ act at points $\mathbf{r}_1, \mathbf{r}_2, \mathbf{r}_3$ where
$\mathbf{F}_1 = 3\mathbf{i} - \mathbf{j} + 2\mathbf{k}$ $\quad \mathbf{r}_1 = 3\mathbf{i} - \mathbf{k}$,
$\mathbf{F}_2 = -\mathbf{i} - 4\mathbf{j} + \mathbf{k}$ $\quad \mathbf{r}_2 = 2\mathbf{i} - 4\mathbf{j}$,
$\mathbf{F}_3 = \mathbf{i} + \mathbf{j} - 2\mathbf{k}$ $\quad \mathbf{r}_3 = -3\mathbf{j} + 5\mathbf{k}$.
When a fourth force $\mathbf{F}_4$ is added the system is in equilibrium. Find $\mathbf{F}_4$ and a vector equation of its line of action.

7) A force $\mathbf{i} - \mathbf{j} + 2\mathbf{k}$ acts at the point with position vector $-\mathbf{i} - \mathbf{j} + \mathbf{k}$. Show that this force is equivalent to an equal force acting through the origin together with a couple. Find the vector moment of the couple.

8) A force $\mathbf{i} + \mathbf{j} - \mathbf{k}$ acts through the origin together with a couple of moment $\mathbf{i} + \mathbf{j} + 2\mathbf{k}$. Prove that the vector moment of this couple is perpendicular to the force and hence show that the couple and force are coplanar. Show that the couple and force are equivalent to a single force and find a vector equation of its line of action.

MULTIPLE CHOICE EXERCISE 4

Instructions for answering these questions are given on page (xi)

TYPE I

1) A force $\mathbf{F}$ acts through the point A. If the position vector of A is $\mathbf{a}$ and the position vector of another point B is $\mathbf{b}$, the vector moment of $\mathbf{F}$ about B is:
(a) $\mathbf{F} \times \mathbf{a}$ (b) $\mathbf{b} \times \mathbf{F}$ (c) $\mathbf{0}$ (d) $(\mathbf{a} - \mathbf{b}) \times \mathbf{F}$ (e) $\mathbf{a} \times \mathbf{F}$.

2) If $\mathbf{a} = \mathbf{i} + \mathbf{j}$, $\mathbf{b} = 2\mathbf{i} - \mathbf{j}$ then $\mathbf{a} \times \mathbf{b}$ is:
(a) $2\mathbf{i} - \mathbf{j}$ (b) $-3\mathbf{k}$ (c) $\mathbf{0}$ (d) $\mathbf{i}$ (e) $\sqrt{10}$.

3) The vertices A, B, C of ΔABC have position vectors $\mathbf{a}, \mathbf{b}, \mathbf{c}$ respectively. The area of Δ ABC is:
(a) $\mathbf{a} \times \mathbf{b}$ (b) $\frac{1}{2}|\mathbf{b} \times \mathbf{c}|$ (c) $\frac{1}{2}|(\mathbf{a} - \mathbf{b}) \times (\mathbf{b} - \mathbf{c})|$ (d) $(\mathbf{b} - \mathbf{a}) \times (\mathbf{a} - \mathbf{c})$.

4) Forces $\mathbf{i} - 2\mathbf{j}$ and $-\mathbf{i} + 2\mathbf{j}$ act through the points whose position vectors are $\mathbf{k}$ and $3\mathbf{k}$. The vector sum of the moments of these forces about O is:
(a) $2(-2\mathbf{i} - \mathbf{j})$ (b) $\mathbf{0}$ (c) $2(\mathbf{i} - 2\mathbf{j})$ (d) $4\mathbf{k}$ (e) $2\mathbf{k}$.

5) A force $\mathbf{F}$ acts along the line whose vector equation is $\mathbf{r} = \mathbf{a} + \lambda\mathbf{b}$. The moment of $\mathbf{F}$ about O is:
(a) $\mathbf{b} \times \mathbf{F}$ (b) $\mathbf{a} \times \mathbf{F}$ (c) $\mathbf{F} \times \mathbf{a}$ (d) $\mathbf{a} \cdot \mathbf{F}$ (e) $\mathbf{F} \times \mathbf{b}$.

6) Forces $\mathbf{i}-2\mathbf{j}+\mathbf{k}$, $2\mathbf{i}+\mathbf{j}-\mathbf{k}$, $3\mathbf{i}$ act through the origin. The resultant of these forces is:
(a) a couple (b) a single force (c) they are in equilibrium
(d) none of these.

TYPE II

7) ABC is a triangle.
(a) The area of $\Delta ABC = \frac{1}{2}|\mathbf{AB} \times \mathbf{BC}|$.
(b) Forces **AB**, **BC**, **CA** acting round the sides of the triangle are in equilibrium.
(c) The vector $\mathbf{BC} \times \mathbf{CA}$ is perpendicular to the plane containing ΔABC.

8) A system of forces $\mathbf{F}_1$, $\mathbf{F}_2$ and $\mathbf{F}_3$ act through the points with position vectors $\mathbf{r}_1, \mathbf{r}_2, \mathbf{r}_3$. The system is in equilibrium.
(a) $\Sigma\, \mathbf{r} \times \mathbf{F} = \mathbf{0}$.
(b) The lines of action of the forces are concurrent.
(c) $\mathbf{F}_1$, $\mathbf{F}_2$ and $\mathbf{F}_3$ are coplanar.

9) **a** and **b** are perpendicular vectors.
(a) $\mathbf{a} \times \mathbf{b} = \mathbf{0}$.
(b) $\mathbf{a} \times \mathbf{b}$, **a**, **b** are mutually perpendicular.
(c) $|\mathbf{a} \times \mathbf{b}| = ab$.

10) A force **F** acts along the line whose vector equation is $\mathbf{r} = \mathbf{a} + \lambda\mathbf{b}$. The moment of **F** about O is:
(a) $\mathbf{a} \times \mathbf{F}$ (b) $(\mathbf{a}+\mathbf{b}) \times \mathbf{F}$ (c) $\mathbf{b} \times \mathbf{F}$.

TYPE III

11) (a) Two couples are equivalent.
(b) Two couples act in the same plane.

12) (a) $\mathbf{a} = \lambda\mathbf{b}$.
(b) $\mathbf{a} \times \mathbf{b} = \mathbf{0}$.

13) (a) A system of forces $\mathbf{F}_1, \mathbf{F}_2, \mathbf{F}_3, \ldots$ is in equilibrium
(b) $\Sigma\, \mathbf{F} = \mathbf{0}$.

14) (a) $\mathbf{a} \times \mathbf{b} = ab\hat{\mathbf{n}}$
(b) $\mathbf{a} \cdot \mathbf{b} = 0$.

15) (a) **a** and **b** are non-parallel vectors and $\mathbf{a} \neq \mathbf{0}$, $\mathbf{b} \neq \mathbf{0}$.
(b) $\mathbf{a} \times \mathbf{b} = \mathbf{b} \times \mathbf{a}$.

TYPE IV

16) Forces $\mathbf{F}_1, \mathbf{F}_2, \mathbf{F}_3, \mathbf{F}_4$ act through points whose position vectors are $\mathbf{r}_1, \mathbf{r}_2, \mathbf{r}_3, \mathbf{r}_4$. Find the resultant of this system.
(a) $\Sigma\, \mathbf{F} = \mathbf{0}$ (b) $\Sigma\, \mathbf{r} = \mathbf{0}$ (c) $\Sigma\, \mathbf{r} \times \mathbf{F} = \mathbf{0}$.

17) A force **F** acting through the point with position vector **a** is equivalent to a force **F** acting through the point with position vector **b** together with a couple. Find the magnitude of the moment of this couple.
(a) $\mathbf{F} = \mathbf{i} + \mathbf{j}$ (b) $\mathbf{a} = 2\mathbf{k}$ (c) $\mathbf{b} = \mathbf{i} - \mathbf{j}$.

18) A particle moves under the action of a constant force **F**. Find the Cartesian equation of the path of the particle.
(a) The particle moves in a straight line.
(b) The acceleration of the particle is $\mathbf{i} - \mathbf{j}$.
(c) The moment of **F** about O is $\mathbf{i} + \mathbf{j}$.

19) ABC is a triangle. Find the direction cosines of a normal to the plane of the triangle.
(a) $\overrightarrow{AB} = \mathbf{i} + \mathbf{j}$.
(b) The position vector of B is $2\mathbf{i} - \mathbf{j} + \mathbf{k}$.
(c) $\overrightarrow{AC} = 2\mathbf{i} - \mathbf{k}$.

TYPE V

20) Any three forces whose lines of action are concurrent are in equilibrium.

21) Three non zero vectors **a**, **b** and **c** are such that $\mathbf{a} \times \mathbf{b} = \mathbf{a} \times \mathbf{c}$, therefore $\mathbf{b} = \mathbf{c}$ or $\mathbf{a} = \lambda(\mathbf{b} - \mathbf{c})$.

22) ABCD is a parallelogram. The unit vector perpendicular to the plane of ABCD is $\mathbf{AB} \times \mathbf{BC}$ divided by the area of ABCD.

23) A system of forces is such that $\Sigma\, \mathbf{r} \times \mathbf{F} = \mathbf{0}$, therefore the forces must be in equilibrium.

MISCELLANEOUS EXERCISE 4

1) Define the scalar product $\mathbf{a} \cdot \mathbf{b}$ and the vector product $\mathbf{a} \wedge \mathbf{b}$ of two vectors **a** and **b**.
The points P, Q, R have co-ordinates (1, 1, 1), (1, 3, 2), (2, 1, 3) respectively, referred to rectangular axes $Oxyz$.
Calculate the products $\overrightarrow{PQ} \cdot \overrightarrow{PR}$, $\overrightarrow{PQ} \wedge \overrightarrow{PR}$ and deduce the values of the cosine of the angle QPR and the area of the triangle PQR. (JMB part question)

2) Define the vector product $\mathbf{a} \wedge \mathbf{b}$ of two vectors **a** and **b**. Three non-collinear points P, Q and R have position vectors **p**, **q** and **r**, respectively, relative to an origin O, not necessarily in the plane PQR. Prove that the area of the triangle PQR is equal to the magnitude of the vector

$$\tfrac{1}{2}(\mathbf{q} \wedge \mathbf{r} + \mathbf{r} \wedge \mathbf{p} + \mathbf{p} \wedge \mathbf{q}).$$

When O does lie in the plane PQR interpret this result in terms of the areas of the triangles OPQ, OQR and ORP, considering the cases (i) O inside the triangle PQR, (ii) O in the region bounded by PR produced and QR produced.

(JMB part question)

3) Define the moment of a force **F** about a point O as a vector product, and show that the value given by your definition depends only on **F**, the position of O and the line of action of the force.
Calculate the moment about the point C(1, 1, 1) of a force of 5 N acting along the line $\overrightarrow{AB}$, where A, B are the points (2, 3, 4), (3, 5, 6) respectively, the distances being measured in metres. (JMB part question)

4) (i) Define the scalar and vector products of two vectors **a**, **b** in terms of a, b and the angle θ between the vectors, explaining carefully any sign conventions used.
Deduce the formula

$$|\mathbf{a} \wedge \mathbf{b}|^2 = a^2b^2 - (\mathbf{a}.\mathbf{b})^2$$

(ii) Find the most general form for the vector **u** when

$$\mathbf{u} \wedge (\mathbf{i}+\mathbf{j}+2\mathbf{k}) = \mathbf{i}-\mathbf{j}$$

(JMB part question)

5) (a) Define the moment of a force **F** about a point A as a vector product. Hence, or otherwise, prove that the moment of the couple formed by the forces **F** and $-$**F** acting in different lines is the same about all points.
(b) A force of unit magnitude acts through the origin O of a rectangular system of axes Ox, Oy and Oz and has equal moments about each of the points (1, 1, 0) and (2, 0, 1). Find the possible values of the components of the force in the directions Ox, Oy and Oz. (JMB)

6) (a) The line AB is the common perpendicular to two skew lines AP and BQ, and C and R are the mid-points of AB and PQ respectively. Prove by vector methods, that CR and AB are perpendicular.
(b) Three vectors **a**, **b** and **c** are such that $\mathbf{a} \neq \mathbf{0}$ and

$$\mathbf{a} \wedge \mathbf{b} = 2\mathbf{a} \wedge \mathbf{c}$$

Show that

$$\mathbf{b} - 2\mathbf{c} = \lambda\mathbf{a},$$

where λ is a scalar.
Given that

$$|\mathbf{a}| = |\mathbf{c}| = 1, \quad |\mathbf{b}| = 4$$

and the angle between **b** and **c** is $\arccos \frac{1}{4}$, show that

$$\lambda = +4 \quad \text{or} \quad -4$$

For each of these cases find the cosine of the angle between **a** and **c**.

(JMB)

7) Find the shortest distance between the straight lines with vector equations:
$\mathbf{r} = -3\mathbf{i} + 5s\mathbf{j} + \mathbf{k}$, $\quad \mathbf{r} = 3\mathbf{i} - 2\mathbf{j} + \mathbf{k} + t(-\mathbf{i} + 2\mathbf{j} - \mathbf{k})$. (U of L)

8) Prove that the lines with vector equations:
$\mathbf{r} = (-\mathbf{i} + 2\mathbf{j}) + a(-\mathbf{i} + 2\mathbf{j} - \mathbf{k})$,
$\mathbf{r} = \mathbf{k} + b(\mathbf{i} + \mathbf{j} - 2\mathbf{k})$,
$\mathbf{r} = \mathbf{i} + \mathbf{j} - \mathbf{k} + c(\mathbf{i} + \mathbf{k})$,
$\mathbf{r} = 2\mathbf{i} + \mathbf{j} + d(3\mathbf{i} - \mathbf{j})$,
in the given order, form a (skew) quadrilateral.
Prove that the ratio of the shortest distances between the two pairs of opposite sides of this quadrilateral is $\sqrt{7}:1$. (U of L)

9) (i) Show that $\mathbf{r} = 2\mathbf{i} + \mathbf{k}$ is a solution of the equation

$$\mathbf{r} \times (\mathbf{i} + \mathbf{j}) = -\mathbf{i} + \mathbf{j} + 2\mathbf{k}$$

Find the general solution of the equation and give a geometrical interpretation.

(ii) The position vector of a point P at time t is given by:

$$\mathbf{r} = (a \cos t)\mathbf{i} + (a \sin t)\mathbf{j} + (at)\mathbf{k}$$

Show that the point P remains at a constant distance from a fixed line. The position vectors of the points Q and R are $(a\mathbf{i} + a\mathbf{j})$ and $(a\mathbf{i} + a\mathbf{j} + a\mathbf{k})$ respectively. Find the times at which the area of the triangle PQR is a minimum. (U of L)

10) Let $\mathbf{i}, \mathbf{j}$ be orthogonal unit vectors in a plane and let

$$\mathbf{F} = X\mathbf{i} + Y\mathbf{j}$$

be a force acting at a point P in the plane whose position vector relative to the origin O is

$$\mathbf{r} = x\mathbf{i} + (-y\mathbf{j})$$

Prove that the force is equivalent to an equal force at O together with a couple. Three variable forces $\mathbf{F}_1, \mathbf{F}_2, \mathbf{F}_3$ act at points with position vectors $\mathbf{0}$, $\mathbf{i} + \mathbf{j}$, $-3\mathbf{i} + 2\mathbf{j}$ respectively and, at time t,

$$\mathbf{F}_1 = 2 \cos t\mathbf{i}, \quad \mathbf{F}_2 = \cos t\mathbf{i} + 2 \sin t\mathbf{j}, \quad \mathbf{F}_3 = 3 \sin t\mathbf{i} + \cos t\mathbf{j}$$

If the system is reduced to a single force $\mathbf{F}$ at O and a couple G, find the values of $\mathbf{F}$ and G. Deduce the equation of the line of action of the resultant and show that this line passes through a fixed point which is independent of t. (Oxford)

11) $\mathbf{F}$ and $k\mathbf{F}$ where k is a scalar, are two parallel forces with lines of action l_1, l_2 at a distance a apart.

(i) For each of the three cases $k < -1$, $-1 < k < 0$, $k > 0$ indicate clearly the position of the single force $(1 + k)\mathbf{F}$ which is equivalent to the given forces. Prove that the moment of this force about each point of the plane containing l_1 and l_2 is the same as the sum of the moments of $\mathbf{F}$ and $k\mathbf{F}$ about the point.

(ii) If $k = -1$, show that the sum of the moments of the two forces is the same about every point in the plane containing l_1 and l_2.

ABCD is a square with sides of length $2a$. A system S of coplanar forces consists of four forces of magnitude $8F$, $6F$, $8F$, $3F$ acting along AB, CB, CD,

AD respectively (the order of the letters indicating the sense), together with a couple of moment G and a force $\mathbf{P}$ with line of action l.

(a) If $G = 0$ and S is equivalent to a couple of moment aF, find the direction and magnitude of $\mathbf{P}$ and the possible positions of l.

(b) If l lies along BC and S is a null system (i.e. if the system S acted on a lamina in its plane, then the lamina would be in equilibrium), find G.

(WJEC)

12) Given that $\mathbf{i}, \mathbf{j}$ are perpendicular unit vectors and that $\mathbf{r}_1 = x_1\mathbf{i} + y_1\mathbf{j}$, $\mathbf{r}_2 = x_2\mathbf{i} + y_2\mathbf{j}$ are any two vectors, the scalar quantity $\mathbf{r}_1 \circ \mathbf{r}_2$ is defined for these two vectors by the equation

$$\mathbf{r}_1 \circ \mathbf{r}_2 = x_1y_2 - x_2y_1$$

Deduce that $\mathbf{r}_2 \circ \mathbf{r}_1 = -\mathbf{r}_1 \circ \mathbf{r}_2$.
If $\mathbf{r}_3 = x_3\mathbf{i} + y_3\mathbf{j}$ is a further vector, prove that

$$(\mathbf{r}_1 + \mathbf{r}_2) \circ \mathbf{r}_3 = (\mathbf{r}_1 \circ \mathbf{r}_3) + (\mathbf{r}_2 \circ \mathbf{r}_3)$$

If a force $\mathbf{F} = X\mathbf{i} + Y\mathbf{j}$ acts at a point P with position vector $\mathbf{r} = x\mathbf{i} + y\mathbf{j}$, prove that its moment about the origin O in an anticlockwise direction is $\mathbf{r} \circ \mathbf{F}$. Deduce that, if $\mathbf{r}' = x'\mathbf{i} + y'\mathbf{j}$ is the position vector of a further point P$'$, the moment of $\mathbf{F}$ about O is equal to the sum of the moments of the force $\mathbf{F}$ acting at P$'$ about O and the force $\mathbf{F}$ acting at P about P$'$. (Oxford)

13) The three forces $\mathbf{F}_1$, $\mathbf{F}_2$ and $\mathbf{F}_3$ are in equilibrium when acting at the points whose position vectors are $\mathbf{r}_1$, $\mathbf{r}_2$ and $\mathbf{r}_3$ respectively. If

$$\begin{aligned} \mathbf{F}_1 &= 5\mathbf{i} + 6\mathbf{j} & \mathbf{r}_1 &= c\mathbf{i} + \mathbf{j} \\ \mathbf{F}_2 &= a\mathbf{i} - 4\mathbf{j} & \mathbf{r}_2 &= 2\mathbf{i} - \mathbf{j} \\ \mathbf{F}_3 &= -6\mathbf{i} + b\mathbf{j} \quad \text{and} & \mathbf{r}_3 &= 3\mathbf{i} + 2\mathbf{j} \end{aligned}$$

calculate the values of the constants a, b and c and the position vector of the point of concurrence of the lines of action of the three forces.
The force $\mathbf{F}_3$ is now reversed, $\mathbf{F}_1$, $\mathbf{F}_2$, $\mathbf{r}_1$, $\mathbf{r}_2$ and $\mathbf{r}_3$ remain unchanged, and a clockwise couple of magnitude 21 units is introduced into the system. Calculate the magnitude and direction of the resultant of this system and find a vector equation of its line of action. (AEB)

14) ABCDEG is a regular hexagon of side a and O is its centre. Forces of magnitude 2, 3, 4, 5 units act along AB, BC, CD, DE respectively and a variable force $\mathbf{F}$ acts at G in the plane of the hexagon. Cartesian axes Ox, Oy are chosen parallel to AB and BD respectively. The resultant of this system of forces will be denoted by $\mathbf{R}$.

(a) If $\mathbf{F} = 0$, show that $|\mathbf{R}| = 7$ and find the Cartesian equation of the line of action of $\mathbf{R}$.

(b) Find the magnitude and direction of $\mathbf{F}$, indicating the direction clearly on a figure, if the line of action of $\mathbf{R}$ is to coincide with OD. Find also the magnitude of $\mathbf{R}$.

(c) If **F** is chosen to make the system of forces reduce to a couple, show that the moment of the couple is of magnitude $10.5\sqrt{3}a$. (WJEC)

15) Forces $-9\mathbf{i}+\mathbf{j}-2\mathbf{k}$, $3\mathbf{i}+2\mathbf{j}-3\mathbf{k}$ and $6\mathbf{i}-3\mathbf{j}+5\mathbf{k}$ act through points with position vectors $-11\mathbf{i}+2\mathbf{j}-5\mathbf{k}$, $\mathbf{i}-4\mathbf{j}+5\mathbf{k}$, and $-8\mathbf{i}+4\mathbf{j}-8\mathbf{k}$ respectively. Prove that these forces are equivalent to a couple, and find the moment of this couple. (U of L)

16) Show that each of the following systems of forces is in equilibrium:

(i) forces $3\overrightarrow{AB}$ and $4\overrightarrow{AC}$, where A, B and C have position vectors $(4\mathbf{i}-\mathbf{k})$, $(4\mathbf{j}+3\mathbf{k})$ and $(7\mathbf{i}-3\mathbf{j}-4\mathbf{k})$ respectively:

(ii) forces $\mathbf{F}_1 = 3\mathbf{i}+4\mathbf{j}+5\mathbf{k}$ acting at the point L $(7\mathbf{i}+9\mathbf{j}+11\mathbf{k})$,
$\mathbf{F}_2 = \mathbf{i}+\mathbf{j}+\mathbf{k}$ acting at the point M $(4\mathbf{i}+4\mathbf{j}+4\mathbf{k})$,
$\mathbf{F}_3 = -4\mathbf{i}-5\mathbf{j}-6\mathbf{k}$ acting at the point N $(5\mathbf{i}+6\mathbf{j}+7\mathbf{k})$.

Find the cosine of the angle between the lines of action of the forces $\mathbf{F}_1$ and $\mathbf{F}_2$. (AEB)

17) The vertices of a tetrahedron ABCD have position vectors **a**, **b**, **c**, **d** respectively, where:

$$\mathbf{a} = 3\mathbf{i}-4\mathbf{j}+\mathbf{k} \qquad \mathbf{b} = 4\mathbf{i}+4\mathbf{j}-2\mathbf{k}$$
$$\mathbf{c} = 4\mathbf{i}+\mathbf{k} \qquad \mathbf{d} = \mathbf{i}-2\mathbf{j}+\mathbf{k}$$

Forces of magnitude 30 and $3\sqrt{13}$ units act along CB and CD respectively. A third force acts at A. If the system reduces to a couple, find the magnitude of this couple and the force at A. Find also a unit vector along the axis of the couple. (U of L)

18) Two forces $\mathbf{F}_1 = 5\mathbf{i}+2\mathbf{j}+7\mathbf{k}$ and $\mathbf{F}_2 = 4\mathbf{i}+\mathbf{j}-\mathbf{k}$ act at points whose position vectors are $\mathbf{r}_1 = -3\mathbf{i}+3\mathbf{j}-2\mathbf{k}$ and $\mathbf{r}_2 = 10\mathbf{i}+7\mathbf{j}+3\mathbf{k}$ respectively. Show that the lines of action of these forces are coplanar.
Find the magnitude of their resultant and the vector and Cartesian equations of the line of action of this resultant. Write down the unit vectors which lie in the lines of action of $\mathbf{F}_1$ and $\mathbf{F}_2$ and hence find the cosine of the angle between these lines of action. (AEB)

CHAPTER 5

PLANES, COORDINATE GEOMETRY IN THREE DIMENSIONS

SCALAR PRODUCT FORM FOR THE VECTOR EQUATION OF A PLANE

A particular plane can be specified in several ways, for example:

(a) One and only one plane can be drawn through three non-collinear points, therefore three given points specify a particular plane.

(b) One and only one plane can be drawn to contain two concurrent lines, therefore two given concurrent lines specify a particular plane.

(c) One and only one plane can be drawn perpendicular to a given direction at a given distance from the origin, therefore the normal to a plane and the distance of the plane from the origin specify a particular plane.

(d) One and only one plane can be drawn through a given point and perpendicular to a given direction, therefore a point on the plane and a normal to the plane specify a particular plane.

There are many other ways of specifying a particular plane but the most useful are described in (c) and (d) as these lead to the same, very simple, form for the vector equation of a plane.

Consider the plane which is at a distance d from the origin and which is perpendicular to the unit vector $\hat{\mathbf{n}}$. ($\hat{\mathbf{n}}$ being directed away from O).

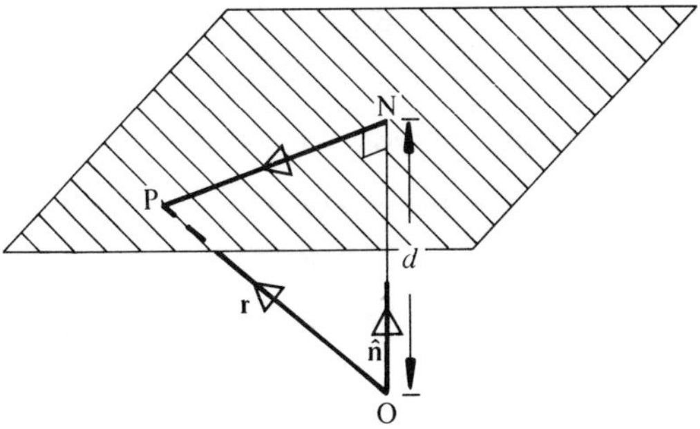

If ON is the perpendicular from the origin to the plane then $\overrightarrow{ON} = d\hat{\mathbf{n}}$.
If P is any point on the plane, NP is perpendicular to ON, and conversely if P is not on the plane, NP is not perpendicular to ON.
Therefore if and only if P is a point on the plane

$$\overrightarrow{NP} \cdot \overrightarrow{ON} = 0 \qquad (1)$$

This equation is called the scalar product form of the vector equation of the plane.
If $\mathbf{r}$ is the position vector of P, $\overrightarrow{NP} = \mathbf{r} - d\hat{\mathbf{n}}$

Therefore (1) $\Rightarrow$ $(\mathbf{r} - d\hat{\mathbf{n}}) \cdot d\hat{\mathbf{n}} = 0$

$$\mathbf{r} \cdot \hat{\mathbf{n}} - d\hat{\mathbf{n}} \cdot \hat{\mathbf{n}} = 0$$

$$\mathbf{r} \cdot \hat{\mathbf{n}} = d \quad \text{as } \hat{\mathbf{n}} \cdot \hat{\mathbf{n}} = 1$$

The equation $\mathbf{r} \cdot \hat{\mathbf{n}} = d$ is the standard form of the vector equation of the plane where:
$\mathbf{r}$ is the position vector of any point on the plane,
$\hat{\mathbf{n}}$ is the unit vector perpendicular to the plane,
d is the distance of the plane from the origin.
Consider the plane which contains the point A whose position vector is a and which is perpendicular to the unit vector $\hat{\mathbf{n}}$.

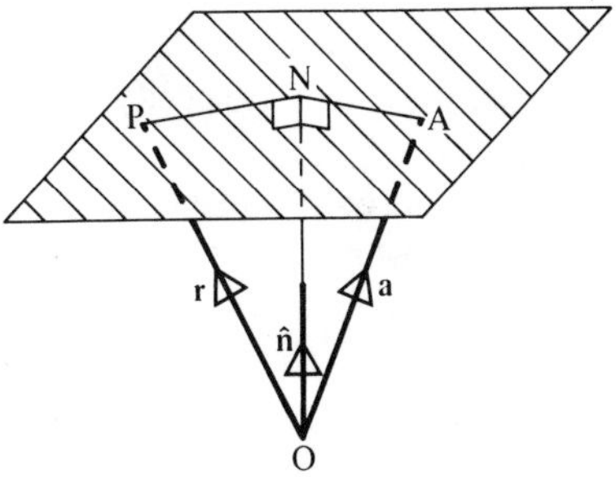

The distance of the plane from the origin is ON and ON = $\mathbf{a} \cdot \hat{\mathbf{n}}$
Therefore the equation of the plane is
$\mathbf{r} \cdot \hat{\mathbf{n}} = \mathbf{a} \cdot \hat{\mathbf{n}}$
where $\mathbf{r}$ is the position vector of any point P on the plane.

The standard form of the vector equation of a plane can be multiplied by any scalar quantity, thus any equation of the form $\mathbf{r}.\mathbf{n} = D$ represents a plane perpendicular to $\mathbf{n}$. Converting to standard form we have $\mathbf{r}.\hat{\mathbf{n}} = \frac{D}{|\mathbf{n}|}$ giving $\frac{D}{|\mathbf{n}|}$ as the distance of the plane from the origin.

THE CARTESIAN EQUATION OF A PLANE

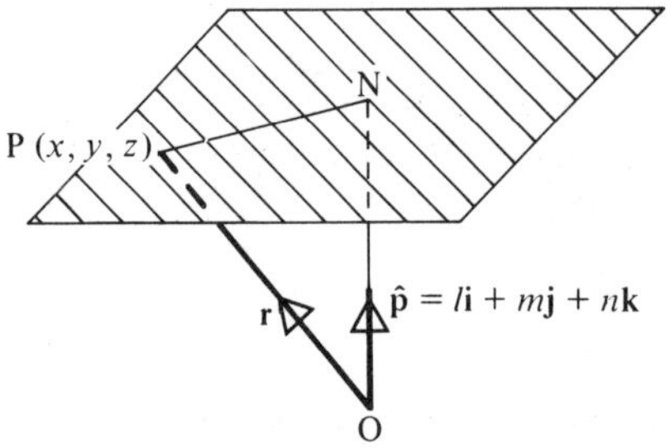

The standard form of the vector equation of the plane shown in the diagram is

$$\mathbf{r}.\hat{\mathbf{p}} = d \qquad (1)$$

If $P(x, y, z)$ is any point on the plane $\mathbf{r} = x\mathbf{i} + y\mathbf{j} + z\mathbf{k}$ and if l, m, n are the direction cosines of $\hat{\mathbf{p}}$ then $\hat{\mathbf{p}} = l\mathbf{i} + m\mathbf{j} + n\mathbf{k}$.

Substituting in (1) $(x\mathbf{i} + y\mathbf{j} + z\mathbf{k}).(l\mathbf{i} + m\mathbf{j} + n\mathbf{k}) = d$

$$\Rightarrow \quad lx + my + nz = d$$

which is the Cartesian equation of the plane.
Therefore if $\mathbf{r}.(A\mathbf{i} + B\mathbf{j} + C\mathbf{k}) = D$ is the vector equation of a plane
then $Ax + By + Cz = D$ is the Cartesian equation of the same plane
where A, B, C are the direction ratios of the normal to the plane. (So there is an easy transfer between these two forms for the equation of a plane.)

THE PARAMETRIC FORM FOR THE VECTOR EQUATION OF A PLANE

Consider the plane which is parallel to the vectors **d** and **e** (**d** not parallel to **e**) and which also contains the point A whose position vector is **a**.

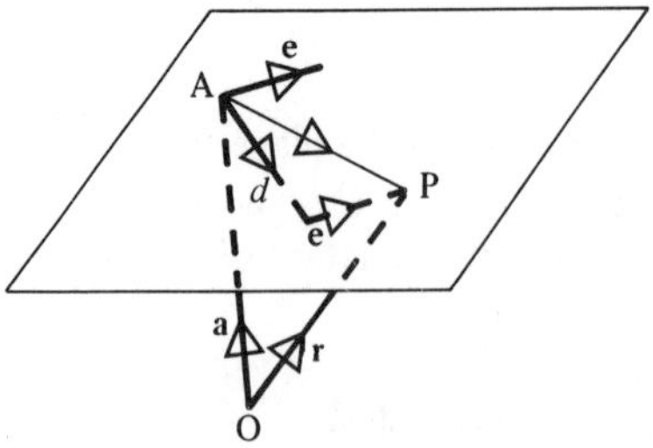

The vectors **d** and **e** determine the orientation of the plane and **a** fixes it in space, so these vectors specify a particular plane.
If P is any point on this plane $\overrightarrow{AP} = \lambda\mathbf{d} + \mu\mathbf{e}$ where λ and μ are independent parameters.

If **r** is the position vector of P

$$\mathbf{r} = \mathbf{a} + \overrightarrow{AP}$$
$$= \mathbf{a} + \lambda\mathbf{d} + \mu\mathbf{e}$$

thus any equation of the form $\mathbf{r} = \mathbf{a} + \lambda\mathbf{d} + \mu\mathbf{e}$, where λ and μ are independent parameters represents the plane parallel to the vectors **d** and **e** and containing the point **a**.
It should be noted that the parametric form is not a unique equation for a particular plane: **a** is any one of an infinite number of points on the plane, and **d** and **e** are only one pair of the infinite set of vectors parallel to the plane.
An interesting variation of the parametric form is found by considering the plane passing through three given points.

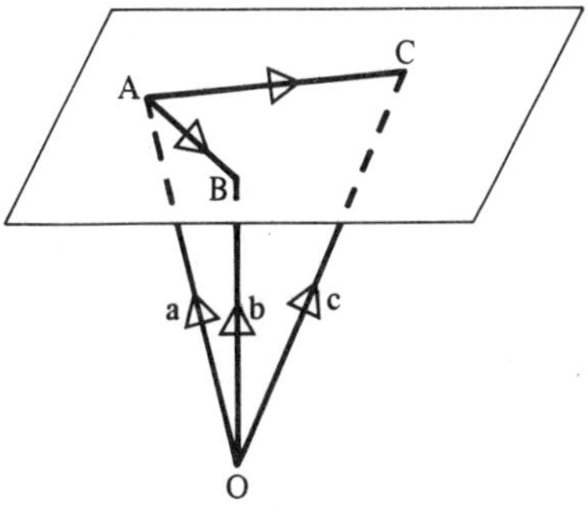

The vector equation of the plane in the diagram can be written as

$$\begin{aligned}\mathbf{r} &= \mathbf{a} + t\overrightarrow{AB} + u\overrightarrow{AC}\\ &= \mathbf{a} + t(\mathbf{b} - \mathbf{a}) + u(\mathbf{c} - \mathbf{a})\\ &= (1 - t - u)\mathbf{a} + t\mathbf{b} + u\mathbf{c}\\ &= s\mathbf{a} + t\mathbf{b} + u\mathbf{c} \quad \text{where } s = 1 - t - u\end{aligned}$$

Therefore any equation of the form $\mathbf{r} = s\mathbf{a} + t\mathbf{b} + u\mathbf{c}$, where $s + t + u = 1$, represents the plane passing through the points **a**, **b**, **c**.
(*Note*: In this form, s, t and u are *not* independent parameters.)

SUMMARY

The *scalar product form* of the vector equation of a plane is

$$\mathbf{r}.\mathbf{n} = D$$

where **n** is a vector perpendicular to the plane and $\dfrac{D}{|\mathbf{n}|}$ is the distance of the plane from the origin.
The *parametric form* of the vector equation of a plane is

$$\mathbf{r} = \mathbf{a} + \lambda\mathbf{b} + \mu\mathbf{c}$$

where **a** is a point on the plane and **b** and **c** are parallel to the plane.
The *Cartesian equation* of a plane is

$$Ax + By + Cz = D$$

where A, B, C are the direction ratios of **n**, the normal to the plane.

TRANSFORMATION OF THE EQUATION OF A PLANE FROM ONE FORM TO ANOTHER

Any one of these forms for the equation of a plane can be transformed into any other form.

Consider the plane whose equation, in scalar product form, is $\mathbf{r}.(2\mathbf{i}+\mathbf{j}-2\mathbf{k}) = 6$
The Cartesian equation of the plane is therefore: $2x+y-2z = 6$
The parametric equation of the plane can be written in terms of any three points on the plane. (Three such points can be found from the Cartesian equation by giving arbitrary values to any two coordinates and finding the related value of the third.)
Thus $(0, 0, -3)$, $(0, 6, 0)$, $(3, 0, 0)$ are points on the plane.
Therefore $\mathbf{r} = \lambda(-3\mathbf{k})+\mu(6\mathbf{j})+\eta(3\mathbf{i})$ where $\lambda+\mu+\eta = 1$ is the parametric equation of the plane.
Substituting $1-\lambda-\mu$ for η gives

$$\mathbf{r} = 3\mathbf{i}-\lambda(3\mathbf{i}+3\mathbf{k})+\mu(-3\mathbf{i}+6\mathbf{j})$$

or
$$\mathbf{r} = 3\mathbf{i}+s(\mathbf{i}+\mathbf{k})+t(-\mathbf{i}+2\mathbf{j})$$

which is the parametric equation of the plane in terms of two independent parameters.
Converting this last equation back into Cartesian form we have:
If $P(x, y, z)$ is any point on the plane

then
$$\left.\begin{aligned} x &= 3+s-t \\ y &= 2t \\ z &= s \end{aligned}\right\}$$
eliminating s and t gives the Cartesian equation
$$2x+y-2z = 6$$

and again this leads automatically to

$$\mathbf{r}.(2\mathbf{i}+\mathbf{j}-2\mathbf{k}) = 6$$

which is the scalar product form.

EXAMPLES 5a

1) Find the vector equation of the plane through the points A(0, 1, 1), B(2, 1, 0), C(−2, 0, 3), (a) in parametric form, (b) in scalar product form.

(a)

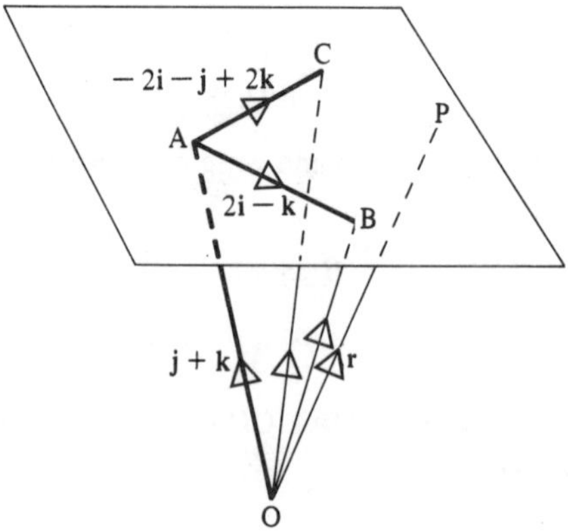

$\overrightarrow{AB}$ and $\overrightarrow{AC}$ are two vectors contained in the plane.
$\overrightarrow{AB} = \overrightarrow{OB}-\overrightarrow{OA} = 2\mathbf{i}-\mathbf{k}$

$\overrightarrow{AC} = \overrightarrow{OC} - \overrightarrow{OA} = -2\mathbf{i} - \mathbf{j} + 2\mathbf{k}$
$\overrightarrow{OA} = \mathbf{j} + \mathbf{k}$ is the position vector of a point on the plane.
Therefore a vector equation of this plane in the form $\mathbf{r} = \mathbf{a} + \lambda\mathbf{d} + \mu\mathbf{e}$

is $$\mathbf{r} = \mathbf{j} + \mathbf{k} + \lambda(2\mathbf{i} - \mathbf{k}) + \mu(-2\mathbf{i} - \mathbf{j} + 2\mathbf{k})$$

or $$\mathbf{r} = (2\lambda - 2\mu)\mathbf{i} + (1 - \mu)\mathbf{j} + (1 - \lambda + 2\mu)\mathbf{k} \qquad (1)$$

(b) If $P(x, y, z)$ is any point on this plane, then from equation (1) we have

$$\left.\begin{aligned} x &= 2\lambda - 2\mu \\ y &= 1 - \mu \\ z &= 1 - \lambda + 2\mu \end{aligned}\right\}$$ eliminating λ and μ gives $x + 2y + 2z = 4$ which is the Cartesian equation of the plane.

Therefore the scalar product form of the equation is

$$\mathbf{r}.(\mathbf{i} + 2\mathbf{j} + 2\mathbf{k}) = 4$$

Alternatively the scalar product form can be found directly from the co-ordinates of A, B and C.

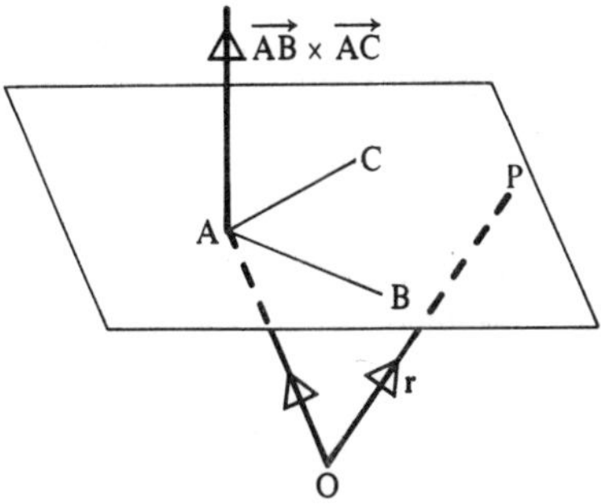

As $\overrightarrow{AB}$ and $\overrightarrow{AC}$ are parallel to the plane, $\overrightarrow{AB} \times \overrightarrow{AC}$ is perpendicular to the plane. Therefore the scalar product form can be written as:

$$\mathbf{r}.(\overrightarrow{AB} \times \overrightarrow{AC}) = \overrightarrow{OA}.(\overrightarrow{AB} \times \overrightarrow{AC})$$

$$\mathbf{r}.(2\mathbf{i} - \mathbf{k}) \times (-2\mathbf{i} - \mathbf{j} + 2\mathbf{k}) = (\mathbf{j} + \mathbf{k}).(2\mathbf{i} - \mathbf{k}) \times (-2\mathbf{i} - \mathbf{j} + 2\mathbf{k})$$

$$\mathbf{r}.(-\mathbf{i} - 2\mathbf{j} - 2\mathbf{k}) = (\mathbf{j} + \mathbf{k}).(-\mathbf{i} - 2\mathbf{j} - 2\mathbf{k})$$

$$\mathbf{r}.(-\mathbf{i} - 2\mathbf{j} - 2\mathbf{k}) = -4$$

or $$\mathbf{r}.(\mathbf{i} + 2\mathbf{j} + 2\mathbf{k}) = 4$$

2) Show that the plane whose vector equation is $\mathbf{r}.(\mathbf{i} + 2\mathbf{j} - \mathbf{k}) = 3$ contains the line whose vector equation is $\mathbf{r} = \mathbf{i} + \mathbf{j} + \lambda(2\mathbf{i} + \mathbf{j} + 4\mathbf{k})$.

The line is contained in the plane if any two points on the line are on the plane. Taking $\lambda = 0$ and $\lambda = 1$ we find that $\mathbf{i} + \mathbf{j}$ and $3\mathbf{i} + 2\mathbf{j} + 4\mathbf{k}$ are two points on the line.

If $\mathbf{r} = \mathbf{i} + \mathbf{j}$ then $\mathbf{r}.(\mathbf{i} + 2\mathbf{j} - \mathbf{k}) = (\mathbf{i} + \mathbf{j}).(\mathbf{i} + 2\mathbf{j} - \mathbf{k}) = 3$
Therefore $\mathbf{i} + \mathbf{j}$ is a point on the plane.
If $\mathbf{r} = 3\mathbf{i} + 2\mathbf{j} + 4\mathbf{k}$ then $\mathbf{r}.(\mathbf{i} + 2\mathbf{j} - \mathbf{k}) = (3\mathbf{i} + 2\mathbf{j} + 4\mathbf{k}).(\mathbf{i} + 2\mathbf{j} - \mathbf{k}) = 3$
Therefore $3\mathbf{i} + 2\mathbf{j} + 4\mathbf{k}$ is a point on the plane.
Therefore the line is contained in the plane.

3) Find the vector equation of the line passing through the point (3, 1, 2) and perpendicular to the plane $\mathbf{r}.(2\mathbf{i} - \mathbf{j} + \mathbf{k}) = 4$.

As $\mathbf{r}.(2\mathbf{i} - \mathbf{j} + \mathbf{k}) = 4$ is the equation of the plane
$2\mathbf{i} - \mathbf{j} + \mathbf{k}$ is perpendicular to the plane and therefore parallel to the required line.
As this line passes through the point $3\mathbf{i} + \mathbf{j} + 2\mathbf{k}$ its equation is:

$$\mathbf{r} = (3\mathbf{i} + \mathbf{j} + 2\mathbf{k}) + \lambda(2\mathbf{i} - \mathbf{j} + \mathbf{k})$$

EXERCISE 5a

1) Find a vector equation of the plane containing the points A, B and C in parametric form and in scalar product form where:
(a) A, B and C are the points (1, 2, − 1), (1, 3, 2), (0, 2, 1) respectively,
(b) the position vectors of A, B and C are $\mathbf{i} + \mathbf{j} - 2\mathbf{k}$, $\mathbf{i} + \mathbf{k}$, $-2\mathbf{i} + \mathbf{j} - 3\mathbf{k}$.

2) Find the vector equation of the following planes in scalar product form:
(a) $\mathbf{r} = \mathbf{i} - \mathbf{j} + \lambda(\mathbf{i} + \mathbf{j} + \mathbf{k}) + \mu(\mathbf{i} - 2\mathbf{j} + 3\mathbf{k})$,
(b) $\mathbf{r} = 2\mathbf{i} - \mathbf{j} + \mathbf{k} + \lambda(\mathbf{i}) + \mu(\mathbf{i} - 2\mathbf{j} - \mathbf{k})$,
(c) $\mathbf{r} = (1 + s - t)\mathbf{i} + (2 - s)\mathbf{j} + (3 - 2s + 2t)\mathbf{k}$.

3) Find the Cartesian equation of the following planes:
(a) $\mathbf{r}.(\mathbf{i} + \mathbf{j} - \mathbf{k}) = 2$,
(b) $\mathbf{r}.(2\mathbf{i} + 3\mathbf{j} - 4\mathbf{k}) = 1$,
(c) $\mathbf{r} = (s - 2t)\mathbf{i} + (3 - t)\mathbf{j} + (2s + t)\mathbf{k}$.

4) Find the vector equation in scalar product form of the plane that contains the lines $\mathbf{r} = (\mathbf{i} + \mathbf{j}) + s(\mathbf{i} + 2\mathbf{j} - \mathbf{k})$ and $\mathbf{r} = (\mathbf{i} + \mathbf{j}) + t(-\mathbf{i} + \mathbf{j} - 2\mathbf{k})$.

5) Find the vector equation in parametric form of the plane that contains the lines $\mathbf{r} = -3\mathbf{i} - 2\mathbf{j} + t(\mathbf{i} - 2\mathbf{j} + \mathbf{k})$, $\mathbf{r} = \mathbf{i} - 11\mathbf{j} + 4\mathbf{k} + s(2\mathbf{i} - \mathbf{j} + 2\mathbf{k})$.

6) Find the vector equation in parametric form of the plane that goes through the point with position vector $\mathbf{i} + \mathbf{j}$ and which is parallel to the lines $\mathbf{r}_1 = 2\mathbf{i} - \mathbf{j} + \lambda(\mathbf{i} + \mathbf{k})$ and $\mathbf{r}_2 = 2\mathbf{j} - \mathbf{k} + \mu(\mathbf{i} - \mathbf{j} + \mathbf{k})$. Is either of these lines contained in the plane?

7) A plane goes through the three points whose position vectors are **a**, **b** and **c**

where
$$\mathbf{a} = \mathbf{i} + \mathbf{j} + 2\mathbf{k}$$
$$\mathbf{b} = 2\mathbf{i} - \mathbf{j} + 3\mathbf{k}$$
$$\mathbf{c} = -\mathbf{i} + 2\mathbf{j} - 2\mathbf{k}.$$

Find the vector equation of this plane in scalar product form and hence find the distance of the plane from the origin.

8) A plane goes through the points whose position vectors are $\mathbf{i} - 2\mathbf{j} + \mathbf{k}$ and $2\mathbf{i} - \mathbf{j} - \mathbf{k}$ and is parallel to the line $\mathbf{r} = \mathbf{i} - \mathbf{j} + \lambda(3\mathbf{i} + \mathbf{j} - 2\mathbf{k})$. Find the distance of this plane from the origin.

9) Two planes Π_1 and Π_2 have vector equations $\mathbf{r}.(2\mathbf{i} + \mathbf{j} - 2\mathbf{k}) = 3$ and $\mathbf{r}.(2\mathbf{i} + \mathbf{j} - 2\mathbf{k}) = 9$. Explain why Π_1 and Π_2 are parallel and hence find the distance between them.

10) Find the vector equation of the line through the origin which is perpendicular to the plane $\mathbf{r}.(\mathbf{i} - 2\mathbf{j} + \mathbf{k}) = 3$.

11) Find the vector equation of the line through the point $(2, 1, 1)$ which is perpendicular to the plane $\mathbf{r}.(\mathbf{i} + 2\mathbf{j} - 3\mathbf{k}) = 6$.

12) Find the vector equation of the plane which goes through the point $(0, 1, 6)$ and is parallel to the plane $\mathbf{r}.(\mathbf{i} - 2\mathbf{j}) = 3$.

13) Find the vector equation of the plane which goes through the origin and which contains the line $\mathbf{r} = 2\mathbf{i} + \lambda(\mathbf{j} + \mathbf{k})$.

THE ANGLE BETWEEN TWO PLANES

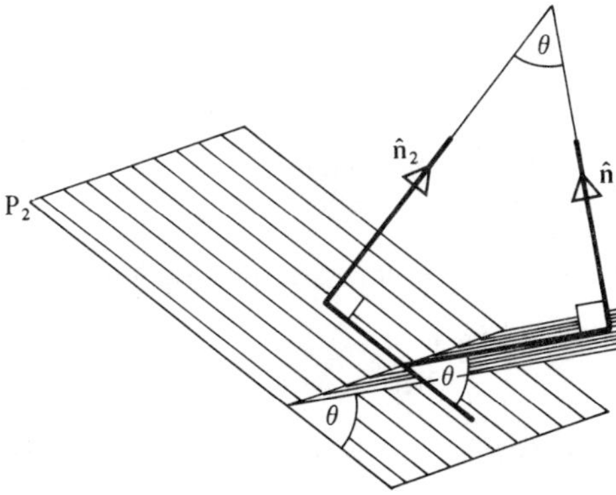

Consider two planes P_1 and P_2 whose vector equations are $\mathbf{r}.\hat{\mathbf{n}}_1 = d_1$ and $\mathbf{r}.\hat{\mathbf{n}}_2 = d_2$.
The angle between P_1 and P_2 is equal to the angle between the normals to P_1 and P_2, i.e. the angle between $\hat{\mathbf{n}}_1$ and $\hat{\mathbf{n}}_2$.
Therefore if θ is the angle between P_1 and P_2,

$$\cos\theta = \hat{\mathbf{n}}_1.\hat{\mathbf{n}}_2$$

Thus *two planes are perpendicular* if $\hat{\mathbf{n}}_1.\hat{\mathbf{n}}_2 = 0$

and *two planes are parallel if* $\qquad \hat{\mathbf{n}}_1 = \hat{\mathbf{n}}_2$.
Thus the angle between the planes whose vector equations are $\mathbf{r}\cdot(\mathrm{i}+\mathrm{j}-2\mathrm{k}) = 3$ and $\mathbf{r}\cdot(2\mathrm{i}-2\mathrm{j}+\mathrm{k}) = 2$ is given by:

$$\cos\theta = \frac{(\mathrm{i}+\mathrm{j}-2\mathrm{k})}{\sqrt{6}}\cdot\frac{(2\mathrm{i}-2\mathrm{j}+\mathrm{k})}{3}$$

$$= \frac{-2}{3\sqrt{6}} = -\frac{\sqrt{6}}{9}$$

This is the cosine of the obtuse angle between the planes.

The acute angle is $\arccos\left(\frac{\sqrt{6}}{9}\right)$.

THE ANGLE BETWEEN A LINE AND A PLANE

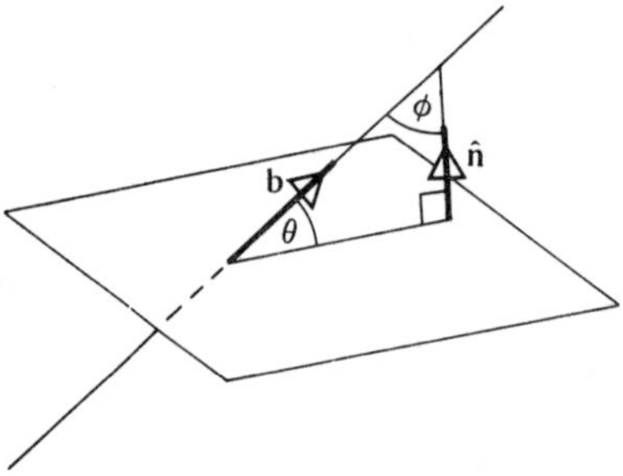

Consider the line $\mathbf{r} = \mathbf{a} + \lambda\mathbf{b}$ and the plane $\mathbf{r}\cdot\hat{\mathbf{n}} = d$
The angle ϕ between the line and the normal to the plane is given by

$$\cos\phi = \frac{\mathbf{b}\cdot\hat{\mathbf{n}}}{|\mathbf{b}|}$$

If θ is the angle between the line and the plane then $\theta = \frac{\pi}{2} - \phi$:

i.e. $\quad \sin\theta = \cos\phi$.
Therefore the angle between the line $\mathrm{r} = \mathbf{a} + \lambda\mathbf{b}$ and the plane $\mathbf{r}\cdot\hat{\mathbf{n}} = d$ is given by:

$$\sin\theta = \frac{\mathbf{b}\cdot\hat{\mathbf{n}}}{|\mathbf{b}|}$$

So the angle θ between the line $\mathbf{r} = (\mathrm{i}+2\mathrm{j}-\mathrm{k}) + \lambda(\mathrm{i}-\mathrm{j}+\mathrm{k})$ and the plane $\mathbf{r}\cdot(2\mathrm{i}-\mathrm{j}+\mathrm{k}) = 4$ is given by

$$\sin\theta = \frac{(\mathrm{i}-\mathrm{j}+\mathrm{k})}{\sqrt{3}}\cdot\frac{(2\mathrm{i}-\mathrm{j}+\mathrm{k})}{\sqrt{6}} = \frac{2\sqrt{2}}{3}$$

Therefore $\qquad \theta = \arcsin\frac{2\sqrt{2}}{3}$

THE DISTANCE OF A POINT FROM A PLANE

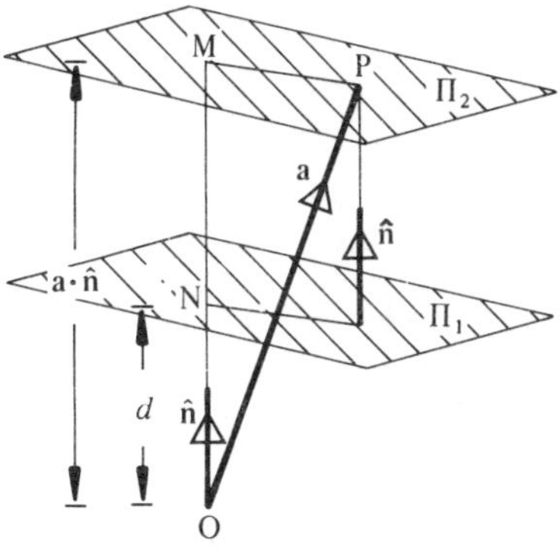

Consider a point P with position vector **a** and a plane Π_1 whose equation is $\mathbf{r}\cdot\hat{\mathbf{n}} = d$.

The equation of the plane Π_2 through P parallel to the plane Π_1 is $\mathbf{r}\cdot\hat{\mathbf{n}} = \mathbf{a}\cdot\hat{\mathbf{n}}$, i.e. the distance OM of this plane from the origin is $\mathbf{a}\cdot\hat{\mathbf{n}}$.

Therefore (assuming that P and O are on opposite sides of Π_1) the distance MN of P from the plane Π_1 is $\mathbf{a}\cdot\hat{\mathbf{n}} - d$.

(If P and O are on the same side of the plane the use of this formula will give a negative result.)

Thus the distance of the point with position vector $\mathbf{i} - 3\mathbf{j} + 3\mathbf{k}$ from the plane Π with equation $\mathbf{r}\cdot(2\mathbf{i} + 3\mathbf{j} - 6\mathbf{k}) = 9$ is given by:

$$(\mathbf{i} - 3\mathbf{j} + 3\mathbf{k})\cdot\frac{(2\mathbf{i} + 3\mathbf{j} - 6\mathbf{k})}{7} - \frac{9}{7} = -\frac{25}{7} - \frac{9}{7} = -\frac{34}{7}$$

The negative sign indicates that the point and the origin are on the same side of the plane.

THE INTERSECTION OF TWO PLANES

Unless two planes are parallel they will contain a common line which is the line of intersection of the two planes.

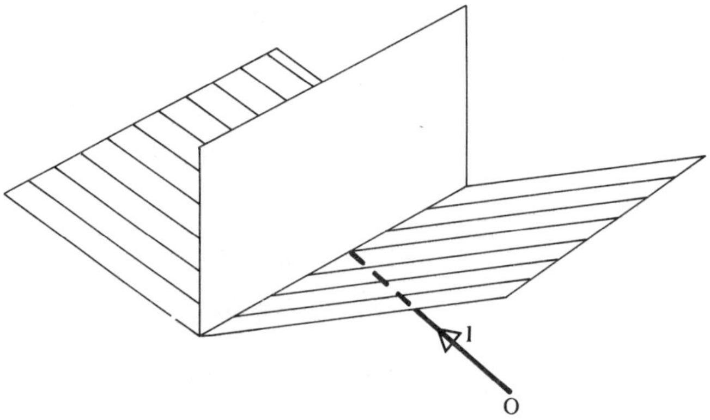

If Π_1 and Π_2 are planes with equations $\mathbf{r}.\hat{\mathbf{n}}_1 = d_1$ and $\mathbf{r}.\hat{\mathbf{n}}_2 = d_2$ respectively, the position vector of any point on the line of intersection must satisfy both equations.

If $\mathbf{l}$ is the position vector of a point on this line then $\mathbf{l}.\hat{\mathbf{n}}_1 = d_1$ and $\mathbf{l}.\hat{\mathbf{n}}_2 = d_2$

Therefore, for any value of k, $\mathbf{l}.\hat{\mathbf{n}}_1 - k\mathbf{l}.\hat{\mathbf{n}}_2 = d_1 - kd_2$

or $\mathbf{l}.(\hat{\mathbf{n}}_1 - k\hat{\mathbf{n}}_2) = d_1 - kd_2$

But the equation $\mathbf{r}.(\hat{\mathbf{n}}_1 - k\hat{\mathbf{n}}_2) = d_1 - kd_2$ represents a plane Π_3 which is such that if any vector $\mathbf{r}$ satisfies the equations of Π_1 and Π_2 it also satisfies the equation of Π_3.

i.e. *Any plane passing through the intersection of the planes* $\mathbf{r}.\hat{\mathbf{n}}_1 = d_1$ *and* $\mathbf{r}.\hat{\mathbf{n}}_2 = d_2$ *has an equation*

$$\mathbf{r}.(\hat{\mathbf{n}}_1 - k\hat{\mathbf{n}}_2) = d_1 - kd_2$$

Conversely, for all real values of k the equation $\mathbf{r}.(\hat{\mathbf{n}}_1 - k\hat{\mathbf{n}}_2) = d_1 - kd_2$ represents the family of planes passing through the line of intersection of the planes $\mathbf{r}.\hat{\mathbf{n}}_1 = d_1$ and $\mathbf{r}.\hat{\mathbf{n}}_2 = d_2$.

This is a particular case of a more general result, viz, if $E_1 = 0$ and $E_2 = 0$ are the equations of two members of a family of curves (or surfaces) then the equation

$$E_1 = kE_2$$

represents, for all real values of k, those members of that family that contain the point (or points) of intersection of E_1 and E_2.

The Line of Intersection of Two Planes

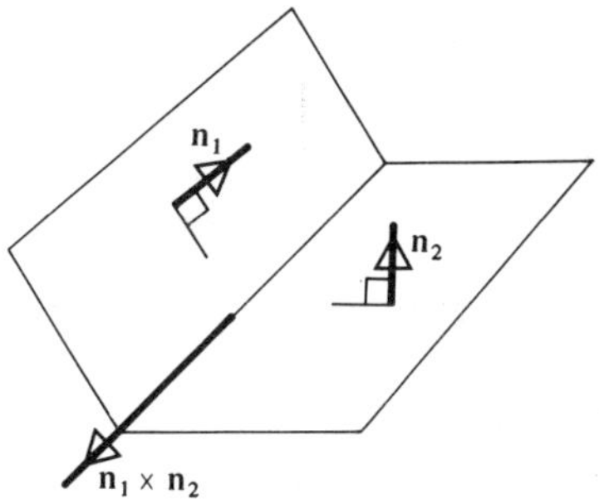

As the line of intersection of two planes

$$\mathbf{r}.\mathbf{n}_1 = D_1$$

$$\mathbf{r}.\mathbf{n}_2 = D_2$$

is contained in both planes it is perpendicular to both $\mathbf{n}_1$ and $\mathbf{n}_2$, i.e. parallel to $\mathbf{n}_1 \times \mathbf{n}_2$.

To find the equation of the line of intersection of two planes, consider the planes:

$$\left.\begin{aligned} \mathbf{r}.(\mathbf{i}+\mathbf{j}-3\mathbf{k}) &= 6 \\ \mathbf{r}.(2\mathbf{i}-\mathbf{j}+\mathbf{k}) &= 4 \end{aligned}\right\} \Rightarrow \begin{aligned} x+y-3z &= 6 \quad (1) \\ 2x-y+z &= 4 \quad (2) \end{aligned}$$

These planes meet where $\quad 3x-2z = 10 \qquad (1)+(2)$

and $\quad 7x-2y = 18 \qquad (1)+3(2)$

i.e. where
$$x = \frac{2z+10}{3} = \frac{2y+18}{7} = \lambda$$

These are the Cartesian equations of the line.
So any point on the line has co-ordinates:

$$x = \lambda, \quad y = \frac{7\lambda-18}{2}, \quad z = \frac{3\lambda-10}{2}$$

or
$$x = 2s, \quad y = 7s-9, \quad z = 3s-5 \quad (s = 2\lambda)$$

Hence the position vector of any point on the line is:

$$\mathbf{r} = -9\mathbf{j}-5\mathbf{k}+s(2\mathbf{i}+7\mathbf{j}+3\mathbf{k})$$

This is the vector equation of the line.
If the equations of the planes are in parametric form it is not necessary to convert these equations into Cartesian form. The line of intersection may be found as follows.
Consider the planes:

$$\mathbf{r} = \mathbf{i}+\mathbf{j}+\lambda(2\mathbf{i}-\mathbf{k})+\mu(\mathbf{i}-\mathbf{j}+\mathbf{k}) \qquad (1)$$

$$\mathbf{r} = 3\mathbf{i}-\mathbf{k}+s(\mathbf{i}-\mathbf{j}+2\mathbf{k})+t(\mathbf{i}+2\mathbf{j}-\mathbf{k}) \qquad (2)$$

rearranging:
$$\mathbf{r} = (1+2\lambda+\mu)\mathbf{i}+(1-\mu)\mathbf{j}+(-\lambda+\mu)\mathbf{k}$$

$$\mathbf{r} = (3+s+t)\mathbf{i}+(-s+2t)\mathbf{j}+(-1+2s-t)\mathbf{k}$$

These planes meet where
$$\begin{aligned} 1+2\lambda+\mu &= 3+s+t \\ 1-\mu &= -s+2t \\ -\lambda+\mu &= -1+2s-t \end{aligned}$$

eliminating λ and μ from these equations gives $3 = 2s+5t$

i.e. the planes meet at all points where $s = \dfrac{3-5t}{2}$

Therefore substituting $\dfrac{3-5t}{2}$ for s in equation (2) gives the vector equation of the line of intersection of the planes:

i.e. $$\mathbf{r} = \tfrac{1}{2}[(9\mathbf{i}-3\mathbf{j}+4\mathbf{k})+t(-3\mathbf{i}+9\mathbf{j}-12\mathbf{k})]$$

EXERCISE 5b

1) Find the cosine of the angle between the two planes whose equations are:
(a) $\mathbf{r}.(\mathbf{i}-\mathbf{j}+3\mathbf{k}) = 3$, $\mathbf{r}.(2\mathbf{i}-\mathbf{j}+2\mathbf{k}) = 5$.
(b) $\mathbf{r} = (\mathbf{i}+\mathbf{j})+\lambda(\mathbf{i}+\mathbf{j}-\mathbf{k})+\mu(2\mathbf{i}-\mathbf{j}+3\mathbf{k})$,
$\mathbf{r} = (\mathbf{i}-2\mathbf{j}+\mathbf{k})+s(2\mathbf{i}+\mathbf{k})+t(\mathbf{i}-2\mathbf{j}-\mathbf{k})$.
(c) $2x+2y-3z = 3$, $x+3y-4z = 6$.

2) Find the sine of the angle between the line and plane whose equations are:
(a) $\mathbf{r} = \mathbf{i}-\mathbf{j}+\lambda(\mathbf{i}+\mathbf{j}+\mathbf{k})$, $\mathbf{r}.(\mathbf{i}-2\mathbf{j}+2\mathbf{k}) = 4$.
(b) $\mathbf{r} = \mathbf{i}-2\mathbf{j}+\mathbf{k}+\lambda(2\mathbf{i}-\mathbf{j})$,
$\mathbf{r} = \mathbf{i}-\mathbf{j}+s(\mathbf{i}+\mathbf{k})+t(\mathbf{j}-\mathbf{k})$.
(c) $\dfrac{x-2}{2} = \dfrac{y+1}{6} = \dfrac{z+3}{3}$, $2x-y-2z = 4$.

3) Find the distance of the point (1, 3, 2) from the following planes:
(a) $\mathbf{r}.(7\mathbf{i}+4\mathbf{j}+4\mathbf{k}) = 9$.
(b) $6x+6y+3z = 8$.
(c) $\mathbf{r} = \mathbf{i}-\mathbf{j}+\mathbf{k}+\lambda(\mathbf{i})+\mu(\mathbf{j}-2\mathbf{k})$.

4) Find the vector equation of the line of intersection of the following pairs of planes.
(a) $\mathbf{r}.(\mathbf{i}-2\mathbf{j}+\mathbf{k}) = 3$, $\mathbf{r}.(3\mathbf{i}+\mathbf{j}-2\mathbf{k}) = 4$.
(b) $\mathbf{r} = (1-\lambda+\mu)\mathbf{i}+(\lambda-\mu)\mathbf{j}+(2-\mu)\mathbf{k}$,
$\mathbf{r} = (2-s)\mathbf{i}+(1-3t)\mathbf{j}+(2s-3t)\mathbf{k}$.

5) Prove that the line $\mathbf{r} = \mathbf{i}-2\mathbf{j}+\lambda(\mathbf{i}-3\mathbf{j}-\mathbf{k})$ is parallel to the intersection of the planes $\mathbf{r}.(\mathbf{i}+\mathbf{j}-2\mathbf{k}) = 2$ and $\mathbf{r}.(2\mathbf{i}+\mathbf{j}-\mathbf{k}) = 0$.

SUMMARY OF RESULTS FOR THREE DIMENSIONAL CARTESIAN GEOMETRY

Points and Distances

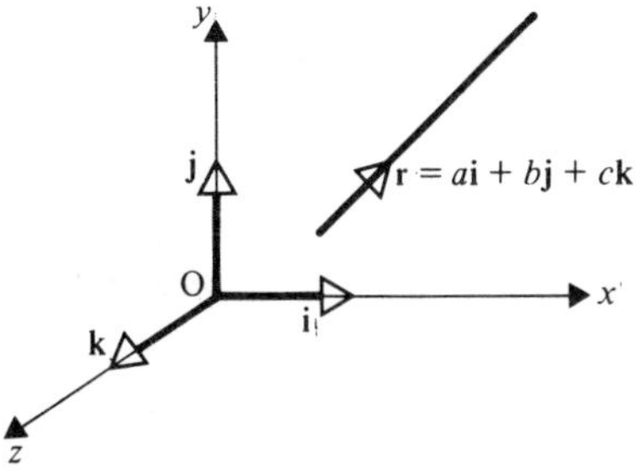

If $\mathbf{r} = a\mathbf{i} + b\mathbf{j} + c\mathbf{k}$ is a vector

$$|\mathbf{r}| = \sqrt{a^2 + b^2 + c^2}$$

If l, m, n are the *direction cosines* of $\mathbf{r}$:

$$l = \frac{a}{|\mathbf{r}|}, m = \frac{b}{|\mathbf{r}|}, n = \frac{c}{|\mathbf{r}|}$$

$$l^2 + m^2 + n^2 = 1$$

$$l:m:n = a:b:c$$

$$\hat{\mathbf{r}} = l\mathbf{i} + m\mathbf{j} + n\mathbf{k}$$

$$\mathbf{r} = r(l\mathbf{i} + m\mathbf{j} + n\mathbf{k})$$

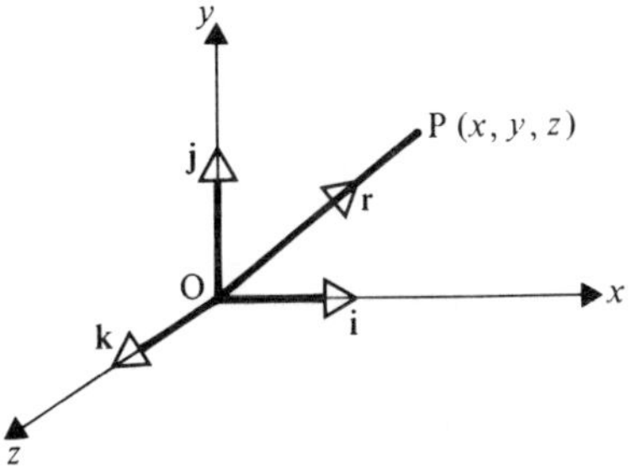

The *position vector* $\mathbf{r}$ of the point $P(x, y, z)$ is given by:

$$\mathbf{r} = x\mathbf{i} + y\mathbf{j} + z\mathbf{k} = \overrightarrow{OP}$$

If points $A(x_1, y_1, z_1)$, $B(x_2, y_2, z_2)$ have position vectors $\mathbf{a}$ and $\mathbf{b}$:

$$\begin{aligned}\overrightarrow{AB} &= \overrightarrow{OB} + \overrightarrow{AO} \\ &= \mathbf{b} - \mathbf{a} \\ &= (x_2 - x_1)\mathbf{i} + (y_2 - y_1)\mathbf{j} + (z_2 - z_1)\mathbf{k} \\ AB &= |\overrightarrow{AB}| \\ &= \sqrt{(x_2 - x_1)^2 + (y_2 - y_1)^2 + (z_2 - z_1)^2}\end{aligned}$$

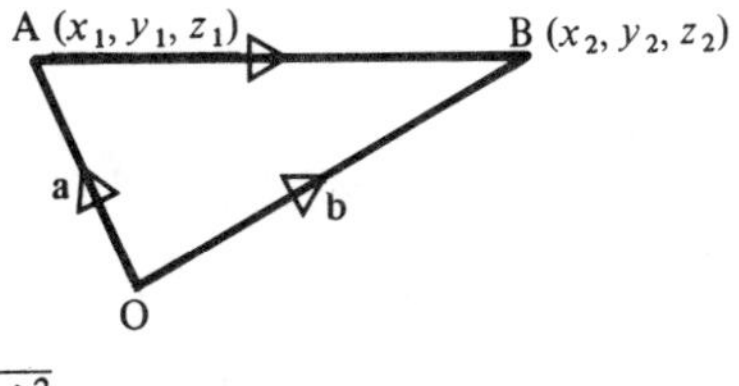

If P divides AB in the ratio $m:n$

$$\overrightarrow{OP} = \frac{n\mathbf{a} + m\mathbf{b}}{m + n}$$

The co-ordinates of P are:

$$\frac{nx_1 + mx_2}{m + n}, \frac{ny_1 + my_2}{m + n}, \frac{nz_1 + mz_2}{m + n}$$

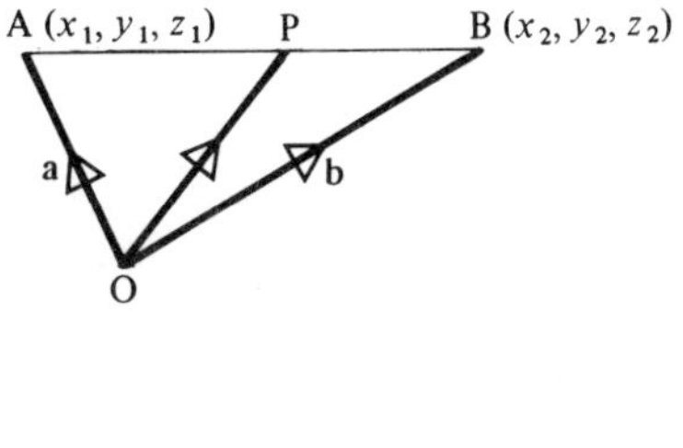

Products and Angles Between Vectors

If **a** and **b** are two vectors inclined to each other at an angle θ

where $\mathbf{a} = a_1\mathbf{i} + a_2\mathbf{j} + a_3\mathbf{k} = a(l\mathbf{i} + m\mathbf{j} + n\mathbf{k})$

and $\mathbf{b} = b_1\mathbf{i} + b_2\mathbf{j} + b_3\mathbf{j} = b(l'\mathbf{i} + m'\mathbf{j} + n'\mathbf{k})$

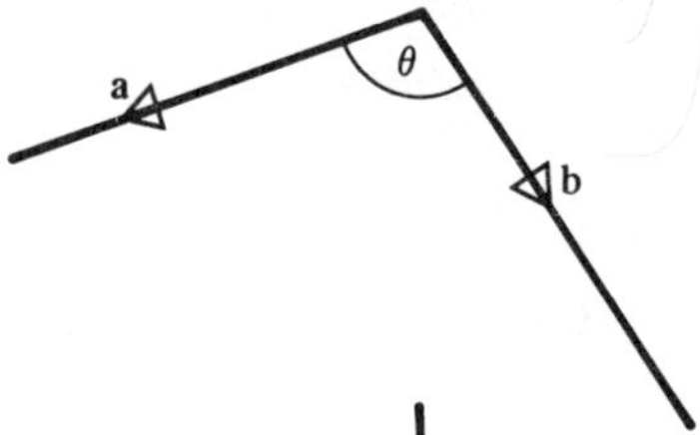

The *scalar product* of **a** and **b** is

$$\mathbf{a}\cdot\mathbf{b} = a_1b_1 + a_2b_2 + a_3b_3 = ab\cos\theta$$

The *angle* θ is given by

$$\cos\theta = \frac{\mathbf{a}\cdot\mathbf{b}}{ab} = \hat{\mathbf{a}}\cdot\hat{\mathbf{b}} = ll' + mm' + nn'$$

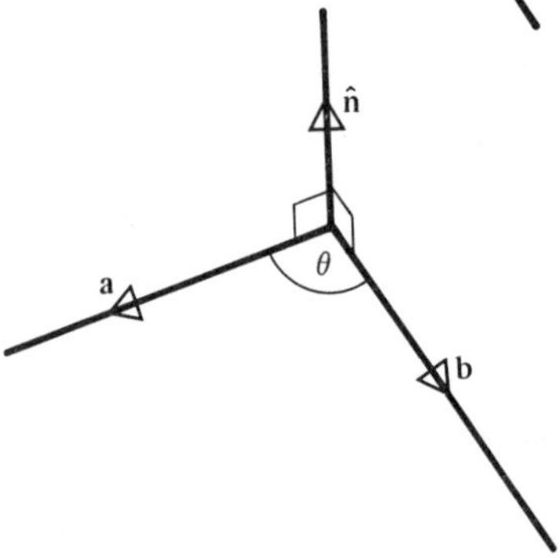

The *vector product* of **a** and **b** is

$$\mathbf{a}\times\mathbf{b} = \begin{vmatrix} i & j & k \\ a_1 & a_2 & a_3 \\ b_1 & b_2 & b_3 \end{vmatrix} = ab\sin\theta\,\hat{\mathbf{n}}$$

a and **b** are *perpendicular* if $\mathbf{a}\cdot\mathbf{b} = 0$
a and **b** are *parallel* if $\mathbf{a}\times\mathbf{b} = \mathbf{0}$

Straight Lines

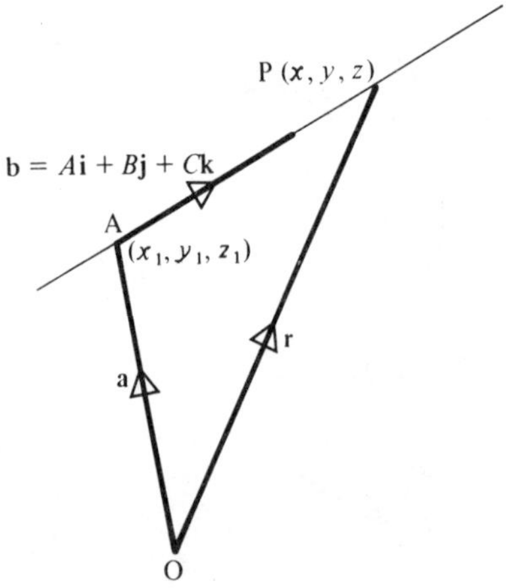

The *vector equation of a straight line* is

$$\mathbf{r} = \mathbf{a} + \lambda\mathbf{b} \qquad (1)$$

where **r** is the position vector of any point on the line,
a is the position vector of a point A on the line,
b is a vector parallel to the line.

If A is the point (x_1, y_1, z_1)
and $\mathbf{b} = A\mathbf{i} + B\mathbf{j} + C\mathbf{k}$

$$(1) \Rightarrow \mathbf{r} = x_1\mathbf{i} + y_1\mathbf{j} + z_1\mathbf{k} + \lambda(A\mathbf{i} + B\mathbf{j} + C\mathbf{k})$$

$$\Rightarrow \frac{x - x_1}{A} = \frac{y - y_1}{B} = \frac{z - z_1}{C} = \lambda$$

$A:B:C = l:m:n$ where l, m, n are the direction cosines of the line. A, B, C are the direction ratios of the line.

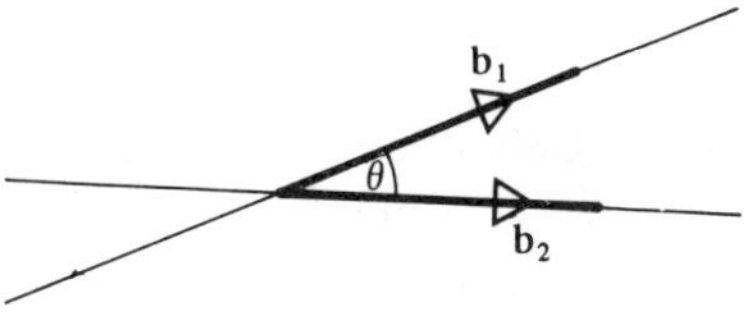

The *angle θ between two lines* with vector equations

$$\left.\begin{aligned}\mathbf{r} &= \mathbf{a}_1 + \lambda\mathbf{b}_1\\ \mathbf{r} &= \mathbf{a}_2 + \lambda\mathbf{b}_2\end{aligned}\right\}\quad \text{is given by}$$

$$\cos\theta = \frac{\mathbf{b}_1 \cdot \mathbf{b}_2}{b_1 b_2}$$

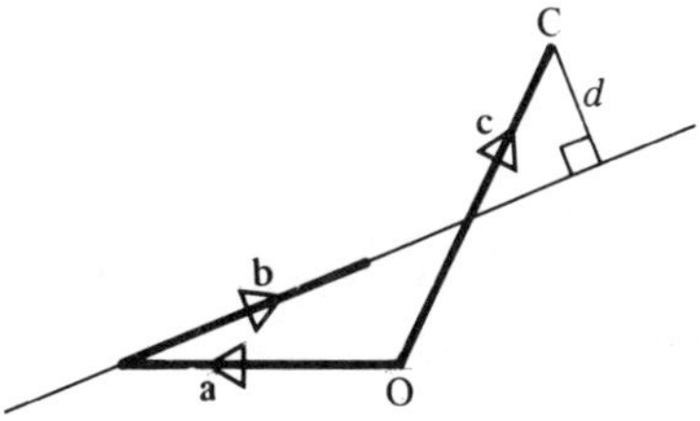

The *distance of the point* C with position vector **c** *from the line whose vector equation is* $\mathbf{r} = \mathbf{a} + \lambda\mathbf{b}$ is

$$\frac{1}{b}\,|(\mathbf{c} - \mathbf{a}) \times \mathbf{b}|$$

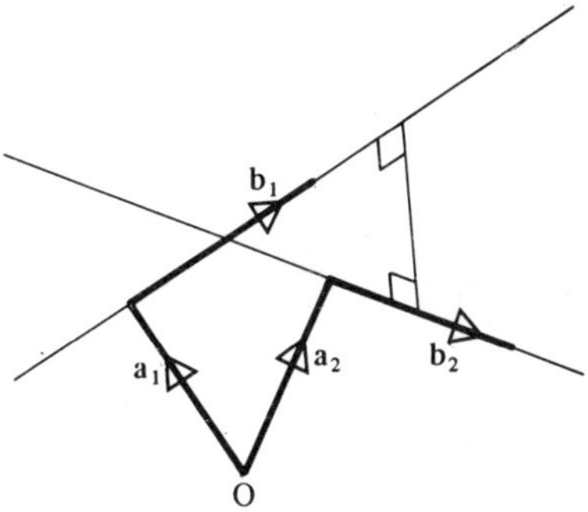

The *shortest distance between a pair of skew lines* whose vector equations are

$$\mathbf{r} = \mathbf{a}_1 + \lambda\mathbf{b}_1$$

$$\mathbf{r} = \mathbf{a}_2 + \lambda\mathbf{b}_2$$

is $$\left|\frac{(\mathbf{a}_1 - \mathbf{a}_2)\cdot(\mathbf{b}_1 \times \mathbf{b}_2)}{|\mathbf{b}_1 \times \mathbf{b}_2|}\right|$$

The *condition that the lines intersect* is

$$(\mathbf{a}_1 - \mathbf{a}_2)\cdot(\mathbf{b}_1 \times \mathbf{b}_2) = 0$$

Planes

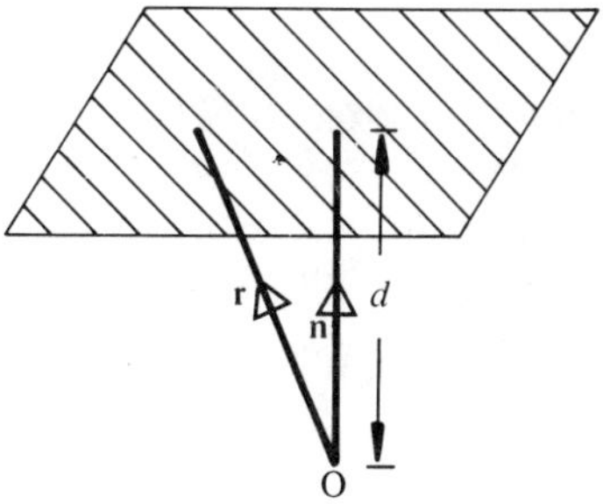

The *vector equation of the plane* can be written as

$$\mathbf{r}\cdot\mathbf{n} = D$$

$$\Rightarrow \mathbf{r}\cdot\hat{\mathbf{n}} = d \qquad (1)$$

where d is the distance of the plane from O and **n** is a vector perpendicular to the plane.

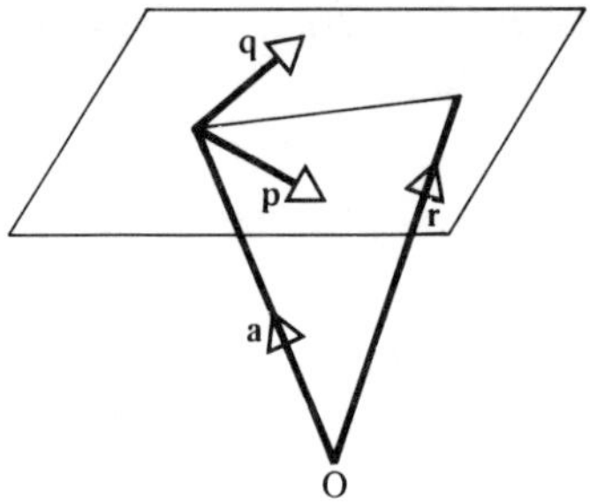

or $\mathbf{r} = \mathbf{a} + \lambda\mathbf{p} + \mu\mathbf{q}$ (2)

where **a** is the position vector of a point on the plane and **p** and **q** are parallel to the plane.

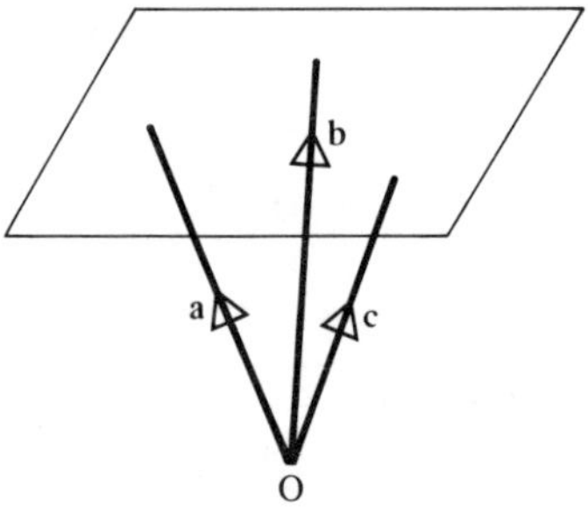

or $\mathbf{r} = (1 - \lambda - \mu)\mathbf{a} + \lambda\mathbf{b} + \mu\mathbf{c}$ (3)

where **a**, **b**, **c** are the position vectors of non-collinear points in the plane. The *Cartesian equation of the plane* is

$$Ax + By + Cz = D$$

where A, B, C are the direction ratios of the normal to the plane.

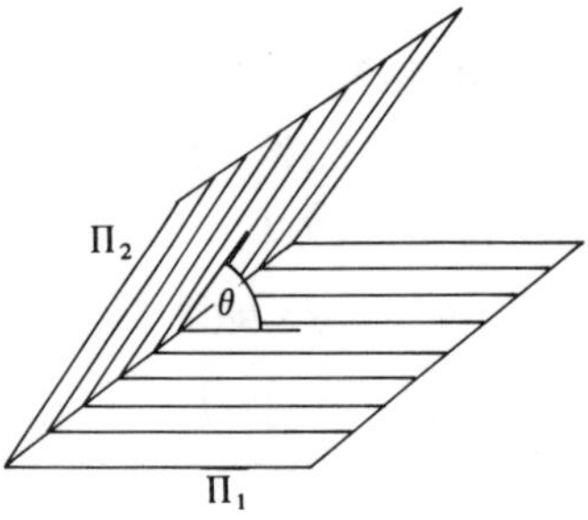

The *angle* θ *between two planes* Π_1 and Π_2 whose vector equations are:

$$\left.\begin{aligned} \mathbf{r}\cdot\hat{\mathbf{n}}_1 &= d_1 \\ \mathbf{r}\cdot\hat{\mathbf{n}}_2 &= d_2 \end{aligned}\right\} \quad \text{is given by}$$

$$\cos\theta = \hat{\mathbf{n}}_1\cdot\hat{\mathbf{n}}_2$$

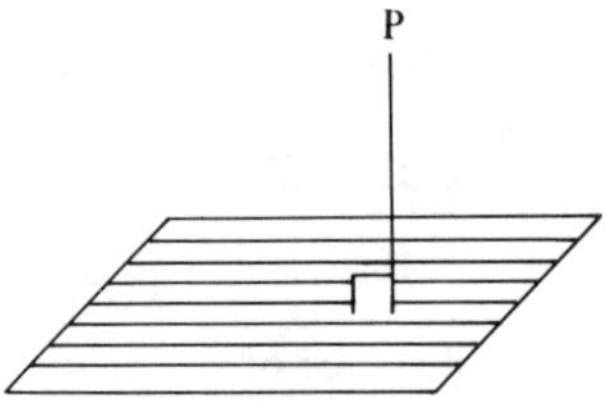

The *distance of the point* P *from the plane* $\mathbf{r}\cdot\hat{\mathbf{n}} = d$
is given by $\mathbf{a}\cdot\mathbf{n} - d$
where **a** is the position vector of P.

Areas and Volumes

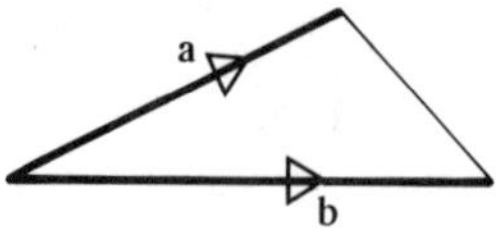

Area of a triangle $= \frac{1}{2}|\mathbf{a} \times \mathbf{b}|$ where **a** and **b** define two sides of the triangle.

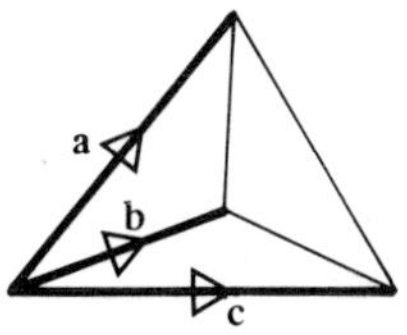

Volume of a tetrahedron $= \frac{1}{6}|(\mathbf{a} \times \mathbf{b} \cdot \mathbf{c})|$ where **a**, **b** and **c** define three concurrent edges.

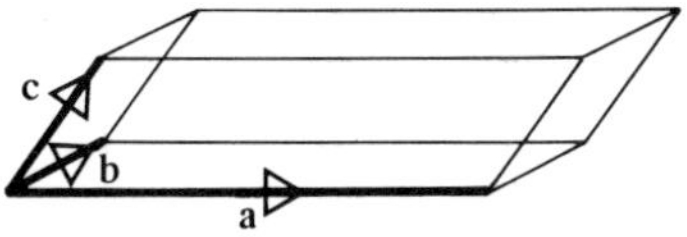

Volume of a parallelepiped $= |\mathbf{a} \times \mathbf{b} \cdot \mathbf{c}|$ where **a**, **b** and **c** define three concurrent edges.

The list of results given above contains several formulae which must be used with caution. The use of a formula, particularly in the case of finding an area or volume, is not always the simplest method of solving a geometric problem. Consideration should first be given to the particular information provided in that problem so that full use can be made of special properties.

EXAMPLES 5c

1) Show that the points P(3, 0, 1), Q(2, 1, −2) lie on opposite sides of the plane Π whose equation is $\mathbf{r}.(2\mathbf{i}-\mathbf{j}+\mathbf{k}) = 3$. Find the co-ordinates of the point of intersection of the plane Π and the line PQ.

Rewriting the equation of Π in the form $\mathbf{r}.\hat{\mathbf{n}} = d$ gives

$$\mathbf{r}\cdot\frac{1}{\sqrt{6}}(2\mathbf{i}-\mathbf{j}+\mathbf{k}) = \frac{3}{\sqrt{6}}$$

The distance of P from Π is

$$\mathbf{OP}.\hat{\mathbf{n}}-d = (3\mathbf{i}+\mathbf{k})\cdot\frac{1}{\sqrt{6}}(2\mathbf{i}-\mathbf{j}+\mathbf{k})-\frac{3}{\sqrt{6}} = \frac{4}{\sqrt{6}}$$

i.e. P and O are on opposite sides of Π.
The distance of Q from Π is

$$\mathbf{OQ}.\hat{\mathbf{n}}-d = (2\mathbf{i}+\mathbf{j}-2\mathbf{k})\cdot\frac{1}{\sqrt{6}}(2\mathbf{i}-\mathbf{j}+\mathbf{k})-\frac{3}{\sqrt{6}} = -\frac{2}{\sqrt{6}}$$

i.e. Q and O are on the same side of Π.
Therefore P and Q are on opposite sides of Π.

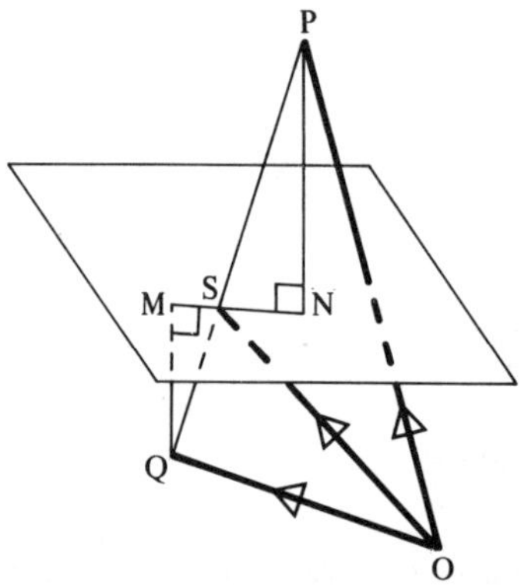

If S is the point of intersection of PQ and Π, PS : SQ = PN : MQ = 2 : 1
Therefore the position vector of S is given by $\frac{1}{3}(\mathbf{OP} + 2\mathbf{OQ})$
Therefore the co-ordinates of S are $(\frac{7}{3}, \frac{2}{3}, -1)$
(Alternatively **OS** can be found by solving simultaneously the equation of Π and the equation of PQ, but this method is longer.)

2) A right circular cone has its vertex at the point (2, 1, 3) and the centre of its plane face at the point (1, − 1, 2). A generator of the cone has equation $\mathbf{r} = (2\mathbf{i} + \mathbf{j} + 3\mathbf{k}) + \lambda(\mathbf{i} - \mathbf{j} - \mathbf{k})$. Find the radius of the base of the cone and hence its volume.

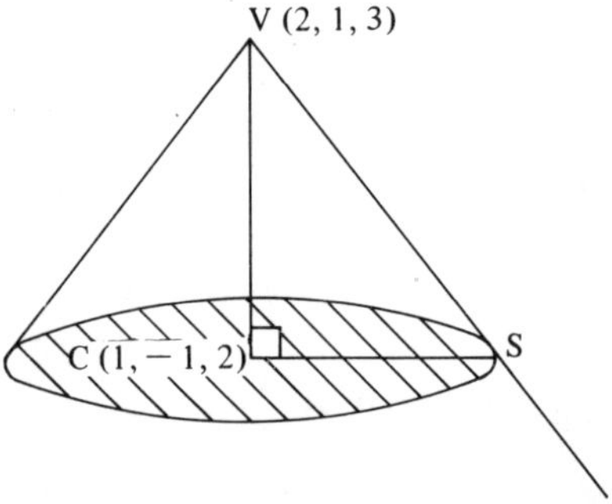

To find the radius of the base we need to find S, the point where the generator VS meets the plane Π containing the base.

Equation of Π:

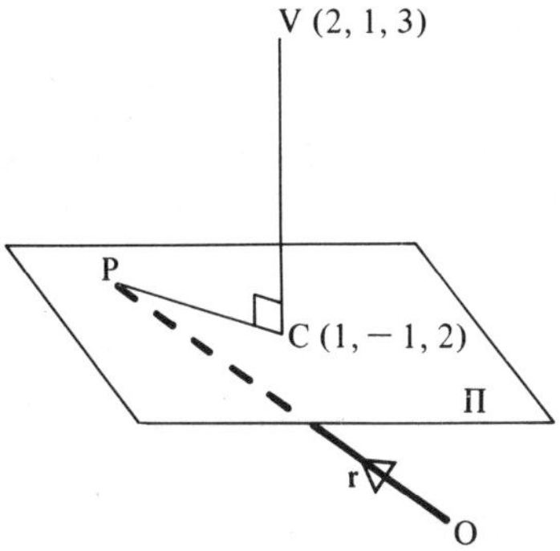

If **r** is the position vector of any point P in the plane Π then $\overrightarrow{PC} \cdot \overrightarrow{VC} = 0$

i.e. $[\mathbf{r} - (\mathbf{i} - \mathbf{j} + 2\mathbf{k})] \cdot (\mathbf{i} + 2\mathbf{j} + \mathbf{k}) = 0$

$$\mathbf{r} \cdot (\mathbf{i} + 2\mathbf{j} + \mathbf{k}) = 1$$

Any point on the given generator has co-ordinates $[(2 + \lambda), (1 - \lambda), (3 - \lambda)]$. The co-ordinates of S also satisfy the equation of Π.

Therefore $\quad [(2 + \lambda)\mathbf{i} + (1 - \lambda)\mathbf{j} + (3 - \lambda)\mathbf{k}] \cdot (\mathbf{i} + 2\mathbf{j} + \mathbf{k}) = 1$

$$\lambda = 3$$

Therefore the co-ordinates of S are $(5, -2, 0)$.
The radius of the base, $CS = \sqrt{[(1 - 5)^2 + (-1 + 2)^2 + (2 - 0)^2]} = \sqrt{21}$.
The volume of the cone is $\frac{1}{3}\pi r^2 h = \frac{1}{3}\pi(21)\sqrt{(1^2 + 2^2 + 1^2)}$

$$= 7\pi\sqrt{6}.$$

3) A tetrahedron has one vertex at O and the other vertices at the points A(2, 0, 0), B(0, 3, 0), C(0, 0, 1). Find the volume of the tetrahedron. This tetrahedron is divided into two parts by a plane Π which contains the line OB and which is inclined at 60° to the face OAB. Find the vector equation of Π in scalar product form and the ratio in which it divides the volume of the tetrahedron OABC.

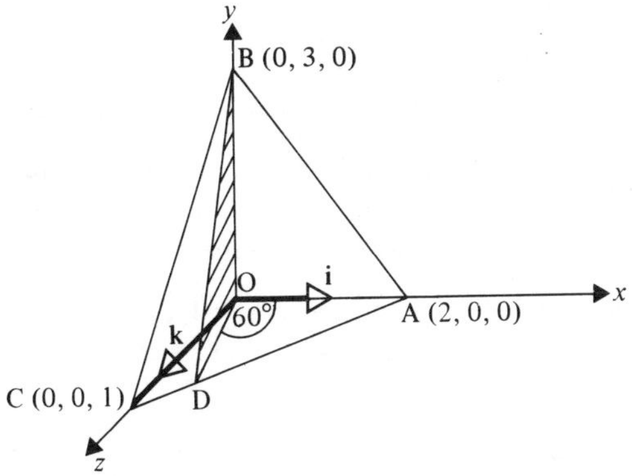

$$\text{Volume of OABC} = (\tfrac{1}{3}\text{ area }\Delta\text{OAC})(\text{OB})$$
$$= \tfrac{1}{3}(1)(3) = 1$$

The plane Π contains the line OB, (which is parallel to **j**)
the line OD, (which is parallel to $\cos 60°\mathbf{i} + \cos 30°\mathbf{k}$ or $\mathbf{i} + \sqrt{3}\mathbf{k}$)
and the origin.

Therefore the equation of Π in parametric form is

$$\mathbf{r} = \lambda\mathbf{j} + \mu(\mathbf{i} + \sqrt{3}\mathbf{k})$$

If (x, y, z) is any point on Π:

$$\left.\begin{aligned} x &= \mu \\ y &= \lambda \\ z &= \sqrt{3}\mu \end{aligned}\right\} \Rightarrow \sqrt{3}x - z = 0$$

Therefore the equation of Π in scalar product form is

$$\mathbf{r}\cdot(\sqrt{3}\mathbf{i} - \mathbf{k}) = 0$$

If D is the point of intersection of Π and AC, OBCD and OABD are tetrahedrons with a common base OBD: therefore their volumes are in the ratio of the distances of C and A from the plane containing OBD (Π).

The distance of C from Π is $\mathbf{k}\cdot\tfrac{1}{2}(\sqrt{3}\mathbf{i} - \mathbf{k}) - 0 = -\tfrac{1}{2}$.

The distance of A from Π is $2\mathbf{i}\cdot\tfrac{1}{2}(\sqrt{3}\mathbf{i} - \mathbf{k}) - 0 = +\sqrt{3}$.

Therefore Π divides the volume of OABC in the ratio $\sqrt{3} : \tfrac{1}{2}$
$= 2\sqrt{3} : 1$

EXERCISE 5c

1) A tetrahedron has one vertex at O and the other vertices at the points A(1, 3, 2), B(1, − 1, 0), C(2, 3, 1). Find the distance of O from the face ABC.

2) Show that the points P(3, 2, − 2), Q(1, 2, 1) are on opposite sides of the plane $\mathbf{r}\cdot(\mathbf{i} - \mathbf{j} - \mathbf{k}) = 2$. Find the position vector of the point of intersection of the line PQ with the plane.

3) A tetrahedron has vertices at the points A(2, − 1, 0), B(3, 0, 1), C(1, − 1, 2), D(− 1, 3, 0). Find the cosine of the angle between the faces ABC and ABD.

4) OABC is one face of a cube, where A and C are the points (1, 4, − 1), (3, 0, 3) respectively. Find the co-ordinates of B. Find also the vector equation of the plane containing the other face of the cube of which AB is an edge.

5) A tetrahedron is bounded by the planes $\mathbf{r}\cdot\mathbf{i} = 0$, $\mathbf{r}\cdot\mathbf{j} = 0$, $\mathbf{r}\cdot\mathbf{k} = 0$, $\mathbf{r}\cdot(2\mathbf{i} - \mathbf{j} + \mathbf{k}) = 4$. Find the co-ordinates of the vertices and the volume of this tetrahedron.

6) A right circular cylinder has its plane faces contained in the planes $\mathbf{r}\cdot(2\mathbf{i} - \mathbf{j} + 2\mathbf{k}) = 10$, $\mathbf{r}\cdot(-2\mathbf{i} + \mathbf{j} - 2\mathbf{k}) = 6$. Find the height of the cylinder.

The lines $\mathbf{r} = (2\mathbf{i} + \mathbf{j}) + \lambda(2\mathbf{i} - \mathbf{j} + 2\mathbf{k})$, $\mathbf{r} = (\mathbf{i} - \mathbf{j} + \mathbf{k}) + \mu(2\mathbf{i} - \mathbf{j} + 2\mathbf{k})$ are generators of the curved surface of the cylinder, passing through opposite ends of a diameter of its plane face. Find the radius of the cylinder and hence its volume.

7) Find the radius of the circle in which the plane $\mathbf{r} \cdot (2\mathbf{i} + \mathbf{j} - 2\mathbf{k}) = 9$ cuts the sphere of radius 5 and centre the origin. Find the volume of the cone of which this circle is the base and whose vertex is the origin.

8) Show that the point C(2, 0, 1) lies in the plane Π whose equation is $\mathbf{r} \cdot (\mathbf{i} - 2\mathbf{j} + 2\mathbf{k}) = 4$.
A right circular cone has its plane face lying in the plane Π and its centre is at C. Find the vector equation of the axis of the cone.
The line $\mathbf{r} = 4\mathbf{i} - 3\mathbf{j} + 5\mathbf{k} + \lambda(\mathbf{i} + \mathbf{j} + 2\mathbf{k})$ is a generator of the cone. Find the co-ordinates of the vertex of the cone and the point where this generator meets the plane Π. Hence find the volume of the cone.

9) A tetrahedron has three of its vertices at the points A(3, 2, 0), B(1, 3, − 1), C(0, 2, 0). Find the unit vector perpendicular to the face ABC. The fourth vertex D is such that $\overrightarrow{DA} \cdot \overrightarrow{AB} = \overrightarrow{DA} \cdot \overrightarrow{AC} = 0$. Find the vector equation of AD. If the volume of the tetrahedron is $3\sqrt{2}$ cubic units and if D is on the same side of the face ABC as the origin, find the co-ordinates of D.

10) Show that the lines

$$\mathbf{r} = (\mathbf{i} + \mathbf{j}) + \lambda(3\mathbf{i} - \mathbf{j} + 5\mathbf{k})$$
$$\mathbf{r} = (3\mathbf{i} + 2\mathbf{j} + 5\mathbf{k}) + \mu(\mathbf{i} - 2\mathbf{j})$$
$$\mathbf{r} = (2\mathbf{i} - \mathbf{j}) + \eta(2\mathbf{i} + \mathbf{j} + 5\mathbf{k})$$

are coplanar and find in parametric form the vector equation of the plane containing them.

11) A right prism has a triangular cross-section. Two of its rectangular faces are contained in the planes $\mathbf{r} \cdot (\mathbf{i} - 2\mathbf{j}) = 0$

$$\mathbf{r} \cdot (3\mathbf{i} - \mathbf{j} + \mathbf{k}) = 4$$

The two edges of the prism which are parallel to the intersection of these two planes pass through the origin and the point (1, 2, − 1) respectively. Find the vector equations of these edges.
Find also the equation of the plane which contains the cross-section of this prism, one of whose vertices is the origin. Find the area of this cross-section.

THE USE OF VECTORS IN GENERAL GEOMETRIC PROBLEMS

Many of the theorems of Euclidean geometry can be proved easily and quickly using vector methods, as illustrated in the following examples.

EXAMPLES 5d

1) Prove that the diagonals of a parallelogram bisect each other.

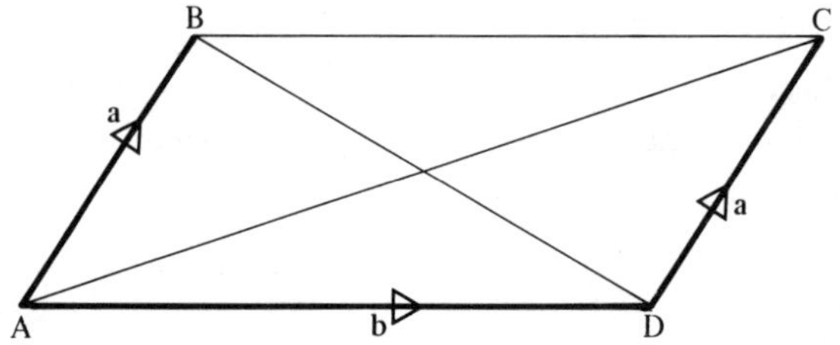

Taking one vertex A as origin and the position vectors of B and D as **a** and **b** respectively:

$$\overrightarrow{AC} = \overrightarrow{AD} + \overrightarrow{DC} = \mathbf{b} + \mathbf{a}$$

The position vector of the mid-point of BD $= \frac{1}{2}(\mathbf{a} + \mathbf{b}) = \frac{1}{2}\overrightarrow{AC}$.
Therefore the diagonals bisect each other.

2) Prove that in any triangle ABC,

$$\frac{\sin A}{a} = \frac{\sin B}{b} = \frac{\sin C}{c}$$

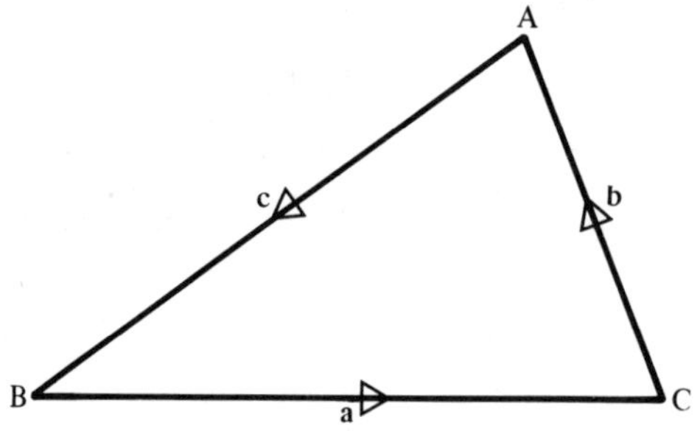

With the notation in the diagram:

$$\begin{aligned} 2\Delta ABC &= |\mathbf{b} \times \mathbf{c}| = bc \sin (180^\circ - A) = bc \sin A \\ &= |\mathbf{a} \times \mathbf{c}| = ac \sin (180^\circ - B) = ac \sin B \\ &= |\mathbf{a} \times \mathbf{b}| = ab \sin (180^\circ - C) = ab \sin C \end{aligned}$$

Therefore $\quad bc \sin A = ac \sin B = ab \sin C$

giving $\quad \dfrac{\sin A}{a} = \dfrac{\sin B}{b} = \dfrac{\sin C}{c}$.

3) Prove that the altitudes of any triangle are concurrent.

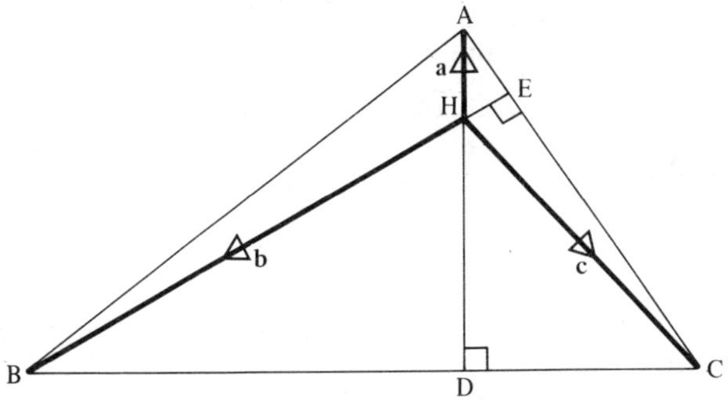

Let the altitudes AD and BE intersect at H.
Taking H as origin, let the position vectors of A, B, C be **a**, **b**, **c** respectively.

Now $\quad \mathbf{HA}\,.\,\mathbf{BC} = 0 \quad$ i.e. $\mathbf{a}\,.\,(\mathbf{c}-\mathbf{b}) = 0 \qquad (1)$

and $\quad \mathbf{HB}\,.\,\mathbf{AC} = 0 \quad$ i.e. $\mathbf{b}\,.\,(\mathbf{c}-\mathbf{a}) = 0 \qquad (2)$

$$(\mathbf{a}\,.\,\mathbf{c}-\mathbf{a}\,.\,\mathbf{b})-(\mathbf{b}\,.\,\mathbf{c}-\mathbf{b}\,.\,\mathbf{a}) = 0 \qquad [(1)-(2)]$$

$$\mathbf{a}\,.\,\mathbf{c}-\mathbf{b}\,.\,\mathbf{c} = 0$$

$$\mathbf{c}\,.\,(\mathbf{a}-\mathbf{b}) = 0$$

Therefore **c** is perpendicular to **a** − **b**, or **HC** is perpendicular to **BA** and so HC is the third altitude.
Therefore the three altitudes are concurrent.

4) Prove that in a skew quadrilateral the joins of the mid-points of opposite edges bisect each other.

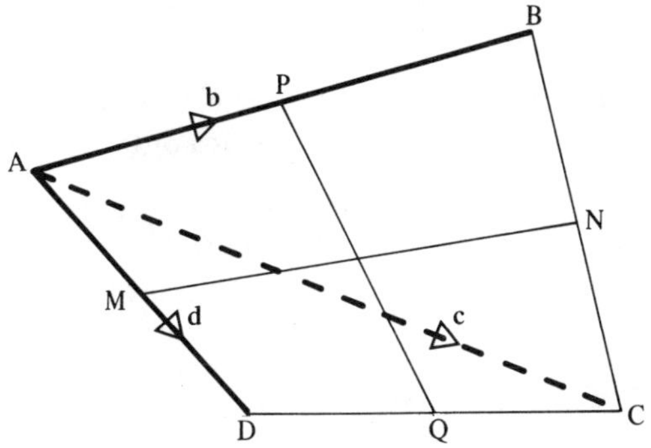

Taking A as origin, the position vectors of B, C and D as **b**, **c**, **d**, and P, N, Q, M as the mid-points of AB, BC, CD, DA respectively we have:

$$\overrightarrow{BC} = \mathbf{c}-\mathbf{b}, \qquad \overrightarrow{DC} = \mathbf{c}-\mathbf{d}$$

Therefore $\overrightarrow{AM} = \frac{1}{2}\mathbf{d}$ and $\overrightarrow{AN} = \mathbf{b}+\frac{1}{2}(\mathbf{c}-\mathbf{b})$
Therefore the position vector of the mid-point of MN is

$$\tfrac{1}{2}(\overrightarrow{AM}+\overrightarrow{AN}) = \tfrac{1}{4}\mathbf{d}+\tfrac{1}{4}\mathbf{b}+\tfrac{1}{4}\mathbf{c}$$

Also $\overrightarrow{AP} = \frac{1}{2}\mathbf{b}$, $\overrightarrow{AQ} = \mathbf{d}+\frac{1}{2}(\mathbf{c}-\mathbf{d})$
Therefore the position vector of the mid-point of PQ is

$$\tfrac{1}{2}(\overrightarrow{AP}+\overrightarrow{AQ}) = \tfrac{1}{4}\mathbf{d}+\tfrac{1}{4}\mathbf{b}+\tfrac{1}{4}\mathbf{c}$$

As the mid-points of PQ and MN have the same position vector, these lines bisect each other.

MISCELLANEOUS EXERCISE 5

In numbers 1–8 give proofs based on vector methods

1) Prove that the line joining the mid-points of two sides of a triangle is parallel to the third side and equal to half of it.

2) Prove that the internal bisectors of the angles of a triangle are concurrent.

3) Prove that the joins of the mid-points of the opposite edges of a tetrahedron bisect each other.

4) Prove that the perpendicular bisectors of the sides of a triangle are concurrent.

5) Prove that the diagonals of a rhombus intersect at right angles.

6) Prove that the lines joining the mid-points of adjacent sides of a skew quadrilateral form a parallelogram.

7) ABCD is a parallelogram and M is the mid-point of AB. Prove that DM and AC cut each other at points of trisection.

8) ABCD is a trapezium and AB and DC are parallel. M is the mid-point of AD. Prove that the area of triangle BMC is half the area of the trapezium.

9) If the position vectors of points P and Q with respect to O as origin are $\mathbf{p}$ and $\mathbf{q}$ respectively, show that the area of the triangle OPQ is $\frac{1}{2}|\mathbf{p}\times\mathbf{q}|$.
The position vectors of the vertices A, B, C of a tetrahedron OABC with respect to O as origin are

$$\overrightarrow{OA} = 2\mathbf{i}-\mathbf{j}, \quad \overrightarrow{OB} = \mathbf{j}+\mathbf{k}, \quad \overrightarrow{OC} = \mathbf{i}+3\mathbf{j}-\mathbf{k}$$

Find the angle between (a) the edges AB, AC, (b) the faces OAB, OAC.
Prove that BC is perpendicular to the plane OAB, and hence prove that the volume of OABC is 3/2. (U of L)

10) Show that the equation of a plane can be expressed in the form $\mathbf{r}.\mathbf{n} = p$. Find the equation of the plane through the origin parallel to the lines $\mathbf{r} = 3\mathbf{i} + 3\mathbf{j} - \mathbf{k} + s(\mathbf{i} - \mathbf{j} - 2\mathbf{k})$ and $\mathbf{r} = 4\mathbf{i} - 5\mathbf{j} - 8\mathbf{k} + t(3\mathbf{i} + 7\mathbf{j} - 6\mathbf{k})$. Show that one of the lines lies in the plane, and find the distance of the other line from the plane. (U of L)

11) Points A, B, C have position vectors **a**, **b**, **c** and λ, μ, ν are variable parameters subject to the condition $\lambda + \mu + \nu = 1$. If the points are not collinear prove that the plane ABC is represented by the equation $\mathbf{r} = \lambda\mathbf{a} + \mu\mathbf{b} + \nu\mathbf{c}$. Prove that the equation of the line of intersection of the two planes:

$$\mathbf{r} = \lambda_1\mathbf{i} + 2\mu_1\mathbf{j} + 3\nu_1\mathbf{k}, \quad \lambda_1 + \mu_1 + \nu_1 = 1$$

and

$$\mathbf{r} = 2\lambda_2\mathbf{i} + \mu_2\mathbf{j} + 2\nu_2\mathbf{k}, \quad \lambda_2 + \mu_2 + \nu_2 = 1$$

can be written in terms of a single parameter t as

$$6\mathbf{r} = (3 + t)\mathbf{i} + 4t\mathbf{j} + 9(1 - t)\mathbf{k}$$

(U of L)

12) A right circular cone has its vertex at the point $(4, -5, 3)$ and the centre of its base at the point $(0, 1, -1)$.
Write down:
(a) equations of the axis of the cone,
(b) the equation of the plane containing the base of the cone.

If the line $\dfrac{x-4}{3} = \dfrac{y+5}{-8} = \dfrac{z-3}{2}$ is a generator of the cone, find the co-ordinates of the point where this generator meets the base, and deduce that the volume of the cone is $6\pi\sqrt{17}$. (U of L)

13) Find the volume of the right triangular prism with rectangular faces in the planes $x = 0$, $y = 0$ and $3x + 2y = 6$, and with triangular faces in the planes $z = 0$ and $z = 4$.
This prism is cut into two portions by the plane $3x + 4y - 6z = 0$. Prove that the ratio of the volumes of the two portions is $1:3$. (U of L)

14) Show that the points $P(5, 5, 3)$ and $Q(-1, 2, -3)$ are on opposite sides of the plane $2x - 3y + 6z = 0$ and find the co-ordinates of the point in which PQ meets the plane. Find the equation of the plane which contains the line PQ and which is perpendicular to the given plane. (U of L)

15) The position vectors of the points A, B, C with respect to the origin O are **a**, **b**, **c** respectively. If OA is perpendicular to BC, and OB is perpendicular to CA, show that OC is perpendicular to AB, and that

$$OA^2 + BC^2 = OB^2 + CA^2 = OC^2 + AB^2.$$

Show that the plane through BC perpendicular to OA meets the plane through AB perpendicular to OC in a line that lies in the plane through OB perpendicular to CA. If this line passes through the centroid of the triangle AOC, show that the angle AOC is $\pi/3$ radians. (U of L)

16) Show that if A, B have position vectors **a**, **b** respectively, and C is the point on AB such that AC:CB $= \alpha:\beta$ $(\alpha+\beta \neq 0)$, where account is taken of sign, then the position vector of C is $(\beta\mathbf{a}+\alpha\mathbf{b})/(\alpha+\beta)$.
ABCD are four points not necessarily in the same plane. P, Q, R, S are points on AB, BC, CD, DA respectively such that AP:PB = BQ:QC = CR:RD = DS:SA = 2:1, and T, U are the mid-points of AC, BD respectively. Show by a *vector method* that the mid-points of PR and of QS both lie on TU, and divide this line into three equal parts. Show further that if V, W, X, Y are the mid-points of PQ, QR, RS, SP respectively, then VX, WY intersect and bisect each other at the mid-point of TU. (WJEC)

17) (i) If AB, BC, CA are the sides of a triangle, show by using the properties of the scalar product that

$$AB^2 = BC^2 + CA^2 - 2CB.CA \cos A\hat{C}B.$$

(ii) If **a**, **b** are unit vectors in the directions OA, OB respectively; **u** is a unit vector in a direction OP such that OP bisects the angle AOB, and **v** is a unit vector in a direction OQ such that O, A, B, Q are coplanar, OQ is at right angles to OA and angle QOB is acute.
(a) Show, by considering $\mathbf{a}.\mathbf{u}$, that $\cos(2\theta) = \cos^2\theta - \sin^2\theta$.
(b) Show, by considering $\mathbf{v}.\mathbf{u}$, that $\sin(2\theta) = 2\sin\theta\cos\theta$.
(WJEC)

18) The vertices A, B, C of a triangle have position vectors **a**, **b**, **c** respectively. Use a vector method to show that the lines joining the vertices to the mid-points of the opposite sides of the triangle are concurrent in the point G (the centroid of the triangle) with position vector $\frac{1}{3}(\mathbf{a}+\mathbf{b}+\mathbf{c})$.
P, Q, R are points on BC, CA, AB respectively such that BP:PC = CQ:QA = AR:RB = 1:2. L, M, N are the mid-points of AP, BQ, CR respectively. Find the position vectors of L, M, N and use a vector method to show that the centroid of LMN coincides with the centroid of ABC. (WJEC)

19) The points A, B have position vectors **a**, **b** respectively with respect to an origin O, and $|\mathbf{a}| = 3$, $|\mathbf{b}| = 2$. By considering the cosines of the angles between the direction of the vector $\mathbf{c} = 2\mathbf{a}+3\mathbf{b}$ and OA, OB, show that the direction of **c** bisects the angle between OA, OB. Hence find the position vector of the point P in which the bisector of the angle AOB meets AB.
If the angle between OA and OB is arc cos $(-\frac{1}{4})$, find the number x if the bisector of the angle OBA is in the direction of the vector $\mathbf{a}+x\mathbf{b}$. (WJEC)

20) (a) The points A, B, C, D (which are not necessarily coplanar) have position vectors **a**, **b**, **c**, **d** respectively, with respect to an origin O, and the mid-points of AB, BC, CD, and DA are P, Q, R and S respectively. Write down the position vectors of P and Q with respect to O and prove that PR and QS bisect each other.
Given that PR is perpendicular to QS prove that

$$|\mathbf{a}|^2 + |\mathbf{c}|^2 - 2\mathbf{a}.\mathbf{c} = |\mathbf{b}|^2 + |\mathbf{d}|^2 - 2\mathbf{b}.\mathbf{d}$$

and interpret this result geometrically.

(b) Three forces of magnitudes kAB, kBC and kCA act along the perpendicular bisectors of the sides AB, BC, CA respectively, of a triangle, ABC and are directed towards the interior of the triangle. Prove that the vector sum of these forces is zero. State and prove the corresponding result for a plane quadrilateral. (JMB)

21) The position vectors of A, B with respect to an origin O are **a**, **b** respectively, where $|\mathbf{a}| = 1$, $|\mathbf{b}| = 3$ and the angle AOB is $\frac{1}{3}\pi$. Show that $\frac{1}{9}\sqrt{3}(3\mathbf{a} - 2\mathbf{b})$ is a unit vector perpendicular to OA.
Find (i) a unit vector perpendicular to AB,
(ii) the position vector of the orthocentre of the triangle OAB (the point of concurrence of the perpendiculars drawn from the vertices to the opposite sides). (WJEC)

22) In a triangle ABC the perpendicular from B to the side AC meets the perpendicular from C to the side AB at H. The position vectors of A, B and C relative to H are **a**, **b** and **c** respectively.
Express $\overrightarrow{CA}$ in terms of **a** and **c** and deduce that $\mathbf{a}.\mathbf{b} = \mathbf{b}.\mathbf{c}$.
Prove that AH is perpendicular to BC. (JMB – part question)

23) In a parallelogram ABCD, X is the mid-point of AB and the line DX cuts the diagonal AC at P. Writing $\overrightarrow{AB} = \mathbf{a}$, $\overrightarrow{AD} = \mathbf{b}$, $\overrightarrow{AP} = \lambda\overrightarrow{AC}$ and $\overrightarrow{DP} = \mu\overrightarrow{DX}$, express AP (i) in terms of λ, **a** and **b**, (ii) in terms of μ, **a** and **b**. Deduce that P is a point of trisection of both AC and DX. (JMB – part question)

CHAPTER 6

INDIRECT IMPACT AND FURTHER WORK ON PROJECTILES

OBLIQUE IMPACT

When two objects collide a pair of equal and opposite impulses act at the moment of impact. If, just before impact, the objects were *not* moving along the line of action of these impulses, the impact is *indirect* or *oblique*.

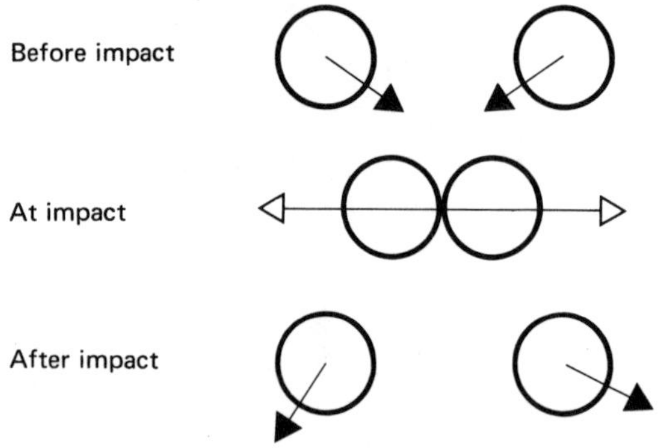

If the impact is elastic (i.e. the objects separate after collision) experimental evidence indicates that *Newton's Law of Restitution* is valid along the line of action of the impulses.

i.e. $$\frac{\text{separation speed along the line of action of impulse}}{\text{approach speed along the line of action of impulse}} = e$$

where e, the coefficient of restitution, is constant for a particular pair of objects irrespective of the magnitude or direction of their velocities before impact.
If the objects do not separate after impact, $e = 0$, and the collision is said to be *inelastic.*
If the relative speed of the objects is not changed by the impact, $e = 1$, and the collision is *perfectly elastic.*
In general $0 \leqslant e \leqslant 1$.
For objects involved in a collision, the *Principle of Conservation of Linear Momentum* applies in certain directions and the following considerations should be borne in mind when using it.

1) When an object is struck, the impulse it receives has a zero component perpendicular to the direction of the blow, therefore *the momentum of an object remains unchanged perpendicular to the impulse it receives.* But in any other direction there is a non-zero impulse component and the *increase in momentum in that direction is equal to the impulse component in the same direction.*

2) When objects that are free to move, collide, the impulses that act occur in equal and opposite pairs so the *total momentum in any direction remains unchanged.* When two smooth spheres collide the impulses act along the line joining their centres.

A Suggested Approach to Problem Solving

(a) Deal with each collision separately.
(b) The solution is usually simplified if the diagrams show velocity components parallel and perpendicular to the impulses.
(c) Draw two diagrams for each collision, the first showing velocity components just before impact, the second showing velocity components just after impact.
(d) Remember that the momentum of *each* object is unchanged in the direction perpendicular to the impulse it receives.

EXAMPLES 6a

1) A smooth sphere is free to move on a horizontal surface. It is projected towards a vertical wall with speed v at an angle of 30° to the wall.
If the coefficient of restitution between the sphere and the wall is $\frac{1}{3}$, find the velocity of the sphere after it hits the wall.

There is no impulse parallel to the wall so the velocity component in this direction is not changed by the impact. This property is incorporated in the diagrams.

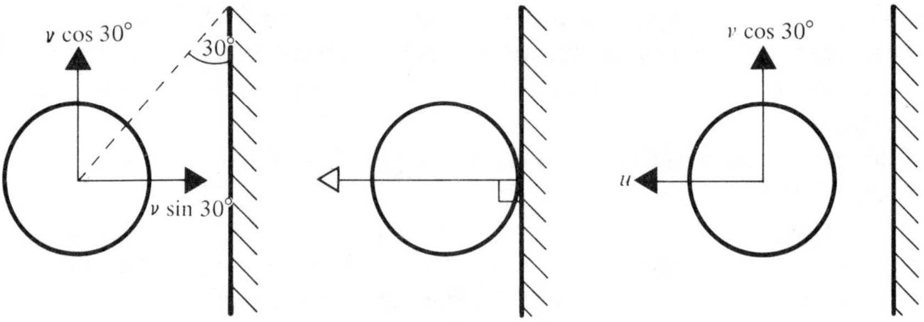

Just before impact At impact Just after impact

Using Newton's Law of Restitution perpendicular to the wall gives

$$\tfrac{1}{3}(v \sin 30°) = u$$

Therefore the magnitude of the velocity after impact is

$$\sqrt{(v^2 \cos^2 30° + \tfrac{1}{9}v^2 \sin^2 30°)} = v\sqrt{7/3}$$

and it is inclined at arctan $\left(\dfrac{\frac{1}{3}v \sin 30°}{v \cos 30°}\right)$ to the wall

$$= \arctan \sqrt{3}/9 \text{ to the wall.}$$

2) Two smooth spheres A and B of equal radii lie on a horizontal table. A has a mass m and B has a mass $2m$. A and B are projected towards each other with velocity vectors $2\mathbf{i} + \mathbf{j}$ and $\mathbf{i} - \mathbf{j}$ and collide when the line joining their centres is parallel to the unit vector $\mathbf{i}$. If the coefficient of restitution between the spheres is $\frac{1}{3}$, find their velocities after impact. Find also the kinetic energy lost at impact.

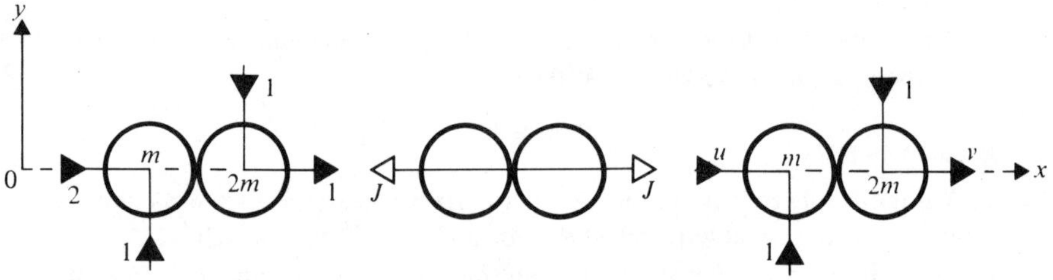

Just before impact Just after impact

Applying the Principle of Conservation of Momentum to each sphere in the direction Oy we get velocity components as shown in the diagrams.
Applying Conservation of Momentum and Newton's Law of Restitution in the direction Ox we get

$$2m + 2m = mu + 2mv \qquad (1)$$

$$\tfrac{1}{3}(2 - 1) = v - u \qquad (2)$$

Hence $\qquad v = \frac{13}{9}, \quad u = \frac{10}{9}$

Therefore the velocity of A after impact is $\qquad \frac{10}{9}\mathbf{i} + \mathbf{j}$

and the velocity of B after impact is $\qquad \frac{13}{9}\mathbf{i} - \mathbf{j}$

The kinetic energy of A before impact is $\qquad \frac{1}{2}m(4 + 1)$

The kinetic energy of B before impact is $\qquad \frac{1}{2}(2m)(1 + 1)$

Kinetic energy of A after impact is $\qquad \frac{1}{2}m(\frac{100}{81} + 1)$

Kinetic energy of B after impact is $\qquad \frac{1}{2}(2m)(\frac{169}{81} + 1)$

Therefore kinetic energy lost in the impact is

$$\tfrac{1}{2}m(4 + 1) + \tfrac{1}{2}(2m)(1 + 1) - \tfrac{1}{2}m(\tfrac{100}{81} + 1) - \tfrac{1}{2}(2m)(\tfrac{169}{81} + 1) = \tfrac{8}{27}m$$

3) A sphere of mass m is moving with velocity vector $4\mathbf{i} - \mathbf{j}$ when it hits a wall. It rebounds from the wall with velocity vector $\mathbf{i} + 3\mathbf{j}$. Find the magnitude and direction of the impulse it receives and the coefficient of restitution between the sphere and the wall.

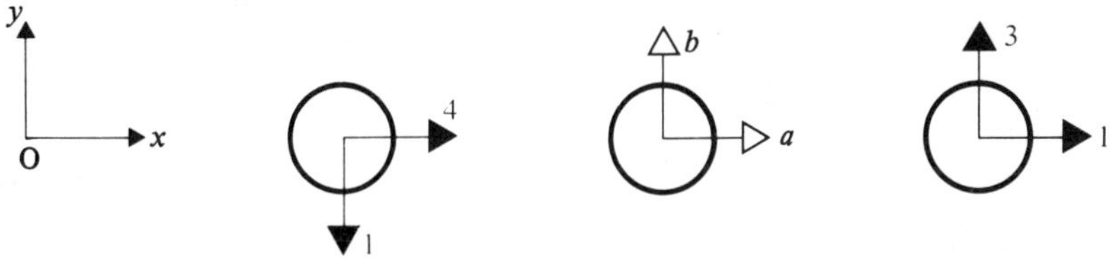

Just before impact $\qquad\qquad$ Just after impact

Let the impulse vector be $a\mathbf{i} + b\mathbf{j}$
As impulse = increase in momentum
in the direction Ox: $\qquad a = m - 4m$
in the direction Oy: $\qquad b = 3m - (-m)$
Therefore the impulse vector is $-3m\mathbf{i} + 4m\mathbf{j}$
The magnitude of the impulse is 5 units and the direction is arctan $(-4/3)$ to Ox.
(To find the coefficient of restitution, e, we apply Newton's Law of Restitution in the direction of the impulse. This requires the velocity components, just before and just after impact, in this direction.)
In the direction of the impulse, the unit vector $\hat{\mathbf{p}}$ is $\frac{1}{5}(-3\mathbf{i} + 4\mathbf{j})$
Therefore the velocity components in the direction of $\hat{\mathbf{p}}$ are:

before impact: $\qquad (4\mathbf{i} - \mathbf{j}) \cdot \frac{1}{5}(-3\mathbf{i} + 4\mathbf{j}) = -16/5$ units

and $\qquad$ after impact: $\qquad (\mathbf{i} + 3\mathbf{j}) \cdot \frac{1}{5}(-3\mathbf{i} + 4\mathbf{j}) = 9/5$ units

Newton's Law of Restitution gives:

$$(\tfrac{9}{5})/(\tfrac{16}{5}) = e$$

therefore $\qquad e = \frac{9}{16}$

4) Two smooth spheres A and B of equal radius but of mass m and M are free to move on a horizontal table. A is projected towards B which is at rest. On impact the line joining their centres makes an angle θ with the velocity of A before impact. If e is the coefficient of restitution between the spheres, find the angle through which A's path is deflected by the impact.

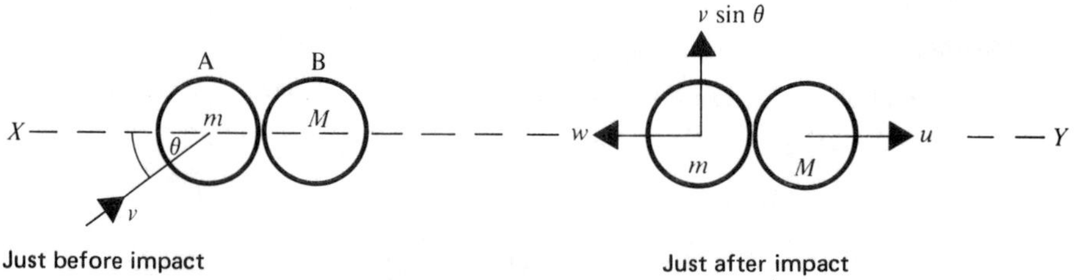

(The momentum both of A and of B is conserved perpendicular to the line of centres XY. Thus B moves off in the direction XY and the velocity component of A perpendicular to XY is unchanged by the impact.) Applying Conservation of Momentum and Newton's Law of Restitution in the direction XY gives

$$mv\cos\theta = Mu - mw \qquad (1)$$

$$ev\cos\theta = u + w \qquad (2)$$

From (1) and (2)

$$w = \frac{v(eM-m)}{M+m}\cos\theta$$

Therefore the velocity of A after impact makes an angle ϕ with YX where

$$\tan\phi = \frac{M+m}{eM-m}\tan\theta$$

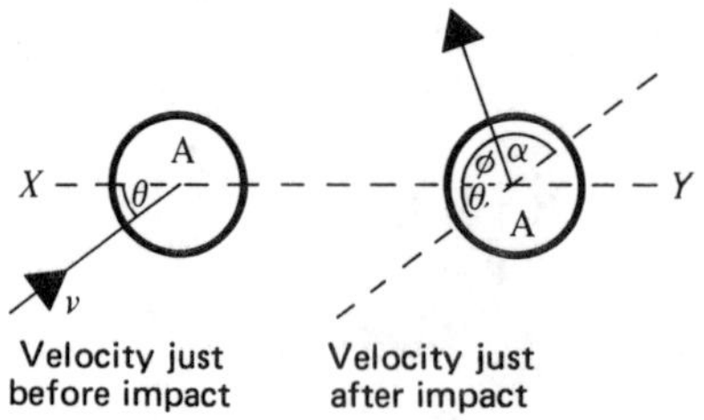

Therefore if α is the angle through which A's path is deflected by the impact then

$$\alpha = 180° - (\theta + \phi)$$

Therefore

$$\tan\alpha = -\tan(\theta+\phi)$$

$$= -\left(\frac{\tan\theta + \tan\phi}{1 - \tan\theta\tan\phi}\right)$$

$$= -\left[\frac{\tan\theta + \dfrac{M+m}{eM-m}\tan\theta}{1 - \dfrac{M+m}{eM-m}\tan^2\theta}\right]$$

$$= \frac{M(1+e)\tan\theta}{(M+m)\tan^2\theta - (eM-m)}$$

(*Note*: this question has been done in general terms but the result should not be taken as a quotable formula for solving similar problems as the relationship between the angles will not necessarily be the same.)

5) Two spheres A and B of equal radius but of masses m and $3m$ move towards each other with velocity vectors $\mathbf{i} + 2\mathbf{j}$ and $-\mathbf{i} + 3\mathbf{j}$ respectively. They collide when the line joining their centres is parallel to $\mathbf{i} - \mathbf{j}$. If $e = \frac{1}{2}$ find the velocity vectors of A and B after impact.

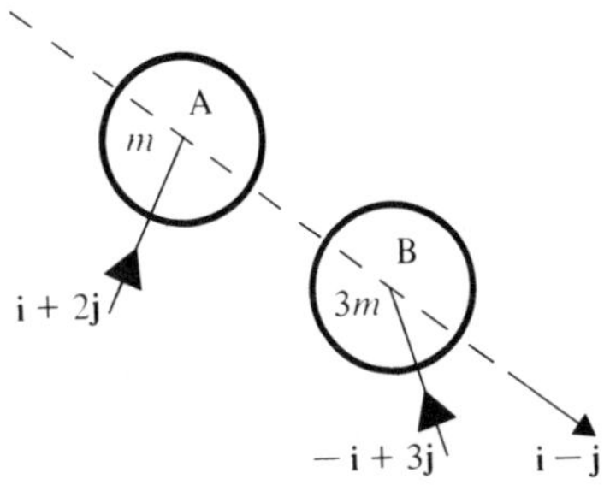

Before impact

We must first find the velocity components (before impact) parallel and perpendicular to the line of centres. Let $\hat{\mathbf{a}}$ be the unit vector parallel to $\mathbf{i} - \mathbf{j}$ and $\hat{\mathbf{b}}$ a unit vector perpendicular to $\mathbf{i} - \mathbf{j}$

Therefore in the direction $\hat{\mathbf{a}}$ the velocity of A has component

$$(\mathbf{i} + 2\mathbf{j}) \cdot \frac{1}{\sqrt{2}}(\mathbf{i} - \mathbf{j}) = -\frac{1}{\sqrt{2}}$$

in the direction $\hat{\mathbf{b}}$ the velocity of A has component

$$(\mathbf{i} + 2\mathbf{j}) \cdot \frac{1}{\sqrt{2}}(\mathbf{i} + \mathbf{j}) = \frac{3}{\sqrt{2}}$$

and in the direction $\hat{\mathbf{a}}$ the velocity of B has component

$$(-\mathbf{i} + 3\mathbf{j}) \cdot \frac{1}{\sqrt{2}}(\mathbf{i} - \mathbf{j}) = -2\sqrt{2}$$

in the direction $\hat{\mathbf{b}}$ the velocity of B has component

$$(-\mathbf{i} + 3\mathbf{j}) \cdot \frac{1}{\sqrt{2}}(\mathbf{i} + \mathbf{j}) = \sqrt{2}$$

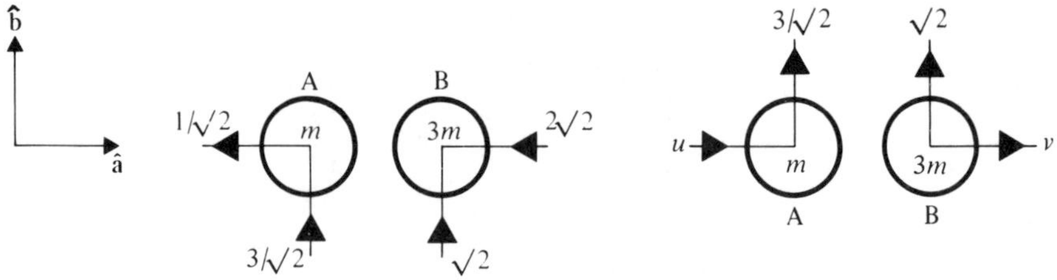

Just before impact Just after impact

Considering the momentum of A and of B perpendicular to the line of centres we get velocity components as shown in the diagram.
Using Conservation of Momentum and Newton's Law of Restitution parallel to the line of centres gives:

$$m\left(-\frac{1}{\sqrt{2}}\right) + 3m(-2\sqrt{2}) = mu + 3mv \qquad (1)$$

$$\frac{1}{2}\left(-\frac{1}{\sqrt{2}} + 2\sqrt{2}\right) = v - u \qquad (2)$$

From (1) and (2) $\quad u = -\dfrac{35\sqrt{2}}{16}, \quad v = -\dfrac{23\sqrt{2}}{16}$

Therefore the velocity of A after impact is

$$-\frac{35\sqrt{2}}{16}\hat{\mathbf{a}} + \frac{3}{\sqrt{2}}\hat{\mathbf{b}} = -\frac{35\sqrt{2}}{16}\left\{\frac{1}{\sqrt{2}}(\mathbf{i}-\mathbf{j})\right\} + \frac{3}{\sqrt{2}}\left\{\frac{1}{\sqrt{2}}(\mathbf{i}+\mathbf{j})\right\}$$

$$= \frac{1}{16}(-11\mathbf{i} + 59\mathbf{j})$$

and the velocity of B after impact is

$$-\frac{23\sqrt{2}}{16}\hat{\mathbf{a}} + \sqrt{2}\,\hat{\mathbf{b}} = -\frac{23\sqrt{2}}{16}\left\{\frac{1}{\sqrt{2}}(\mathbf{i}-\mathbf{j})\right\} + \sqrt{2}\left\{\frac{1}{\sqrt{2}}(\mathbf{i}+\mathbf{j})\right\}$$

$$= \frac{1}{16}(-7\mathbf{i} + 39\mathbf{j})$$

EXERCISE 6a

1) A smooth sphere is projected along horizontal ground and collides obliquely with a vertical wall. It hits the wall when moving at 3 ms^{-1} at an angle of 30° to the wall. Find the velocity of the sphere just after impact with the wall if (a) $e = \frac{1}{2}$, (b) $e = 1$, (c) $e = 0$.

2) A smooth sphere travelling on horizontal ground impinges obliquely on a vertical wall and rebounds at right angles to its original direction of motion. If the sphere is moving at 60° to the wall before impact, find the value of e.

3) Two smooth spheres A and B of equal radius and mass are moving on a horizontal table with velocity vectors $\mathbf{i}+2\mathbf{j}$, $-3\mathbf{i}+\mathbf{j}$ respectively and collide when the line joining their centres is parallel to $\mathbf{i}$. Find the velocity vectors of A and B after the impact if (a) $e=\frac{1}{2}$, (b) $e=1$, (c) the collision is inelastic.

4) Two smooth spheres A and B of equal radius and mass lie on a horizontal surface. B is at rest and A is projected towards B with velocity vector $4\mathbf{i}+3\mathbf{j}$ and they collide when their line of centres is parallel to the vector $\mathbf{i}$. If B moves off with speed 3 units, find the value of e and the velocity vector of A after impact.

5) Two smooth spheres X and Y of equal radius and mass lie on a horizontal billiard table ABCD. X lies at rest and Y is projected towards X with speed u. The spheres collide when their line of centres is parallel to the edge AB. After the impact Y moves directly towards the cushion edge AB and after collision with the cushions both spheres are moving with equal speed. If e, the coefficient of restitution, is the same for all impacts, find e and the direction in which Y was moving before the first collision.

6) A smooth sphere X of mass m lies at rest at the centre of a billiard table ABCD. A second smooth sphere Y of equal radius but mass $2m$ is at rest at the mid-point of the edge BC. If Y is to be projected towards X so that after collision X moves towards the pocket at A what must be the direction of the line of centres at impact? If $e=\frac{1}{5}$ and Y is projected with speed u and X moves off with speed $u/10$, find the direction in which Y was projected.

7) Two identical smooth spheres are moving on a horizontal table with velocity vectors $3\mathbf{i}+4\mathbf{j}$ and $-\mathbf{i}+\mathbf{j}$ and collide when the line joining their centres is parallel to the vector $\mathbf{i}$. If the coefficient of restitution between the spheres is $\frac{1}{2}$, find the velocity vectors of the spheres after impact. Find also the ratio of the magnitudes of the velocities, before and after impact, of the spheres relative to each other. If, at this instant of impact, the centres of the spheres are 2 units of distance apart, find the distance between their centres 1 unit of time later.

(U of L)

8) A smooth uniform ball travelling along a smooth horizontal table collides with a second smooth uniform ball of the same mass and radius which is at rest on the table. At the moment of impact the line of centres makes an angle of 30° with the direction in which the first ball is moving. If the coefficient of restitution between the balls is e, show that the first ball is deflected by the impact through an angle θ, where

$$\tan\theta = \frac{(1+e)\sqrt{3}}{5-3e}$$

(Oxford)

9) Three spheres A, B and C, of the same radius and of mass km, m and $\frac{1}{2}m$ respectively, are free to move on a horizontal plane. The coefficient of restitution between the spheres is $\frac{3}{4}$. Sphere A, moving with speed v, strikes directly

sphere B which is at rest. Sphere A has its velocity halved in magnitude and reversed in direction, and sphere B moves off with speed u. Find the value of k and express v in terms of u.

Sphere B carries on to strike sphere C which is at rest. When impact occurs the line of centres of the spheres is at an angle α to the path of B. After impact the sphere C moves off with speed u. Show that $\cos\alpha = 6/7$ and find the speed of B after this impact. (All frictional forces may be neglected.) (AEB)

10)

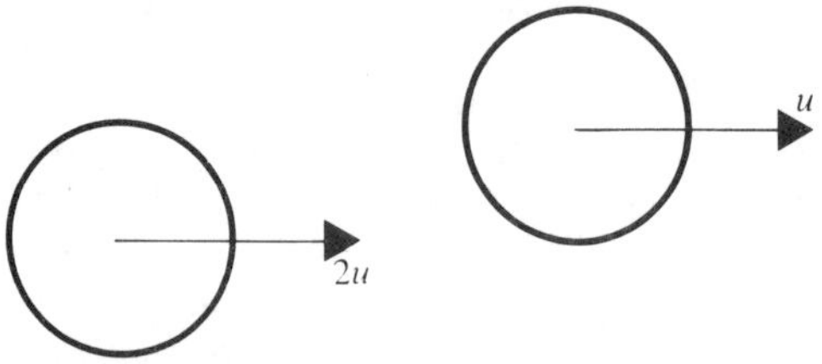

Two uniform smooth spheres, each of mass m and radius a, collide when moving on a horizontal plane. Before impact the spheres are moving with speeds $2u$ and u as shown in the figure, their centres moving in parallel lines which are at a distance $6a/5$ apart. The coefficient of restitution between the spheres is $\frac{1}{2}$. Find the speeds of the spheres after impact and show that the angle between their paths is then approximately 27°. (U of L)

11) A smooth sphere of mass m sliding on a horizontal plane collides obliquely with a sphere of mass $2m$ and of equal radius at rest on the plane. At the moment of impact the velocity u of the moving sphere makes an angle α with the line of centres, and after impact the speed of the heavier sphere is $(2/5)u\cos\alpha$. Find the coefficient of restitution between the spheres.
If the relative velocity of the spheres after impact is at right angles to their relative velocity before impact, find the value of $\tan\alpha$. (U of L)

12) A, B, C are three points on a smooth horizontal table. AB = BC and angle ABC = 120°. A small smooth sphere, P, is projected from A along AB with speed u and strikes an identical sphere Q which is stationary at B. After the collision the sphere Q collides with another sphere R at C; the sphere R was projected from A (along AC) at the same time as P. Show that the speed of projection of R was $\dfrac{\sqrt{3}(1+e)u}{5+e}$ where e is the coefficient of restitution between P and Q.
Given that the mass of R is three times the mass of Q and that $e = \frac{1}{3}$ for all the collisions show that, if the line of centres for the collision between R and Q is AC, the direction of the motion of Q after this collision makes an angle $\arctan\dfrac{2}{3\sqrt{3}}$ with AC produced. (SU)

13) A small sphere of mass 3 grammes is moving with velocity $3\mathbf{j}$ centimetres per second. As a result of a blow this particle is suddenly made to move with velocity

$(3\mathbf{i} - \mathbf{j})$ centimetres per second. Calculate the magnitude and the direction of the impulse received by the particle.
When moving with velocity $(3\mathbf{i} - \mathbf{j})$ centimetres per second the sphere collides with a second sphere of equal radius and mass 2 grammes which is moving with velocity $(2\mathbf{i} + \mathbf{j})$ centimetres per second. At the instant of collision the line of centres of the two spheres is in the direction $\mathbf{i} + \mathbf{j}$ and the coefficient of restitution between them is $\frac{1}{2}$. Calculate:
(i) the velocity vectors of the spheres after impact,
(ii) the kinetic energy lost due to the collision. (AEB)

14) A red ball is stationary on a rectangular billiard table OABC. It is then struck by a white ball of equal mass and equal radius with velocity $u(-2\mathbf{i} + 11\mathbf{j})$ where $\mathbf{i}$ and $\mathbf{j}$ are unit vectors along OA and OC respectively. After impact the red and white balls have velocities parallel to the vectors $-3\mathbf{i} + 4\mathbf{j}$, $2\mathbf{i} + 4\mathbf{j}$ respectively. Prove that the coefficient of restitution between the two balls is $\frac{1}{2}$.
(U of L)

15) Show that the vectors $\mathbf{p}_1 = 3\mathbf{i} + 4\mathbf{j}$ and $\mathbf{p}_2 = 4\mathbf{i} - 3\mathbf{j}$ are at right angles.
If $\mathbf{n}$ and $\mathbf{t}$ are unit vectors in the directions $\mathbf{p}_1$ and $\mathbf{p}_2$ respectively, express each of the vectors $\mathbf{v}_1 = 3\mathbf{i}$ and $\mathbf{v}_2 = \mathbf{i} + \mathbf{j}$ in the form $\mathbf{v} = a\mathbf{n} + b\mathbf{t}$ where a and b are scalar constants and state the values of these constants in each case.
Two billiard balls A_1 and A_2 of equal mass collide; at the instant of collision A_1 and A_2 have velocity vectors $\mathbf{v}_1$ and $\mathbf{v}_2$ respectively, as given above, and $\mathbf{n}$ is the unit vector along the line of centres. If the coefficient of restitution between the billiard balls is $\frac{1}{2}$, find their velocity vectors after impact in terms of $\mathbf{n}$ and $\mathbf{t}$.
Hence find their velocity vectors after impact in terms of $\mathbf{i}$ and $\mathbf{j}$. (AEB)

16) A rectangular billiard table is a long and b wide $(a > b)$. The coefficient of restitution when a particle hits any side cushion is e. If a smooth particle hits a cushion and comes off in a direction at right angles to its original direction, show that the angles made by the two directions with the cushion are $\operatorname{arccot}\sqrt{e}$ and $\arctan\sqrt{e}$.
A smooth particle is projected from a point A on a shorter side. It moves on a rectangular path, hitting each side cushion in turn, and returns to A. Show that A divides the shorter side in the ratio $(a\sqrt{e} - be):(b - a\sqrt{e})$. (Cambridge)

17) A uniform cube of mass $4m$ rests on a rough horizontal plane. The coefficient of friction between the cube and the plane is 0.4. A small sphere, mass m, is projected along the table so as to strike the cube at the centre of a lower edge with speed u. The direction of motion of the particle makes an angle of $30°$ with the edge of the cube. The coefficient of restitution between the sphere and the cube is 0.5.
(i) Show that the initial speed of the cube is $0.15u$.
(ii) Find the distance that the cube moves before coming to rest.
(iii) Show that the sphere is deflected through an angle of $\arctan(3\sqrt{3}/7)$ on impact. (SUJB)

FURTHER WORK ON PROJECTILES

In Volume One, general equations for the motion of a projectile were derived with reference to horizontal and vertical axes. The components of acceleration, velocity and displacement in these directions are simple but in certain problems their use does not lead to the easiest solution.

EXAMPLE

A particle is projected with an initial velocity of 20 ms^{-1} at 75° to the horizontal from a point O at the bottom of a plane inclined at 45° to the horizontal.
The particle hits the plane again at a point A, where OA is a line of greatest slope.
Find (a) the time of flight,
(b) the range on the plane,
(c) the greatest height above the plane reached by the particle.

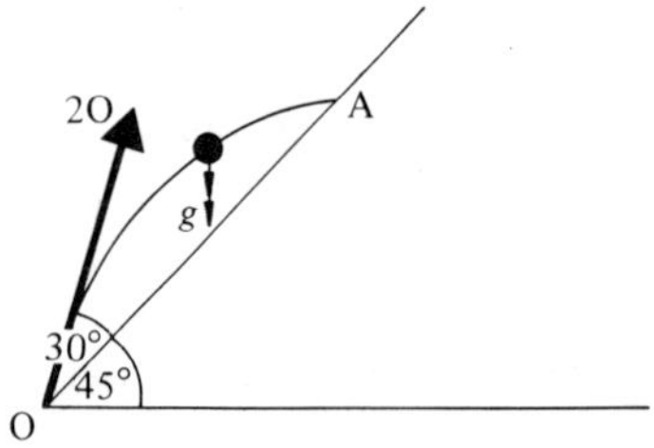

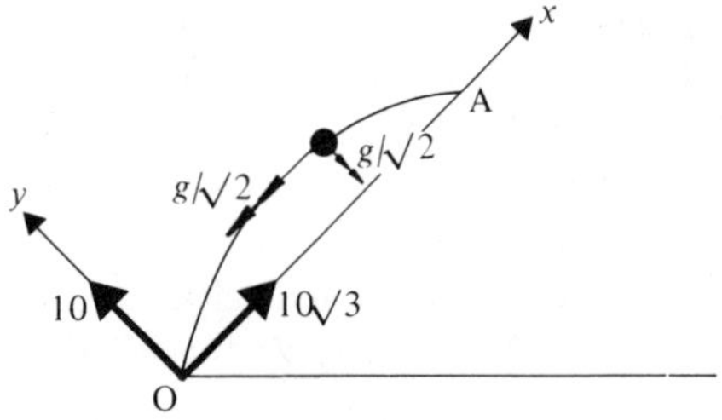

Parts (a) and (b) both require information associated with the displacement of the particle either parallel or perpendicular to the plane. The time of flight is the time taken to travel from O to A, and at A the displacement of the particle perpendicular to the plane is zero. The range on the plane is the distance OA, so in this problem it is logical to consider components parallel and perpendicular to OA.
Taking axes as shown:

$$\ddot{y} = \frac{-g}{\sqrt{2}} \qquad \ddot{x} = \frac{-g}{\sqrt{2}}$$

Therefore $\dot{y} = 10 - \left(\frac{g}{\sqrt{2}}\right)t$ (1), $\dot{x} = 10\sqrt{3} - \left(\frac{g}{\sqrt{2}}\right)t$ (2)

and $$y = 10t - \frac{1}{2}\left(\frac{g}{\sqrt{2}}\right)t^2 \quad (3), \quad x = 10\sqrt{3}t - \frac{1}{2}\left(\frac{g}{\sqrt{2}}\right)t^2 \quad (4)$$

(a) When the particle hits the plane (at A) $y = 0$.
Using (3), this occurs when

$$10t - \left(\frac{g}{2\sqrt{2}}\right)t^2 = 0$$

i.e. when $$t = 0 \text{ or } t = 2\sqrt{2} \qquad (\text{taking } g = 10\,\text{ms}^{-2})$$

Therefore the time of flight is $2\sqrt{2}$ seconds.

(b) The range on the plane is the distance OA, so we require the value of x when $t = 2\sqrt{2}$.

From (4) $$x = 10\sqrt{3}(2\sqrt{2}) - \frac{g}{2\sqrt{2}}(2\sqrt{2})^2$$

$$x = 20\sqrt{2}(\sqrt{3} - 1)$$

Therefore the range up the plane is $20\sqrt{2}(\sqrt{3} - 1)$ m.

(c) The greatest height above the plane, although a vertical displacement, can also be found easily from the equations already derived.

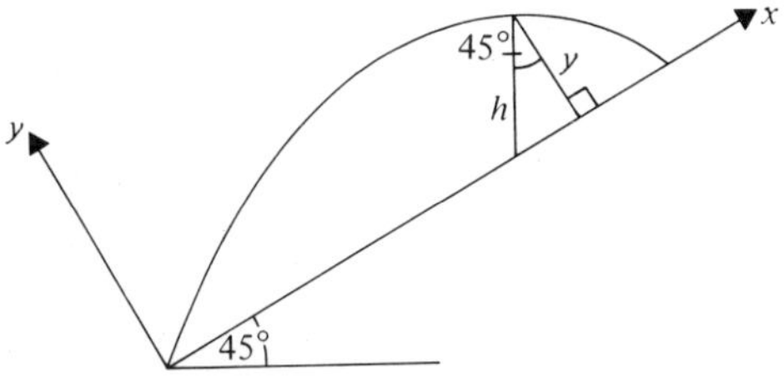

At any time t, the height h above the plane is given by

$$\begin{aligned} h &= y \sec 45^\circ \\ &= \{10t - (g/2\sqrt{2})t^2\}\sqrt{2} \\ &= 10\sqrt{2}t - 5t^2 \end{aligned}$$

Therefore h is maximum when $t = \sqrt{2}$ and the maximum value of h is 10 m.
Note: The maximum height above the plane and the maximum displacement from the plane (i.e. h and y) occur at the same time, and this time is equal to half the time of flight.

RANGE ON AN INCLINED PLANE

Most problems that require components of velocity and displacement in directions other than vertical or horizontal concern particles projected from a point on an inclined plane.
Consider a particle projected with speed u from a point on a plane inclined at an angle α to the horizontal, where the initial velocity of the particle is inclined at an angle β to the line of greatest slope of the plane. The particle can be projected either up or down the plane and the results for these two cases must be analysed separately.

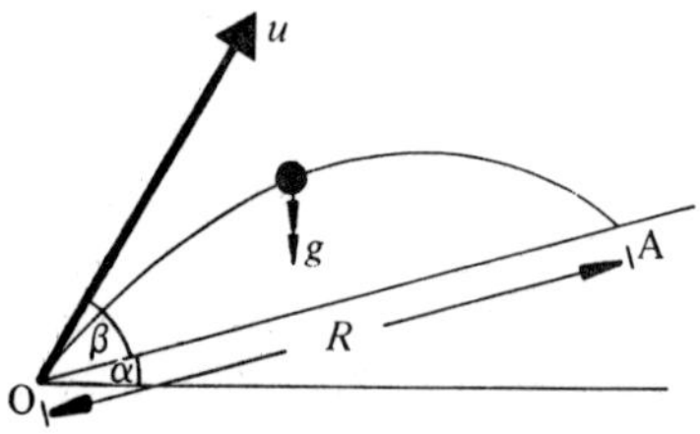

If the particle is projected up the plane, taking axes as shown:

$$\ddot{y} = -g\cos\alpha$$

$$\dot{y} = u\sin\beta - gt\cos\alpha \qquad (1)$$

$$y = ut\sin\beta - \tfrac{1}{2}gt^2\cos\alpha \qquad (2)$$

and $\ddot{x} = -g\sin\alpha$

$$\dot{x} = u\cos\beta - gt\sin\alpha \qquad (3)$$

$$x = ut\cos\beta - \tfrac{1}{2}gt^2\sin\alpha \qquad (4)$$

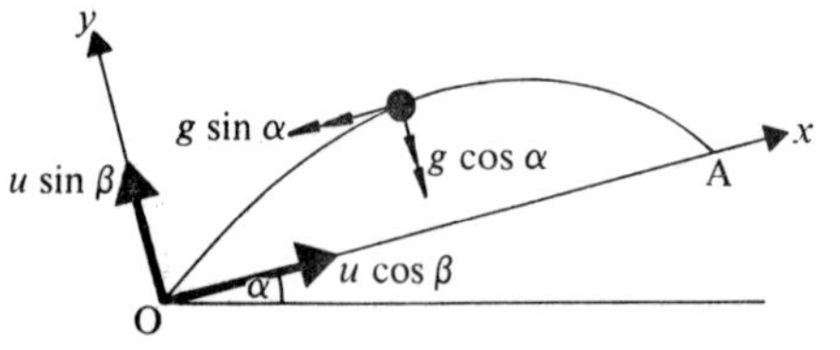

The time of flight is the value of t when $y = 0$ (when the particle is at A).
From (2) this occurs when

$$ut\sin\beta - \tfrac{1}{2}gt^2\cos\alpha = 0$$

Therefore $t = 0$ or $t = \dfrac{2u\sin\beta}{g\cos\alpha}$

The *range up the plane (R)* is the distance OA. i.e. the value of x when $t = \dfrac{2u\sin\beta}{g\cos\alpha}$.

Substituting this value of t in (4) gives

$$R = u\cos\beta\left(\frac{2u\sin\beta}{g\cos\alpha}\right) - \left(\frac{g\sin\alpha}{2}\right)\left(\frac{2u\sin\beta}{g\cos\alpha}\right)^2$$

$$= \frac{2u^2\sin\beta\cos(\beta+\alpha)}{g\cos^2\alpha}$$

Maximum Range up the plane (R_{max})
Rearranging the last equation gives:

$$R = \frac{u^2[\sin(2\beta+\alpha)-\sin\alpha]}{g\cos^2\alpha}$$

For a given value of u, R is a maximum when $\sin(2\beta+\alpha)=1$.

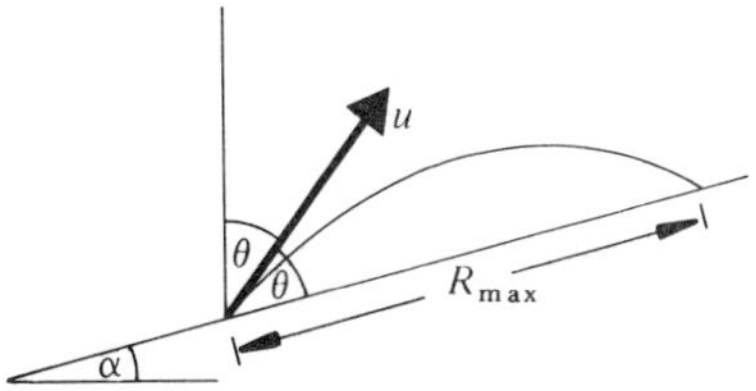

$$\text{Therefore } R_{max} = \frac{u^2(1-\sin\alpha)}{g\cos^2\alpha}$$

and this occurs when $2\beta+\alpha = 90°$

or $\beta = 45° - \dfrac{\alpha}{2}$

i.e. *the range up the plane is maximum when the angle of projection bisects the angle between the upward slope of the plane and the vertical.*

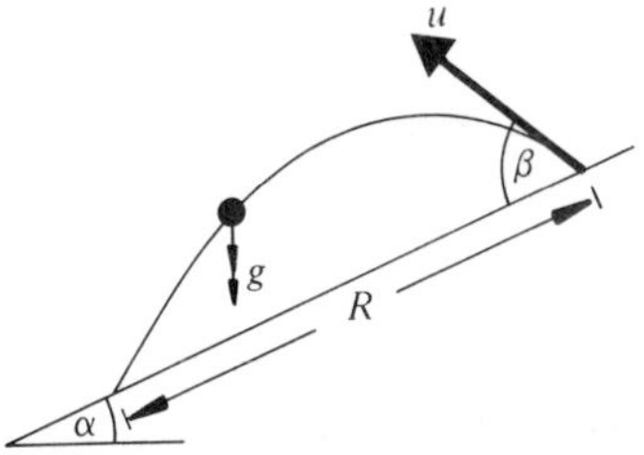

If the particle is projected down the plane, taking axes as shown:

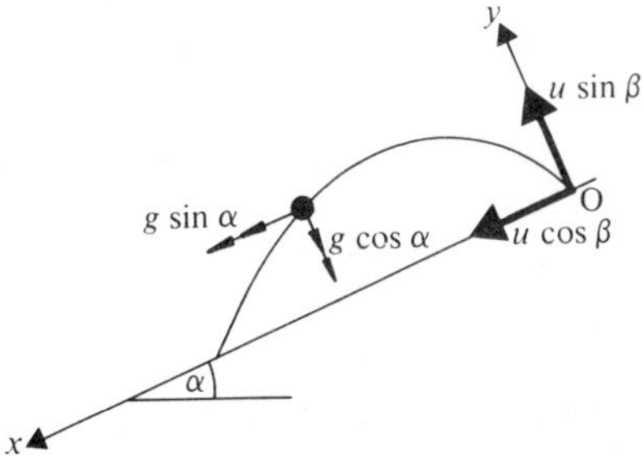

$$\ddot{y} = -g\cos\alpha$$

$$\dot{y} = u\sin\beta - gt\cos\alpha \qquad (1)$$

$$y = ut\sin\beta - \tfrac{1}{2}gt^2\cos\alpha \qquad (2)$$

$$\ddot{x} = g\sin\alpha$$

$$\dot{x} = u\cos\beta + gt\sin\alpha \qquad (3)$$

$$x = ut\cos\beta + \tfrac{1}{2}gt^2\sin\alpha \qquad (4)$$

From (2), *the time of flight* $= \dfrac{2u\sin\beta}{g\cos\alpha}$

Substituting this value of t in (4) gives the *range down the plane*.

i.e.

$$R = u\cos\beta\left(\frac{2u\sin\beta}{g\cos\alpha}\right) + \frac{g\sin\alpha}{2}\left(\frac{2u\sin\beta}{g\cos\alpha}\right)^2$$

$$= \frac{2u^2\sin\beta\cos(\beta-\alpha)}{g\cos^2\alpha}$$

Maximum Range down the plane
Rearranging the last equation gives:

$$R = \frac{u^2[\sin(2\beta - \alpha) + \sin\alpha]}{g\cos^2\alpha}$$

For a given value of u, R is maximum when $\sin(2\beta - \alpha) = 1$

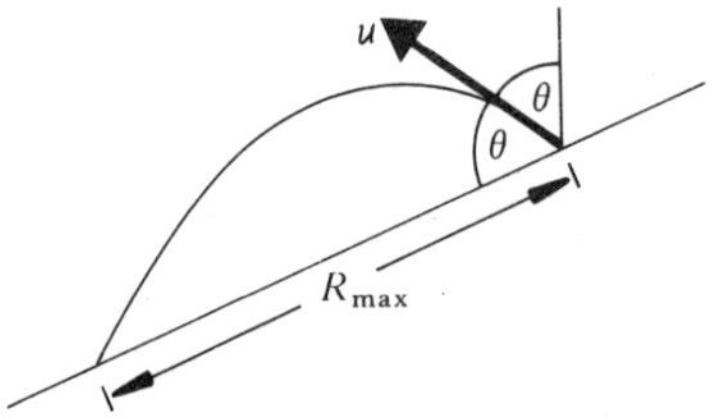

Therefore $R_{max} = \dfrac{u^2(1 + \sin\alpha)}{g\cos^2\alpha}$

and this occurs when $2\beta - \alpha = 90°$

or $\beta = 45° + \dfrac{\alpha}{2}$

i.e. *the range down the plane is maximum when the angle of projection bisects the angle between the downward slope of the plane and the vertical.*

SUMMARY

There are several considerations that should be borne in mind when solving problems concerning projectiles, some of which are covered in Volume One. A summary of the main points is set out below.

1. While a projectile is in flight it has a constant acceleration g vertically downward.
2. When a particle is projected from a point on a plane, the maximum range on that plane is achieved when the angle of projection bisects the angle between the slope of the plane and the vertical. In the special case of a horizontal plane this angle of projection becomes 45°.
   Think carefully about the information given and required in a particular problem. If that information is associated with velocity or displacement in directions other than vertical or horizontal, taking axes in those directions will usually make the solution easier.
   Velocity and displacement components in the chosen directions should be derived afresh for each problem. This may be done by finding first the acceleration components in the desired directions and then using the relevant equations for motion with constant acceleration.

EXAMPLES 6b

1) A particle is projected from a point O on a plane inclined at an angle α to the horizontal. The particle hits the plane at right angles at A, where OA is a line of greatest slope, O being lower than A. If the particle is projected at an angle θ to the horizontal, find the relationship between θ and α.

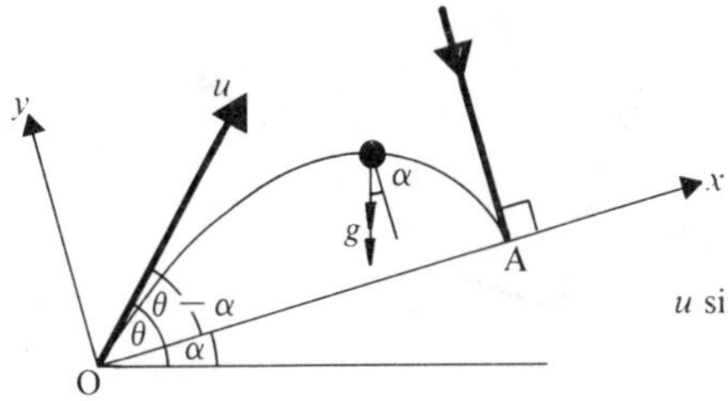

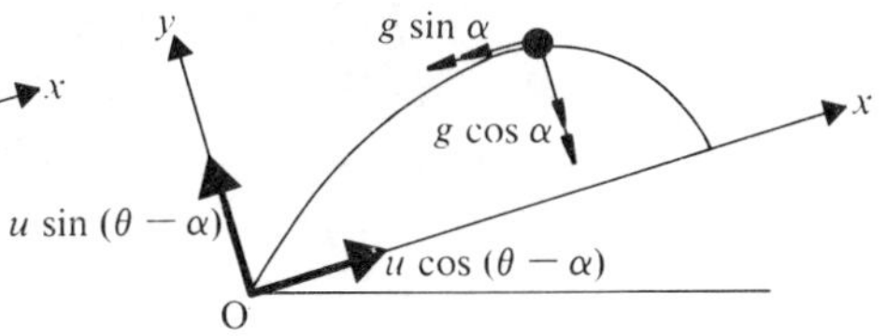

When the particle hits the plane at A, its displacement from the plane is zero; also as it is moving at right angles to the plane at the instant of impact its velocity component parallel to the plane is zero.
Taking axes as shown:

$$\ddot{y} = -g\cos\alpha \qquad \ddot{x} = -g\sin\alpha$$

$$\dot{y} = u\sin(\theta-\alpha) - gt\cos\alpha \qquad \dot{x} = u\cos(\theta-\alpha) - gt\sin\alpha$$

$$y = ut\sin(\theta-\alpha) - \tfrac{1}{2}gt^2\cos\alpha \qquad x = ut\cos(\theta-\alpha) - \tfrac{1}{2}gt^2\sin\alpha$$

When the particle hits the plane $y = 0$ and $\dot{x} = 0$, giving

$$ut\sin(\theta-\alpha) - \tfrac{1}{2}gt^2\cos\alpha = 0$$

$$u\cos(\theta-\alpha) - gt\sin\alpha = 0$$

Eliminating t from these two equations gives:

$$u\sin(\theta-\alpha) - \frac{g\cos\alpha}{2}\left\{\frac{u\cos(\theta-\alpha)}{g\sin\alpha}\right\} = 0$$

$$2\sin\alpha\sin(\theta-\alpha) - \cos\alpha\cos(\theta-\alpha) = 0$$

$$2\tan\alpha\tan(\theta-\alpha) - 1 = 0$$

$$\frac{\tan\theta - \tan\alpha}{1+\tan\theta\tan\alpha} = \frac{1}{2\tan\alpha}$$

$$2\tan\alpha\tan\theta - 2\tan^2\alpha = 1 + \tan\theta\tan\alpha$$

$$\tan\theta = \frac{1+2\tan^2\alpha}{\tan\alpha}$$

$$\tan\theta = \cot\alpha + 2\tan\alpha$$

2) A particle is projected at an angle arctan 2 to the horizontal from a point O on a smooth plane inclined at an angle arctan $\frac{1}{2}$ to the horizontal. The particle hits the plane again at A and rebounds. If OA is a line of greatest slope and O is lower than A, find the range of values of e, the coefficient of restitution between the particle and the plane, if the particle continues to move up the plane after the impact.

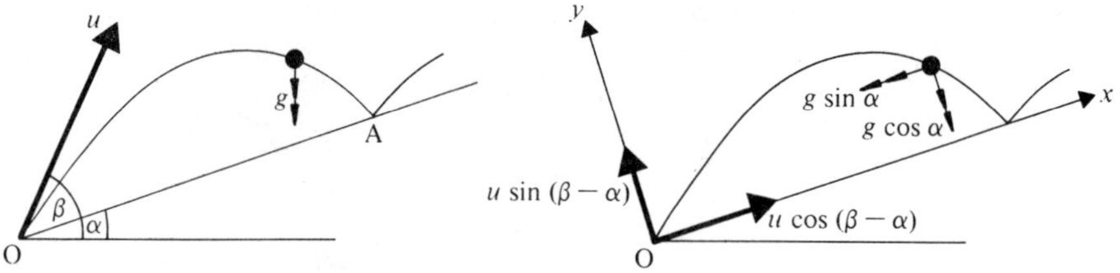

The velocity immediately before impact must be found before we can analyse what happens after the impact.
Taking axes shown, where $\tan \alpha = \frac{1}{2}$ and $\tan \beta = 2$:

$$\ddot{y} = -g\cos\alpha \qquad \ddot{x} = -g\sin\alpha$$

$$\dot{y} = u\sin(\beta-\alpha) - gt\cos\alpha \qquad \dot{x} = u\cos(\beta-\alpha) - gt\sin\alpha$$

$$y = ut\sin(\beta-\alpha) - \tfrac{1}{2}gt^2\cos\alpha$$

The particle hits the plane when $y = 0$

i.e. when $\quad ut\sin(\beta-\alpha) - \frac{1}{2}gt^2\cos\alpha = 0$

$$t = \frac{2u\sin(\beta-\alpha)}{g\cos\alpha}$$

Therefore when the particle hits the plane

$$\dot{y} = u\sin(\beta-\alpha) - g\cos\alpha\left(\frac{2u\sin(\beta-\alpha)}{g\cos\alpha}\right) = -u\sin(\beta-\alpha)$$

$$\dot{x} = u\cos(\beta-\alpha) - g\sin\alpha\left(\frac{2u\sin(\beta-\alpha)}{g\cos\alpha}\right) = u[\cos(\beta-\alpha) - 2\tan\alpha\sin(\beta-\alpha)]$$

We can now investigate the effect of the impact:

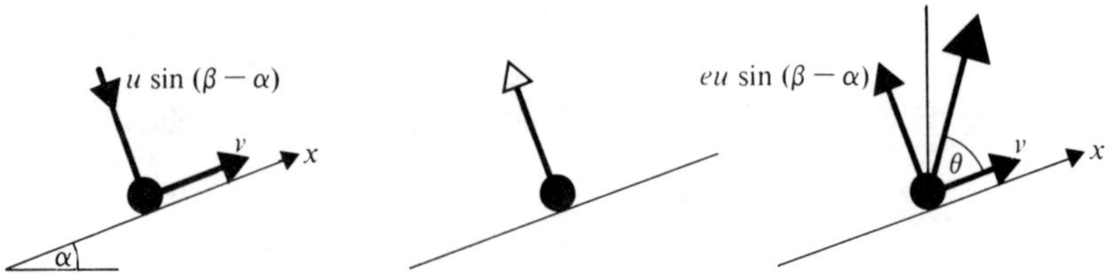

(i) Just before impact (ii) At impact (iii) Just after impact

The impulse received by the particle is in the direction Oy, so the velocity component in the direction Ox is not changed by the impact. The velocity component, after impact, in the direction Oy is found by using Newton's Law of Restitution. Thus the velocity components immediately after impact are as shown in diagram (iii) where

$$v = u[\cos(\beta-\alpha) - 2\tan\alpha\sin(\beta-\alpha)].$$

Therefore the resultant velocity, after impact, makes an angle θ with Ox where

$$\tan\theta = \frac{eu\sin(\beta-\alpha)}{u[\cos(\beta-\alpha)-2\tan\alpha\sin(\beta-\alpha)]}$$

$$= \frac{e\tan(\beta-\alpha)}{1-2\tan\alpha\tan(\beta-\alpha)}$$

Now $\tan\alpha = \frac{1}{2}$, $\tan\beta = 2$, therefore $\tan(\beta-\alpha) = \frac{3}{4}$

therefore $\tan\theta = 3e$

If the particle is to move up the plane after impact then

$$\theta < (90^\circ - \alpha)$$

(i.e. the direction of the resultant velocity after impact must be to the right of the vertical through A in diagram (iii)).

thus $\tan\theta < \tan(90^\circ - \alpha)$

$$3e < 2$$

$$e < \tfrac{2}{3}$$

Also $e \geqslant 0$

Therefore $0 \leqslant e < \frac{2}{3}$

3) A particle is projected from a point A, whose position vector is $2\mathbf{i} + 3\mathbf{j}$, with initial velocity vector $20\mathbf{i} + 30\mathbf{j} + 20\mathbf{k}$, where $\mathbf{i}$ and $\mathbf{k}$ are horizontal and $\mathbf{j}$ is the unit vector in the direction of the upward vertical. Find a vector equation for the path of the particle.

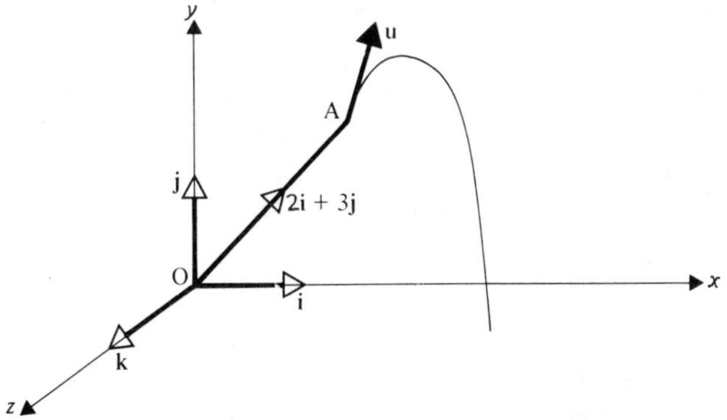

While the particle is in flight it has a constant acceleration g vertically downward

Therefore the acceleration vector at any time t is

$$\mathbf{a} = -g\mathbf{j}$$

and the velocity vector (by integration) is

$$\mathbf{v} = -gt\mathbf{j} + \mathbf{c}$$

When $t = 0$, $\mathbf{v} = 20\mathbf{i} + 30\mathbf{j} + 20\mathbf{k}$

therefore $$\mathbf{v} = 20\mathbf{i} + (30 - gt)\mathbf{j} + 20\mathbf{k}$$

The position vector at any time t is

$$\mathbf{r} = 20t\mathbf{i} + (30t - \tfrac{1}{2}gt^2)\mathbf{j} + 20t\mathbf{k} + \mathbf{c}_1$$

When $t = 0$, $\mathbf{r} = 2\mathbf{i} + 3\mathbf{j}$

therefore $$\mathbf{r} = (20t + 2)\mathbf{i} + (30t - \tfrac{1}{2}gt^2 + 3)\mathbf{j} + 20t\mathbf{k}$$

This is a vector equation of the path of the projectile.

EXERCISE 6b

(Take $g = 10\,\mathrm{ms}^{-2}$.)

1) A particle is projected with an initial velocity of $30\,\mathrm{ms}^{-1}$ at $45°$ to the horizontal from a point O on an inclined plane. The particle hits the plane again at a point A where OA is inclined at $30°$ to the horizontal. Find the time for which the particle is in the air and the distance OA if:
(a) A is higher than O,
(b) A is lower than O.

2) A particle is projected with an initial speed of $100\,\mathrm{ms}^{-1}$ from a point O on a plane inclined at $30°$ to the horizontal. The plane containing the path passes through a line of greatest slope of the inclined plane. Find the maximum range of the particle (a) up the plane, (b) down the plane.

3) A particle is projected from the origin O with initial velocity vector $20\mathbf{i} + 20\mathbf{j}$ where $\mathbf{i}$ is horizontal and $\mathbf{j}$ vertically upward. The path of the particle crosses the line with vector equation $\mathbf{r} = \lambda(3\mathbf{i} + \mathbf{j})$ at the point A. Write down the vector equation of the path of the projectile and hence or otherwise find the distance OA and the time taken for the particle to reach A.

4) A particle is projected with an initial velocity of $30\,\mathrm{ms}^{-1}$ at $\arctan \frac{4}{3}$ to the horizontal from a point O on a plane inclined at $\arctan \frac{1}{2}$ to the horizontal. The path of the particle lies in the vertical plane through a line of greatest slope of the inclined plane. Find the greatest height above the plane reached by the particle if the particle is projected (a) up the plane, (b) down the plane.

5) A particle is projected with an initial velocity vector $50\mathbf{i} + 30\mathbf{j} + 50\mathbf{k}$ from the point with position vector $5\mathbf{i} - 3\mathbf{k}$, where $\mathbf{i}$ and $\mathbf{k}$ are horizontal and $\mathbf{j}$ is

vertically upward. Find vector expressions for the velocity and position of the particle at any time t. If the xz plane is ground level find the time of flight of the particle. Find also the vector equation of the plane that contains the path of the particle.

6)

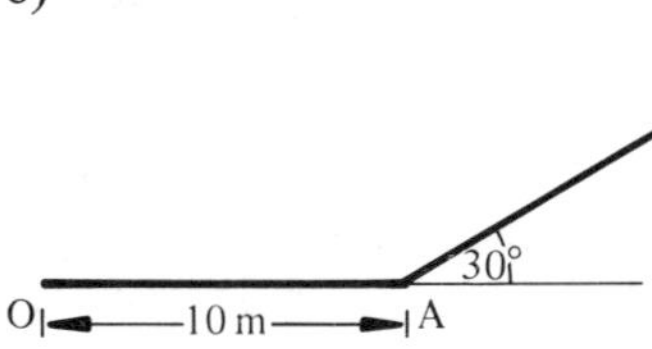

The diagram shows a sloping embankment AB where OA is level ground. A particle is projected towards the slope from O with initial velocity $30\ \text{ms}^{-1}$ at $45°$ to the horizontal. Find the time of flight of the particle and the distance from A of the point at which it hits the slope AB.

7) A particle is projected from a point O on level ground towards a smooth vertical wall 50 m from O and hits the wall. The initial velocity of the particle is $30\ \text{ms}^{-1}$ at $45°$ to the horiztonal and the coefficient of restitution between the particle and the wall is e. Find the distance from O of the point at which the particle hits the ground again if:
(a) $e = 0$, (b) $e = 1$, (c) $e = \frac{1}{2}$.

8) A particle is projected from a point O on a smooth plane, which is inclined at $\arctan \frac{1}{3}$ to the horizontal, and strikes the plane again at A where OA is a line of greatest slope and A is higher than O. The initial velocity of the particle is $40\ \text{ms}^{-1}$ at arctan 3 to the horizontal. Determine in which direction the particle moves after striking the plane at A if:
(a) the impact is inelastic, (b) the impact is perfectly elastic.

9) A particle is projected with speed u from a point on a plane which is inclined at an angle β to the horizontal. The particle is projected at an angle α to the horizontal in a vertical plane through a line of greatest slope of the plane. Show that the range up the plane is

$$\frac{2u^2 \sin(\alpha - \beta) \cos \alpha}{g \cos^2 \beta}.$$

Deduce that the range up the plane is a maximum when $\alpha = \frac{1}{4}\pi + \frac{1}{2}\beta$.
If the *maximum* range down the plane is twice the maximum range up the plane, find the angle β. (U of L)

10) A particle is projected with velocity V, at an angle of elevation α to the horizontal, from a point on a plane inclined at angle β ($< \alpha$) to the horizontal. The path of the particle is in a vertical plane through a line of greatest slope of the plane. If R_1 and R_2 are the respective maximum ranges when the particle is fired up the plane and down the plane, show that

(i) $R_1 = \dfrac{V^2}{g(1 + \sin \beta)}$, (ii) $\sin \beta = \dfrac{R_2 - R_1}{R_2 + R_1}$. (AEB)

11) A projectile is fired with initial speed $\sqrt{(2gh)}$ from a point O on a plane, which is inclined at an angle θ to the horizontal, in a direction such that the range down the plane is a maximum. Show that this range is $2h/(1 - \sin\theta)$, and that the direction bisects the angle between the plane and the vertical.
Show that the highest point of the trajectory is vertically above a point S on the plane such that OS $= h$. (U of L)

12) Ox and Oy are axes fixed, respectively, horizontally and vertically upwards. A particle is projected from O at time $t = 0$, under gravity, with velocity $\mathbf{u}$. Show that the position vector of the particle at time t is $\mathbf{r}$ where $\mathbf{r} = \mathbf{u}t + \frac{1}{2}\mathbf{g}t^2$ where $\mathbf{g}$ is the acceleration due to gravity (assumed constant) and resistances are ignored. Draw a vector triangle to illustrate the above vector equation.
Q is the point with position vector $\frac{1}{2}t\mathbf{u}$ and P is the position of the particle of time t. Show that **QP** is a vector in the same direction as the velocity of the particle at P.
A particle is projected from a point O of a plane inclined at an angle α to the *vertical*. The angle of projection is θ with the vertical and the plane of motion contains a line of greatest slope of the plane. When the particle hits the plane (at a point higher than O) it is moving horizontally. Prove, using the above results, or otherwise, that $\tan\alpha = 2\tan\theta$. (SU)

13) A perfectly elastic particle is projected from a point O on a fixed smooth plane which is inclined at an angle α to the horizontal. The velocity of projection is in the vertical plane through the line of greatest slope through O and makes an angle θ with the upwards line of slope. By using coordinate axes along the perpendicular to this line, or otherwise, prove that, if $\tan\theta = \frac{1}{2}\cot\alpha$, the particle will retrace its path to O after bouncing on the plane. (Oxford)

14) The line OA is inclined at an angle α to the horizontal, with O lower than A and with OA $= 2a$. A particle is projected from O to pass through A. Its initial velocity v makes an angle θ with OA, and its velocity at A is at right angles to OA. Show that
(a) $v^2 \sin\theta \cos(\theta + \alpha) = ga\cos^2\alpha$,
(b) $2\tan\theta\tan\alpha = 1$.
If the perpendicular distance of the particle from OA is a when its velocity is parallel to OA, obtain another relation between $\tan\theta$ and $\tan\alpha$ and deduce that θ equals $\pi/4$ radians. (U of L)

15) Prove, with the usual notation for motion with uniform acceleration, the formulae

$$v = u + ft, \quad s = ut + \tfrac{1}{2}ft^2.$$

A particle projected from a point A on a smooth plane of inclination α strikes the plane at B, where AB is a line of greatest slope with B higher than A. The velocity of projection is V at an angle θ with the plane. Prove that the time of flight to B is $(2V\sin\theta)/(g\cos\alpha)$.

If the particle rebounds in a vertical direction at B, show that $\cot\theta \cot\alpha = 2 + e$, where e is the coefficient of restitution. (JMB)

16) Two particles of equal mass are projected at the same instant and with the same speed $\sqrt{(gl)}$ from points A and B, distant l apart, on a line of greatest slope of a plane inclined at 30° to the horizontal. A is at a level higher than B. The particle at A is projected horizontally towards B and the particle at B is projected at an angle of 60° above the horizontal towards A. Prove that the particles will collide and that if they coalesce, the combined mass will begin to move in a direction inclined at 30° below the horizontal. (U of L)

17) A vehicle is moving with constant acceleration kg up a slope of inclination α, the floor of the vehicle being parallel to the slope. Show that if a particle is falling freely inside the vehicle its acceleration relative to the vehicle makes an angle β with the floor, where

$$\cot\beta = k\sec\alpha + \tan\alpha.$$

A particle is projected inside the vehicle from a point A on the floor, its initial velocity relative to the vehicle being V at an angle θ ($>\beta$) with the floor, as shown in the diagram. It strikes the ceiling at B, where AB is perpendicular to the floor and AB $= h$. By considering motion parallel to the slope, show that the time of flight is

$$\frac{2V\cos\theta}{g\cos\alpha\cot\beta},$$

and deduce that

$$h = \frac{2V^2\cos\theta\tan\beta\sin(\theta-\beta)}{g\cos\alpha\cos\beta}$$

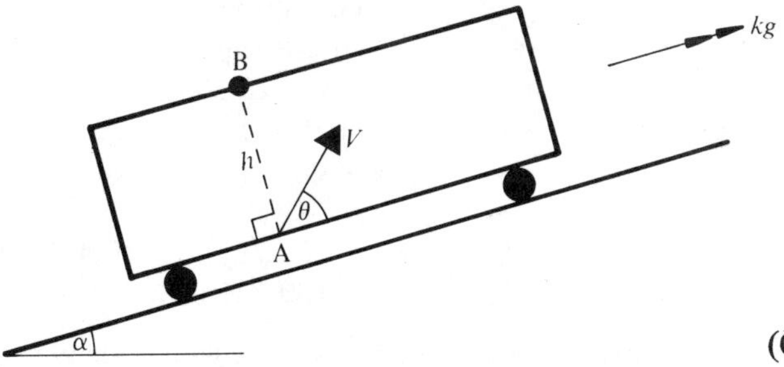

(Cambridge)

MULTIPLE CHOICE EXERCISE 6

(Instructions for answering these questions are given on page xi)

TYPE I

1) A particle is projected from a point on a plane which is inclined at 20° to the horizontal. The vertical plane containing the path of the projectile also contains a line of greatest slope of the plane. For the maximum range down the plane

the angle of projection (measured from the horizontal) is:
(a) 60° (b) 80° (c) 55° (d) 35° (e) 45°.

2)

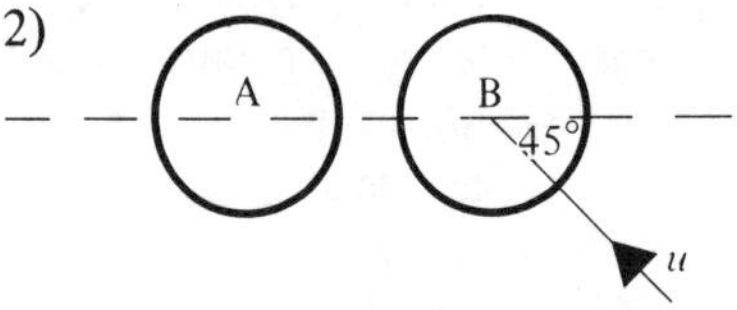

The diagram shows the velocities just before collision of two smooth spheres of equal radius and mass. The impact is perfectly elastic. The velocities just after impact are:

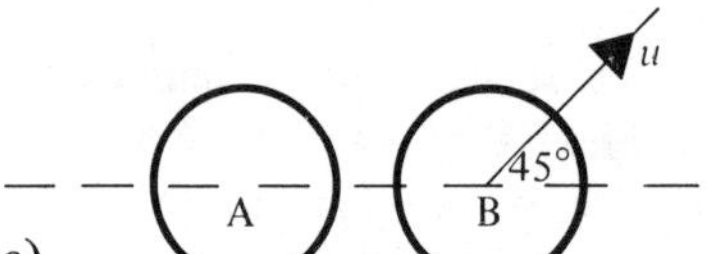

a)

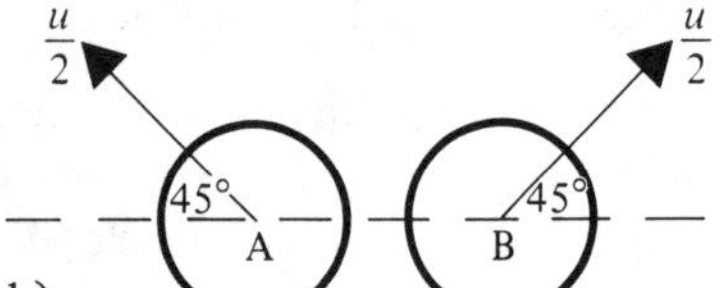

b)

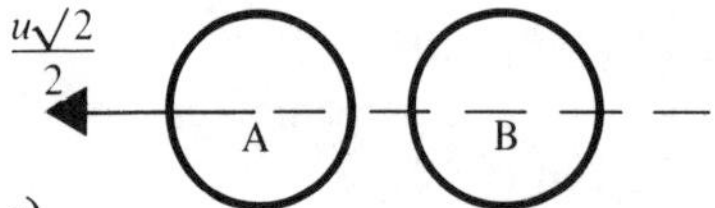

c)

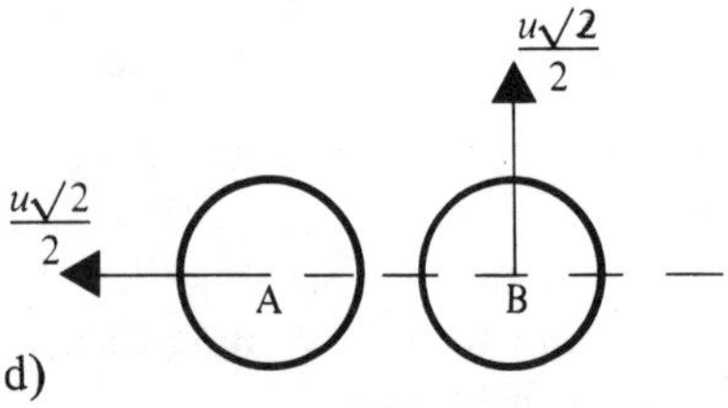

d)

3)

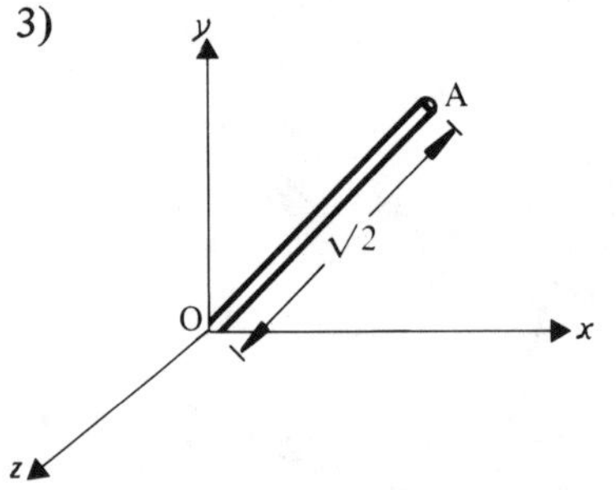

OA is the barrel of a gun which is inclined at 45° to the horizontal (xz plane), and it rotates about Oy in the sense z to x with constant angular velocity 3 rad s^{-1}. As OA passes through the xy plane (Oy is vertically upward) the gun fires a shell with speed 200 ms^{-1} relative to the barrel. The initial velocity vector of the shell is:

(a) $100\sqrt{2}\mathbf{i} + 100\sqrt{2}\mathbf{j} - 3\mathbf{k}$ (b) $100\sqrt{2}\mathbf{i} + 100\sqrt{2}\mathbf{j}$ (c) $100\mathbf{i} + 100\mathbf{j} + 3\mathbf{k}$
(d) $100\mathbf{i} + 3\mathbf{j} + 100\mathbf{k}$ (e) $100\sqrt{2}\mathbf{i} + 100\sqrt{2}\mathbf{j} + 3\mathbf{k}$

4) A smooth sphere is moving on a horizontal surface with velocity vector $3\mathbf{i} + \mathbf{j}$ immediately before it hits a vertical wall. The wall is parallel to the vector $\mathbf{j}$ and the coefficient of restitution between the wall and sphere is $\frac{1}{3}$. The velocity vector of the sphere after it hits the wall is:
(a) $\mathbf{i} + \mathbf{j}$ (b) $3\mathbf{i} - \frac{1}{3}\mathbf{j}$ (c) $-\mathbf{i} + \mathbf{j}$ (d) $\mathbf{i} - \mathbf{j}$ (e) $-\mathbf{i} - \frac{1}{3}\mathbf{j}$.

5) A particle which is projected from a point on an inclined plane strikes the plane and rebounds. After impact the particle retraces its original path. The coefficient of restitution between the particle and the plane is:
(a) 0 (b) -1 (c) $\frac{1}{2}$ (d) 1 (e) $\frac{1}{3}$.

TYPE II

6) A smooth sphere moving on a horizontal surface collides indirectly with a vertical wall. The coefficient of restitution between the sphere and the wall is $e\ (<1)$.

(a) The momentum of the sphere perpendicular to the wall is unchanged by the impact.
(b) The component of velocity parallel to the wall before impact is equal to the component of velocity parallel to the wall after impact.
(c) There is no loss in the kinetic energy of the sphere due to the impact.
(d) If u is the component of velocity perpendicular to the wall before impact and v is the component of velocity in the same direction after impact then $eu + v = 0$.

7)

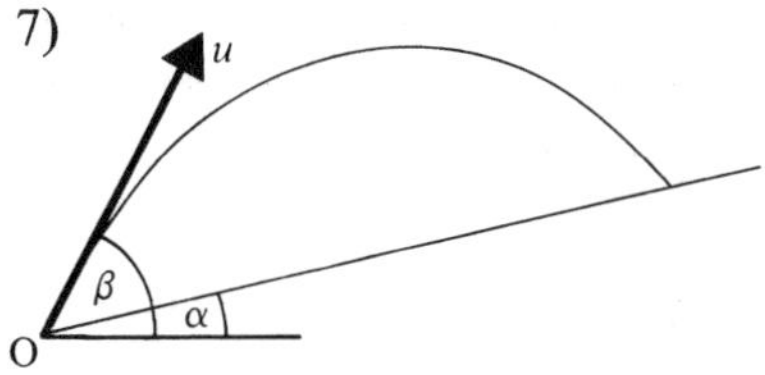

The diagram shows the vertical plane containing the path of a particle which is projected from a point O on an inclined plane.

(a) The path of the projectile is a parabola.
(b) If T is the time of flight, the particle is at its greatest height above the plane after an interval $\dfrac{T}{2}$.
(c) If the range on the plane is maximum then $\beta = \dfrac{\pi}{4} + \dfrac{\alpha}{2}$.
(d) The kinetic energy of the particle is maximum as it leaves O.

8) Two spheres A and B of equal mass are free to move on a smooth horizontal surface. A and B move towards each other with velocity vectors $a\mathbf{i} + b\mathbf{j}$ and $c\mathbf{i} + d\mathbf{j}$ respectively and collide when the line joining their centres is parallel to $\mathbf{i}$. After impact A and B have velocity vectors $p\mathbf{i} + q\mathbf{j}$ and $r\mathbf{i} + s\mathbf{j}$ respectively. The coefficient of restitution between the spheres is $e\ (<1)$.

(a) $b = q$ (b) $c = r$ (c) $a + c = p + r$ (d) $ea = p$.

9)

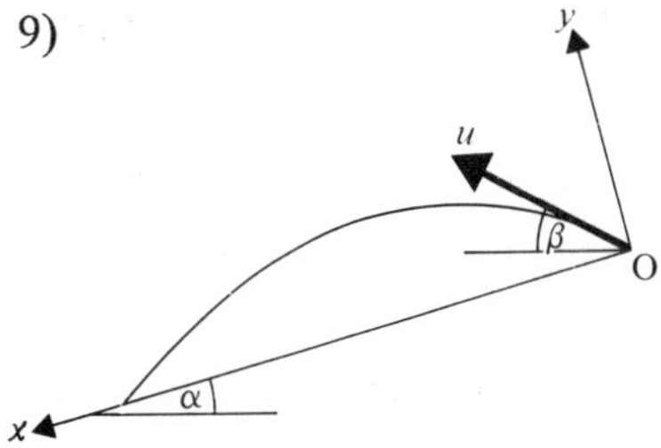

From a point O on a plane inclined at α to the horizontal a particle is projected down the plane with initial velocity u at an angle β to the horizontal. Taking axes as shown:

(a) $\ddot{x} = -g \sin \alpha$ (b) $\ddot{y} = -g \cos \alpha$ (c) $\ddot{x} = g \cos \alpha$ (d) $\ddot{y} = g \sin \alpha$.

TYPE III

10) (a) A smooth sphere moving on a horizontal surface strikes a vertical wall indirectly.
(b) A smooth sphere moving on a horizontal surface strikes a vertical wall and rebounds in a direction perpendicular to the wall.

11) Two smooth spheres A and B move towards each other on a horizontal surface and collide.
(a) A and B move directly towards each other.
(b) A is brought to rest by the impact.

12)

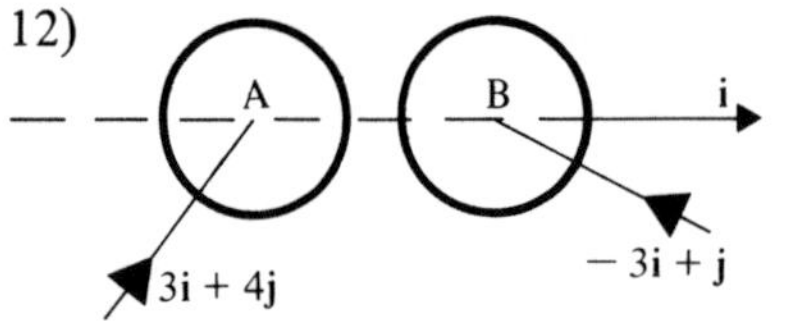

Two spheres A and B of equal radius and mass move towards each other with velocities as shown and collide when the line joining their centres is parallel to **i**.

(a) $e = \frac{1}{2}$.
(b) The velocities of A and B after impact are $-\frac{3}{2}\mathbf{i} + 4\mathbf{j}$ and $\frac{3}{2}\mathbf{i} + \mathbf{j}$ respectively.

TYPE IV

13) A particle is projected from a point O on an inclined plane. Find the maximum range of the particle up the plane.
(a) The plane is inclined at 20° to the horizontal and a line of greatest slope lies in the vertical plane containing the path of the projectile.
(b) The initial speed of the particle is $30\ \text{ms}^{-1}$.
(c) The mass of the particle is 0.2 kg.

14) Two smooth spheres A and B of equal radius move towards each other on a horizontal table and collide. Find the velocity of each sphere after impact.
(a) The initial velocities of A and B are $3\mathbf{i} + 2\mathbf{j}$ and $2\mathbf{i} - 3\mathbf{j}$ respectively.
(b) When A and B collide the line joining their centres is parallel to $\mathbf{i} + \mathbf{j}$.
(c) The coefficient of restitution between A and B is $\frac{1}{2}$.

15) A particle is projected from a point O on a smooth inclined plane and hits the plane again at a point A and rebounds. Find the distance from O of the point at which the particle hits the plane for the third time.
(a) OA is inclined at 15° to the horizontal with A higher than O.
(b) The initial velocity of the particle is $40\ \text{ms}^{-1}$ at 40° to the horizontal.
(c) Immediately after hitting the plane at A the velocity of the particle is inclined at an angle of 20° to the horizontal.

16) Two smooth spheres are moving on a horizontal table and collide. Find the coefficient of restitution between the spheres.
(a) The spheres are of equal mass and each has a radius *a*.

(b) Before impact the centres of the spheres are moving along parallel paths distant $a/2$ apart.
(c) After impact the angle between their paths is 30°.

17) A particle is projected from a point O on an inclined plane and hits the plane again at a point A. Find the angle of projection of the particle.
(a) OA is inclined at α to the horizontal with A lower than O.
(b) The distance OA is l.
(c) The speed of the particle as it hits the plane at A is u.

TYPE V

18) When a perfectly elastic impact occurs between a moving object and a fixed object there is no loss in kinetic energy.

19) When two objects, both of which are free to move, collide and coalesce their total momentum is unchanged by the collision.

20) When two objects which are both free to move, collide, their total mechanical energy is always unchanged by the collision.

CHAPTER 7

MOTION OF A PARTICLE IN A STRAIGHT LINE. DAMPED HARMONIC OSCILLATIONS

The way in which a particle moves depends on several factors, viz. the nature of the forces acting on the particle, the intitial conditions (i.e. velocity, position etc.) and the mass of the particle, which may or may not be constant.
The basic relationship between the applied forces and the motion of the body is expressed in Newton's Second Law of Motion. So far this law has been used in the form which applies only to a body of constant mass. Some moving objects however have a variable mass (e.g. rockets) so we will now restate this law in a more general form which applies also to a body of variable mass.

NEWTON'S SECOND LAW OF MOTION

When an external force is applied to a body the rate of increase of momentum produced is directly proportional to the applied force.
Using the SI system of units, the constant of proportion is unity; hence the rate of increase of momentum is equal to the force which produces it.
Consider a body moving so that:
at time t, the resultant force acting on the body is $\mathbf{F}$
the velocity of the body is $\mathbf{v}$
the mass of the body is m
hence the momentum of the body is $m\mathbf{v}$.

at time $t + \delta t$, the velocity of the body is $\mathbf{v} + \delta\mathbf{v}$
the mass of the body is $m + \delta m$
hence the momentum of the body is $(m + \delta m)(\mathbf{v} + \delta\mathbf{v})$.

The increase in momentum in the interval δt is

$$(m + \delta m)(\mathbf{v} + \delta\mathbf{v}) - m\mathbf{v}$$

The impulse of the force acting on the body in the same interval of time δt is approximately $\mathbf{F}\,\delta t$.

Hence $$\mathbf{F}\,\delta t \simeq (m + \delta m)(\mathbf{v} + \delta\mathbf{v}) - m\mathbf{v}$$

Therefore $$\mathbf{F} \simeq m\frac{\delta\mathbf{v}}{\delta t} + \mathbf{v}\frac{\delta m}{\delta t} + \frac{\delta m\delta\mathbf{v}}{\delta t}$$

As $\delta t \to 0$, this equation becomes

$$\mathbf{F} = m\frac{d\mathbf{v}}{dt} + \mathbf{v}\frac{dm}{dt} \tag{1}$$

In deriving this equation it is assumed that the increase in the mass of the body (δm) in the interval of time δt had no velocity of its own before becoming part of the body. However this is not always the case; a body may gain mass by coalescing with particles which, prior to joining the body concerned, have a velocity of their own, e.g. a rain drop formed by condensing water vapour, where the water vapour has a velocity separate from that of the drop being formed. Alternatively a body may lose mass by ejecting matter from it, e.g. a rocket burning fuel.

Hence if δm has a velocity $\mathbf{u}$ immediately prior to joining the body, the increase in momentum in the interval of time δt is

$$(m + \delta m)(\mathbf{v} + \delta\mathbf{v}) - m\mathbf{v} - \mathbf{u}\,\delta m$$

and $$\mathbf{F}\,\delta t \simeq m\,\delta\mathbf{v} + (\mathbf{v} - \mathbf{u})\,\delta m + \delta m\,\delta\mathbf{v}$$

therefore $$\mathbf{F} = m\frac{d\mathbf{v}}{dt} + (\mathbf{v} - \mathbf{u})\frac{dm}{dt} \tag{2}$$

where $\mathbf{v} - \mathbf{u}$ is the velocity of the mass increment relative to the body.

Note: If the mass of the body is constant, $\dfrac{dm}{dt} = 0$ and both equations (1) and (2) reduce to

$$\mathbf{F} = m\mathbf{a} \tag{3}$$

This is the form in which we have already used this law when considering a body of constant mass.

Basic Equation of Motion

When we apply the appropriate form of Newton's Second Law to analyse the motion of a particle under the action of known forces, we derive a differential equation which is called the *basic equation of motion* of the particle.

Motion of a Particle in a Straight Line under the Action of Known Forces

(Motion in a straight line is sometimes called rectilinear motion. Care must be taken not to confuse the term 'linear motion' with motion in a straight line. Linear motion means motion in a line, which may or may not be straight). Consider a particle of constant mass m moving along a straight line Ox in the direction Ox under the action of a constant force km directed towards O and a resisting force mkv^2 where v is the speed of the particle.

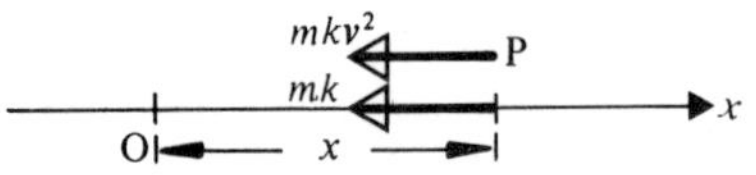

Forces acting

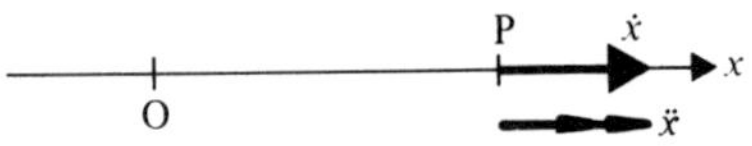

Velocity and acceleration

When the particle is at P where its displacement from O is x the resultant force acting on it, in the direction Ox, is

$$-(km + mk\dot{x}^2)$$

Using Newton's Second Law of Motion gives

$$-(km + mk\dot{x}^2) = m\ddot{x}$$

or

$$\ddot{x} + k\dot{x}^2 + k = 0$$

This is the basic equation of motion for the particle, and it is a second order differential equation in x and t as it has a term in $\dfrac{d^2x}{dt^2}$. Before the motion can be analysed further a complete solution of this equation must be found, involving two stages of integration and therefore two constants of integration. These constants can be evaluated if two facts are known about the conditions of the particle at some given time (e.g. the values of x and $\dot{x}$ when $t = 0$).

Note that this equation of motion is valid only when the particle moves in the sense Ox. To analyse the motion in the direction xO another equation of motion is necessary as the resisting force does not automatically change sign with $\dot{x}$ (as $\dot{x}^2$ is always positive).

In many problems where a particle is moving in a straight line under the action of known forces, the basic equation of motion is a differential equation. The solution of these equations will not always be shown in full in the text since the methods available for solving such differential equations can be found in standard text books on Pure Mathematics.

EXAMPLES 7a

1) A particle moves in a straight line Ox under the action of a force which gives it a constant acceleration of 5 ms^{-2} away from O in a medium which retards it at the rate of $2v$ when its speed is v. Derive the equation of motion for this particle.

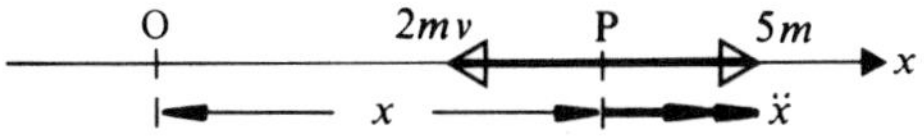

Let the mass of the particle be m

Hence the accelerating force $= 5m$

and the retarding force $= -2m\dot{x}$

Therefore the resultant force in the direction Ox is

$$5m - 2m\dot{x}$$

and $$5m - 2m\dot{x} = m\ddot{x} \qquad (\mathbf{F} = ma)$$

so the equation of motion is $\ddot{x} + 2\dot{x} - 5 = 0$

Note that this equation is valid when the particle is moving in either sense along Ox as the retarding force $(-2m\dot{x})$ automatically changes sign when the direction of motion changes.

2) An engine of mass M is accelerating on the level with a constant power H, against resistances which are proportional to the square of its speed. Find the equation of motion of the engine.

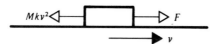

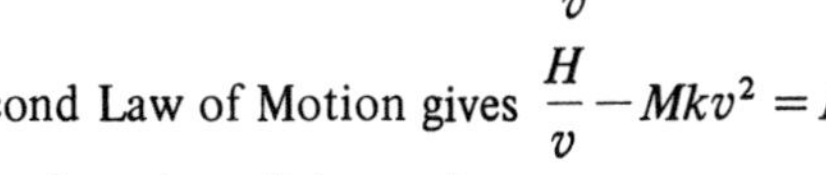

When the speed of the engine is v, the driving force F is $\dfrac{H}{v}$.

Therefore the resultant force on the engine is $\dfrac{H}{v} - Mkv^2$.

Using Newton's Second Law of Motion gives $\dfrac{H}{v} - Mkv^2 = M\dfrac{dv}{dt}$.

This is the equation of motion of the engine.

3) A particle moves along a straight line Ox such that its displacement from O at time t is x where

$$\frac{d^2x}{dt^2} + 3\frac{dx}{dt} + 6 = 0$$

Describe a possible physical situation which would give rise to such an equation of motion.

If, when $t = 1$, $\dfrac{dx}{dt} = 2$ and $x = 0$ find the speed when $t = 0$. Sketch the graph of x against t and hence give a brief description of the motion of the particle.

Rearranging the equation of motion gives $\ddot{x} = -3\dot{x} - 6$

or $$-3m\dot{x} - 6m = m\ddot{x}$$

Thus, from Newton's second law of motion, the resultant force acting on the particle is $-3m\dot{x} - 6m$. This is made up of a constant force in the direction xO and a force directly proportional to the speed of the particle in the direction opposite to the direction of motion.

One physical situation to which this equation of motion applies could be: a particle which, with an initial velocity in the direction Ox, is acted on by a constant braking force and is moving in a medium which offers a resistance proportional to its speed. (No doubt the reader can think of many other alternatives which would give rise to the given equation of motion.)

Writing $\dot{x}$ as v and $\ddot{x}$ as $\dfrac{dv}{dt}$, the equation of motion is

$$\frac{dv}{dt} + 3v + 6 = 0$$

Separating the variables gives

$$\int_2^v \frac{dv}{v+2} = \int_1^t -3dt$$

Therefore $$\ln\left(\frac{v+2}{4}\right) = -3t + 3$$

or $$v = 4e^{3-3t} - 2 \qquad (1)$$

Therefore when $t = 0$, $$v = 4e^3 - 2$$

Replacing v by $\dfrac{dx}{dt}$ in (1) gives

$$\frac{dx}{dt} = 4e^{3-3t} - 2$$

Integrating again with respect to time we have

$$\Big[x\Big]_0^x = \Big[-\tfrac{4}{3}e^{3-3t} - 2t\Big]_1^t$$

$$x = \tfrac{10}{3} - \tfrac{4}{3}e^{3-3t} - 2t$$

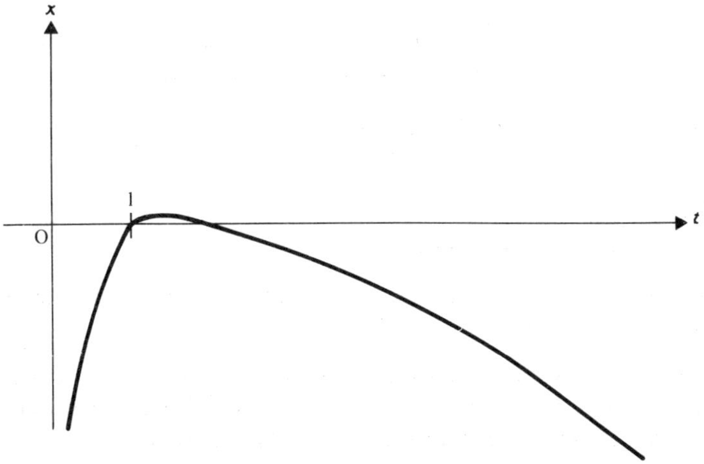

From the graph we see that the particle initially moves in the positive sense along Ox, passing through O when $t = 1$ and returns to O, continuing thereafter to move in the negative sense along Ox.

4) A rocket, with an initial mass of 1000 kg, is launched vertically upwards under gravity. The rocket burns fuel at the rate of 10 kg per second. The burnt matter is ejected vertically downwards with a speed of 2000 ms^{-1} relative to the rocket. If burning ceases after one minute find the maximum velocity of the rocket.
(Take g as constant at 10 ms^{-2}).

The only external force acting on the rocket is its weight vertically downwards. At time t the mass of the rocket is $(1000 - 10t)$ kg
and the weight is $(1000 - 10t)g$ N
Taking the vertically upward direction as positive and using Newton's second law of motion gives

$$-(1000-10t)g = (1000-10t)\frac{dv}{dt} + 2000(-10)$$

$$1000 + 10t = (100-t)\frac{dv}{dt}$$

The maximum velocity is reached when burning ceases.

$$\int_0^v dv = \int_0^{60} \frac{1000+10t}{100-t}\,dt$$

$$= -10\int_0^{60} dt + 10\int_0^{60} \frac{200}{100-t}\,dt$$

$$v_{max} = -600 - \left[2000 \ln (100 - t)\right]_0^{60}$$
$$= -600 - 2000 \ln 40 + 2000 \ln 100$$
$$= 2000 \ln 2.5 - 600$$

EXERCISE 7a

(Take $g = 10\ \text{ms}^{-2}$ when necessary).

In questions 1–5 derive the equation of motion from the physical situation described but do not solve the equation.

1) A particle of mass m is moving along a straight line Ox in the direction Ox under the action of two forces, one proportional to its displacement from O and directed towards O and the other a resistance which is constant.

2) A particle of mass m is projected vertically upward under gravity in a medium which offers a resistance kv when the speed is v.

3) A particle of mass m is attached to one end of a light elastic string of natural length a and modulus mg, the other end of the string is attached to a fixed point O on a smooth horizontal surface. The particle is released from rest at a horizontal distance $a + l$ from O, the subsequent motion taking place in a medium which offers a resistance mkv when the speed of the particle is v.

4) A vehicle of mass M is moving on a level road with constant power H against a resistance of Mkv^2 when the speed is v.

5) A rocket is moving vertically upward against gravity whose mass at time t is $(M - mt)$ and which expels burnt fuel at a speed u vertically downward relative to the rocket.

In questions 6–8 suggest a possible physical situation which would result in the following equations of motion.

6) $\dfrac{d^2x}{dt^2} + k\dfrac{dx}{dt} = 0.$

7) $\dfrac{d^2x}{dt^2} + n^2x = 0.$

8) $\dfrac{d^2x}{dt^2} + kv + g = 0.$

9) A particle of mass m moves along a straight line Ox under the action of a constant force $4m$ towards O and a retarding force of $3mv$ when the speed is v. If when $t = 0, x = 0$ and $\dot{x} = 3$ find the displacement from O when the particle first comes to rest.

10) A particle of mass m slides from rest down a straight groove inclined at 30° to the horizontal against resistances which are equal to $5mv^2$ when the speed of the particle is v. Derive the equation of motion of this particle, taking x as the displacement from the initial position. Find the speed of the particle after it has been moving for $\frac{1}{5}$ s and show that the particle has a terminal velocity of 1 ms^{-1}.

11) A car of mass 1000 kg is accelerating on a level road with constant power of 25 kW against resistances of magnitude ten times the speed of the car. Derive the equation of motion and find the distance covered when the car accelerates from rest to a speed of 25 ms^{-1}.

12) A rocket of initial mass M has a mass $M(1 - \frac{1}{3}t)$ at time t. The rocket is launched from rest vertically upwards under gravity and expels burnt fuel at a speed u relative to the rocket vertically downward. Find the speed and height above the launching pad when $t = 1$.

DAMPED HARMONIC MOTION

There are many forms of oscillatory motion, the most straight forward being simple harmonic motion (which is analysed in Volume One) but this is a form which rarely occurs naturally. A more familiar type of oscillatory motion is that performed by, for example, the end of a tuning fork after it has been struck. The amplitude of this motion gets progressively smaller and the oscillations eventually cease. This is an example of a form of damped oscillations.

Forces Causing Damped Oscillations

A particle will perform linear Simple Harmonic Motion when the resultant force acting on it is proportional to its displacement from a fixed point and is directed towards that fixed point. If, in addition, there is a force resisting the motion of the particle, the oscillations may be damped or the nature of the motion may be completely changed. This resisting force can take many forms, but we shall confine our analysis to resisting forces which are directly proportional to the speed of the particle.

The Equation of Motion of a Particle Moving in a Straight Line under the Action of a Force directed towards a Fixed Point O on that Line and Proportional to the Displacement of the Particle from O and a Resisting Force Proportional to its Speed

Consider a particle of mass m moving along a straight line Ox under the action of a force mn^2x directed towards O when the displacement of the particle from O is x and a resisting force $2mkv$ where v is the speed of the particle, and k and n are constants.

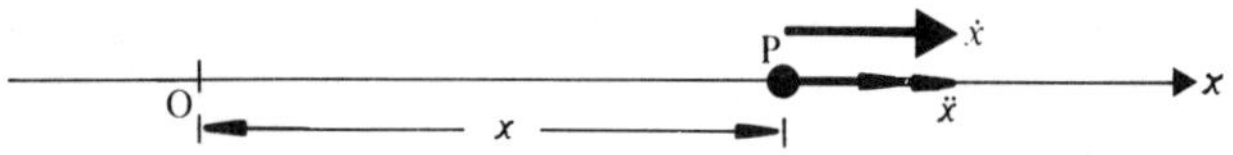

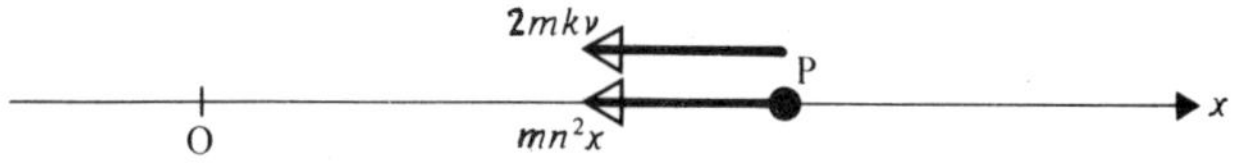

The resultant force in the direction Ox is

$$-mn^2x - 2mk\dot{x}$$

Newton's second law of motion gives

$$m\ddot{x} = -mn^2x - 2mk\dot{x}$$

$$\frac{d^2x}{dt^2} + 2k\frac{dx}{dt} + n^2x = 0 \qquad (1)$$

This is the basic equation of motion for any particle moving in a straight line under the action of the forces described above.
To analyse the motion more fully, equation (1) must be integrated. It is a second order linear differential equation and its solution depends on the nature of the roots of the equation

$$p^2 + 2kp + n^2 = 0 \qquad (2)$$

1. *Real distinct roots* i.e. $k^2 > n^2$
   The complete solution of (1) is

$$x = Ae^{-kt+t\sqrt{(k^2-n^2)}} + Be^{-kt-t\sqrt{(k^2-n^2)}} \qquad (3)$$

2. *Equal roots* i.e. $k^2 = n^2$
   In this case the complete solution of (1) has the form

$$x = e^{-kt}(A + Bt) \qquad (4)$$

3. *Imaginary roots* i.e. $k^2 < n^2$
   In this case the complete solution of (1) has the form

$$x = Ce^{-kt}\cos\left[\sqrt{(n^2 - k^2)}t + \epsilon\right] \qquad (5)$$

where A, B, C, ϵ are constants of integration.

(For a full analysis of the solution of second order linear differential equations refer to a standard text book on Pure Mathematics.)
Therefore equations (3), (4) and (5) represent the three possible complete solutions to equation (1) and we can now determine what form the motion takes in each of these three cases. This is most easily done by reference to a sketch graph of x against t.

CASE 1. $x = Ae^{-kt+\sqrt{(k^2-n^2)}t} + Be^{-kt-\sqrt{(k^2-n^2)}t}$

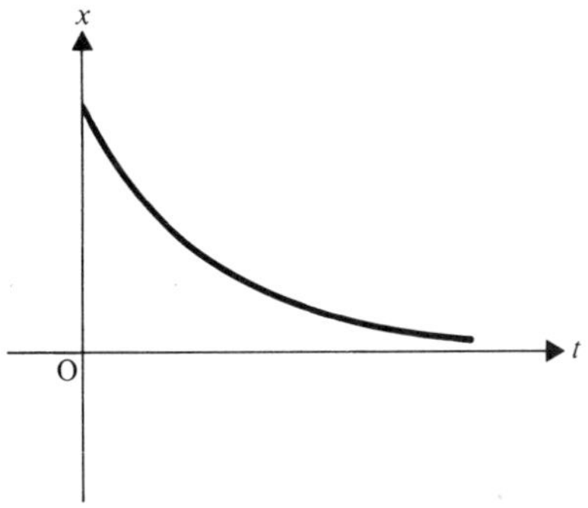

This motion is not oscillatory as there is no value of t for which $x = 0$ and no value of t for which $\dot{x} = 0$.

CASE 2. $x = e^{-kt}(A + Bt)$

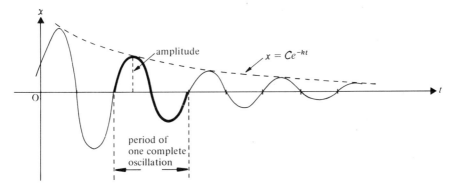

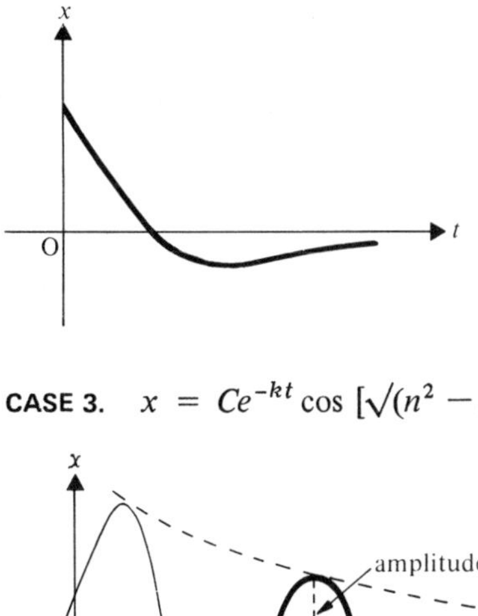

Again this motion is not oscillatory as there is only one finite value of t for which $x = 0$, and only one finite value of t for which $\dot{x} = 0$.

CASE 3. $x = Ce^{-kt} \cos [\sqrt{(n^2 - k^2)}t + \epsilon]$

This motion is oscillatory and is called *damped harmonic motion*.
The oscillations die away as t increases. The function $x = Ce^{-kt}$ forms an upper boundary for the curve and e^{-kt} is called the damping factor.
One complete oscillation is the path travelled by the particle in the interval of time between passing in the same direction through O on successive occasions.

The *period of an oscillation* is the time taken to complete one oscillation.
The *amplitude of an oscillation* is the maximum value of x attained in that oscillation.
Consider a particle which moves along a straight line Ox such that its displacement x from O at time t is given by

$$x = 2e^{-t}\cos(3t + \pi/3) \tag{1}$$

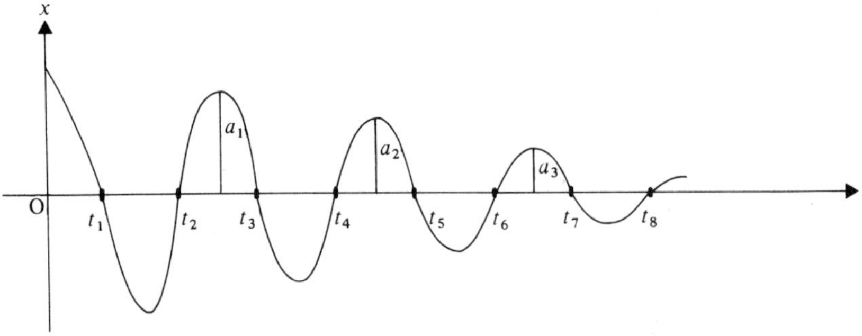

Taking complete oscillations as the path travelled by the particle in the intervals of time from t_2 to t_4, t_4 to t_6 etc. each oscillation begins as the particle passes through O when moving in the positive sense. (We could equally well use the intervals of time t_1 to t_3 etc.)

PERIOD OF THE OSCILLATIONS

When $x = 0$, $\cos(3t + \pi/3) = 0$

i.e. $3t = (2N-1)\pi/2 - \pi/3 \quad (N = 1, 2, 3, \ldots)$

Thus the particle passes through O at times

$$\tfrac{1}{3}(\pi/2 - \pi/3), \tfrac{1}{3}(3\pi/2 - \pi/3), \tfrac{1}{3}(5\pi/2 - \pi/3), \ldots$$

Hence the successive times when the particle passes through O are separated by *equal intervals* of $\pi/3$ seconds.
Therefore the *period of these oscillations is constant* and equal to $2\pi/3$ seconds.

AMPLITUDE

The stationary values of x occur when $\dot{x} = 0$.
From (1)

$$\dot{x} = -2e^{-t}\cos(3t + \pi/3) - 6e^{-t}\sin(3t + \pi/3)$$

Therefore when $\dot{x} = 0$ $\qquad \tan(3t + \pi/3) = -\tfrac{1}{3}$

Thus the stationary (maximum *and* minimum) values of x occur when

$$3t = [N\pi - \alpha], \quad N = 1, 2, 3, \ldots \quad \text{and} \quad \alpha = \arctan\left(\tfrac{1}{3}\right) + \pi/3$$

Now the maximum values of x occur when N is even (i.e. when $\arctan\left(-\frac{1}{3}\right)$ is an angle in the fourth quadrant), giving:

$$t = \tfrac{1}{3}(2n\pi - \alpha) \quad \text{and} \quad \cos(3t + \pi/3) = 1/\sqrt{10}, \quad n = 1, 2, 3, \ldots$$

Therefore successive values of the amplitude are

$$a_1 = \frac{2}{\sqrt{10}} e^{-(2\pi-\alpha)/3}, \; a_2 = \frac{2}{\sqrt{10}} e^{-(4\pi-\alpha)/3}, \; \ldots, a_n = \frac{2}{\sqrt{10}} e^{-(2n\pi-\alpha)/3}$$

From this we see that $\dfrac{a_n}{a_{n+1}} = e^{-2\pi/3}$

i.e. the ratio of the amplitudes of successive oscillations is constant.
Therefore the *amplitude of successive oscillations decreases with time in geometric progression* of common ratio e^{-T} where T is the period of the oscillations.
The properties which this example demonstrates are, in fact, general properties of damped harmonic motion. This can be proved from the general equation for this type of motion, $x = Ce^{-t} \cos(t\sqrt{(n^2 - k^2)} + \epsilon)$, giving results which are summarised below.

SUMMARY

When a particle of mass m moves in a straight line Ox under the action of a force mn^2x directed towards O and a resisting force $2mkv$, the equation of motion is

$$\frac{d^2x}{dt^2} + 2k\frac{dx}{dt} + n^2x = 0$$

Provided that $k^2 < n^2$, the complete solution of this equation is in the form

$$x = Ce^{-kt} \cos(t\sqrt{(n^2 - k^2)} + \epsilon)$$

where C and ϵ are constants of integration.
This equation represents damped harmonic oscillations, and e^{-kt} is called the damping factor. The period of these oscillations is constant and equal to

$$\frac{2\pi}{\sqrt{(n^2 - k^2)}}$$

The amplitude of the oscillations decreases with time in geometric progression.
If $k^2 \geqslant n^2$, the motion is not oscillatory.

EXAMPLES 7b

1) A particle moving in a straight line Ox has a displacement x from O at time t where x satisfies the equation

$$\frac{d^2x}{dt^2} + 4\frac{dx}{dt} + 13x = 0$$

Show that the motion of the particle is oscillatory. If, when $t = 0, x = 3$ and $\dot{x} = -6$, show that the particle oscillates about O with a constant period.

Comparing
$$\frac{d^2x}{dt^2} + 4\frac{dx}{dt} + 13x = 0 \qquad (1)$$

with
$$\frac{d^2x}{dt^2} + 2k\frac{dx}{dt} + n^2x = 0$$

we see that $k = 2$ and $n^2 = 13$, i.e. $k^2 < n^2$.
Therefore the complete solution of (1) is

$$x = Ce^{-2t}\cos(3t + \epsilon) \qquad (2)$$

where C and ϵ are constants.
Therefore (1) represents the equation of motion of an oscillating particle.
From (2),

$$\dot{x} = -2Ce^{-2t}\cos(3t + \epsilon) - 3Ce^{-2t}\sin(3t + \epsilon) \qquad (3)$$

Substituting $t = 0$ and $x = 3$ into (2) gives

$$3 = C\cos\epsilon \qquad (4)$$

Substituting $t = 0$ and $\dot{x} = -6$ into (3) gives

$$-6 = -2C\cos\epsilon - 3C\sin\epsilon \qquad (5)$$

(5) ÷ (4) gives
$$-2 = -2 - 3\tan\epsilon$$
$$3\tan\epsilon = 0$$
$$\epsilon = 0$$

Therefore
$$C = 3$$

Therefore (2) becomes

$$x = 3e^{-2t}\cos 3t \qquad (6)$$

When $x = 0$,

$$3e^{-2t}\cos 3t = 0$$
$$\cos 3t = 0$$
$$3t = (2N - 1)\pi/2$$
$$t = (2N - 1)\pi/6, \quad N = 1, 2, 3, \ldots$$

This shows that the particle passes through O at successive times separated by equal intervals of $\pi/6$ s, therefore the particle oscillates about O with a constant period of $\pi/3$ s.

2) A particle of mass m is attached to one end of a light elastic string of natural length l and modulus $2mg$. The other end of the string is attached to a fixed point O. With the string vertical, the particle is pulled a distance l below the equilibrium position and released. The subsequent motion takes place in a medium which offers a resistance $mv\sqrt{(g/l)}$ when the speed is v. Show that the particle performs damped harmonic motion about the equilibrium position without the string going slack.

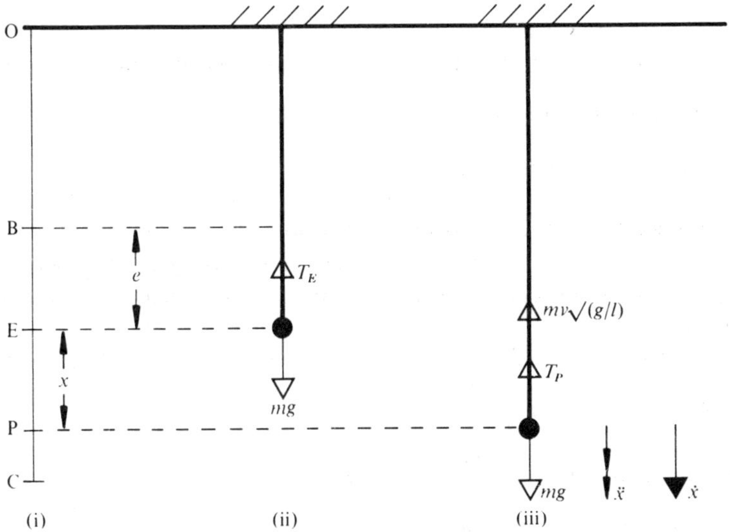

In diagram (i) OB is the natural length of the string,
E is the equilibrium position of the particle,
P is a general position with displacement x from E,
C is the initial position of the particle.

In diagram (ii) $T_E = \dfrac{2mge}{l}$ (Hooke's Law)

As the particle is in equilibrium, $T_E = mg$ and $e = l/2$
In diagram (iii) where the extension is $l/2 + x$

$$T_P = \frac{2mg}{l}(l/2 + x)$$

The resultant downward force on the particle is:

$$mg - \frac{2mg}{l}(l/2 + x) - m\sqrt{\frac{g}{l}}\,\dot{x}$$

Applying Newton's Law in this direction gives

$$mg - \frac{2mg}{l}(l/2 + x) - m\sqrt{\frac{g}{l}}\,\dot{x} = m\ddot{x}$$

or
$$\ddot{x} + \sqrt{\frac{g}{l}}\,\dot{x} + \frac{2g}{l}x = 0 \tag{1}$$

Comparing with
$$\ddot{x} + 2k\dot{x} + n^2 x = 0$$

we see that $k^2 < n^2$

Therefore the complete solution of (1) is

$$x = Ce^{-t\sqrt{(g/4l)}} \cos [t\sqrt{(7g/4l)} + \epsilon] \tag{2}$$

where C and ϵ are constants of integration.

Therefore the motion of this particle is damped harmonic motion with period $2\pi\sqrt{\frac{4l}{7g}}$ provided that the string does not go slack.

Now the particle is initially released from rest at C.

Therefore when $t = 0, \dot{x} = 0$ and $x = l$ showing that the first maximum value of x is l. (x is maximum when the particle is at its lowest position.)

If the string is to remain taut in the subsequent motion the greatest height of the particle above E must not exceed $l/2$. i.e. the next stationary value of x must not numerically exceed $l/2$.

Now the next stationary value of x occurs when a time equal to half the period has elapsed i.e. when $t = \pi\sqrt{\frac{4l}{7g}}$

Substituting $t = 0$ and $t = \pi\sqrt{\frac{4l}{7g}}$ into (2) gives:

first maximum value of $x = C\cos\epsilon = l$.

first minimum value of $x = -Ce^{-\pi/\sqrt{7}}\cos\epsilon = -e^{-\pi/\sqrt{7}}l$.

Therefore the maximum height above E to which the particle rises is

$$l/e^{\pi/\sqrt{7}} = l/3.27 \qquad \text{i.e.} < l/2$$

Therefore the string does not go slack in the subsequent motion.

EXERCISE 7b

1) A particle moves along a straight line Ox such that its displacement x from O at time t is given by

(a)
$$\ddot{x} + 3\dot{x} + 4x = 0$$

(b)
$$\frac{d^2x}{dt^2} + 4\frac{dx}{dt} + 4 = 0$$

(c)
$$\ddot{x} + 2\sqrt{\frac{g}{l}}\,\dot{x} + \frac{3g}{l}x = 0$$

Determine in each case whether the particle performs damped harmonic motion and, if it does, write down the period of the motion.

2) A particle moves along a straight line Ox such that at time t its displacement from O is x where

$$\ddot{x} + 2\dot{x} + 3x = 0$$

Show that the particle performs damped harmonic motion. If when $t = 0, x = 0$, $\dot{x} = 3$, find the value of x when $t = 2$.

3) A particle moves along a straight line Ox such that its displacement x from O at time t is given by

$$x = 2e^{-t} \cos 3t.$$

Prove that the particle performs oscillations about O with a constant period and that the amplitude of the motion decreases in geometric progression.
Find the common ratio of this geometric progression.

4) A light elastic string of natural length $2a$ and modulus mg has a particle of mass m attached to its mid-point. One end of the string is attached to a point A and the other end to a point B, both on a smooth horizontal table where AB $= 4a$. The particle is held at a point C where ACB is a straight line and AC $= a$, and released. The subsequent motion takes place in a medium which offers a resistance $mv\sqrt{g/2a}$ when the speed of the particle is v. Derive the equation of motion of this particle and determine whether the motion is damped harmonic.

5) A light elastic string of natural length $2a$ and modulus $2mg$ has a particle of mass m attached to its mid-point. One end of the string is fixed to a point A, the other end is fixed to a point B, a distance $4a$ vertically below A. Find the equilibrium position of the particle.
The particle is held at a distance $3a$ below A and released from rest in a medium which offers a resistance of $mv\sqrt{g/a}$ when the speed of the particle is v.
Show that the particle performs damped harmonic motion about the equilibrium position and find the maximum height to which the particle rises above the equilibrium position.

MISCELLANEOUS EXERCISE 7

1) A small bead of mass m is threaded on a fixed horizontal straight smooth wire. It is acted on by a force of constant magnitude F directed towards a point A which is fixed at a distance h from the wire. If O is the point of the wire nearest to A and x is the distance of the bead from O at time t prove that the equation of motion is

$$m\ddot{x} = -Fx/(x^2 + h^2)^{1/2}$$

Prove that, if the speed of the bead at O is V, the greatest value of x is

$$\left\{h\frac{mV^2}{F} + \frac{1}{4}\left(\frac{mV^2}{F}\right)^2\right\}^{1/2}$$

(Oxford)

2) A particle of unit mass is moving in a horizontal straight line so that its displacement x from a fixed point of the line at time t satisfies the differential equation

$$4\frac{d^2x}{dt^2} + 12\frac{dx}{dt} + 13x = 0$$

Describe a physical situation which could give rise to such an equation of motion.
Solve the equation given that $x = 4$ and $dx/dt = 6$ when $t = 0$. Show from the form of the solution that the motion is oscillatory with a constant period, and that the amplitudes of successive oscillations decrease in geometric progression.
(U of L)

3) A particle of mass m is attached to one end, B, of a light spring, AB, of natural length l and modulus mln^2. At time $t = 0$ the spring and particle are lying at rest on a smooth horizontal table, with the spring straight but unstretched. The end A is then moved in a straight line in the direction BA with constant acceleration f, so that, after t seconds, its displacement in this direction from its initial position is $\frac{1}{2}ft^2$. Show that the displacement, x, of the particle at time t in the direction BA from its initial position satisfies the equation

$$\frac{d^2x}{dt^2} + n^2x = \tfrac{1}{2}n^2ft^2.$$

By using the substitution $y = \frac{1}{2}ft^2 - x$, or otherwise, find the value of x at time t and show that the tension in the spring never exceeds $2mf$. (General solutions of differential equations may be quoted.) (JMB)

4) A particle of mass m is suspended from a fixed point by a spring of natural length l and modulus $5mn^2l$. When in motion it is resisted by a force of magnitude $2mn$ times its speed. Initially the particle is hanging in equilibrium and it is then projected vertically downwards with speed V. If x is the displacement downwards at time t from the equilibrium position, show that

$$\frac{d^2x}{dt^2} + 2n\frac{dx}{dt} + 5n^2x = 0$$

Find x as a function of t and sketch the graph of this function. Show that the particle is instantaneously at rest when $nt = \frac{1}{2}k\pi + \alpha$, where k is a non-negative integer and α is the acute angle such that $\tan 2\alpha = 2$.
(General solutions of differential equations may be quoted.) (JMB)

5) A particle of unit mass is tied to one end of a light elastic string, of natural length a and modulus $2an^2$, and the other end of the string is attached to a fixed point O. The particle is released from rest at a point a distance $2a$ vertically below O. When the particle is moving with velocity v, the air resistance is $2nv$. Prove that, when the extension of the string is x,

$$\frac{d^2x}{dt^2} + 2n\frac{dx}{dt} + 2n^2x = g$$

Prove also that the particle first comes to rest before the string becomes slack provided that

$$2an^2 < g(e^{\pi} + 1)$$ (Oxford)

6) A particle of mass m is attached to one end of an elastic string of modulus mg and natural length l. The other end of the string is attached to a fixed point O. The particle is held below O so that the string is vertical and of length $3l$, and is then released. The subsequent motion takes place in a medium which offers a resistance equal to $(3g/l)^{1/2}mv$, where v is the speed of the particle. Show that the extension x of the string at time t satisfies the differential equation

$$l\frac{d^2x}{dt^2} + (3gl)^{1/2}\frac{dx}{dt} + gx = gl$$

Prove that the string does not become slack, and that the particle is next at rest when its depth below O is $l(2 - e^{-\sqrt{3}\pi})$. (U of L)

7) A rocket vehicle uses fuel at a rate μ, so that

$$dm/dt = -\mu,$$

where m is the vehicle's mass at time t. The fuel used is expelled backwards from the vehicle with a constant velocity V relative to the vehicle. The vehicle starts from rest at time $t = 0$ with a total mass m_0 and reaches a final speed v_1 at a time t_1, the mass at that time being m_1. If no external forces act on the vehicle and v denotes the speed at time t, use the conservation of linear momentum to show that

$$m\,dv/dt = \mu V$$

and deduce that

$$v_1 = V\log_e(m_0/m_1).$$

If, for a particular vehicle, $m_0 = 10{,}000$ kg, $V = 2{,}500$ m s^{-1}, $v_1 = 5{,}000$ m s^{-1}, $t_1 = 200$ s and μ is constant, find numerical values for μ, m_1 and the greatest and least acceleration of the vehicle during the period of powered flight.

(WJEC)

8) A particle of mass m is attached to one end of a light elastic string of natural length a and modulus $\frac{25}{16}mn^2a$ $(n > 0)$. The other end of the string is fixed to a point A on a horizontal plane. The particle is held at rest on the plane with string extended to a length $2a$ and then released. During the ensuing motion the particle experiences a resistance of $\frac{3}{2}mn$ times its speed. If at time t after the release of the particle the length of the string is $a + x$ $(0 \leqslant x \leqslant a)$ prove that

$$\frac{d^2x}{dt^2} + \frac{3}{2}n\frac{dx}{dt} + \frac{25}{16}n^2x = 0$$

Deduce that the string slackens at $t = T$, where $\tan nT = -\frac{4}{3}$. Prove also that the particle does not reach A. (JMB)

9) A clock has a simple pendulum of length l with a bob of mass m. The effect of air resistance is to produce a force on the bob of magnitude $2km$ times its speed, where $k < \sqrt{(g/l)}$. Show that, when the pendulum is swinging freely with small amplitude, its angular displacement θ from the downward vertical at time t (measured from an instant when $\theta = 0$) is given approximately by $\theta = Ae^{-kt} \sin nt$, where $n = \sqrt{[(g/l) - k^2]}$ and A is a constant.
Every time θ passes through the value zero the clock's mechanism applies an impulse which causes the pendulum's angular speed to increase by ω. Use the above formula for the motion in the interval between the impulse at $t = 0$ and the next impulse to determine the value of A, in terms of k, n and ω, for which the angular speed immediately after every impulse is the same. Hence show that, in a steady motion of this type, the maximum value of θ is

$$\frac{\omega e^{-k\tau} \sin n\tau}{n(1 - e^{-\pi k/n})}$$

where $\tan n\tau = n/k$. (JMB)

CHAPTER 8

CENTRE OF MASS AND ITS MOTION. STABILITY OF EQUILIBRIUM

CENTRE OF MASS

Consider a set of n particles whose masses are $m_1, m_2, m_3, \ldots, m_p, \ldots, m_n$, and whose position vectors relative to an origin O are $\mathbf{r}_1, \mathbf{r}_2, \mathbf{r}_3, \ldots, \mathbf{r}_p, \ldots, \mathbf{r}_n$. *The centre of mass of the set of particles is defined as the point with position vector* $\mathbf{r}_M$ where

$$\mathbf{r}_M = \frac{\sum_{p=1}^{n} m_p \mathbf{r}_p}{\sum_{p=1}^{n} m_p}$$

EXAMPLE

Find the position vector of the centre of mass of three particles A, B and C where the mass and position vector of each particle is given in the following table.

Particle	Mass	Position vector
A	2 units	$\mathbf{i} + 4\mathbf{j} - 7\mathbf{k}$
B	5 units	$3\mathbf{i} - 2\mathbf{j} + \mathbf{k}$
C	3 units	$\mathbf{i} - 6\mathbf{j} + 13\mathbf{k}$

The position vector $\mathbf{r}_M$ of the centre of mass of A, B and C is given by

$$\mathbf{r}_M = \frac{2(\mathbf{i} + 4\mathbf{j} - 7\mathbf{k}) + 5(3\mathbf{i} - 2\mathbf{j} + \mathbf{k}) + 3(\mathbf{i} - 6\mathbf{j} + 13\mathbf{k})}{2 + 5 + 3}$$

$$= \frac{(2 + 15 + 3)\mathbf{i} + (8 - 10 - 18)\mathbf{j} + (-14 + 5 + 39)\mathbf{k}}{10}$$

Hence $\quad \mathbf{r}_M = 2\mathbf{i} - 2\mathbf{j} + 3\mathbf{k}$

We shall now show that the position of the centre of mass of a system of particles does not depend on the choice of origin.
Suppose that, relative to an origin O, the centre of mass of a set of particles is at a point C with position vector $\mathbf{r}_M$ and that, relative to an origin O_1, the centre of mass is at a point C_1 with position vector $\mathbf{s}_M$. The position vector of O_1 relative to O is $\mathbf{d}$.

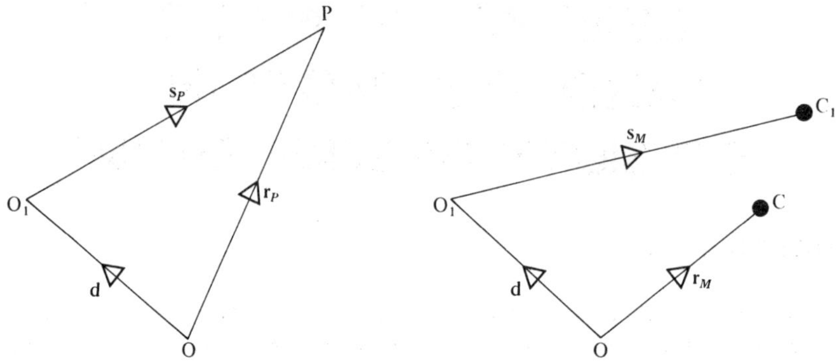

For each particle $\qquad \mathbf{r}_P = \mathbf{d} + \mathbf{s}_P$

For the set of particles $\qquad \mathbf{r}_M = \dfrac{\Sigma\, m_P \mathbf{r}_P}{\Sigma\, m_P}$

$$= \frac{\Sigma\, m_P(\mathbf{d} + \mathbf{s}_P)}{\Sigma\, m_P}$$

$$= \frac{\mathbf{d}\, \Sigma\, m_P + \Sigma\, m_P \mathbf{s}_P}{\Sigma\, m_P}$$

$$= \mathbf{d} + \frac{\Sigma\, m_P \mathbf{s}_P}{\Sigma\, m_P}$$

But, by definition, the position vector of the centre of mass relative to O_1 is given by

$$\mathbf{s}_M = \frac{\Sigma\, m_P \mathbf{s}_P}{\Sigma\, m_P}$$

Hence $\qquad \mathbf{r}_M = \mathbf{d} + \mathbf{s}_M$

Thus C and C_1 coincide

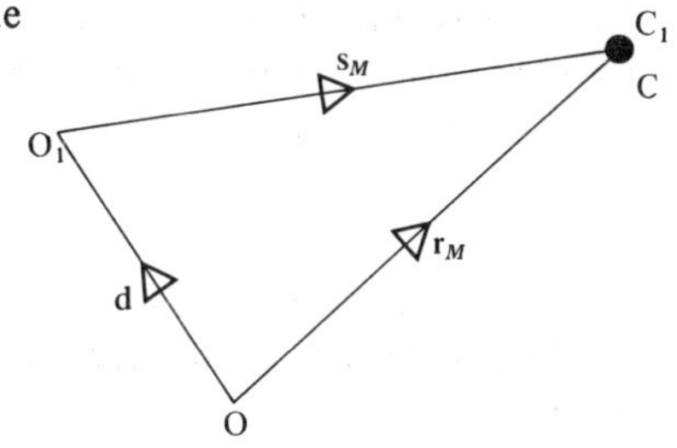

DISTINCTION BETWEEN CENTRE OF MASS AND CENTRE OF GRAVITY

The position of the centre of mass of a system depends only upon the mass and position of each constituent particle,

i.e.
$$\mathbf{r}_M = \frac{\Sigma\, m_p \mathbf{r}_p}{\Sigma\, m_p} \tag{1}$$

The location of G, the centre of gravity of the system, depends however upon the moment of the gravitational force acting on each particle in the system (about any point, the sum of the moments for all the constituent particles is equal to the moment for the whole system concentrated at G).
Hence, if $\mathbf{g}_P$ is the acceleration vector due to gravity of a particle P, the position vector $\mathbf{r}_G$ of the centre of gravity of the system is given by:

$$\mathbf{r}_G \times \sum m_P \mathbf{g}_P = \sum (\mathbf{r}_P \times m_P \mathbf{g}_P) \tag{2}$$

Now if the system is in a *uniform gravitational field* so that the accleration vector due to gravity ($\mathbf{g}$) is constant, then (2) becomes

$$\mathbf{r}_G \times \sum m_P \mathbf{g} = \sum (\mathbf{r}_P \times m_P \mathbf{g})$$

Hence
$$\mathbf{r}_G \sum m_P = \sum \mathbf{r}_P m_P$$

and
$$\mathbf{r}_G = \frac{\Sigma\, \mathbf{r}_P m_P}{\Sigma\, m_P} = \mathbf{r}_M$$

In this case, therefore, the centre of gravity and the centre of mass coincide.
If, however, the gravitational field is *not* uniform and $\mathbf{g}_P$ is not constant then, in general, equation (2) cannot be simplified and $\mathbf{r}_G \neq \mathbf{r}_M$.
Thus, for a system of particles in a uniform gravitational field, the centre of mass and the centre of gravity are identical points but *in a variable gravitational field, the centre of mass and the centre of gravity are, in general, two distinct points.*
The calculation of the centre of gravity of a system of particles in a non-uniform gravitational field is beyond the scope of this book, but it is important that the

distinction between centre of mass and centre of gravity is understood.
In this book it will be assumed, unless otherwise stated, that all analysis is carried out in a uniform gravitational field.

EXERCISE 8a

1) Four particles of masses 4, 1, 2 and 5 kg are placed at points with position vectors $\mathbf{i}+\mathbf{j}+3\mathbf{k}$, $4\mathbf{i}-3\mathbf{j}+\mathbf{k}$, $5\mathbf{i}-\mathbf{j}$ and $2\mathbf{j}-\mathbf{k}$ respectively. Find the position vector of their centre of mass.

2) Particles of masses $3m$, $2m$, $5m$ and $2m$ are placed at points with position vectors $\mathbf{a}$, $\mathbf{b}$, $3\mathbf{a}$ and $4\mathbf{b}$ respectively. Find the position vector of their centre of mass.

3) Three particles at points A, B and C have a centre of mass at the point with co-ordinates $(1, 2, 3)$. The mass at $A(4, -1, 2)$ is 2 kg and the mass at $B(-6, 1, -5)$ is 1 kg. If C is the point $(p, q, 8)$, calculate the values of p and q and the mass at C.

4) Find the centre of gravity of three particles of masses 2, 2 and 3 kg if they are at points with co-ordinates $(1, 0, 2)$, $(3, 1, -4)$ and $(2, -3, -1)$ respectively. (Assume that the particles are in a uniform gravitational field.)

5) OABC is a parallelogram. The position vectors relative to O of A and C are $\mathbf{a}$ and $\mathbf{c}$ respectively. Find the centre of mass of particles placed at O, A, B and C whose masses are 1, 2, 3 and 4 kg respectively.

6) Particles of masses 3, 3 and 1 unit are placed at points with position vectors $a\mathbf{i}+2\mathbf{j}+2\mathbf{k}$, $6\mathbf{i}+b\mathbf{k}$ and $c\mathbf{j}-3\mathbf{k}$ respectively. If their centre of mass is at the point with position vector $3\mathbf{i}+\mathbf{j}-3\mathbf{k}$, evaluate a, b and c. If the particle of mass 1 unit is then replaced by one of mass 3 units, where is the new centre of mass?

PROPERTIES OF THE MOTION OF THE CENTRE OF MASS

In this section of the work we shall analyse the motion of the centre of mass of a set of particles.
Consider a typical particle P of mass m_P which, at time t has a position vector $\mathbf{r}_P$, a velocity vector $\mathbf{v}_P$, an acceleration vector $\mathbf{a}_P$ and is under the action of a single force $\mathbf{F}_P$. Initially the position and velocity vectors of P are $\mathbf{R}_P$ and $\mathbf{V}_P$. Similarly $\mathbf{r}_M$, $\mathbf{v}_M$ and $\mathbf{a}_M$ are the position, velocity and acceleration vectors of the centre of mass at time t, its initial position and velocity vectors being $\mathbf{R}_M$ and $\mathbf{V}_M$.

Acceleration

By definition
$$\mathbf{r}_M = \frac{\Sigma\, m_P\mathbf{r}_P}{\Sigma\, m_P}$$

or
$$\mathbf{r}_M \sum m_P = \sum m_P \mathbf{r}_P$$

Differentiating twice with respect to time we have

$$\frac{d^2\mathbf{r}_M}{dt^2} \sum m_P = \sum m_P \frac{d^2\mathbf{r}_P}{dt^2}$$

or
$$\mathbf{a}_M \sum m_P = \sum m_P \mathbf{a}_P \tag{1}$$

Now applying Newton's Law to the motion of one particle,

$$\mathbf{F}_P = m_P \mathbf{a}_P$$

So for the whole set of particles

$$\sum \mathbf{F}_P = \sum m_P \mathbf{a}_P \tag{2}$$

From (1) and (2) we see that

$$\sum \mathbf{F}_P = \left(\sum m_P\right) \mathbf{a}_M$$

But $\Sigma\, \mathbf{F}_P$ is the resultant force acting on the system, $\Sigma\, m_P$ is the total mass of the system and $\mathbf{a}_M$ is the acceleration of the centre of mass.
Thus the acceleration of the centre of mass of a system is the same as that of a particle whose mass is the total mass of the system, acted upon by the resultant of the forces acting on the system.

EXAMPLE

Find the acceleration vector of the centre of mass of particles A, B, C and D whose masses are 1, 2, 3 and 4 units respectively and which are moving under the action of force vectors $\mathbf{i} - \mathbf{j} + 3\mathbf{k}$, $4\mathbf{i} - 3\mathbf{j}$, $2\mathbf{j} + \mathbf{k}$ and $5\mathbf{i} - 4\mathbf{k}$ respectively.

The resultant force $\mathbf{F}$ is given by

$$\begin{aligned}\mathbf{F} &= (\mathbf{i} - \mathbf{j} + 3\mathbf{k}) + (4\mathbf{i} - 3\mathbf{j}) + (2\mathbf{j} + \mathbf{k}) + (5\mathbf{i} - 4\mathbf{k}) \\ &= 10\mathbf{i} - 2\mathbf{j}\end{aligned}$$

The total mass of the system is 10 units. Hence the acceleration vector $\mathbf{a}_M$ of the centre of mass is given by

$$\begin{aligned}\mathbf{a}_M &= \text{resultant force/total mass} \\ &= (10\mathbf{i} - 2\mathbf{j})/10\end{aligned}$$

Note: Since in this problem the constituent particles are moving under the action of constant forces, the acceleration of the centre of mass also is constant.

Momentum

The impulse applied to each particle in the system is equal to the change in its momentum, thus

$$\mathbf{I}_P = m_P\mathbf{v}_P - m_P\mathbf{V}_P$$

Hence for the whole system

$$\sum \mathbf{I}_P = \sum m_P\mathbf{v}_P - \sum m_P\mathbf{V}_p \quad (3)$$

Now for the centre of mass

$$\mathbf{r}_M \sum m_P = \sum m_P\mathbf{r}_P$$

Differentiating with respect to time gives

$$\frac{d\mathbf{r}_M}{dt} \sum m_P = \sum m_P \frac{d\mathbf{r}_P}{dt}$$

But at time t, $\quad \dfrac{d\mathbf{r}_M}{dt} = \mathbf{v}_M$ and $\dfrac{d\mathbf{r}_P}{dt} = \mathbf{v}_P$

and when $t = 0$ $\quad \dfrac{d\mathbf{r}_M}{dt} = \mathbf{V}_M$ and $\dfrac{d\mathbf{r}_P}{dt} = \mathbf{V}_P$

Therefore at time t $\quad \mathbf{v}_M \sum m_P = \sum m_P\mathbf{v}_P$

and initially $\quad \mathbf{V}_M \sum m_P = \sum m_P\mathbf{V}_P$

Subtracting $\quad (\mathbf{v}_M - \mathbf{V}_M) \sum m_P = \sum m_P\mathbf{v}_P - \sum m_P\mathbf{V}_P \quad (4)$

Equation (3) shows that the resultant impulse acting on the system is equal to the change in the resultant momentum of the set of particles.
From equation (4) we see that *the change in resultant momentum of the system is the same as the change in momentum of a particle, of mass equal to the total mass of the system, placed at the centre of mass.*

EXAMPLE

Force vectors $2\mathbf{i} + 5\mathbf{j} - \mathbf{k}$, $\mathbf{i} - \mathbf{j} + 3\mathbf{k}$ and $3\mathbf{i} + \mathbf{j}$ act respectively on three particles of masses 4, 2 and 3 units which are initially at rest. Find the resultant momentum of the system after 4 seconds. Find the impulse of each force and show that the vector sum of these impulses is equal to the change in the momentum of a particle of mass 9 units at the centre of mass.

$$\begin{aligned}\text{The resultant force } \mathbf{F} &= (2\mathbf{i} + 5\mathbf{j} - \mathbf{k}) + (\mathbf{i} - \mathbf{j} + 3\mathbf{k}) + (3\mathbf{i} + \mathbf{j}) \\ &= 6\mathbf{i} + 5\mathbf{j} + 2\mathbf{k}\end{aligned}$$

The total mass $= 4 + 2 + 3 = 9$ units

Now the motion of the centre of mass is the same as the motion of a particle of mass 9 units subjected to the force **F**.
Hence, applying Newton's Law

$$\mathbf{F} = 9\mathbf{a}_M$$

Integrating with respect to time gives

$$(6\mathbf{i} + 5\mathbf{j} + 2\mathbf{k})t = 9\mathbf{v}_M + \mathbf{V}$$

When $t = 0, \quad \mathbf{v}_M = \mathbf{0} \quad \text{so} \quad \mathbf{V} = \mathbf{0}$

Therefore $(6\mathbf{i} + 5\mathbf{j} + 2\mathbf{k})t = 9\mathbf{v}_M$

So when $t = 4$ $\quad \mathbf{v}_M = \frac{4}{9}(6\mathbf{i} + 5\mathbf{j} + 2\mathbf{k})$

The resultant momentum of the system is equal to the momentum of a particle of mass 9 units at the centre of mass, i.e. $9\mathbf{v}_M$.

Thus the resultant momentum is $24\mathbf{i} + 20\mathbf{j} + 8\mathbf{k}$

Using $\mathbf{I} = \int \mathbf{F}dt$ for each particle, where $\mathbf{I}$ is the impulse vector, we have,

$$\mathbf{I}_1 = \int_0^4 (2\mathbf{i} + 5\mathbf{j} - \mathbf{k})dt = 8\mathbf{i} + 20\mathbf{j} - 4\mathbf{k}$$

$$\mathbf{I}_2 = \int_0^4 (\mathbf{i} - \mathbf{j} + 3\mathbf{k})dt = 4\mathbf{i} - 4\mathbf{j} + 12\mathbf{k}$$

$$\mathbf{I}_3 = \int_0^4 (3\mathbf{i} + \mathbf{j})dt = 12\mathbf{i} + 4\mathbf{j}$$

$$\begin{aligned}\text{Resultant impulse} &= (8\mathbf{i} + 20\mathbf{j} - 4\mathbf{k}) + (4\mathbf{i} - 4\mathbf{j} + 12\mathbf{k}) + (12\mathbf{i} + 4\mathbf{j}) \\ &= 24\mathbf{i} + 20\mathbf{j} + 8\mathbf{k} \\ &= 9\mathbf{v}_M \qquad \text{Q.E.D.}\end{aligned}$$

Work Done and Change in Kinetic Energy

Consider a set of particles moving under the action of a set of constant forces. The work done when a force $\mathbf{F}_P$ is applied to one particle P of mass m_P and moves it from its initial position vector $\mathbf{R}_P$ to a general position vector $\mathbf{r}_P$.
The work done by $\mathbf{F}_P = \mathbf{F}_P \cdot (\mathbf{r}_P - \mathbf{R}_P)$
The total work done by all the forces acting on the system of particles is therefore given by

$$\sum \mathbf{F}_P \cdot (\mathbf{r}_P - \mathbf{R}_P) = \sum \mathbf{F}_P \cdot \mathbf{r}_P - \sum \mathbf{F}_P \cdot \mathbf{R}_P$$

Now consider the action of a force $\Sigma\, \mathbf{F}_P$ on a particle of mass $\Sigma\, m_P$ at the centre of mass, causing a displacement from the initial position vector $\mathbf{R}_M$ to a general position vector $\mathbf{r}_M$.

The work done by the force $\Sigma\, \mathbf{F}_P$ is given by

$$\left(\sum \mathbf{F}_P\right) . (\mathbf{r}_M - \mathbf{R}_M) = \left(\sum \mathbf{F}_P\right) . \left(\frac{\Sigma\, m_P \mathbf{r}_P}{\Sigma\, m_P} - \frac{\Sigma\, m_P \mathbf{R}_P}{\Sigma\, m_P}\right)$$

In general this is *not* equal to $\Sigma\, \mathbf{F}_P . \mathbf{r}_P - \Sigma\, \mathbf{F}_P . \mathbf{R}_P$

So *the total work done by the individual forces displacing the individual particles is not, in general, equal to the work done by the resultant force displacing a particle of mass $\Sigma\, m_P$ at the centre of mass of the system.*

If $\mathbf{F}_P$ is the resultant force acting on the particle P, the change in kinetic energy of P is equal to the work done by $\mathbf{F}_p$.

It then follows immediately from the last section that:

the total change in kinetic energy of all the individual particles is not in general, equal to the change in kinetic energy of a particle of mass $\Sigma\, m_p$ at the centre of mass of the system.

Clearly the two conclusions just reached are 'negative' results i.e. they make no constructive contribution to the solution of those problems which require the calculation of the work done by a number of forces acting on a system of particles, or the change in kinetic energy so caused.

Nevertheless they are important results, since they show that it is invalid to use the motion of the centre of mass of a system of *independent* particles subjected to *random* forces, in evaluating the total work done or the change in kinetic energy of the system as a whole.

EXAMPLE

Two forces $\mathbf{F}_1$ and $\mathbf{F}_2$ act respectively on particles A and B which are initially at rest.

If $\mathbf{F}_1 = 2\mathbf{i} + \mathbf{j} - 3\mathbf{k}$, $\mathbf{F}_2 = 4\mathbf{i} + \mathbf{k}$, A is of mass 3 kg and B is of mass 5 kg, find the change in kinetic energy of A and B in the first 12 seconds of motion.

Find also the change in kinetic energy of a particle of mass 8 kg, initially at rest, and subjected to a force $\mathbf{F}_1 + \mathbf{F}_2$. (Use SI units throughout.)

For particle A, the acceleration vector $\mathbf{a}_1$ is given by

$$\mathbf{a}_1 = \mathbf{F}_1/3$$

Hence $$\mathbf{v}_1 = \tfrac{1}{3} \int \mathbf{F}_1\, dt = \tfrac{1}{3}(2\mathbf{i} + \mathbf{j} - 3\mathbf{k})t + \mathbf{V}_1$$

But $\mathbf{v}_1 = \mathbf{0}$ when $t = 0$, so $\mathbf{V}_1 = \mathbf{0}$

Therefore, after 12 seconds,

$$\mathbf{v}_1 = \tfrac{12}{3}(2\mathbf{i} + \mathbf{j} - 3\mathbf{k})$$

$$|\mathbf{v}_1|^2 = 16[2^2 + 1^2 + (-3)^2] = 16 \times 14$$

$$(\mathrm{KE})_A = \tfrac{1}{2}(3)(16 \times 14) = 336 \text{ J}$$

Similarly for particle B,

$$\mathbf{a}_2 = \mathbf{F}_2/5$$

$$\mathbf{v}_2 = \int \mathbf{a}_2\,dt = \tfrac{1}{5}(4\mathbf{i}+\mathbf{k})t + \mathbf{0}$$

Hence, after 12 seconds,

$$\mathbf{v}_2 = \tfrac{12}{5}(4\mathbf{i}+\mathbf{k})$$

$$|\mathbf{v}_2|^2 = (\tfrac{12}{5})^2[4^2 + 1^2] = \tfrac{144}{25} \times 17$$

$$(\text{KE})_B = \tfrac{1}{2}(5)(\tfrac{144}{25} \times 17) = 244.8 \text{ J}$$

The change in kinetic energy of A and B together is therefore 580.8 J
Considering now a force $\mathbf{F}_1 + \mathbf{F}_2$ acting on a mass of (3 + 5) kg, causing an acceleration vector **a** we have,

$$\mathbf{a} = (\mathbf{F}_1 + \mathbf{F}_2)/8 = \tfrac{1}{8}(6\mathbf{i}+\mathbf{j}-2\mathbf{k})$$

$$\Big[\mathbf{v}\Big]_{\mathbf{0}}^{\mathbf{v}} = \int_0^t \mathbf{a}\,dt = \frac{t}{8}(6\mathbf{i}+\mathbf{j}-2\mathbf{k}) + \mathbf{0}$$

After 12 seconds $\quad \mathbf{v} = \tfrac{12}{8}(6\mathbf{i}+\mathbf{j}-2\mathbf{k})$

$$|\mathbf{v}|^2 = (\tfrac{3}{2})^2[6^2 + 1^2 + (-2)^2] = \tfrac{9}{4} \times 41$$

$$\text{KE} = \tfrac{1}{2}(8)(\tfrac{9}{4} \times 41) = 369 \text{ J}$$

Note: This example verifies that the total change in kinetic energy of the particles is not equal to the change in kinetic energy of the total mass subjected to the resultant force.

MOTION OF THE CENTRE OF MASS IN A UNIFORM GRAVITATIONAL FIELD

If the only forces acting on a set of particles are the weights of those particles then (using the notation introduced earlier in this chapter)

$$\mathbf{F}_P = m_P\mathbf{g}$$

Thus the total work done by all the forces in a specified time is given by

$$\sum \mathbf{F}_P \cdot (\mathbf{r}_P - \mathbf{R}_P) = \sum m_P\mathbf{g} \cdot (\mathbf{r}_P - \mathbf{R}_P)$$

$$= \mathbf{g} \cdot \left(\sum m_P\mathbf{r}_P - \sum m_P\mathbf{R}_P \right)$$

Now the work done when a force $\Sigma\,\mathbf{F}_P$ acts on a mass $\Sigma\, m_P$ at the centre of mass is given by

$$\left(\sum \mathbf{F}_P\right) . (\mathbf{r}_M - \mathbf{R}_M) = \left(\sum m_P \mathbf{g}\right) . \left(\frac{\Sigma\, m_P \mathbf{r}_P - \Sigma\, m_P \mathbf{R}_P}{\Sigma\, m_P}\right)$$

$$= \mathbf{g} . \left(\sum m_P \mathbf{r}_P - \sum m_P \mathbf{R}_P\right)$$

In this case we see that the total work done by the individual forces *is* equal to the work done by the resultant force applied to the total mass concentrated at the centre of mass.

Note: In this case all the forces act in the same direction so the scalar sum of the work they all do is consistent with the work done by the resultant force (a vector sum) in that direction.

Now in a uniform gravitation field, the centre of mass and centre of gravity of a system coincide.

Therefore the properties of the linear motion of a system of particles moving under gravity in a uniform gravitational field correspond in all respects to the motion produced when the resultant force acts on the total mass concentrated at the centre of gravity.

CENTROID

The centroid of an object is a geometric centre. For instance the centroid of a circle is the point of intersection of two diameters; the centroid of a parallelogram is the point of intersection of its diagonals; the centroid of a triangle is the point of concurrence of its medians.

In certain circumstances the centre of mass or the centre of gravity of the object happens to lie at the centroid, but this is not always the case. If the mass of the body is uniformly distributed, the centre of mass and the centroid coincide. But when the mass is unevenly distributed throughout the body it is most unlikely that the centroid and the centre of mass will be the same point.

For example, the centroid of a triangle PQR is at C as shown in diagram (i)

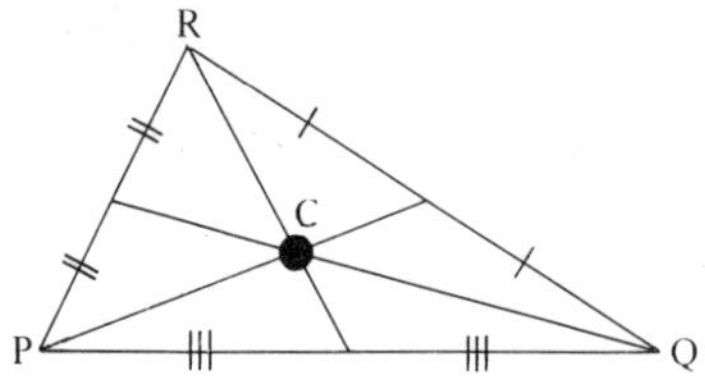

(i)

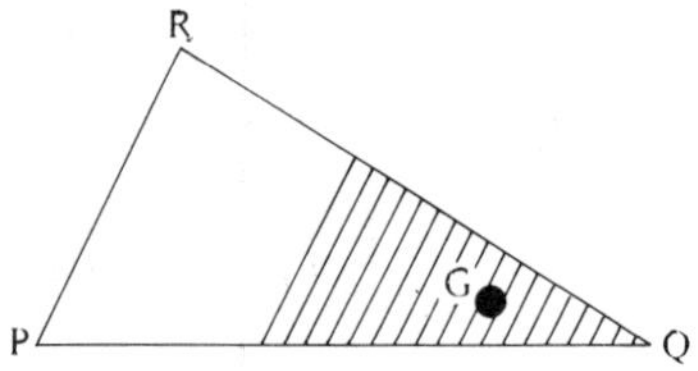

(ii)

But if the mass of a triangular lamina PQR is heavily concentrated in the shaded section as shown in diagram (ii) the centre of mass will be approximately at point G.

It is therefore misleading to use the term centroid when referring to centre of mass or to centre of gravity. Unfortunately this is a fairly common practice and the student must decide with care what is really meant by centroid whenever this term is encountered. Amongst the variety of expressions which include the word *centroid*, the two that follow are fairly frequently used and again they require thoughtful interpretation.

(i) The centre of mass of a set of particles of various masses may be called the *weighted centroid*.

(ii) If each particle of a set has an *associated number* (n, say) then, using the usual notation, the point with position vector $\dfrac{\Sigma\, \mathbf{r}_p \mathbf{n}_p}{\Sigma\, n_p}$ may be called the *centroid of a set of particles with associated numbers*.

SUMMARY

1. The position vector of the centre of mass of a set of particles is given by:

$$\mathbf{r}_M = \frac{\Sigma\, m_P \mathbf{r}_P}{\Sigma\, m_P}$$

2. The acceleration of the centre of mass is the same as the acceleration of a particle of mass $\Sigma\, m_P$ whose position vector is $\mathbf{r}_M$ subjected to a force $\Sigma\, \mathbf{F}_P$.
3. The change in the resultant momentum of the system is the same as the change in momentum of a particle of mass $\Sigma\, m_P$ at the centre of mass, under the action of a force $\Sigma\, \mathbf{F}_P$.
4. In general the total work done by the individual forces in displacing the individual particles is *not* equal to the work done by the resultant force acting on the concentrated mass at the centre of mass.
   Further, the total change in kinetic energy of all the particles is *not*, in general, equal to the change in kinetic energy of the concentrated mass subjected to the resultant force.

EXERCISE 8b

1) Forces $\mathbf{F}_1$, $\mathbf{F}_2$ and $\mathbf{F}_3$ act on particles A, B and C whose masses are m_1, m_2 and m_3 respectively. Find the acceleration of the centre of mass of the particles if:

(i) $\mathbf{F}_1 = 3\mathbf{i} + 7\mathbf{j} - \mathbf{k}$, $\quad \mathbf{F}_2 = 2\mathbf{i} - \mathbf{j} + 6\mathbf{k}$, $\quad \mathbf{F}_3 = \mathbf{i} + 4\mathbf{j} + 2\mathbf{k}$,
$m_1 = 2$ units, $\quad m_2 = 5$ units, $\quad m_3 = 3$ units.

(ii) $\mathbf{F}_1 = \mathbf{i} - 2\mathbf{k}$, $\quad \mathbf{F}_2 = 3\mathbf{i} + 4\mathbf{j}$, $\quad \mathbf{F}_3 = 2\mathbf{i} - \mathbf{j} - \mathbf{k}$,
$m_1 = m_2 = m_3 = m$.

2) Forces $\mathbf{F}_1 = 4\mathbf{i} + 3\mathbf{j} - \mathbf{k}$ and $\mathbf{F}_2 = 3\mathbf{i} - 5\mathbf{j} + 8\mathbf{k}$ act on two particles of masses 2 kg and 1 kg respectively. The particles are initially at rest at points with position vectors $\mathbf{i} + \mathbf{j} + \mathbf{k}$ and $2\mathbf{j} - 7\mathbf{k}$ respectively. Find:

(i) the initial position vector of the centre of mass,
(ii) the acceleration of the centre of mass,
(iii) the total momentum of the particles after 2 seconds.

3) ABCD is a square framework of rigid light rods. Particles of masses m, $2m$, $3m$ and $4m$ are attached at A, B, C and D. A force of magnitude $3F$ is applied at A towards B and a force of magnitude $4F$ is applied at B towards C. Find the linear acceleration of the centre of mass. If the mass at D is removed what effect does this have on (a) the magnitude (b) the direction of the acceleration of the centre of mass?

4) Two particles of mass 1 kg lie at rest at points P, Q on a smooth horizontal surface. A force of magnitude 2 N is applied to the particle at P in a direction inclined at $60°$ to PQ. Find the accleration of the centre of mass of the particles. Find also the total momentum and the total kinetic energy of the particles after 5 seconds. If the force of magnitude 2 N is now removed and replaced by two forces each of magnitude 1 N in the same direction as the original force, one applied to P and one applied to Q, in what way does this affect the motion of the centre of mass of the particles?

5) At time t, two particles of equal mass 3 kg have velocity vectors

$$(2+3t)\mathbf{i}+(-1+t)\mathbf{j} \quad \text{and} \quad (3-4t)\mathbf{i}+(2+5t)\mathbf{j}.$$

Determine the acceleration vector of their centre of mass and the total work done on the particles in the first two seconds.

6) A, B and C are points with co-ordinates $(4, 0)$, $(4, 3)$ and $(0, 3)$ respectively relative to an origin O. Forces of magnitudes 6, 10 and 4 N begin to act respectively on particles of equal mass at A, B and C, each force being directed towards O. Determine the initial position vector of the centre of mass of the particles and state the direction in which the centre of mass begins to move.

7) Three particles of masses 3, 3 and 4 units are at rest on a smooth horizontal plane at points A, B and C with position vectors $\mathbf{i}+4\mathbf{j}$, $5\mathbf{i}+10\mathbf{j}$ and $3\mathbf{i}+2\mathbf{j}$ respectively.
Write down the position vector of the centre of mass of the particles.
If constant forces of magnitudes $3\sqrt{17}$, $10\sqrt{5}$ and $\sqrt{13}$ units respectively now act on the particles in the directions $\overrightarrow{OA}$, $\overrightarrow{OB}$ and $\overrightarrow{OC}$ find the magnitude of the initial acceleration of the centre of mass. Find the total work done by the forces at the instant when the total momentum vector of the particles is $48\mathbf{i}+102\mathbf{j}$.

STABILITY OF EQUILIBRIUM

Consider first a smooth bead threaded on to a wire whose shape is shown in the diagram and which is fixed in a vertical plane.

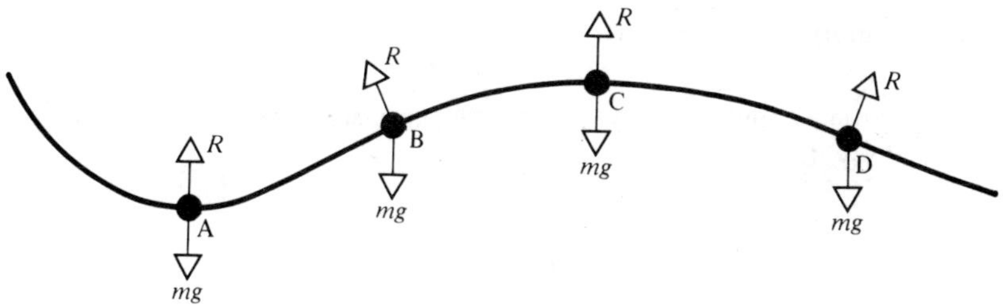

If the bead is placed at A or at C it will rest in equilibrium (mg and R are equal and opposite in these positions).

When placed at points such as B or D however the bead is not in equilibrium.

When the bead is slightly displaced from A, its natural tendency is to return to A which is called a *position of stable equilibrium*. When the bead is slightly displaced from C however, it will continue to move further from C which is called a *position of unstable equilibrium*.

It is interesting to notice that:

(i) the total mechanical energy of the bead remains constant as it moves along the wire, since the normal reaction R between the bead and the wire is always perpendicular to the direction of motion of the bead and hence R does no work,

(ii) the potential energy of the bead has a minimum value at A and a maximum value at C.

The above example illustrates the concept of stable and unstable equilibrium. We will now investigate a more general case.

Consider a system whose total mechanical energy remains constant (i.e. a conservative system).

Then $$\text{PE} + \text{KE} = \lambda \quad \text{(a constant)}$$

where the term PE includes all the potential energy of the system whether gravitational or elastic.

But $$\text{KE} = \tfrac{1}{2}mv^2$$

So $$\text{PE} = \lambda - \tfrac{1}{2}mv^2$$

Differentiating with respect to time gives

$$\frac{d(\text{PE})}{dt} = 0 - mv\frac{dv}{dt}$$

Now in a position of equilibrium the resultant force acting on the system is zero. Therefore the resultant acceleration is also zero.

i.e. $$\frac{dv}{dt} = 0$$

So, for equilibrium, $\dfrac{d(\text{PE})}{dt} = 0$

If the potential energy of the system can be expressed in terms of any single variable, θ say, then

$$\frac{d(\text{PE})}{dt} = \frac{d\theta}{dt}\frac{d(\text{PE})}{d\theta}$$

Thus if $\qquad \dfrac{d(\text{PE})}{d\theta} = 0 \quad$ it follows that $\dfrac{d(\text{PE})}{dt} = 0$

and the system is in equilibrium.
Hence any conservative system whose potential energy can be expressed in terms of a single variable θ, is in equilibrium in positions for which $\dfrac{d(\text{PE})}{d\theta} = 0$.

Stable Equilibrium

When a body is slightly displaced in any direction from a position of equilibrium E, it will tend to return to E if E is below the displaced position. If this condition is satisfied, the potential energy at E has a minimum value.
So an equilibrium position where $\theta = \alpha$ is stable if $\dfrac{d^2(\text{PE})}{d\theta^2} > 0$
(*Note*: The condition given above is the conclusion of an intuitive argument rather than a proof, which is too difficult at this stage. However the condition so derived is correct and can be used with confidence.)
So the conditions for stable equilibrium are

$$\frac{d(\text{PE})}{d\theta} = 0 \quad \text{and} \quad \frac{d^2(\text{PE})}{d\theta^2} > 0$$

Conversely a position for which

$$\frac{d(\text{PE})}{d\theta} = 0 \quad \text{and} \quad \frac{d^2(\text{PE})}{d\theta^2} < 0$$

is a position of unstable equilibrium.

EXAMPLES 8c

1) Two uniform smooth rods AB and BC, each of length $2a$ and mass m, are smoothly jointed at B. They rest symmetrically in a vertical plane over a beam of width $2b$. If the rods are not horizontal, show that there is a position of equilibrium only if $b \leqslant a$.

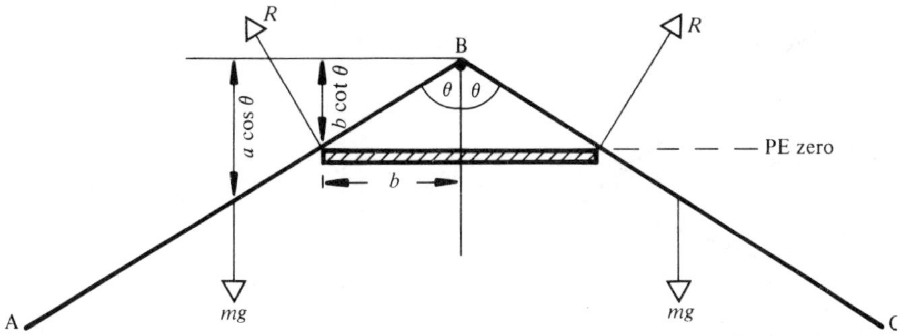

Let the rods be equally inclined to the vertical at a general angle θ.
(As θ varies, the two reactions, R, do no work so the system is conservative and we can use the potential energy method to investigate the equilibrium of the rod.)
In general the potential energy of the rod is given by:

$$\text{PE} = -2mg(a\cos\theta - b\cot\theta)$$

Hence
$$\frac{d(\text{PE})}{d\theta} = -2mg(-a\sin\theta + b\operatorname{cosec}^2\theta)$$
$$= 2mg(a\sin\theta - b\operatorname{cosec}^2\theta) \qquad (1)$$

There are equilibrium positions where $\dfrac{d(\text{PE})}{d\theta} = 0$

i.e. where
$$a\sin\theta - b\operatorname{cosec}^2\theta = 0$$

or
$$\sin^3\theta = \frac{b}{a}$$

This equation has a real solution only if $\dfrac{b}{a} \leqslant 1$

In this case $\dfrac{d(\text{PE})}{d\theta} = 0$ when $\theta = \arcsin\left(\dfrac{b}{a}\right)^{1/3}$

Therefore the rod is in equilibrium when $\theta = \arcsin\left(\dfrac{b}{a}\right)^{1/3}$ if $b \leqslant a$.

2) A uniform rod AB of mass m and length $2l$ is smoothly jointed at A to a fixed point. A light elastic string of length l and modulus of elasticity mg connects the end B to a point C distant $2l$ vertically above A. Show that there are two positions in which the rod can rest in equilibrium with the string taut and determine whether the equilibrium at these positions is stable or unstable.

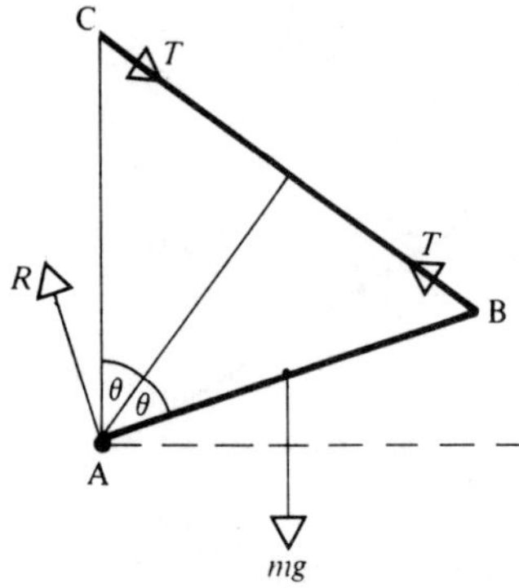

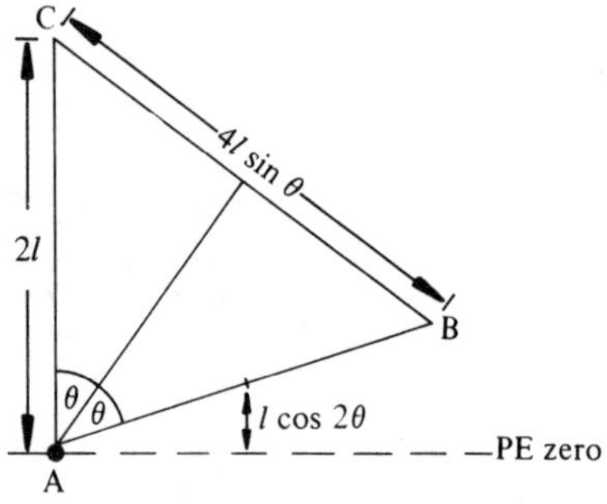

(This is a conservative system since the external reaction at the hinge does no work as the rod rotates about A. So it is valid to use the potential energy method to investigate the equilibrium of the rod.)
Consider the rod in a general position where the angle CAB = 2θ

In this position $$\text{PE} = mgl\cos 2\theta + \frac{mg}{2l}(4l\sin\theta - l)^2$$

Hence $$\frac{1}{mgl}\frac{d(\text{PE})}{d\theta} = -2\sin 2\theta + (4\sin\theta - 1)4\cos\theta$$

$$= 6\sin 2\theta - 4\cos\theta \qquad (1)$$

Therefore $$\frac{1}{mgl}\frac{d^2(\text{PE})}{d\theta^2} = 12\cos 2\theta + 4\sin\theta \qquad (2)$$

There are equilibrium positions where $\dfrac{d(\text{PE})}{d\theta} = 0$

From equation (1) $$6\sin 2\theta - 4\cos\theta = 0$$

$$4\cos\theta(3\sin\theta - 1) = 0$$

Hence the rod is in equilibrium when

$$\cos\theta = 0 \quad \text{or} \quad \sin\theta = \tfrac{1}{3}$$

i.e. $$\theta = \pi/2 \quad \text{or} \quad \arcsin\tfrac{1}{3}$$

Now the stability of equilibrium depends upon whether the PE is maximum or minimum, i.e. it depends upon the sign of $\dfrac{d^2(\text{PE})}{d\theta^2}$.
Referring to equation (2) we have

θ	$\pi/2$	$\arcsin \frac{1}{3}$
$\dfrac{d^2(\text{PE})}{d\theta^2}$	$-ve$	$+ve$
PE	Maximum	Minimum
State of equilibrium	Unstable	Stable

So the rod is in equilibrium, with the string taut, in two positions:
(i) with B vertically below A (an unstable position),
(ii) with the rod inclined at 2 arcsin $\frac{1}{3}$ to the upward vertical (a stable position).
(*Note*: The *positions* of equilibrium could have been determined by the standard consideration of three forces in equilibrium.)

3) A uniform rod AB of mass $12m$ is smoothly hinged at A to a fixed point. A light inextensible string attached to B passes over a small smooth peg at a point C and carries at its other end a particle of mass m. AC is horizontal, AC = AB and the angle CAB is θ. Find the value of θ for which the rod is in a position of stable equilibrium.

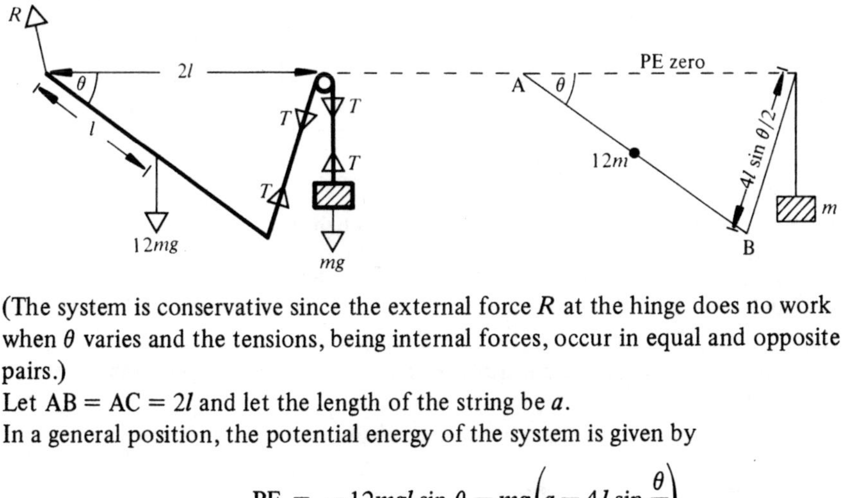

(The system is conservative since the external force R at the hinge does no work when θ varies and the tensions, being internal forces, occur in equal and opposite pairs.)
Let AB = AC = $2l$ and let the length of the string be a.
In a general position, the potential energy of the system is given by

$$\text{PE} = -12mgl\sin\theta - mg\left(a - 4l\sin\frac{\theta}{2}\right)$$

$$\frac{1}{mg}\frac{d(\text{PE})}{d\theta} = -12l\cos\theta + 2l\cos\frac{\theta}{2}$$

$$\frac{1}{mg}\frac{d^2(\text{PE})}{d\theta^2} = +12l\sin\theta - l\sin\frac{\theta}{2}$$

There is an equilibrium position if, for some value of θ, $\dfrac{d(\text{PE})}{d\theta} = 0$.

Hence $$12l\cos\theta = 2l\cos\frac{\theta}{2}$$

$$6\left(2\cos^2\frac{\theta}{2} - 1\right) = \cos\frac{\theta}{2}$$

The solutions of this quadratic equation are

$$\cos\frac{\theta}{2} = \frac{3}{4} \quad \text{or} \quad -\frac{2}{3}$$

The corresponding values of $\cos\theta$ are:

$$\cos\frac{\theta}{2} = \frac{3}{4} \Rightarrow \cos\theta = \frac{1}{8}$$

$$\cos\frac{\theta}{2} = -\frac{2}{3} \Rightarrow \cos\theta = -\frac{1}{9}$$

Checking for stability in equation (2) we have:

θ	$\arccos\frac{1}{8}$	$\arccos(-\frac{1}{9})$
$\dfrac{d^2(\text{PE})}{d\theta^2}$	$+ve$	$-ve$
PE	minimum	maximum
State of equilibrium	stable	unstable

Thus there is a position of stable equilibrium when $\theta = \arccos\frac{1}{8}$.

4) Three equal uniform rods PQ, QR, RS each of length a are smoothly jointed at Q and R. The rods rest symmetrically with PQ and RS supported by two smooth pegs which are distant $(a + b)$ apart in a horizontal line.

(i) If $2a > 3b$ show that there are two equilibrium positions and investigate their stability.

(ii) If $2a = 3b$ show that there is only one position of equilibrium.

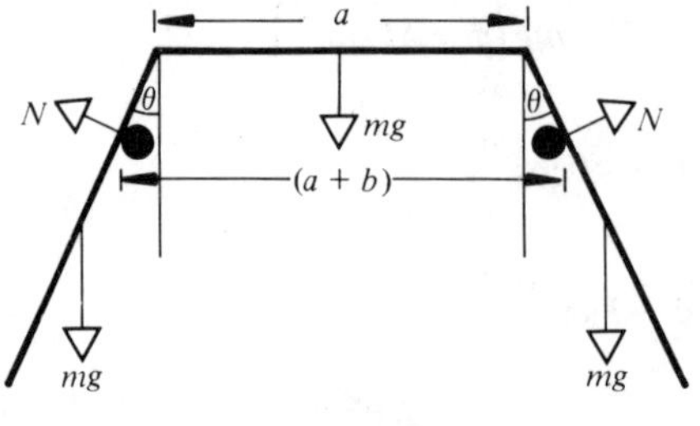

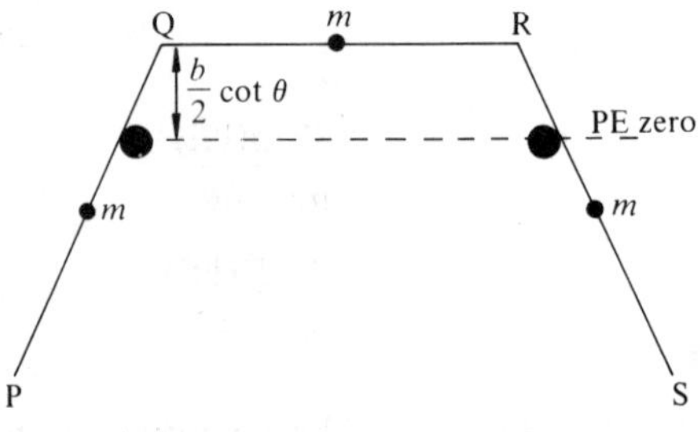

(When the angle between the inclined rods and the vertical varies the forces at the pegs do no work, so the system is conservative.)
Let θ be the angle between PQ (or RS) and the upward vertical in a general position.
The total PE of the three rods is then given by:

$$PE = mg\frac{b}{2}\cot\theta - 2mg\left(\frac{a}{2}\cos\theta - \frac{b}{2}\cot\theta\right)$$

$$\frac{2}{mg}\frac{d(PE)}{d\theta} = -b\,\mathrm{cosec}^2\theta + 2a\sin\theta - 2b\,\mathrm{cosec}^2\theta$$

$$= 2a\sin\theta - 3b\,\mathrm{cosec}^2\theta \qquad (1)$$

$$\frac{2}{mg}\frac{d^2(PE)}{d\theta^2} = 2a\cos\theta + 6b\,\mathrm{cosec}^2\theta\cot\theta \qquad (2)$$

For equilibrium $\frac{d(PE)}{d\theta} = 0$

Hence from (1) $2a\sin\theta = 3b\,\mathrm{cosec}^2\theta$

$$\sin^3\theta = 3b/2a$$

(i) If $2a > 3b$, $\sin\theta < 1$, so there are two positions of equilibrium given by $\theta = \alpha$ and $\theta = \pi - \alpha$ where α is the acute angle $\arcsin(3b/2a)^{1/3}$.
Now the stability of equilibrium depends upon whether the potential energy is maximum or minimum.
Putting $\theta = \alpha$ in equation (2); $\cos\alpha$, $\mathrm{cosec}\,\alpha$ and $\cot\alpha$ are all positive so $\frac{d^2(PE)}{d\theta^2}$ is positive.
Hence when $\theta = \alpha$ the rods are in a position of stable equilibrium.
Putting $\theta = \pi - \alpha$ in equation (2); $\cos(\pi - \alpha)$ and $\cot(\pi - \alpha)$ are negative but $\mathrm{cosec}^2(\pi - \alpha)$ is positive so $\frac{d^2(PE)}{d\theta^2}$ is negative.
Hence when $\theta = \pi - \alpha$ the rods are in a position of unstable equilibrium.

(ii) If $2a = 3b$, the condition for equilibrium is $\sin^3\theta = 1$.
Hence the only angle giving a position of equilibrium is $\theta = \pi/2$.

(*Note*: In each of the examples above, the level of some *fixed* object was chosen as the potential energy zero level.)

EXERCISE 8c

In this exercise use potential energy methods to investigate positions of equilibrium and their stability.

1) Two uniform smooth heavy rods AB and BC, each of length $2a$, are smoothly jointed together at B. If they are placed symmetrically in a vertical plane over a fixed cylinder of radius a show that the rods are in equilibrium if $\cos^3\theta = \sin\theta$ where θ is the inclination of each rod to the horizontal.

2) Four equal uniform heavy rods are smoothly jointed to form a rhombus ABCD. A and C are connected by a light elastic string whose natural length is half the length of each rod and whose modulus of elasticity is twice the weight of a rod. Find the angle BAC when the framework hangs in equilibrium from A.

3)

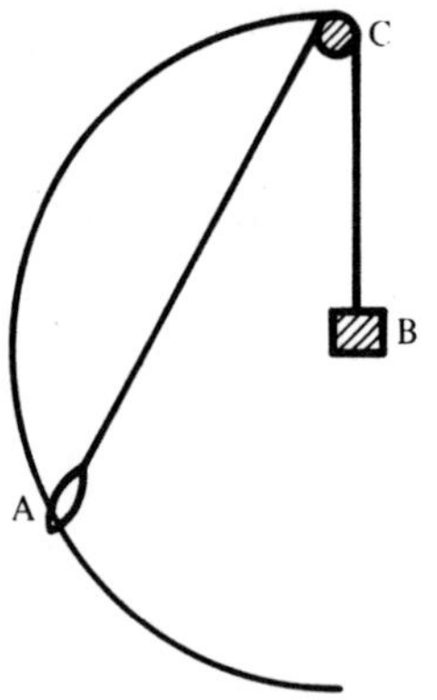

A is a smooth ring of mass m which is threaded on to a wire in the shape of a semi-circle in a vertical plane.
The bead is connected by a light inextensible string to a particle B also of mass m. The string passes over a small smooth pulley at the highest point of the wire.

If the length of the string exceeds the diameter of the circle, show that there are two positions of equilibrium for A and B and investigate their stability.

4)

A heavy circular disc of radius a hangs so that it touches a smooth vertical wall and is attached to the wall as shown by a light inextensible string AB of length $2a$.
Show that there is only one position of equilibrium and that it is stable.

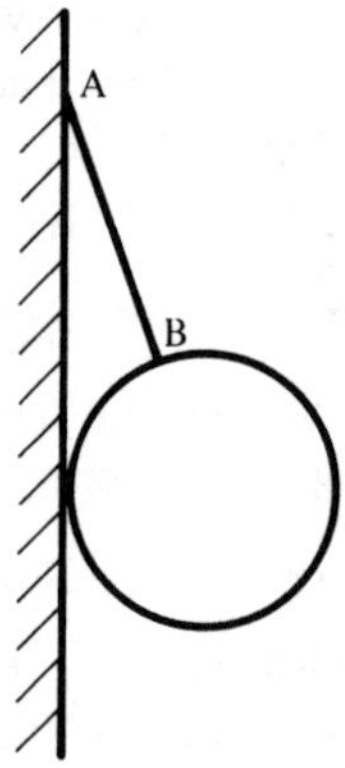

5)

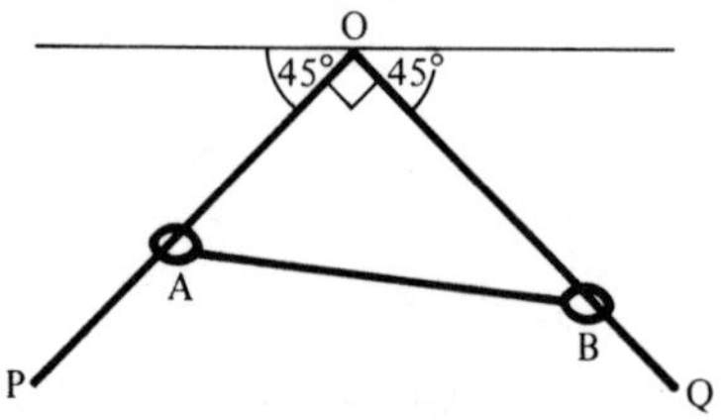

OP and OQ are two fixed smooth wires and angle POQ is a right angle. AB is a uniform heavy rod with a smooth light ring at each end. The rings are threaded on to OP and OQ as shown. Find the position in which AB is in stable equilibrium.
Is there an unstable position?

6)

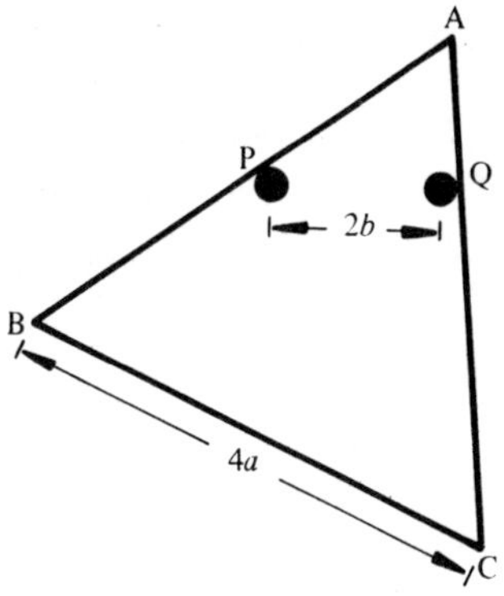

ABC is an equilateral triangular framework of three equal uniform heavy rods each of length $4a$. The framework rests over two smooth fixed pegs P and Q distant $2b$ apart in a horizontal line. Investigate the positions of equilibrium of the framework if
(i) $2b > a$ (ii) $2b < a$.

7)

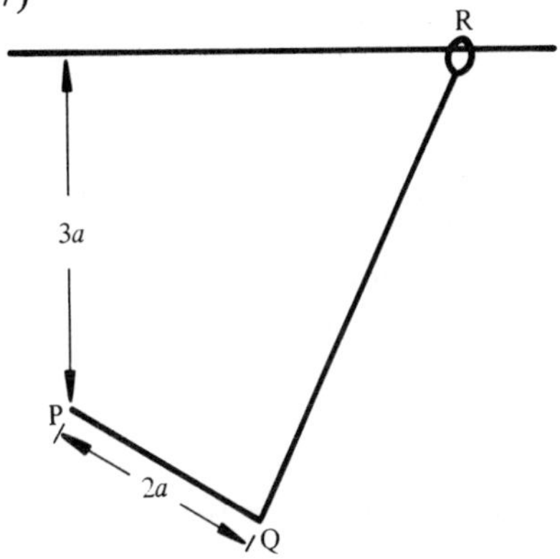

In the diagram PQ is a uniform rod of mass $2m$ and length $2a$. The end P is smoothly pivoted to a fixed point. A light elastic string of natural length a and modulus mg connects Q to a small smooth ring R which slides on a horizontal wire which is distant $3a$ vertically above P. Find the positions of equilibrium and investigate their stability.

MISCELLANEOUS EXERCISE 8

1) Define the centre of mass of n particles of masses $m_1, m_2, \ldots, m_n$, placed at points whose position vectors with respect to an origin O are $\mathbf{r}_1, \mathbf{r}_2, \ldots, \mathbf{r}_n$, respectively, and show that it is independent of the choice of origin.
Find the position vector of the centre of mass of particles of masses 4, 3, 2, 3 units at rest at the points

$$\mathbf{i}+\mathbf{j}, \quad 2\mathbf{i}-\mathbf{j}, \quad 2\mathbf{i}+\mathbf{j}, \quad 2\mathbf{i}+3\mathbf{j}$$

respectively. If each mass is acted upon by a force directed towards the origin and proportional to its distance from the origin, find the direction of the initial acceleration of the centre of mass. (U of L)

2) Show that, if $\mathbf{F}$ is the resultant of the external forces acting on a system of particles with centroid G, then the motion of the point G is the same as if the force $\mathbf{F}$ acted on a particle of suitable mass placed at G.
Particles of masses 6, 5, 2, 3 units are fixed to the vertices of a light square framework, the position vectors of the vertices being $a\mathbf{i}, a\mathbf{j}, -a\mathbf{i}, -a\mathbf{j}$ respectively.

Each particle is acted on by a force represented by the vector drawn from the particle to the point with position vector $3a\mathbf{i} + 4a\mathbf{j}$. Find the position vector of the centroid of the particles and the direction in which it will begin to move.

(U of L)

3) Two particles P_1 and P_2, of masses m_1 and m_2 respectively, are free to move on a smooth horizontal plane. Let $\mathbf{i}, \mathbf{j}$ be perpendicular unit vectors in the plane and let

$$\mathbf{r}_1 = x_1\mathbf{i} + y_1\mathbf{j}, \quad \mathbf{r}_2 = x_2\mathbf{i} + y_2\mathbf{j}$$

be the position vectors of P_1, P_2 relative to the origin. Show that the position vector $\mathbf{r}$ and the velocity $\mathbf{v}$ of the centre of mass G of the particles are given by

$$\mathbf{r} = \frac{m_1\mathbf{r}_1 + m_2\mathbf{r}_2}{m_1 + m_2}$$

and

$$\mathbf{v} = \frac{m_1\mathbf{v}_1 + m_2\mathbf{v}_2}{m_1 + m_2}$$

respectively where $\mathbf{v}_1 = \dot{x}_1\mathbf{i} + \dot{y}_1\mathbf{j}$ and $\mathbf{v}_2 = \dot{x}_2\mathbf{i} + \dot{y}_2\mathbf{j}$ are the velocities of P_1 and P_2.

Show that the velocity of particle P_1 relative to the point G is

$$\mathbf{V}_1 = \frac{m_2(\mathbf{v}_1 - \mathbf{v}_2)}{m_1 + m_2}$$

and write down the velocity $\mathbf{V}_2$ of the particle P_2 relative to G.

If T is the kinetic energy of the pair of particles P_1 and P_2, prove that

$$T = \tfrac{1}{2}(m_1 + m_2)\mathbf{v}.\mathbf{v} + \tfrac{1}{2}m_1\mathbf{V}_1.\mathbf{V}_1 + \tfrac{1}{2}m_2\mathbf{V}_2.\mathbf{V}_2$$

(Oxford)

4) Three particles are at rest on a smooth horizontal plane, their masses being 2, 2, 3 units and their position vectors respectively being

$$3\mathbf{i} + 4\mathbf{j}, \quad \mathbf{i} + \mathbf{j}, \quad \mathbf{i} - \mathbf{j}$$

Find the position vector of their centre of mass G.

Each mass is acted on by a constant force directed away from the origin, the magnitudes of the forces being 5, 1, 7 units respectively. Find the magnitude of the acceleration of G. Find also the resultant linear momentum of the particles at the instant when the total work done by the forces is 150/7 units. (U of L)

5) Two particles of mass m and $2m$ are attached to the ends of an elastic string of natural length $2a$ which would be doubled in length by a tension $4mg$. The particles are held at rest at a distance $3a$ apart on a smooth horizontal plane and are then released. Explain why the centre of mass of the system remains at rest during the motion of the particles. Find how far each particle travels before they collide.

From the moment of release the particle of mass $2m$ travels a distance x in time

t. Show that, for $x \leqslant a/3$,

$$\frac{d^2x}{dt^2} = g(a-3x)/a$$

Find the speed of each particle immediately before impact. (U of L)

6) A particle of mass 3 units is acted on by the forces

$$\mathbf{F}_1 = 2\mathbf{i}+3\mathbf{j}, \quad \mathbf{F}_2 = 3\mathbf{j}+4\mathbf{k}, \quad \mathbf{F}_3 = \mathbf{i}+2\mathbf{k}$$

and initially it is at rest at the point $\mathbf{i}-\mathbf{j}-\mathbf{k}$. Find the position and the momentum of the particle after 2 seconds. Find also the work done on the particle in this time. (U of L)

7) The position vectors of the vertices A, B, C of a triangle are respectively

$$4\mathbf{i}+2\mathbf{j}+2\mathbf{k}, \quad 2\mathbf{i}+\mathbf{j}+4\mathbf{k}, \quad -3\mathbf{i}+6\mathbf{j}-6\mathbf{k}$$

Find the position vector of G, the centroid of the triangle.
Forces $3\overrightarrow{AB}$ and $2\overrightarrow{AC}$ act along AB and AC. Find the magnitude of their resultant and the position vector of the point P in which the line of action of this resultant meets BC. Show that the area of the triangle GAP is $\frac{1}{2}\sqrt{5}$ and hence find the moment of the resultant about an axis through G perpendicular to the plane ABC. (U of L)

8) A particle of mass m kg is acted upon by a constant force of $21m$ newtons in the direction of the vector $3\mathbf{i}+2\mathbf{j}-6\mathbf{k}$. Initially the particle is at the origin moving with speed 18 m/s in the direction of the vector $7\mathbf{i}-4\mathbf{j}+4\mathbf{k}$. Find the position vector of the particle 4 seconds later and find the work done by the force in this period of 4 seconds. (U of L)

9) (i) Masses $5m$, $3m$ and $2m$ are situated at points with position vectors $2\mathbf{i}+2\mathbf{k}$, $-8\mathbf{i}-12\mathbf{j}+5\mathbf{k}$ and $3\mathbf{i}+3\mathbf{j}+7\mathbf{k}$ respectively. Find the position vector of the centroid of these point masses.

(ii) A particle with position vector $2\mathbf{i}$ is kept in equilibrium by the following five forces:
$\mathbf{P} = 7\mathbf{i}-15\mathbf{j}+\mathbf{k}$, $\quad \mathbf{Q} = 5\mathbf{i}-8\mathbf{j}-7\mathbf{k}$,
$\mathbf{R}_1$ acting along the line $\mathbf{r} = 2\mathbf{i}+a(3\mathbf{i}+4\mathbf{k})$,
$\mathbf{R}_2$ acting along the line $\mathbf{r} = 2\mathbf{i}+b(2\mathbf{i}-\mathbf{j}+2\mathbf{k})$,
$\mathbf{R}_3$ acting along the line $\mathbf{r} = 2\mathbf{i}+c(-3\mathbf{i}+4\mathbf{j})$.
Find the magnitudes of $\mathbf{R}_1$, $\mathbf{R}_2$ and $\mathbf{R}_3$. (U of L)

10) A uniform rod AB is of weight W and length $2a$. The upper end A is attached to a small light ring which can slide on a fixed smooth vertical wire; the end B is attached to a small light ring which can slide on a fixed smooth horizontal wire which intersects the vertical wire in O. One end of a light inextensible string is attached to the rod at a point C where $BC = \frac{1}{2}a$; the string passes over a small smooth pulley at O and, at its other end, carries a weight $\frac{1}{2}W$ which hangs freely.

Find the potential energy of the system when $\mathrm{O\hat{A}B} = \theta$.
Show that there is a position of equilibrium given by $\theta = 30^\circ$ and determine whether it is stable or unstable. (Cambridge)

11) A uniform rod OA of weight W and length $2a$ is freely pivoted at O; to the end A is attached a light inextensible string which passes over a small smooth pulley at B, at a distance $4a$ vertically above O. To the other end of the string is attached a weight λW hanging freely. Show that, when $\mathrm{A\hat{O}B} = \theta$, the potential energy V of the system is

$$Wa[\cos\theta + 2\lambda\sqrt{(5-4\cos\theta)}] + \text{constant}$$

Hence or otherwise show that, in addition to the equilibrium positions given by $\theta = 0^\circ$ and $\theta = 180^\circ$, there is another position of equilibrium for a value of θ between 0° and 180°, provided that $\frac{1}{4} < \lambda < \frac{3}{4}$.
By expanding V in powers of θ, or otherwise, determine whether the equilibrium position given by $\theta = 0$ is stable or unstable when

$$\text{(i) } \lambda = 1, \quad \text{(ii) } \lambda = \tfrac{1}{8}, \quad \text{(iii) } \lambda = \tfrac{1}{4}$$

[Assume throughout that the string is of such length that it never leaves the pulley.] (Cambridge)

12) A uniform straight rod OA of mass m and length $2a$ is free to move in a vertical plane about its end O, which is fixed. B and C are fixed points vertically above O and at distances $b(> 2a)$ and $b + c$ $(c > 0)$ from O respectively. A light string of natural length c and modulus of elasticity λ passing through a small smooth ring at B has one end attached to C and the other to a point D of the rod at a distance d from O. Assuming that $mgac < \lambda bd$, find all the positions of equilibrium of the rod and investigate their stability. (WJEC)

13) A uniform rod AB, of mass m and length $2a$, is free to rotate in a vertical plane about the end A which is fixed. The end B is attached to one end of a light elastic string of natural length a and modulus $mg/(2\sqrt{3})$, whose other end is attached to a fixed point O, at the same horizontal level as A. If $\mathrm{OA} = 2a$, and the angle between AB and the downward vertical at A is $(90^\circ - 2\theta)$, show that, apart from an additive constant, the potential energy of the system is

$$\frac{-mga}{\sqrt{3}}(\sqrt{3}\sin 2\theta + 2\cos 2\theta + 2\sin\theta)$$

provided that the string is not slack.
Hence or otherwise, *verify* that there is a position of equilibrium in which $\theta = 30^\circ$ and determine whether this position is stable or unstable.
Show also that there is another position of equilibrium in which $\theta = \alpha$, where $90^\circ < \alpha < 120^\circ$. (Cambridge)

14) A light rod AB can turn freely in a vertical plane about a smooth hinge at A and carries a mass m hanging from B. A light string of length $2a$ fastened to the

rod at B passes over a smooth peg at a point C vertically above A and carries a mass km at its free end. If $AC = AB = a$, find the range of values of k for which equilibrium is possible with the rod inclined to the vertical.
Given that equilibrium is possible with the rod horizontal find the value of k.
If the rod is slightly disturbed when horizontal and in equilibrium, determine whether it will return to the horizontal position or not. (U of L)

15) A uniform rod OA of length $2a$ and weight $8W$ is freely pivoted to a fixed support at O. A light elastic string of natural length a and modulus nW has one end tied to A and the other to a small ring which is free to slide on a smooth horizontal straight wire passing through a point at a height $7a$ above O. Show that when the rod makes an angle θ with the upward vertical at O, and the string is vertical, the potential energy of the system is

$$2Wa[n\cos^2\theta - (6n-4)\cos\theta] + \text{constant}$$

Obtain the positions of equilibrium and discuss their stability in each of the following cases: (i) $n > 1$, (ii) $n = \frac{4}{5}$, (iii) $n < \frac{1}{2}$. (Cambridge)

16) A uniform circular hoop of weight W and radius a is free to rotate in a vertical plane about a horizontal axis through a fixed point O on its circumference. A light inextensible string AB of length $a\sqrt{7}$ is attached at one end A to the point on the circumference of the hoop diametrically opposite to O and at the other end B carries a small smooth ring of weight $\frac{1}{2}W$ which is free to slide on a smooth vertical wire which lies in the plane of the hoop and at a distance $3a$ from O. Write down an expression for the potential energy of the system when OA makes an acute angle θ with the horizontal and prove that there exists a single position of equilibrium which occurs when $\theta = \frac{1}{3}\pi$.
Prove that this position is stable. (Oxford)

17) A uniform rod OA of mass m and length $2a$ is freely pivoted at a fixed point O and has a light smooth ring attached at A. A light elastic string of natural length $4a$ and modulus nmg $(n > 0)$ is threaded through the ring and its ends are attached to fixed points B and C where B, O, C lie on a horizontal line and $BO = OC = 2a$. Show that when the rod lies in the vertical plane containing BOC and makes an angle $\theta(-\frac{1}{2}\pi < \theta < \frac{1}{2}\pi)$ with the downward vertical at O, the potential energy V of the system (apart from an additive constant) is $mgaf(\theta)$ where

$$f(\theta) = (2n-1)\cos\theta - 4n\sqrt{2}\cos\frac{\theta}{2}$$

Hence show that there are two positions of equilibrium in which the rod is inclined to the vertical, A being below the level of O, provided that $n > 1 + \dfrac{1}{\sqrt{2}}$.
If $n = (3+\sqrt{6})/2$ determine the stability of these positions and also of the equilibrium position given by $\theta = 0$. (Cambridge)

18) Particles of mass $m_i (i = 1, 2, \ldots, n)$ have position vectors $\mathbf{r}_i$ relative to an origin O, and $\mathbf{r}'_i$ relative to an origin O′. The points G, G′ are defined by

$$M\mathbf{OG} = \sum_{i=1}^{n} m_i \mathbf{r}_i \quad \text{and} \quad M\mathbf{O'G'} = \sum_{i=1}^{n} m_i \mathbf{r}'_i$$

where $M = \sum_{i=1}^{n} m_i =$ total mass of the particles.

Prove that G and G′ coincide.

If the particles have position vectors $\mathbf{R}_i$ relative to an origin X, prove that there is a unique position of X (*the centre of mass* of the particles) for which

$$\sum_{i=1}^{n} m_i \mathbf{R}_i = 0$$

(i) A bead is free to move on a smooth wire which is bent into a circle and fixed with its plane horizontal. The circumference of the circle is divided into n equal parts by points $A_1, A_2, \ldots, A_n$. At any point P on the wire the bead is acted upon simultaneously by forces $k\mathbf{PA}_1, k\mathbf{PA}_2, \ldots, k\mathbf{PA}_n$. Prove that the bead can rest in equilibrium at any point of the wire.

(ii) A set of n particles of mass $m_i (i = 1, 2, \ldots, n)$ placed at the points B_i have centre of mass G_1. A second set of particles of mass $m_i (i = 1, 2, \ldots, n)$ placed at the points C_i have centre of mass G_2. Prove that

$$m_1 \mathbf{B}_1\mathbf{C}_1 + m_2 \mathbf{B}_2\mathbf{C}_2 + \ldots + m_n \mathbf{B}_n\mathbf{C}_n = M\mathbf{G}_1\mathbf{G}_2$$

where $M = \sum_{i=1}^{n} m_i$. (Oxford)

CHAPTER 9

UNITS AND DIMENSIONS

UNITS

Most of the quantities that occur in the solution of mechanics problems require a unit for their measurement.
Some of these units are *fundamental*, that is each can be chosen in an arbitrary manner and is quite independent of the others.
The quantities whose units are fundamental are mass, length and time.
Other quantities are measured in units which are derived from the fundamental ones. For instance the unit of speed is one unit of length described in one unit of time. If we denote unit mass by M, unit length by L and unit time by T, a *dimensional symbol* can be formed for the units of non-fundamental quantities.
The dimensional symbol for the unit of speed, for example, is L/T or LT^{-1}.

DIMENSIONS

The *dimensions* of any physical quantity in terms of mass, length and time are indicated by the indices attached to M, L and T.
For example:

1. The unit of acceleration is a unit increase in speed in one unit of time, i.e. the metre per second per second.
   The dimensional symbol for this unit is therefore LT^{-2}.
2. The unit of force imparts a unit of acceleration to a unit of mass, i.e. the kilogramme metre per second per second. This is too clumsy an expression to use for a unit of force so it is given the simple name newton.
   The dimensional symbol, however, is MLT^{-2} and this cannot be simplified in any way.
3. Work is the product of a force and a distance so its dimensions are given by

multiplying the dimensions of force by the dimension of distance, i.e. $(MLT^{-2})(L) = ML^2T^{-2}$.

4. An angle is measured in a unit defined by the ratio of unit arc length to unit radius, so its dimensional symbol is LL^{-1}, i.e. L^0. An angle is therefore said to be *dimensionless*.
   Further examples of dimensionless quantities are coefficient of friction and coefficient of restitution.
5. Moment of inertia has been defined as $\Sigma\, mr^2$ so the unit in which it is measured is the kilogramme metre squared and its dimensional symbol is ML^2.
   Quantities of different dimensions cannot either be collected together or equated to each other. Hence, whenever an equation is formed in the solution of a mechanics problem, every term must have the same dimensions. Checking this property is a useful aid to accuracy.

EXAMPLES 9a

1) Find the dimensions of impulse and momentum and hence verify the relationship Impulse = Change in Momentum.

The unit of impulse is a unit force acting for a unit of time, so its dimensional symbol is $(MLT^{-2})(T)$

i.e. MLT^{-1}

The unit of momentum is a unit mass moving with unit velocity.
The dimensional symbol is therefore $(M)(LT^{-1})$,

i.e. MLT^{-1}

The dimensions of impulse and momentum are identical.
Impulse = Change in Momentum is therefore a dimensionally correct relationship.

2) If the period of oscillation of a simple pendulum can be assumed to depend on nothing other than the length l of the string, the mass m of the bob and the acceleration g due to gravity. Use the theory of dimensions to determine a formula for the period.

If the period depends on mass, length and acceleration its dimensional symbol is made up of

$$(M^{\alpha})(L^{\beta})(LT^{-2})^{\gamma}$$

where α, β and γ are unknown constants.
But the dimensional symbol of a periodic time must be T, i.e. $M^0L^0T^1$.
Hence $M^0L^0T^1 \equiv M^{\alpha}L^{(\beta+\gamma)}T^{-2\gamma}$
Therefore $\alpha = 0$, $\gamma = -\frac{1}{2}$ and $\beta = \frac{1}{2}$.
So the structure of the dimensional symbol of the period is

$$M^0L^{1/2}(LT^{-2})^{-1/2}$$

The formula for calculating the period of oscillation of the pendulum can therefore be written as

$$\text{Period} = k\sqrt{\frac{l}{g}}$$

(*Note*: The value of the numerical constant k cannot be found by considering dimensions since k is dimensionless.)

3) The formula

$$v^2 = \frac{3\,mgl(1-\cos\theta)}{M(1+\sin^2\theta)}$$

is obtained as the solution of a problem. Use dimensions to find whether this is a reasonable solution (v is a velocity, m, M are masses, l is a length and g is gravitational acceleration).

Considering the dimensions of the left-hand side we have,

$$(LT^{-1})^2 = L^2T^{-2}$$

On the right-hand side, the following terms are dimensionless: $3, 1, \cos\theta, \sin^2\theta$. So the dimensions of the right-hand side are:

$$\frac{(M)(LT^{-2})(L)}{(M)} = L^2T^{-2}$$

As the dimensions of both sides of the equation are the same, the formula is a reasonable solution of the problem.
(*Note*: It is not possible to say that the formula is exactly correct because the dimensionless terms may not be numerically accurate).

EXERCISE 9a

1) Find the dimensional symbol for the following quantities:
Kinetic Energy, Momentum, Torque, Angular Velocity, Angular Acceleration, Area, Volume, Weight, Impulse, Power, Modulus of Elasticity.

2) A particle of mass m is attached to the end of a light string of length l. The other end of the string is attached to a fixed point on a smooth table. The particle is travelling in a horizontal circle on the table with angular velocity ω. Assuming that the tension in the string depends only upon m, l and ω find an expression for this tension.

3) Show that the relationship

$$\text{Work done} = \text{Change in Kinetic Energy}$$

is dimensionally correct.

In questions 4–12 there are errors in some of the equations. By considering dimensions determine which equations are definitely wrong.

The symbols used in these questions are explained below:

m	mass	θ	angle
F	force	$\dot{\theta}, \omega$	angular velocity
J	impulse	$\ddot{\theta}$	angular acceleration
C	torque	I	moment of inertia
u, v	velocity	λ	modulus of elasticity
l, r	length	μ	coefficient of friction

4) $5\mu g\theta = r\omega^2$

5) $C = I\dot{\theta}$

6) $J/\omega = mgl^2/C$

7) $I = \dfrac{mr}{v}(ru - l\omega)$

8) $mlg^2 = \lambda v$

9) $l^2 = v^2 r/g + \mu^2 r^2$

10) $F = 2\omega^2 l^2/7\mu g$

11) $Cv = I\dot{\theta}$

12) $J = mv - I\ddot{\theta}$

CHAPTER 10

ROTATION

One of the most fundamental characteristics possessed by an object is its intrinsic reluctance to accept a change in its state of motion, i.e. its inertia.
A body whose linear motion (translation) is changing has a linear acceleration. In order to produce such a change it is necessary to apply a force.
The same force applied to different bodies, however, produces different linear accelerations, indicating that each body has an individual amount of *linear inertia* which controls the degree of change in linear motion.
The measure of a body's linear inertia is better known as *mass*.
The relationship between mass, force and the linear acceleration produced was recognised by Newton whose second law can, for a body of constant mass, be expressed in the familiar form

$$F = ma$$

On the other hand the state of motion of a body can undergo a change in rotation if a torque is applied.
The resulting angular acceleration depends in part on the magnitude of the applied torque.
Experimental evidence shows, however, that the same torque applied to different bodies produces different angular accelerations, indicating that each body has an individual amount of *rotational inertia* which controls the degree of change in rotation.
The measure of a body's rotational inertia is called *moment of inertia* and it is represented by the symbol I.
At this stage we are not in a position to specify the actual constitution of a body's moment of inertia. Experiments can be carried out, however, which show that when a torque is applied to a body causing it to rotate about an axis then:
1. for a particular body free to rotate about a specific axis, the angular acceleration is proportional to the torque applied.

2. the angular acceleration produced by a particular torque changes if:
   (a) the mass of the body is altered,
   (b) the axis of rotation is changed,
   (c) a body of equal mass but different size or shape is used.

These results suggest that the moment of inertia of a body is a function of the mass of the body, the distribution of that mass (e.g. size and shape) and the position of the axis of rotation.
The theoretical analysis of a rotating body will be seen to confirm this hypothesis.

ROTATION OF A RIGID BODY ABOUT A FIXED AXIS

A rigid body is made up of a large number of particles whose relative displacements are fixed.
Consider a rigid body which is freely rotating about a fixed axis and let P be a typical constituent particle, of mass m and distant r from the axis of rotation. The diagram below shows a cross-section of the body perpendicular to the axis, which passes through A.
Because the constituent particles do not move relative to one another, every particle has at any instant the same angular velocity $\dot{\theta}$, the same angular acceleration $\ddot{\theta}$ and is describing a circle about A as centre.

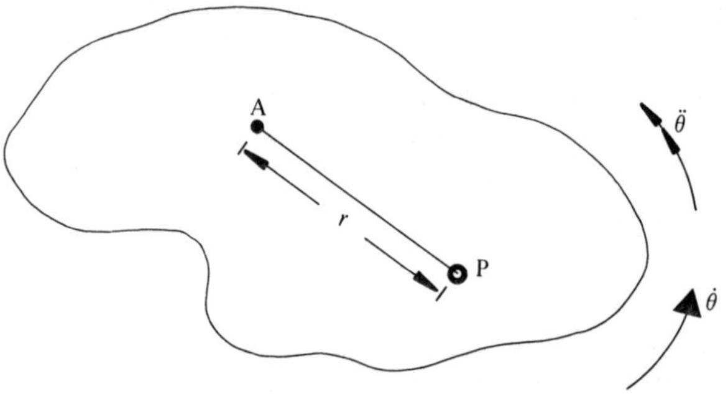

In the direction perpendicular to AP, the linear velocity and acceleration components of P are $r\dot{\theta}$ and $r\ddot{\theta}$ respectively (P also has a linear acceleration component $r\dot{\theta}^2$ in the direction PA).
Now suppose that P is under the action of a force whose components parallel and perpendicular to AP (i.e. normal and tangential) are F_N and F_T.

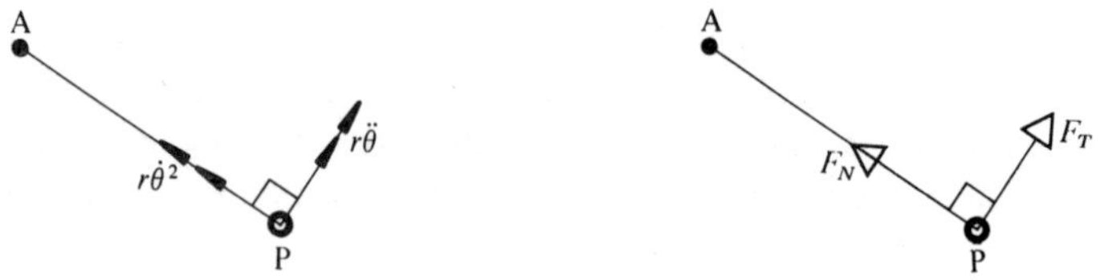

If we now apply Newton's Law to the particle in the tangential direction we have:

$$F_T = m(r\ddot{\theta})$$

Therefore $$F_T r = mr^2\ddot{\theta}$$

But $F_T r$ is the moment of F_T about the axis of rotation and F_N has no moment about this axis.
Now for the whole body

$$\sum F_T r = \sum mr^2\ddot{\theta}$$

But $\Sigma F_T r$ is the resultant torque about the axis of rotation, C say.

Thus $$C = \left(\sum mr^2\right)\ddot{\theta} \qquad (1)$$

In this equation Σmr^2 is a quantity which depends solely on the mass of the body and the distribution of the mass about the axis of rotation, i.e. it has the properties of moment of inertia.
Then, defining the value of moment of inertia as the quantity Σmr^2, we have

$$I = \sum mr^2$$

and equation (1) becomes

$$C = I\ddot{\theta}$$

This relationship is, in effect, Newton's Second Law adapted to rotation, as can be seen by comparing

$$F = M\ddot{x}$$

and $$C = I\ddot{\theta}$$

since:

F is necessary to produce change in translation,
C is necessary to produce change in rotation.

M is the intrinsic control over change in translation,
I is the intrinsic control over change in rotation.

$\ddot{x}$ is the resulting linear acceleration in a straight line,
$\ddot{\theta}$ is the resulting angular acceleration.

PROPERTIES OF A ROTATING BODY

Consider a rigid body which is free to rotate about a smooth fixed axis and whose angular velocity at any instant is represented by $\dot{\theta}$.

A cross-section of the body, perpendicular to the axis of rotation, is shown in the diagram. The axis passes through A and P is a typical constituent particle of mass m and distant r from A.

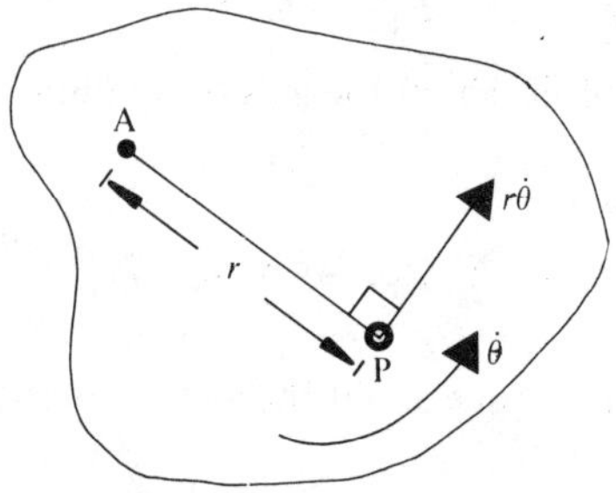

The linear velocity of P is $r\dot{\theta}$ perpendicular to AP as shown.

Kinetic Energy of Rotation

For P, Kinetic Energy $= \frac{1}{2}m(r\dot{\theta})^2$

So for the whole body

Kinetic Energy $= \sum \frac{1}{2}mr^2\dot{\theta}^2$

But every constituent particle has the same angular velocity $\dot{\theta}$.
Hence the kinetic energy of the rotating body is given by

$$\text{Kinetic Energy of rotation} = \frac{1}{2}\left(\sum mr^2\right)\dot{\theta}^2$$
$$= \tfrac{1}{2}I\dot{\theta}^2$$

This quantity has dimensions $(ML^2)(T^{-1})^2$ and it is therefore dimensionally equal to linear kinetic energy.
Consequently linear kinetic energy and rotational kinetic energy can be added together in an equation and can also be measured in the same unit, the joule.

Moment of Momentum

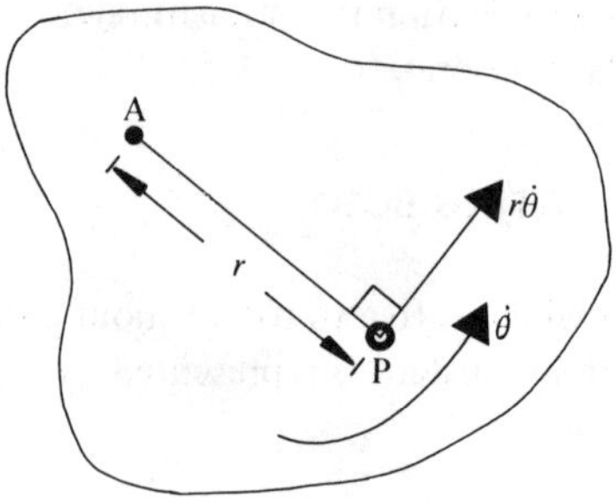

For one particle P,

$$\text{Linear momentum} = m(r\dot{\theta})$$

Then the *moment of the momentum* of P about the axis through A is given by:

$$\text{Moment of momentum} = (mr\dot{\theta})r$$
$$= mr^2\dot{\theta}$$

Again $\dot{\theta}$ is common to all constituent particles.
So for the whole body we have:

$$\text{Total moment of momentum} = \left(\sum mr^2\right)\dot{\theta}$$
$$= I\dot{\theta}$$

Hence the moment of momentum of a rotating body is $I\dot{\theta}$.
An alternative name for this quantity is *angular momentum* and it is measured in a rather clumsy unit, the $\text{kg m}^2\,\text{s}^{-1}$.
The dimensional symbol for this unit is ML^2T^{-1} or $(MLT^{-1})L$ which can be seen to have the same dimensions as linear momentum times distance.
Hence, in forming an equation, the angular momentum of a rotating body can be added to the moment of the linear momentum of another body.

SUMMARY

1. Moment of inertia is the name given to a body's intrinsic control over its change in rotation.
2. The value of a body's moment of inertia, I, is defined by $I = \Sigma\, mr^2$.
3. When a torque C acts on a body with moment of inertia I, producing an angular acceleration $\ddot{\theta}$, then:
$$C = I\ddot{\theta}$$
4. The kinetic energy of a body whose angular velocity is $\dot{\theta}$ is $\frac{1}{2}I\dot{\theta}^2$.
5. The angular momentum, or moment of momentum of that body is $I\dot{\theta}$.

PROBLEMS

Detailed methods of evaluating the moment of inertia of a variety of bodies are explained in Chapter 11. At present, whenever a specific moment of inertia is required in the solution of a problem, it will be stated and used without proof. In some problems the rigid body concerned will be found to have a constant angular acceleration ($\ddot{\theta} = \alpha$). The motion of such a body can be analysed using the standard formulae for motion of this type. A summary of these formulae, which were derived in Volume One, is given below.

If
$$\omega_1 = \text{initial angular velocity} \quad (\text{rad s}^{-1})$$
$$\omega_2 = \text{final angular velocity} \quad (\text{rad s}^{-1})$$
$$\theta = \text{angular displacement} \quad (\text{rad})$$
$$\alpha = \text{angular acceleration} \quad (\text{rad s}^{-2})$$
$$t = \text{time} \quad (\text{s})$$

Then
$$\omega_2 = \omega_1 + \alpha t$$
$$\theta = \omega_1 t + \tfrac{1}{2}\alpha t^2$$
$$\theta = \omega_2 t - \tfrac{1}{2}\alpha t^2$$
$$\theta = (\omega_1 + \omega_2)t/2$$
$$2\alpha\theta = \omega_2^{\,2} - \omega_1^{\,2}$$

EXAMPLES 10a

1) A door swings open with an angular speed of 0.5 rad s^{-1}. If the door has a moment of inertia about its hinges of 160 kg m^2, find its angular momentum.

$$\text{Angular momentum} = I\dot{\theta}$$

But $I = 160\,\text{kg m}^2$ and $\dot{\theta} = 0.5\,\text{rad s}^{-1}$;

Hence the angular momentum of the door is given by:

$$I\dot{\theta} = (160)(0.5)$$
$$= 80\,\text{kg m}^2\,\text{s}^{-1}$$

2) The moment of inertia of a flywheel about its axis is 20 kg m^2. When it is stationary, a constant torque of 40 Nm is applied to the flywheel. Find its kinetic energy after three seconds assuming the flywheel has smooth bearings. (A flywheel is either a circular disc or a circular rim which can rotate about an axis through its centre perpendicular to the flywheel.)

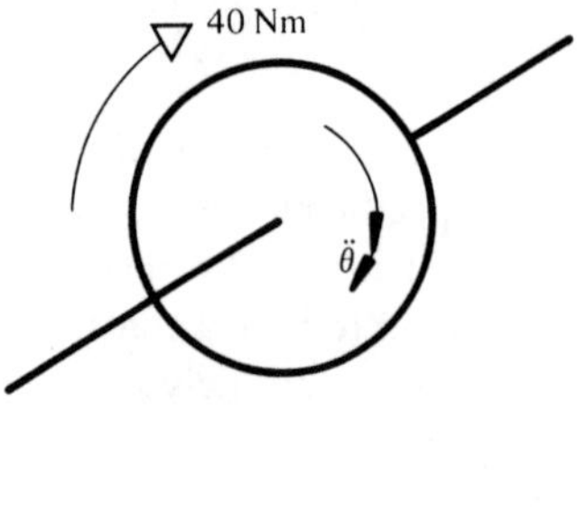

Let $\ddot{\theta}$ represent the angular acceleration of the flywheel at any instant and let ω be the angular velocity after three seconds.

Using $C = I\ddot{\theta}$ we have

$$40 = 20\ddot{\theta}$$

Hence $\ddot{\theta} = 2$

Therefore the flywheel has a constant angular acceleration of 2 rads^{-2} and its motion can be investigated using the standard equations for motion of this type. The flywheel starts from rest so, after three seconds,

$$\omega = \ddot{\theta}t = (2)(3)$$

Hence the flywheel has an angular velocity of 6 rads^{-1} and therefore possesses kinetic energy given by:

$$\begin{aligned} \text{KE} &= \tfrac{1}{2}I\omega^2 \\ &= \tfrac{1}{2}(20)(6)^2 \text{ J} \\ &= 360 \text{ J} \end{aligned}$$

3) A crate of mass 100 kg is dragged across smooth horizontal ground by a light rope whose other end is being wrapped round a pulley of radius 0.4 m. The pulley has a fixed smooth vertical axle and its moment of inertia about this axis is 24 kg m^2.
If the crate moves with an acceleration of 0.2 ms^{-2}, find the magnitude of the torque driving the pulley.
Find also the total kinetic energy of the moving system when the speed of the crate is 2 ms^{-1}.

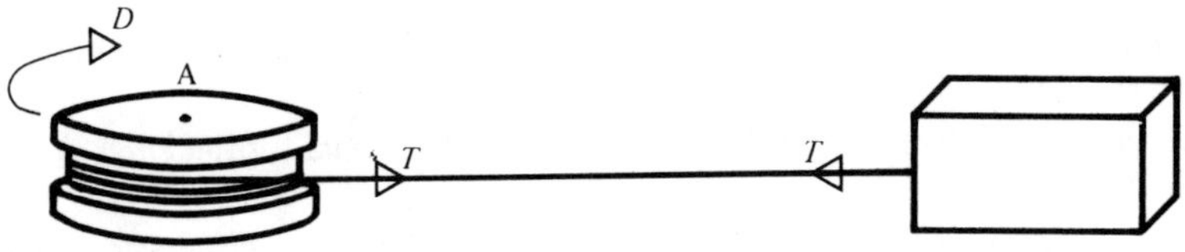

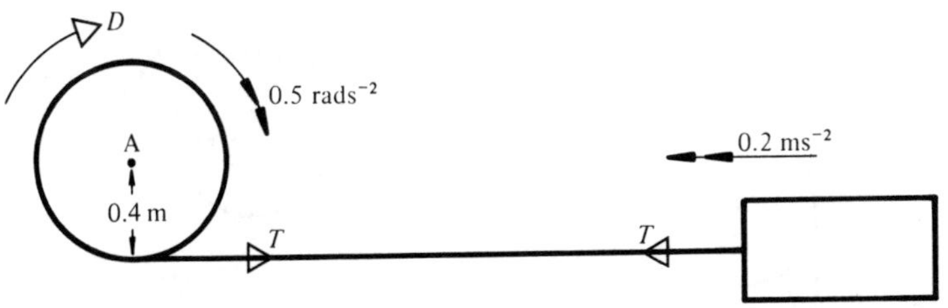

Let D be the torque driving the pulley.
(The linear acceleration of the crate is equal to the linear acceleration of the rope on the rim of the pulley. So the angular acceleration of the pulley is given by dividing the linear acceleration of the crate by the radius of the pulley, i.e. $\ddot{\theta} = 0.2/0.4 = 0.5$ rads^{-2}.)
Using $F = ma$ for the motion of the crate:

$$\leftarrow \qquad T = (100)(0.2) \qquad (1)$$

Using $C = I\ddot{\theta}$ for the rotation of the pulley

$$\text{A}\circlearrowright \quad D - 0.4T = 24(0.5) \qquad (2)$$

Eliminating T from (1) and (2) gives:

$$D - 8 = 12$$

Hence $$D = 20\ \text{Nm}$$

(When the linear speed of the crate is 2 ms^{-1}, this is also the speed of the rope as it touches the pulley so the angular speed of the pulley is 2/0.4 rads^{-1}, i.e. 5 rads^{-1}.)

The total kinetic energy is made up of two parts:
(a) the linear kinetic energy of the crate ($\frac{1}{2}mv^2$),
(b) the rotational kinetic energy of the pulley ($\frac{1}{2}I\dot{\theta}^2$).

$$\text{KE of crate} = \tfrac{1}{2}(100)(2)^2\ \text{J} = 200\ \text{J}$$

$$\text{KE of pulley} = \tfrac{1}{2}(24)(5)^2\ \text{J} = 300\ \text{J}$$

Thus the total kinetic energy of the system is 500 J.

4) A cylinder of radius r is free to rotate about its axis which is horizontal. A light string hangs over the pulley and carries a particle P of mass m at one end and a particle Q of mass km at the other end. The string is rough enough not to slip on the pulley.
When the system is released from rest a constant frictional torque $\frac{1}{5}mgr$ acts on the pulley.
If the acceleration of the particles when $k = 3$ is twice their acceleration when $k = 2$ find the moment of inertia of the pulley.

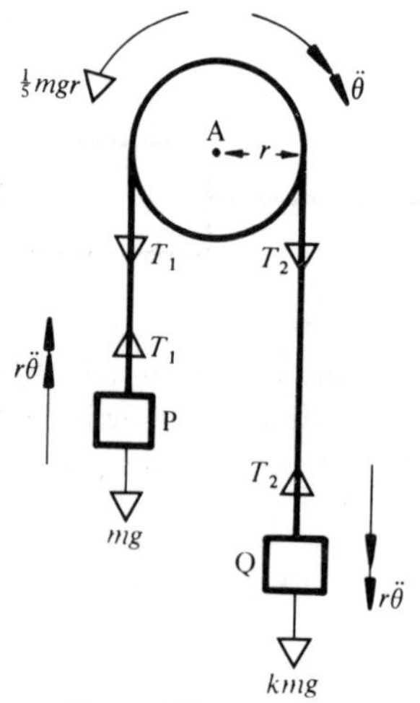

General Diagram

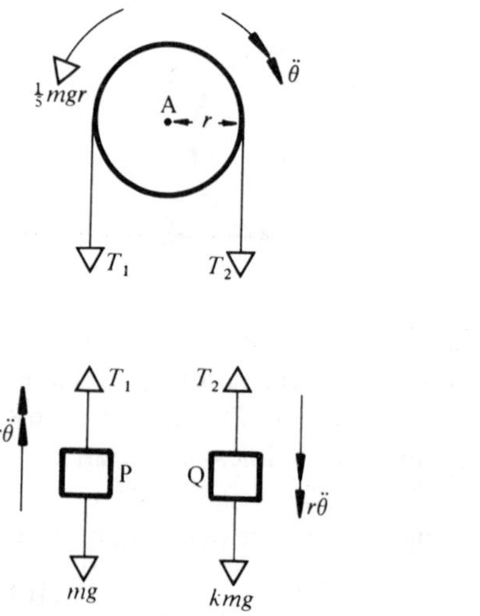

Diagram showing separately the forces acting on the cylinder and on each particle.

(When a string passes over a rough pulley, the tensions in the string on either side of the pulley are different.)

Let the moment of inertia of the pulley be I.
Using $F = ma$ for the motion of P and Q and
using $C = I\ddot{\theta}$ for the motion of the pulley we have:

For P $\uparrow$ $$T_1 - mg = m(r\ddot{\theta}) \quad (1)$$

For Q $\downarrow$ $$kmg - T_2 = km(r\ddot{\theta}) \quad (2)$$

For pulley A $\circlearrowright$ $$T_2 r - T_1 r - \tfrac{1}{5}mgr = I\ddot{\theta}$$

i.e. $$T_2 - T_1 - \tfrac{1}{5}mg = \frac{I\ddot{\theta}}{r} \quad (3)$$

Adding (1), (2) and (3) $$\left(k - \frac{6}{5}\right)mg = \left(kmr + mr + \frac{I}{r}\right)\ddot{\theta}$$

Now when $k = 3$, $$\ddot{\theta}_3 = \frac{9}{5}mg\Big/\left(4mr + \frac{I}{r}\right)$$

And when $k = 2$, $$\ddot{\theta}_2 = \frac{4}{5}mg\Big/\left(3mr + \frac{I}{r}\right).$$

But $$\ddot{\theta}_3 = 2\ddot{\theta}_2$$

Therefore $$\frac{9}{5}mg\Big/\left(4mr + \frac{I}{r}\right) = \frac{8}{5}mg\Big/\left(3mr + \frac{I}{r}\right)$$

Hence $$9(3mr + I/r) = 8(4mr + I/r)$$

$$I/r = 5mr$$

$$I = 5mr^2$$

Thus the moment of inertia of the pulley is $5mr^2$.

EXERCISE 10a

1) A flywheel whose moment of inertia is $50\ \text{kg m}^2$ is rotating at $4\ \text{rad s}^{-1}$. Find its kinetic energy and angular momentum.

2) A flywheel loses kinetic energy amounting to 640 J when its angular speed falls from $7\ \text{rad s}^{-1}$ to $3\ \text{rad s}^{-1}$. What is the moment of inertia of the flywheel?

3) Find the kinetic energy of rotation of the earth, taking its moment of inertia as $98 \times 10^{36}\ \text{kg m}^2$.

4) A string is wound round a pulley of radius 0.3 m which is free to rotate about a smooth vertical axis. When the string is pulled horizontally with a constant force of 48 N, the angular velocity of the pulley increases from $2\ \text{rad s}^{-1}$ to $6\ \text{rad s}^{-1}$ in five seconds, find the moment of inertia of the pulley.

5) A uniform circular flywheel whose moment of inertia is 64 kg m^2 rotates under the action of a constant couple. The flywheel starts from rest and after turning through 200 revolutions its angular speed is 20 revolutions per second. Find the moment of the couple.

6) An engine is running at 160 revolutions per minute and its flywheel has a moment of inertia of 600 kg m^2. When the engine is switched off the flywheel continues to rotate for 40 seconds before coming to rest. Find the magnitude of the constant retarding torque acting on the flywheel.

7) A particle of mass m is attached to one end of a light rough string which is wrapped several times round a cylinder of radius r. The cylinder can rotate about its axis of symmetry which is horizontal, and its moment of inertia about that axis is $4mr^2$.
Assuming that the string is rough enough not to slip on the cylinder find the angular acceleration of the cylinder when the system is released from rest.

8) A string is wrapped round a cylindrical shaft of radius 0.1 m and moment of inertia 200 kg m^2, and is pulled with a constant force of 400 N for five seconds. If there is a frictional couple of 15 Nm acting on the cylinder find the angular acceleration of the cylinder and the angle through which it turns in the five seconds (from rest).

9) Two particles of masses m_1 and m_2 are connected by a light string which passes over a pulley of radius r and moment of inertia I. If the string does not slip on the pulley find the difference between the tensions on the two sides of the pulley when the system is released from rest.
Find also the force which the string exerts on the pulley.

10)

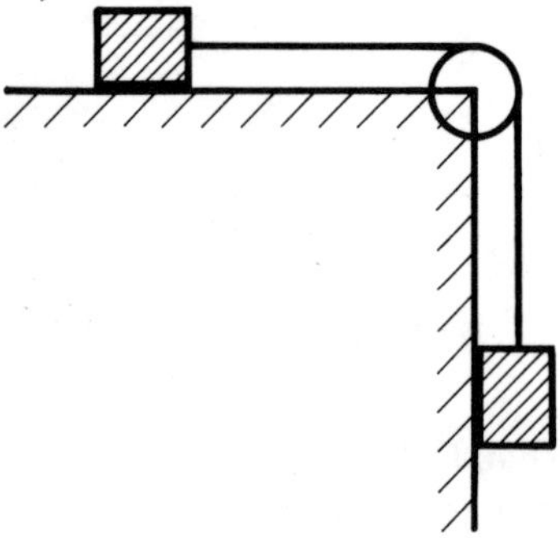

The diagram shows two particles of equal mass m connected by a rough light string which passes over a pulley at the edge of a smooth horizontal table. If the pulley is free to rotate about its axis and has a moment of inertia $4mr^2$ where r is its radius, find its angular acceleration when the system is released from rest.

MISCELLANEOUS EXERCISE 10

1) A flywheel is brought to rest by a constant retarding torque. From the instant this torque is applied, the flywheel is observed to make 200 revolutions in the

first minute and 120 revolutions in the next minute. Calculate how many more revolutions the wheel makes before coming to rest and the time taken to stop the wheel. (AEB)

2) A uniform solid cylinder of mass $2m$ and radius r is free to rotate about its axis which is fixed and horizontal. A light inextensible string is wrapped round the surface of the cylinder several times and one end of the string is attached to the cylinder. The other end of the string is attached to a particle of mass m which hangs vertically downwards. (For the cylinder $I = mr^2$.)
If the system is released from rest, prove that when the particle has fallen a distance r the angular velocity of the cylinder is $\sqrt{(g/r)}$. Find the time taken for the cylinder to complete one revolution from rest. (U of L)

3) A flywheel is being brought to rest by a constant retarding torque. The flywheel is observed to do n_1 revolutions in the first half-minute and n_2 revolutions in the next half-minute. Find:
(a) the angular retardation of the wheel,
(b) how many more revolutions the wheel does before coming to rest. (U of L)

4) Two particles of masses M and m are connected by a light inextensible string. The string passes over a pulley of radius a smoothly mounted on a horizontal axis and the particles hang freely. The string does not slip on the pulley and the moment of inertia of the pulley about its axis is na^2. Find the angular acceleration of the pulley while the system is in motion, and prove that the strings exert on the pulley a vertical force of magnitude

$$\frac{(M+m)n+4Mm}{M+m+n}g$$

(Oxford)

5) A thin uniform circular disc, of mass M, radius a and centre C, is smoothly hinged to a fixed pivot at a point A on its circumference and is free to rotate in its own plane. When this plane is horizontal a constant coplanar couple G will bring the disc to rest from an angular velocity ω in exactly two revolutions. Show that $G = 3Ma^2\omega^2/(16\pi)$ and find the time taken in thus coming to rest. If the couple continues to act after bringing the disc momentarily to rest, find the further time taken and the number of revolutions completed whilst the disc reaches an angular velocity -2ω.
[The moment of inertia of the disc about the given axis is $\frac{3}{2}Ma^2$] (U of L)

6) A uniform square lamina ABCD is free to rotate about a fixed vertical axis which coincides with the diagonal AC. The lamina is given an initial angular velocity ω_0 and, under the action of a constant driving torque G against a constant frictional torque T, completes 10 revolutions in the first second and 20 revolutions in the next second. Show that $\omega_0 = 10\pi$ rad/s.
The constant driving torque G is then removed and the lamina is brought to rest by the frictional torque T which has been constant throughout the motion. The

lamina is thus brought to rest in a further 15 revolutions. Find T in terms of G. Find also in terms of G, the moment of inertia of the square about AC.

7) A cylinder, of radius r, is free to rotate about its axis which is horizontal. A light inextensible string is wrapped round the cylinder and has a load attached to its free end. With a load of W it is found that the load falls from rest a distance l in one second and a load of $4W$ similarly falls a distance $2l$ in one second. Assuming the string does not slip and the frictional couple opposing the motion of the cylinder remains constant, calculate:
(i) the moment of inertia of the cylinder about its axis,
(ii) the magnitude of the frictional couple.

8) A light inextensible string is connected at its ends to two particles of masses m_1 and m_2 $(m_1 > m_2)$ and passes over a uniform circular pulley of radius a which can rotate freely about a fixed horizontal axis through its centre. The particles hang freely and the system is released from rest. If the pulley is sufficiently rough to prevent the string slipping, find the acceleration of either particle. (For the pulley $I = \frac{1}{2}Ma^2$.)
The heavy pulley is now replaced by a light smooth one and both particles have their masses increased by the same amount, m.
If the acceleration of the particles is the same as it was in the first case, find an expression for m.

9) A heavy pulley, which may be regarded as a uniform circular disc of mass $6m$, centre O and radius a, can turn freely in a vertical plane about a fixed horizontal axis through O. A light inextensible string passes over the pulley and particles of mass $2m$ and $3m$ are attached at its free ends. The system is released from rest and the string does not slip on the pulley. Show that the heavier particle falls with acceleration $g/8$. (For the pulley $I = 3ma^2$.)
After time t a constant frictional couple is applied to the pulley and in consequence the system comes to rest again in a *further* time t.
Assuming that the lighter particle does not reach the pulley throughout the motion, calculate:
(i) the total distance covered by the heavier particle,
(ii) the magnitude of the constant frictional couple.

(U of L)

10) A rigid body can turn freely about a smooth fixed vertical axis and the moment of inertia of the body about this axis is I. A constant couple of magnitude N acts on the body in a plane perpendicular to the axis of rotation. Find (a) the time taken for the angular velocity of the body to increase from ω_1 to ω_2, (b) the angle through which the body rotates in this time.
A uniform circular disc whose moment of inertia is 20 kgm^2 rotates, under the action of a constant couple, about a fixed axis through its centre perpendicular to its plane. If the speed of rotation changes from 10 revolutions per minute to

35 revolutions per minute in 15 seconds, find the moment of the couple acting on the disc.
Find also the angular acceleration and the number of revolutions made by the disc in this time.

11) The diagram shows particles A and B, each of mass M, connected by a light inextensible string which passes over a smooth light fixed pulley C. The particle A is on a plane inclined at an angle α to the horizontal, the coefficient of friction between A and the plane being μ. Between A and the pulley the string is parallel to a line of greatest slope of the plane and B is hanging freely. Find the acceleration with which the particle B descends.
The pulley C is now replaced by a heavy one D, of the same radius a, which is rough enough to prevent the string slipping over it, all other conditions remaining unchanged. As a result the acceleration of B is halved. Assuming that the bearings of D are frictionless, find the moment of inertia of D about its axis. If the change of pulley also causes a 10% increase in the tension in the vertical portion of the string, show that

$$\mu = \frac{2 - 3\sin\alpha}{3\cos\alpha}$$

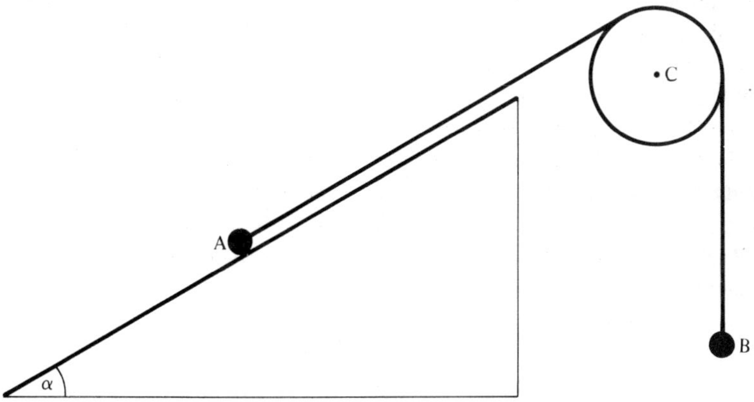

(U of L)

12) When a uniform solid cylinder of radius a and mass M is rotating about its fixed horizontal axis there is a constant resisting torque of magnitude G. A light inextensible string has one end attached to the curved surface of the cylinder and is wound several times round this surface. Hanging from the free end of this string is a particle of mass km. When the system is released from rest the particle has an acceleration $\frac{1}{4}g$ if $k = 1$ and an acceleration $\frac{1}{2}g$ if $k = 2$. Find M and G in terms of a, m and g. Find the acceleration of the particle when $k = 4$.
(The moment of inertia of the cylinder is $\frac{1}{2}Ma^2$.) (U of L)

13) A uniform circular pulley, of radius r and mass m, is in a vertical plane and is free to turn about a horizontal axis passing through its centre, the moment of inertia of the pulley about this axis being $\frac{mr^2}{2}$. A light inextensible string passes over the pulley and hangs vertically on each side, supporting a particle of mass m on one side and a particle of mass $2m$ on the other. The system is released from rest at time $t = 0$, and, in the subsequent motion, the string does not slip on the pulley and the rotation of the pulley is resisted by a frictional couple of constant moment G. If θ is the angular displacement of the pulley at time t, prove that

$$7mr^2 \frac{d^2\theta}{dt^2} = 2(mgr - G)$$

When the pulley has turned through an angle α, the string is completely detached from the pulley without any impulse. If the pulley then turns through a further angle β before being brought to rest by the frictional couple, show that

$$G = \frac{mgr\alpha}{\alpha + 7\beta}$$

(JMB)

14) A uniform solid smooth circular cylinder of mass M and radius a is free to rotate about its axis which is fixed and horizontal. One end of a light inextensible string of length $2\pi a$ is attached to the curved surface of the cylinder at a point A, level with the axis of the cylinder. The string passes over the cylinder and to its other end is attached a particle P of mass M. The system is held at rest with the string taut and in a vertical plane which is perpendicular to the axis of the cylinder. The particle P is then raised to the point B on the cylinder diametrically opposite to the point A. The system is now released from rest so that P falls vertically. When the string becomes taut again the cylinder starts to rotate with angular speed ω. Show that

$$\omega = \sqrt{\left(\frac{8g\pi}{9a}\right)}$$

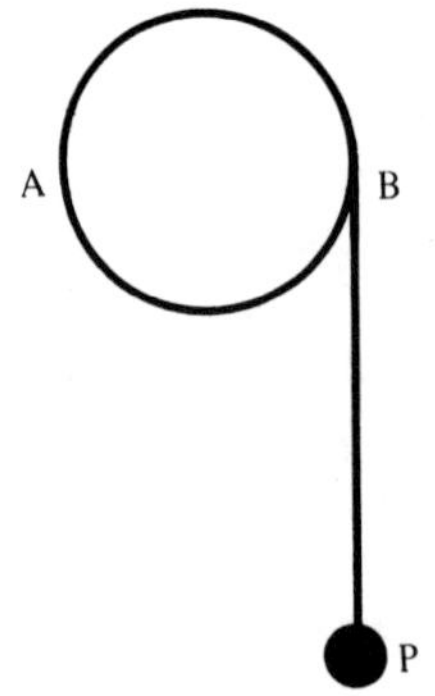

When the particle has fallen a further distance πa the angular speed of the cylinder is ω_1. Show that $2\omega_1{}^2 = 5\omega^2$ and find the tension in the string during this part of the motion.
(The moment of inertia of the cylinder is $\frac{1}{2}Ma^2$.) (U of L)

CHAPTER 11

CALCULATION OF MOMENT OF INERTIA

When a rigid body rotates about a fixed axis the moment of inertia of that body about that axis is $\Sigma\, mr^2$ where m is the mass of a constituent particle of the body and r is the distance of that particle from the axis of rotation.
As moment of inertia is the product of mass and the square of a distance the *unit* in which it is measured is one *kilogram metre squared* ($\mathrm{kg\,m^2}$). It should also be noted that $\Sigma\, mr^2$ is a scalar quantity.
Thus the moment of inertia of a body depends on two properties:
(a) the position of the axis of rotation,
(b) the distribution of the mass about that axis.
It must be emphasised that it is inadequate to refer simply to 'the moment of inertia of a body' – this term is meaningless unless the axis of rotation is also specified.

CALCULATION

The evaluation of $\Sigma\, mr^2$ for a particular body is done in one of two ways. If the body consists of a finite number of particles (e.g. two particles fixed at the ends of a light rod), this summation can be done by simple addition. In the case of a solid body the summation may have to be done by integration.
Several examples follow which are important because they illustrate the way in which standard results are built up from results already obtained. They also deal with objects that occur frequently in problems.

EXAMPLES 11a

1) Three light rods, each of length $2l$, are joined together to form a triangle. Three particles A, B, C of mass m, $2m$, $3m$ are fixed to the vertices of the triangle. Find the moment of inertia of the resulting body about:
(a) an axis through A perpendicular to the plane ABC,
(b) an axis passing through A and the midpoint of BC.

(a)

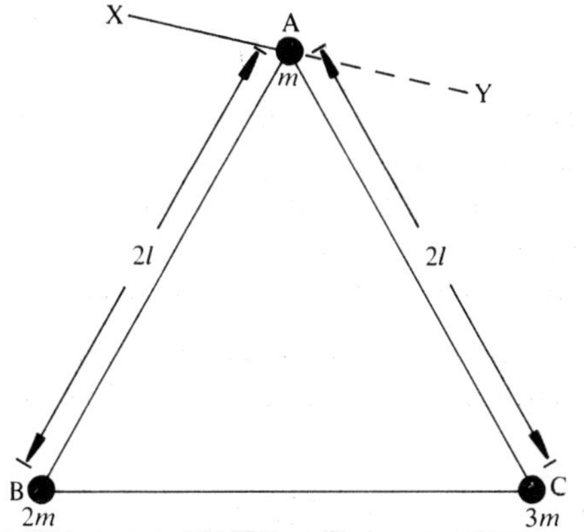

B is distant $2l$ from the axis XY
so the moment of inertia of B (I_B) about XY is $2m(2l)^2$
similarly I_C about XY is $3m(2l)^2$
and I_A about XY is $m(0)^2$

Therefore the moment of inertia of the body about XY is

$$2m(2l)^2 + 3m(2l)^2 + m(0)^2$$

$$= 20ml^2$$

(b)

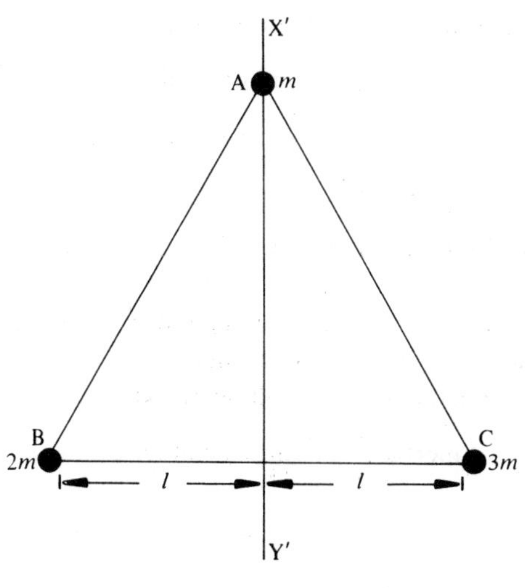

I_A about X′Y′ $= m(0)^2$

I_B about X′Y′ $= 2m(l)^2$

I_C about X′Y′ $= 3m(l)^2$

Therefore the moment of inertia of the body about X′Y′ is

$$m(0)^2 + 2m(l)^2 + 3m(l)^2$$
$$= 5ml^2$$

(This example emphasises that the moment of inertia of a body is not unique but depends on the axis of rotation.)
The moment of inertia of any object which consists of a *finite* number of particles can be found in a similar way.
We shall now consider bodies which have a continuous distribution of mass.

2) Find the moment of inertia of a uniform ring of mass M and radius a about an axis through its centre and perpendicular to the plane of the ring.

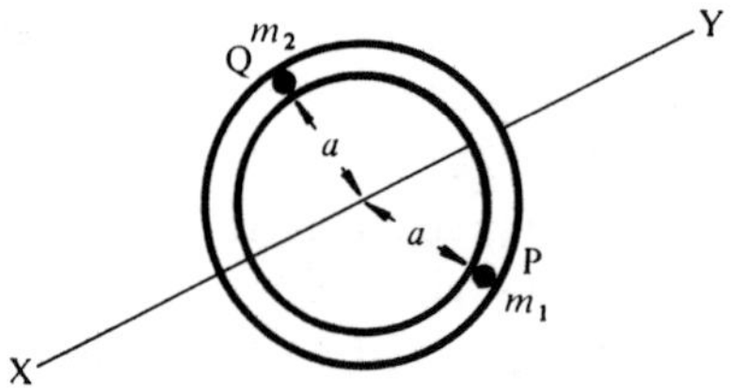

Consider the ring to be made up of particles whose masses are $m_1, m_2, m_3, \ldots$.
The moments of inertia of these particles about XY are

$$m_1a^2, m_2a^2, m_3a^2, \ldots$$

Therefore the moment of inertia of the ring about XY is

$$m_1a^2 + m_2a^2 + m_3a^2 + \ldots$$
$$= (m_1 + m_2 + m_3 + \ldots)a^2$$
$$= Ma^2$$

as the total mass M of the ring is the sum of the masses of its constituent particles.
This result is particularly useful as a 'building block' for finding the moments of inertia of more complex bodies which can be considered as being made up of ring-like elements as illustrated in the following examples.

3) Find the moment of inertia of a hollow cylinder of mass M and radius a about its axis.

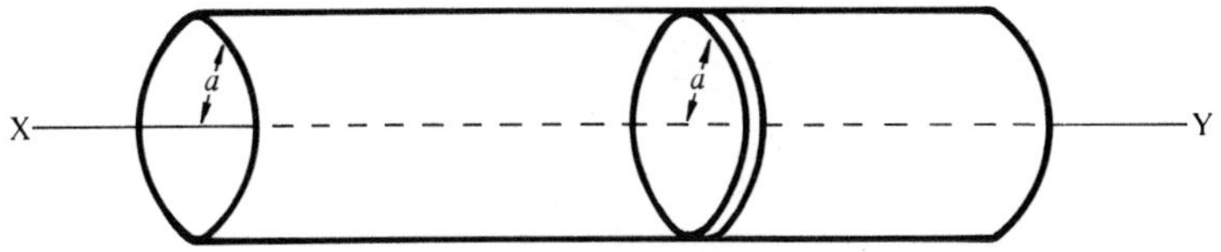

As we now know the moment of inertia of a ring about an axis through its centre perpendicular to its plane we can find the moment of inertia of the cylinder by dividing it into rings perpendicular to its axis and summing their moments of inertia about the axis of the cylinder.
Thus considering a typical such ring of mass m_r.
The moment of inertia of this ring about XY is $m_r a^2$ as it has a radius a.
Therefore the moment of inertia of the cylinder is

$$\sum m_r a^2 = a^2 \sum m_r \quad \text{(as each ring has a radius } a\text{)}$$

$$= Ma^2$$

Any hollow object whose surface is a surface of revolution (i.e. formed by rotating a portion of a curve about the x-axis) can be divided into cylinder-like elements.

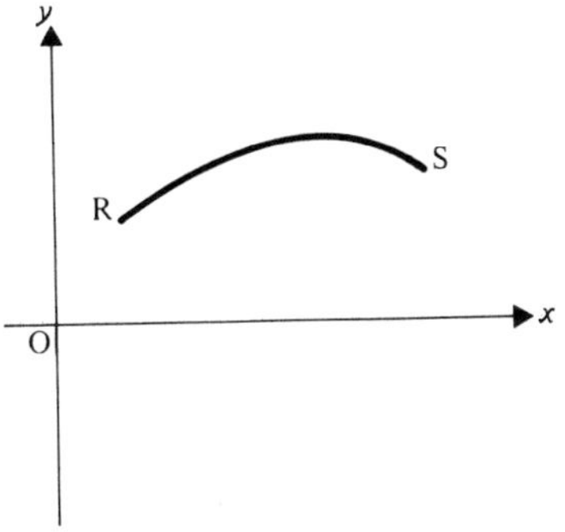

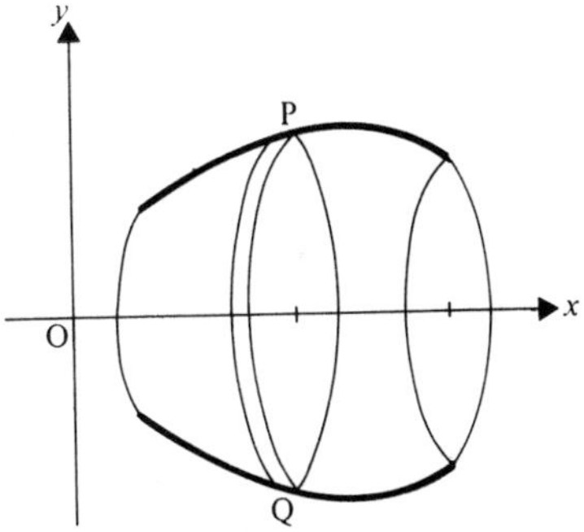

Consider the hollow body formed by rotating the curve RS about the x-axis. This body can be considered as being made up of elements (of which PQ is a typical element) which are approximately cylinders. Using the result of Example 2 the moment of inertia of such an element about Ox can be written down. The mass and radius of such an element depend on its position within the body, so the expression for its moment of inertia about Ox is a function of its position. The necessary summing to find the moment of inertia of the whole body about Ox therefore has to be done by integration. This general method is illustrated in the next example.

4) Find the moment of inertia of a uniform hollow sphere of radius a and mass M about a diameter.

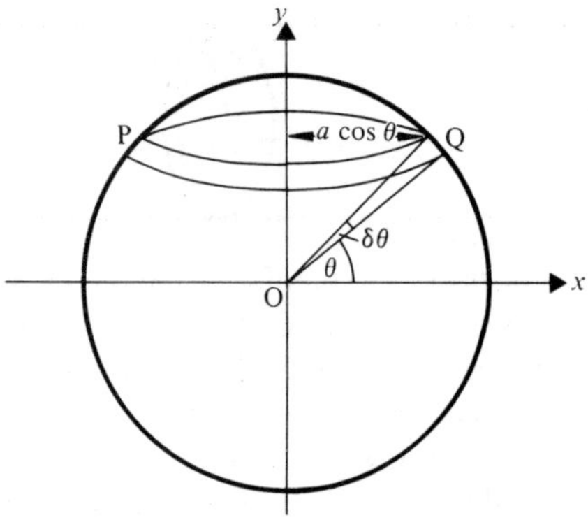

If the sphere is divided into elements by parallel cuts perpendicular to Oy, each element is approximately a cylinder.
The moment of inertia about Oy of each such cylindrical element is its mass multiplied by the square of its radius.
Consider a typical element PQ.
The radius of this 'cylinder' is $a \cos \theta$ and its width is $a\delta\theta$.
If m is the mass per unit area of the sphere, the mass of the element PQ is approximately

$$m(2\pi a \cos \theta)(a\delta\theta)$$

Therefore the moment of inertia of the element PQ about Oy

$$\simeq (2\pi ma^2 \cos \theta\delta\theta)(a \cos \theta)^2$$

Therefore the moment of inertia of the sphere about Oy is

$$\int_{-\pi/2}^{\pi/2} 2\pi ma^4 \cos^3\theta \, d\theta$$

$$= 2\pi ma^4 \int_{-\pi/2}^{\pi/2} \cos \theta(1 - \sin^2\theta) \, d\theta$$

$$= 2\pi ma^4 \left[\sin \theta - \tfrac{1}{3} \sin^3\theta\right]_{-\pi/2}^{\pi/2}$$

$$= 2\pi ma^4 (\tfrac{4}{3})$$

$$= \frac{8\pi ma^4}{3}$$

$$= \frac{2Ma^2}{3} \qquad \text{as } M = (4\pi a^2)m$$

(*Note*: do not take the element as approximately a cylinder of radius x and width δx as this gives a bad approximation to the surface area.)

5) Find the moment of inertia of a uniform disc of mass M and radius a about an axis through the centre of the disc perpendicular to the plane of the disc.

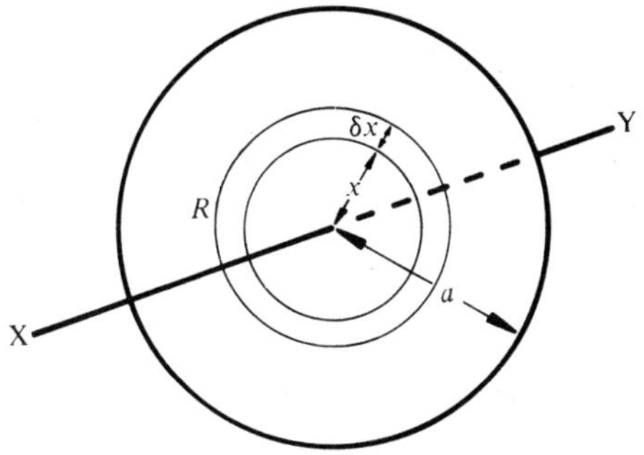

If the disc is divided into annuli each of width δx then, if δx is small, each such annulus is approximately a ring.

Note: an annulus is a circular lamina with a concentric hole in it.

Consider a typical such 'ring' R of radius x.
If m is the mass per unit area of the disc, the mass of the ring is m times its area which is approximately $m(2\pi x\ \delta x)$ (taking the area as approximately the area of a strip of length $2\pi x$ and width δx).
Therefore the moment of inertia of the ring about XY $\simeq m(2\pi x\ \delta x)(x)^2$.
Therefore the moment of inertia I of the disc about XY $\simeq \Sigma\ 2\pi m x^3\,\delta x$.

i.e. $I \simeq 2\pi m \sum x^3\ \delta x$

$$I = 2\pi m \int_0^a x^3\,dx = 2\pi m \left[\frac{x^4}{4}\right]_0^a = \frac{m\pi a^4}{2}$$

$$= \frac{Ma^2}{2} \qquad \text{as } \pi a^2 m = M$$

This is another very useful result as it enables us to find the moment of inertia of bodies which can be divided into 'disc-like' elements. In particular, any solid of revolution can be divided into such elements perpendicular to its axis. The next example illustrates this.

6) Find the moment of inertia of a uniform solid right circular cone of radius a, height h, and mass M about its axis.

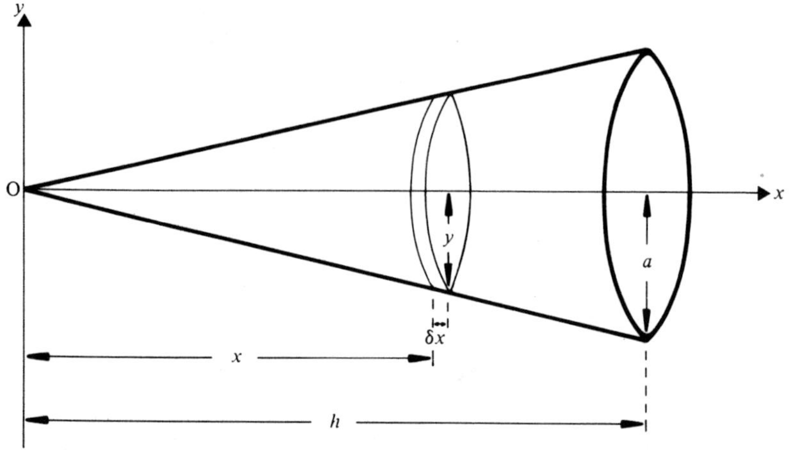

If the cone is divided into elements by cuts parallel to its base, each such element is approximately a disc.

Taking axes as shown, consider a typical such element, E, distant x from O.

The volume of $E \simeq \pi y^2\, \delta x$.

The mass of $E \simeq \pi y^2 m\, \delta x$ where m is the mass per unit volume of the cone.

If δx is small E is approximately a disc.

Therefore the moment of inertia of E about Ox $\simeq \frac{1}{2}(\pi y^2 m\, \delta x)y^2$

Therefore the moment of inertia of the cone about Ox, $I = \int_0^h \frac{1}{2}\pi y^2 m\, dx$

$$= \tfrac{1}{2}\pi m \int_0^h y^4\, dx$$

From similar triangles $y = \dfrac{xa}{h}$.

$$I = \tfrac{1}{2}\pi m \int_0^h \frac{x^4 a^4}{h^4}\, dx = \left(\frac{\pi m a^4}{2h^4}\right)\left[\frac{x^5}{5}\right]_0^h = \frac{\pi m a^4 h}{10}$$

$$= \frac{3Ma^2}{10} \qquad \text{as } M = \tfrac{1}{3}m\pi a^2 h$$

Another very useful 'building block' is a rod, as several bodies can be divided into 'rod-like' elements, so in the next example we find the moment of inertia of a rod about three different axes.

7) Find the moment of inertia of a uniform rod of mass M and length $2l$ about an axis:

(a) through the centre of the rod and perpendicular to the rod,

(b) through the centre of the rod and inclined at an angle θ to the rod,

(c) parallel to the rod and distant d from it.

(a)

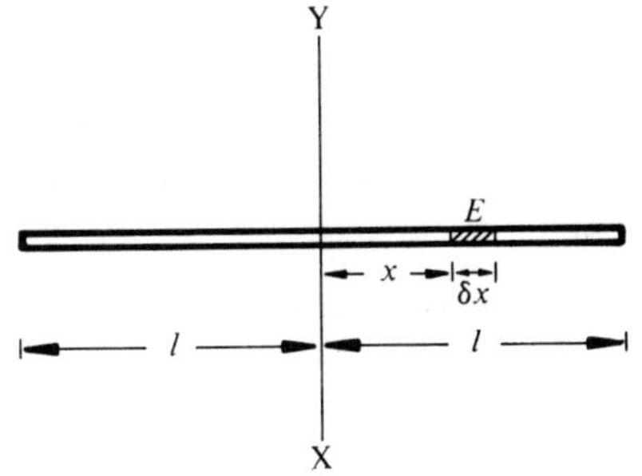

If the rod is divided into elements each of length δx, and the mass per unit length of the rod is m then, if δx is small, each such element can be considered as approximately a particle of mass $m\delta x$.

Considering a typical such element E, distant x from XY:

The moment of inertia of E about XY $= m\delta x\,(x)^2$

Therefore the moment of inertia of the rod about XY, $I = \int_{-l}^{l} mx^2\,dx$

$$I = m\left[\frac{x^3}{3}\right]_{-l}^{l} = \frac{2ml^3}{3} = \frac{Ml^2}{3} \qquad \text{as } M = 2lm$$

(b)

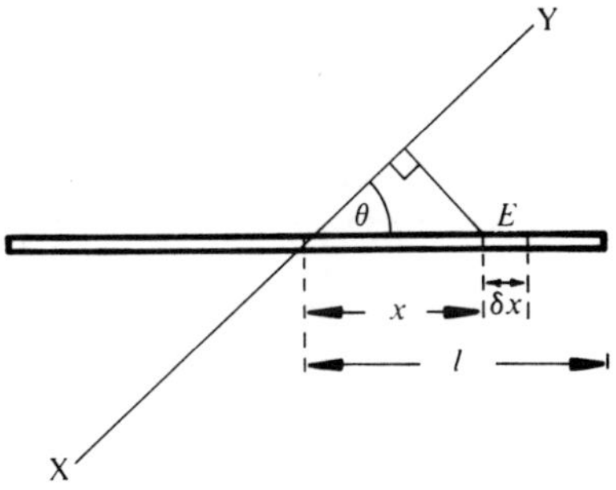

Dividing the rod into elements as in (a) and considering the element E again, the mass of E, as before, is $m\delta x$.

The distance of E from XY is $x \sin\theta$.

Therefore the moment of inertia of E about XY $= m\delta x\,(x\sin\theta)^2$

Therefore the moment of inertia of the rod about XY $= \int_{-l}^{l} mx^2 \sin^2\theta\,dx$

$$= m\sin^2\theta\left[\frac{x^3}{3}\right]_{-l}^{l} = \frac{Ml^2\sin^2\theta}{3}$$

(c)

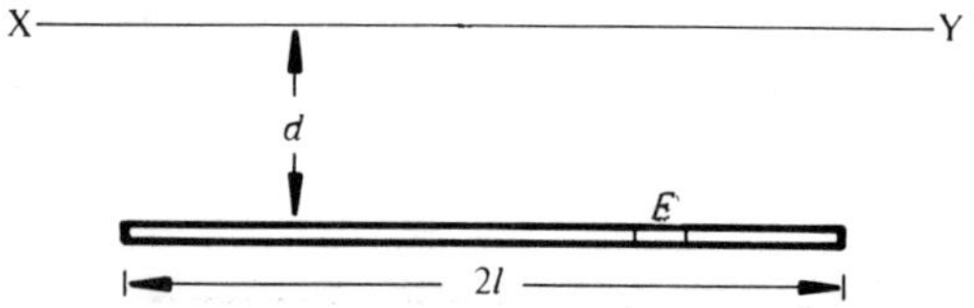

In this case every element of the rod is the same distance from the axis XY. Considering the element E of mass m

the moment of inertia of E about XY $= md^2$

Therefore the moment of inertia of the rod about XY $= \sum md^2$

$$= d^2 \sum m$$

$$= Md^2$$

We are now in a position to find the moments of inertia about certain axes of bodies that can be divided into rod-like elements.

8) Find the moment of inertia of a uniform lamina in the shape of a rectangle ABCD of mass M, where AB $= 2a$ and BC $= 2b$, about an axis through the mid-points of AB and CD.

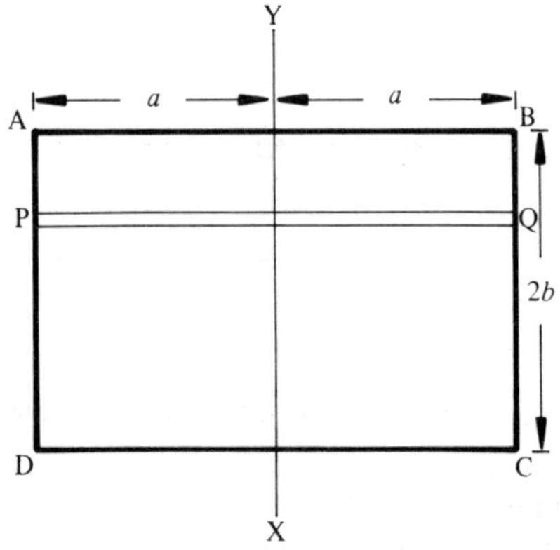

If the lamina is divided into strips parallel to AB, then each strip is approximately a rod of length $2a$.
Considering a typical such rod PQ, the moment of inertia of PQ about XY is $(m)\dfrac{a^2}{3}$ where m is the mass of PQ.
Therefore the moment of inertia of the lamina ABCD about XY is

$$\sum (m) \frac{a^2}{3}$$

$$= \frac{a^2}{3} \sum (m) \qquad \text{as each such rod is of length } 2a$$

$$= \frac{Ma^2}{3}$$

EXERCISE 11a

1) Four particles, A, B, C and D of mass 2 kg, 5 kg, 6 kg, 3 kg respectively are rigidly joined together by light rods to form a rectangle ABCD, where AB = $2a$ and BC = $4a$. Find the moment of inertia of this system of particles about an axis along (a) AB (b) BC (c) AC (d) through the midpoints of AB and CD.

2) Find the moment of inertia of a uniform hollow cone of mass M, base radius a and height h, about its axis.

3) Find the moment of inertia of a uniform solid cylinder of radius a and mass M about its axis.

4) Find the moment of inertia of a uniform solid sphere of radius a and mass M about a diameter, and deduce the moment of inertia of a uniform solid hemisphere of radius a and mass M' about its axis.

5) ABCD is a uniform rectangle of mass M where AB = $2a$ and BC = $2b$. Find the moment of inertia of ABCD about the edge AB by dividing the rectangle into rod-like elements parallel to AB.

6) Find the moment of inertia of a uniform lamina in the form of an equilateral triangle of mass M and side $2l$ about a median of the triangle.

7) Find, by integration, the moment of inertia of a uniform rod of mass M and length $2l$ about an axis through one end of the rod which is (a) perpendicular to the rod (b) inclined at an angle θ to the rod.

8) Use the result obtained in question 7(b) to find the moment of inertia of a uniform lamina in the form of an equilateral triangle of side $2a$ and mass M, about one side.

9) A uniform solid is formed by rotating the portion of the curve $y^2 = 4ax$ between $x = 0$ and $x = 2a$ about the x-axis. If the mass of the solid is M and the mass per unit volume is m, find a relationship between M and m and then find the moment of inertia of the solid about the x-axis.

10) A uniform hollow frustum of a cone has a mass M. The radii of its circular ends are a and $2a$ and the distance between them is h. Find the moment of inertia of the frustum about its axis.

COMPOUND BODIES

Consider a ring of radius a and mass M with a particle of mass M' attached to a point on its circumference.

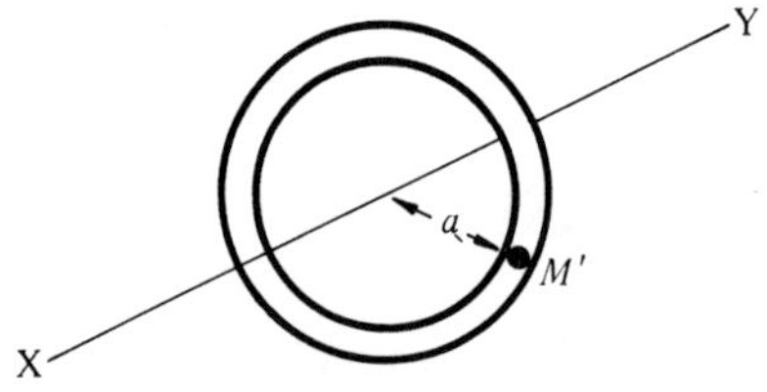

If the resulting body is free to rotate about an axis through the centre of the ring perpendicular to the plane of the ring, then
the moment of inertia of the body about XY

$= (m_1 + m_2 + \ldots) a^2 + M'a^2$ where $m_1, m_2, \ldots$ are the masses of the constituent particles of the ring.

$= Ma^2 + M'a^2$

= moment of inertia of the ring about XY plus the moment of inertia of the particle about XY.

In general if two solid bodies A and B are rigidly joined together, and the resulting body is free to rotate about an axis XY and, if I_{A+B} is the moment of inertia of the compound body about XY, where I_A and I_B are the moments of inertia of A and B respectively about XY, then

$$\begin{aligned} I_{A+B} &= \sum mr^2 \text{ summed over all the particles of the compound body} \\ &= \sum mr^2 \text{ summed over all the particles of A} \\ &\quad + \sum mr^2 \text{ summed over all the particles of B} \\ &= I_A + I_B \end{aligned}$$

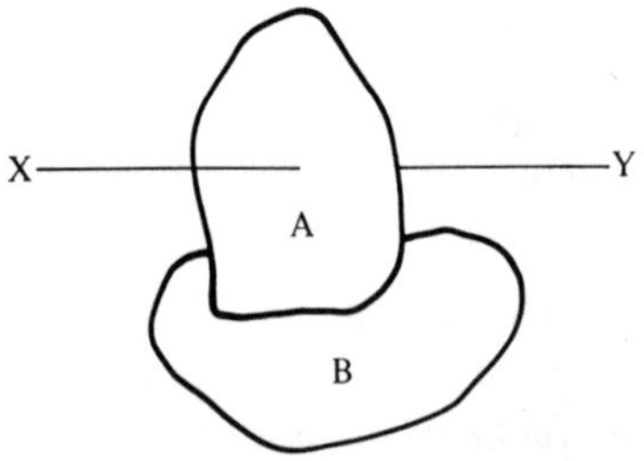

This result can obviously be extended to cover any number of bodies rigidly joined together to form one compound body. Thus the *moment of inertia of a compound body about an axis is the sum of the moments of inertia of the separate objects forming the compound body about the same axis.*

EXAMPLES 11b

1) Three uniform rods, each of length $2l$ and mass M are rigidly joined at their ends to form a triangular framework. Find the moment of inertia of the framework about an axis passing through the midpoints of two of its sides.

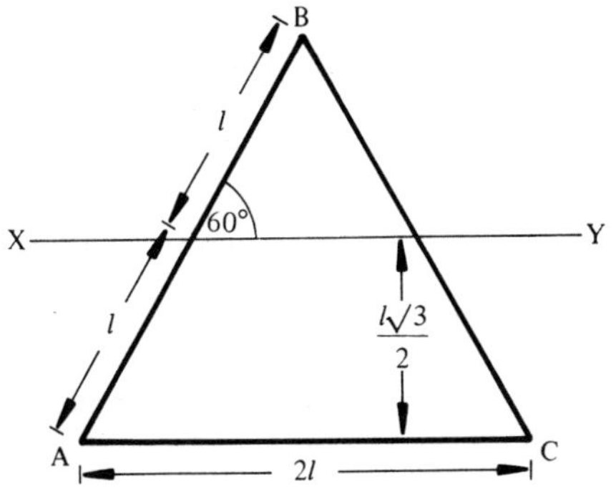

Using the results of Example 11a, No. 7:

The moment of inertia of the rod AB about XY $= \frac{Ml^2}{3}\sin^2 60°$

$$= \frac{Ml^2}{4}$$

The moment of inertia of the rod BC about XY $= \frac{Ml^2}{4}$

The moment of inertia of the rod AC about XY $= M\left(\frac{l\sqrt{3}}{2}\right)^2$

$$= \frac{3Ml^2}{4}$$

Therefore the moment of inertia of the framework about XY

$$= \frac{Ml^2}{4} + \frac{Ml^2}{4} + \frac{3Ml^2}{4}$$

$$= \frac{5Ml^2}{4}$$

NON-UNIFORM BODIES

All the bodies we have considered so far have been uniform, although some of the results obtained are true also if the body concerned is not uniform. This is so in the case of bodies whose constituent particles are all at the same distance from the axis of rotation:

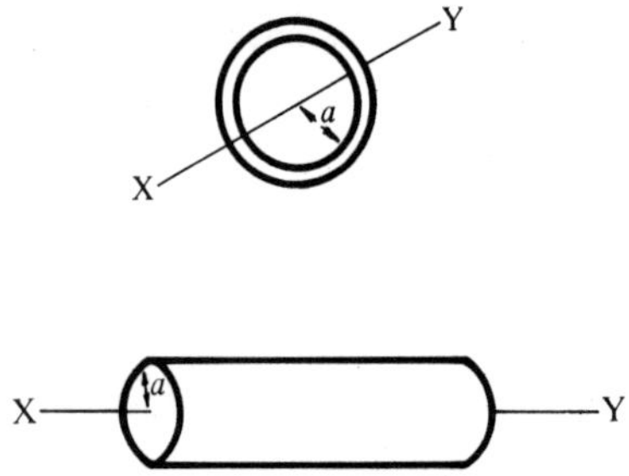

The moment of inertia of the ring (mass M) about XY is Ma^2 whether the ring is uniform or not.
Similarly the moment of inertia of the *hollow* cylinder (mass M) about XY is Ma^2 irrespective of whether the cylinder is uniform or not.

It must be noted that these results would *not* be true of any other axis.
In general, the mass of an element of a non-uniform body is a function of its position in the body, so the evaluation of $\Sigma\, mr^2$ has to be done by integration.

EXAMPLES 11b (continued)

2) A semicircular lamina of radius a has a mass per unit area which is mx at a point distant x from O, the midpoint of its straight edge. If M is the total mass of the lamina, find, in terms of M and a, its moment of inertia about an axis through O, perpendicular to the lamina.

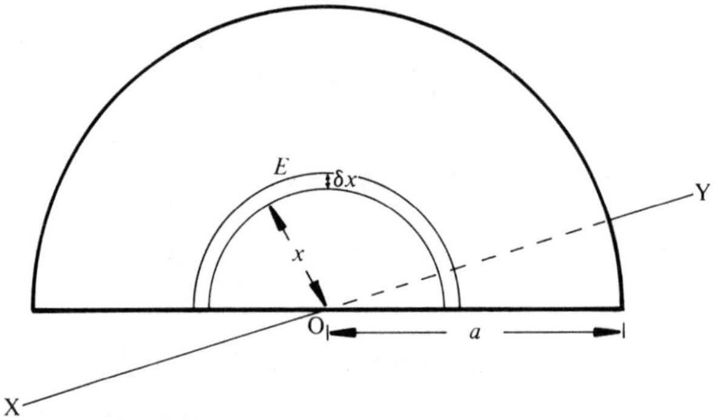

If the lamina is divided into semicircular ring-like elements, each constituent particle is the same distance from O.
Considering a typical such element E,

the area of E $\simeq \pi x\, \delta x$

Therefore the mass of $E \simeq (\pi x\, \delta x)mx = \pi m x^2\, \delta x$

Therefore I_E about XY $\simeq (\pi m x^2\, \delta x)x^2$ as each particle of E is distant x from O

Therefore the moment of inertia of the lamina about XY

$$= \int_0^a \pi m x^4 \, dx$$

$$= \pi m \left[\frac{x^5}{5}\right]_0^a$$

$$= \frac{\pi m a^5}{5}$$

As the mass of the element E is $\pi m x^2 \, \delta x$

the mass of the lamina $M = \int_0^a \pi m x^2 \, dx$

Therefore $M = \dfrac{\pi m a^3}{3}$

Therefore the moment of inertia of the lamina about XY $= \dfrac{3Ma^2}{5}$.

EXERCISE 11b

1) A uniform rod of mass M and length $2l$ has a particle of mass $\frac{1}{2}M$ fixed to one end. Find the moment of inertia of this system about an axis through the centre of the rod and perpendicular to the rod.

2) A uniform ring of radius a and mass M has two particles each of mass M attached to it. Find the moment of inertia of the resulting body about an axis through the centre of the ring and perpendicular to the ring.

3) A uniform solid consists of a solid cylinder of radius a and length l surmounted by a solid hemisphere of radius a. If the total mass of the solid is M find the moment of inertia of the solid about its axis.

4) A triangular framework ABC consists of three uniform rods each of mass M where $AB = BC = 2l$, $AC = 2\sqrt{2}l$. Find the moment of inertia of the framework about an axis passing through the midpoints of AB and BC.

5)

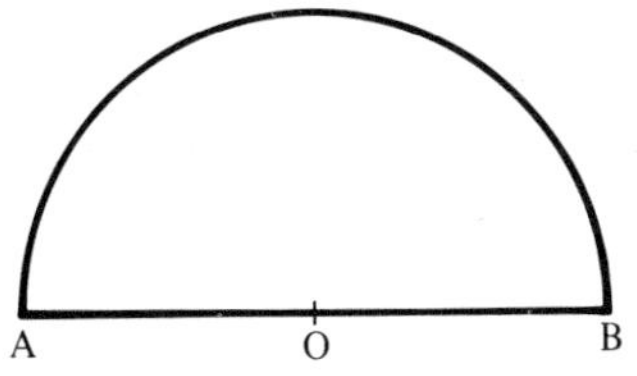

A uniform wire of mass M is bent into the form shown in the diagram where the arc AB is a semicircle of radius a and AOB is a straight line. Find the moment of inertia of this framework about an axis through O (the mid-point of AB) and perpendicular to the plane of the framework.

6) A non-uniform rod AB has a mass M and length $2l$. The mass per unit length of the rod is mx at a point of the rod distant x from A. Find the moment of inertia of this rod about an axis perpendicular to the rod passing (a) through A (b) through the midpoint of AB.

7) A non-uniform solid sphere of radius a and mass M has a mass mx per unit volume at all points distant x from its centre. By dividing the sphere into elements which are approximately hollow spheres, find the moment of inertia of the sphere about a diameter. (Use the result of Examples 11a, No. 4 for the moment of inertia of the element.)

RADIUS OF GYRATION

The moment of inertia of any rigid body about a specified axis can be expressed in the form Mk^2 where M is the mass of the body – this is the same as the moment of inertia of a particle of mass M distant k from the axis and k is called the *radius of gyration* of the body about that axis.
Consider, for example, a uniform rod of mass M and length $2l$ rotating about an axis through its centre and perpendicular to the rod. If I is the moment of inertia of the rod about this axis then:

$$I = \frac{Ml^2}{3} = M\left(\frac{l}{\sqrt{3}}\right)^2$$

Therefore $$k = \frac{l}{\sqrt{3}} = \frac{l\sqrt{3}}{3}$$

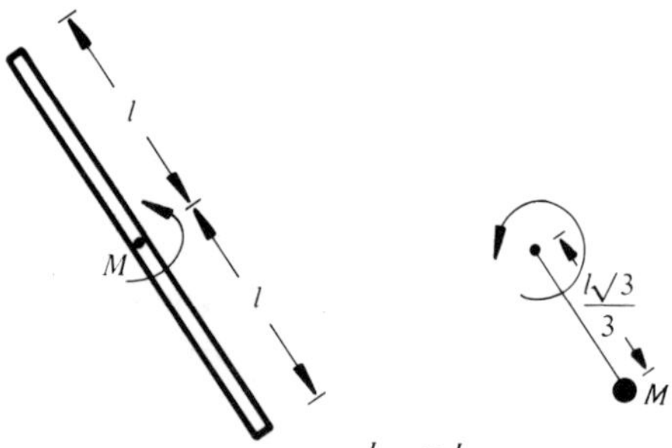

$I_{\text{rod}} = I_{\text{particle}}$

Many students are tempted to replace a rotating rigid body by a particle of equal mass at the centre of gravity, but as can be seen from the above example this does *not* give the correct result for its moment of inertia.

Second moment of Area and Volume

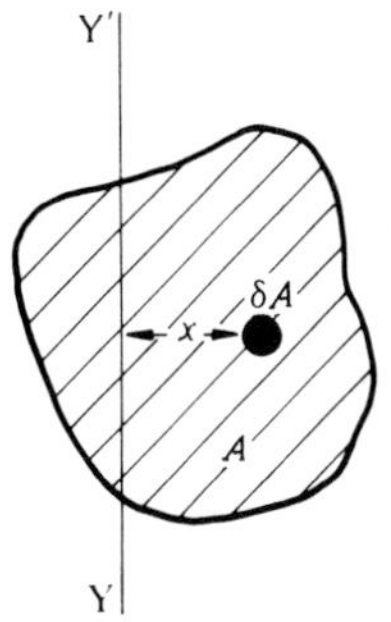

If δA is a small element of an area A, the second moment of the area A about an axis YY$'$ is defined as $\Sigma x^2 \delta A$ where x is the distance of δA from YY$'$. If the area A represents a uniform lamina of mass m per unit area, the moment of inertia of the lamina about YY$'$ is $\Sigma x^2 (m \delta A) = m \Sigma x^2 \delta A =$ m times second moment of its area about YY$'$.
Thus the second moment of area is calculated in the same way as the moment of inertia of a lamina except that the mass per unit area is not included.

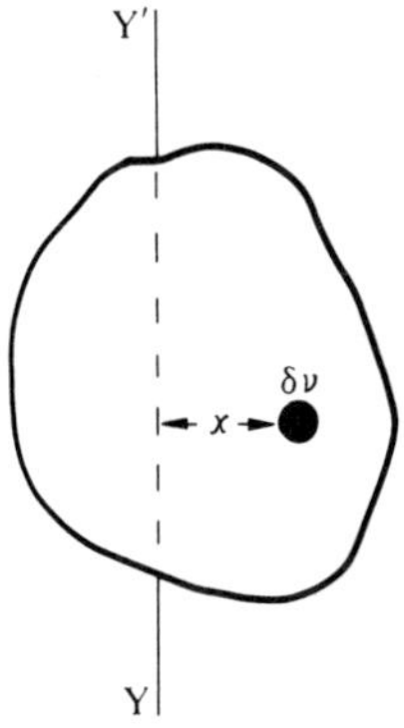

If δv is a small element of a volume v, the second moment of the volume about YY$'$ is $\Sigma x^2 \delta v$ where x is the distance of δv from YY$'$.
If v represents a uniform solid of mass m per unit volume, the moment of inertia of that solid about YY$'$ is $\Sigma x^2 (m \delta v) = m \Sigma x^2 \delta v = m$ times second moment of volume about YY$'$.
Thus the second moment of volume is calculated in the same way as the moment of inertia of a solid except that the mass per unit volume is not included.

PARALLEL AXIS THEOREM

Up to this point we have calculated the moment of inertia of a body about an axis which passes through its centre of mass. If the moment of inertia of a body about an axis through its centre of mass is known, the following theorem gives a very easy method of finding its moment of inertia about any parallel axis: *If the moment of inertia of a body of mass M about an axis through its centre of mass is Mk^2 its moment of inertia about a parallel axis distant d from the first axis is $M(k^2 + d^2)$.*

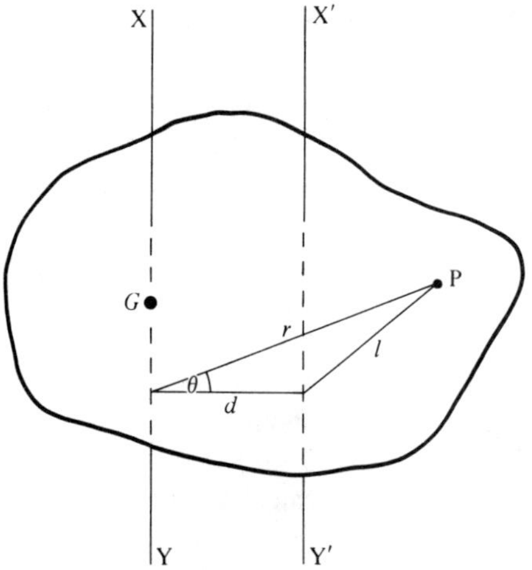

Let XY be an axis passing through the centre of mass G of the body and X'Y' be a parallel axis distant d from XY.
Consider a typical constituent particle P of mass m where P is distant l from X'Y' and distant r from XY.
The moment of inertia of P about X'Y' is

$$ml^2 = m(r^2 + d^2 - 2rd \cos \theta) \quad \text{(cosine rule)}$$

Therefore the moment of inertia of the whole body about X'Y' is

$$\sum m(r^2 + d^2 - 2rd \cos \theta)$$

$$= \sum mr^2 + \sum md^2 - \sum 2mrd \cos \theta$$

$$= Mk^2 + d^2 \sum m - 2d \sum mr \cos \theta$$

Now $mr \cos \theta$ is the moment of the mass of the particle P about the axis XY.
Therefore $\Sigma\, mr \cos \theta = 0$ as XY passes through G, the centre of mass.
Therefore the moment of inertia of the body about X'Y' $= Mk^2 + Md^2$

$$= M(k^2 + d^2).$$

EXAMPLE

The moment of inertia of the uniform rod in the diagram about the axis XY is

$$\frac{Ml^2}{3}.$$

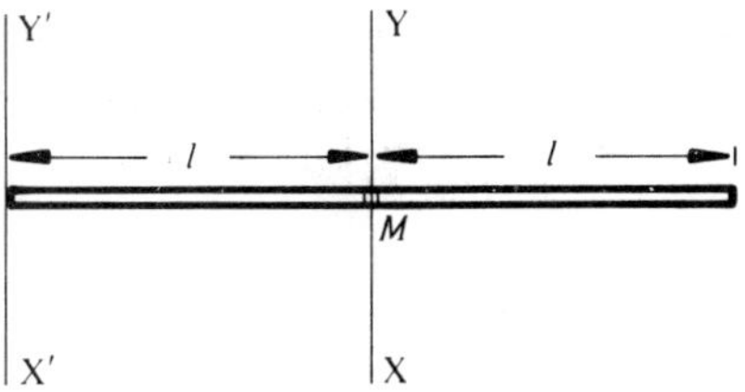

Using the parallel axis theorem the moment of inertia of the rod about the axis X′Y′ is $M\left(\frac{l^2}{3} + l^2\right)$

$$= \frac{4Ml^2}{3}$$

It must be emphasised that this theorem gives the relationship between the moments of inertia of a body about two parallel axis, *one of which passes through the centre of mass of the body*. It is possible, given the moment of inertia of a body about an axis not through the centre of mass, to find the moment of inertia about a parallel axis again not through the centre of mass, but the calculation will have to be done in two stages.
Consider the body shown. If M is the mass of the body and its moment of inertia about XY is I then

$$I = I_G + Md^2$$

Therefore $I_G = I - Md^2$

Also $I_{X'Y'} = I_G + Ma^2$

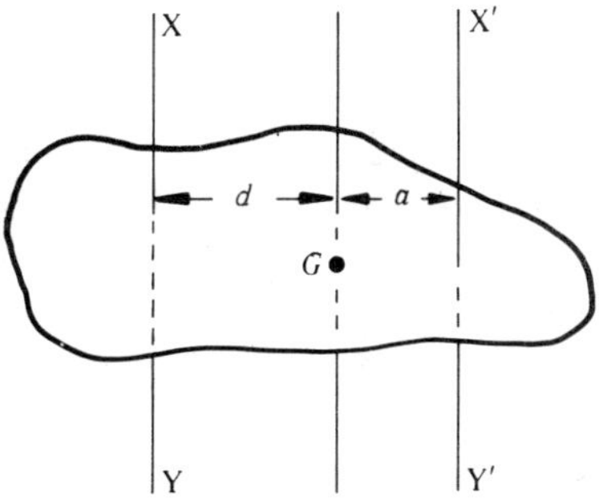

Therefore the moment of inertia of the body about X′Y′

$$= (I - Md^2) + Ma^2$$

$$= I + M(a^2 - d^2)$$

PERPENDICULAR AXES THEOREM

The following theorem is useful for calculating the moment of inertia of any plane object (lamina, ring, triangular framework etc.) about certain axes:

If the moments of inertia of a plane body about two perpendicular axes (Ox and Oy) in the plane of the body are I_x and I_y respectively, the moment of inertia of the body about a third axis Oz which is perpendicular to the first two is $I_x + I_y$.

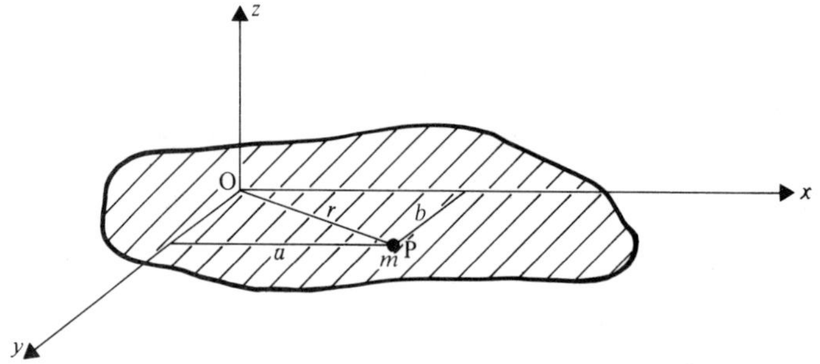

If P is a constituent particle of the lamina and is of mass m, the moment of inertia of P about Oz is

$$mr^2 = m(b^2 + a^2) \quad \text{(Pythagoras' theorem)}$$

Therefore the moment of inertia of the lamina about Oz is

$$\sum mr^2 = \sum mb^2 + \sum ma^2$$

$$= I_x + I_y$$

(*Note*: This theorem *cannot* be applied to three dimensional rigid bodies. Also the three axes under consideration must be mutually perpendicular *and* concurrent, although none of them need pass through the centre of mass of the body.)

We can now use this theorem to extend some of the results already obtained for plane bodies.

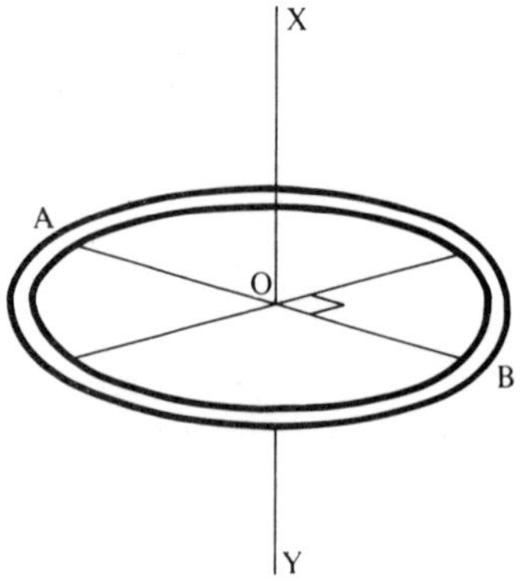

The moment of inertia of a uniform ring of mass M and radius a about XY is Ma^2. If the moment of inertia of the ring about any diameter is I_d then by considering two perpendicular diameters and using the perpendicular axis theorem

$$Ma^2 = I_d + I_d.$$

Therefore the moment of inertia of the ring about a diameter is $\frac{1}{2}Ma^2$.
Similarly the moment of inertia of a uniform disc of mass M and radius a about a diameter is $\frac{1}{4}Ma^2$.

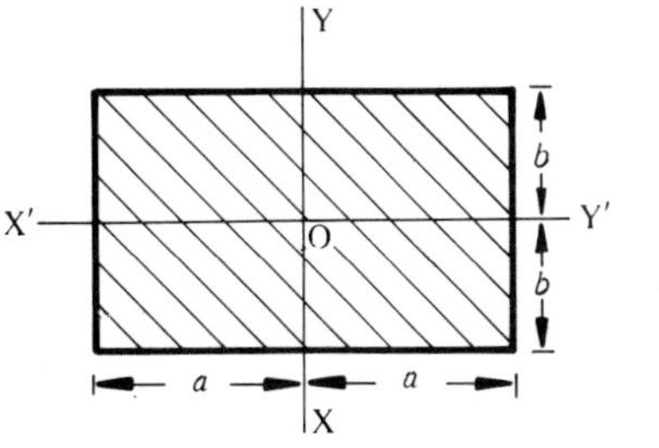

The moment of inertia of the uniform rectangular lamina of mass M shown in the diagram is $\frac{1}{3}Ma^2$ about the axis XY and $\frac{1}{3}Mb^2$ about the axis X'Y'.

Therefore the moment of inertia of the lamina about an axis perpendicular to its plane and passing through its centre of mass is $\frac{1}{3}M(a^2 + b^2)$.
The following examples show how either or both of these theorems can be used to calculate other moments of inertia.

EXAMPLES 11c

1) Find the moment of inertia of a uniform disc of mass M and radius a about a tangent.

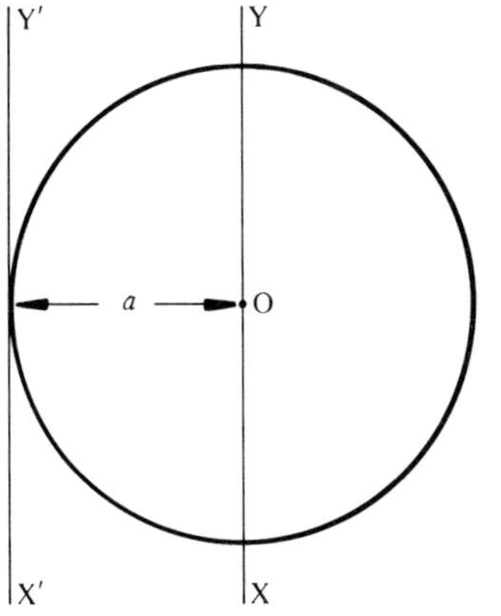

From the result above, the moment of inertia of the disc about the diameter XY is $\frac{1}{4}Ma^2$.
Therefore using the parallel axis theorem the moment of inertia of the disc about the tangent X'Y' is

$$M(\tfrac{1}{4}a^2 + a^2) = \tfrac{5}{4}Ma^2$$

2) Find the moment of inertia of a uniform cube of mass M and edge $2a$ about an axis along one edge.

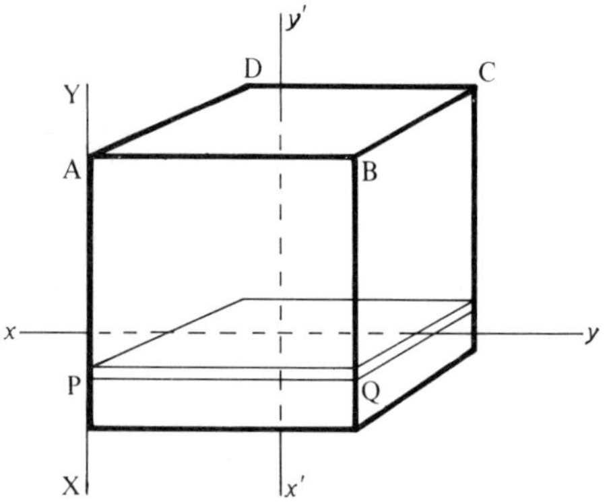

If the cube is divided into elements by cuts parallel to the face ABCD, each element is approximately a square lamina of side $2a$.
Considering a typical such element PQ of mass m:
The moment of inertia of PQ about xy is $\frac{1}{3}ma^2$.
Therefore the moment of inertia of PQ about $x'y'$ is $\frac{2}{3}ma^2$
(perpendicular axis theorem),
and the moment of inertia of PQ about XY is $m(\frac{2}{3}a^2 + 2a^2)$
(parallel axis theorem).
Therefore the moment of inertia of the cube about XY is

$$\sum m(\tfrac{2}{3}a^2 + 2a^2)$$

$$= \sum m \frac{8a^2}{3}$$

$$= \frac{8a^2}{3} \sum m$$

$$= \frac{8Ma^2}{3}$$

3) Find the moment of inertia of a uniform solid cone of radius a, height h and mass M about an axis through its vertex parallel to a diameter of the base.

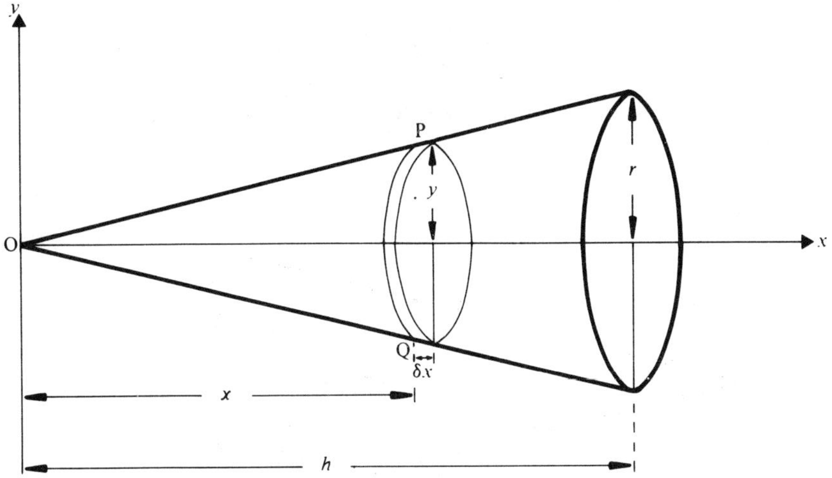

If the cone is divided into elements by cuts parallel to the base, each such element is approximately a disc.
Considering a typical element PQ of radius y and thickness δx:
If m is the mass per unit volume of the cone, the mass of the element PQ is approximately $(\pi y^2 \, \delta x)m$.

Therefore $I_{(\text{element})}$ about Ox $\simeq \frac{1}{2}(\pi y^2 m \, \delta x)y^2$

$I_{(\text{element})}$ about PQ $\simeq \frac{1}{4}(\pi y^2 m \, \delta x)y^2$ (using the perpendicular axis theorem)

$I_{(\text{element})}$ about Oy $\simeq (\pi y^2 m \, \delta x)(\frac{1}{4}y^2 + x^2)$ (using the parallel axis theorem)

$$\simeq \pi m(\tfrac{1}{4}y^4 + y^2x^2)\,\delta x.$$

Therefore $I_{(\text{cone})}$ about Oy $= \int_0^h \pi m(\frac{1}{4}y^4 + y^2x^2)dx$

but $\dfrac{x}{h} = \dfrac{y}{r} \Rightarrow y = \dfrac{rx}{h}$

therefore
$$I_{(\text{cone})} = \int_0^h \pi m \left(\frac{r^4}{4h^4} + \frac{r^2}{h^2}\right) x^4 \, dx$$

$$= \frac{\pi m r^2}{h^2}\left(\frac{r^2}{4h^2} + 1\right)\left[\frac{x^5}{5}\right]_0^h$$

$$= \frac{m\pi r^2 h}{20}(r^2 + 4h^2)$$

$$= \frac{3M}{20}(r^2 + 4h^2) \quad \text{as } M = \tfrac{1}{3}\pi r^2 hm.$$

This result can be extended to find the moment of inertia of the cone about a diameter of the base.

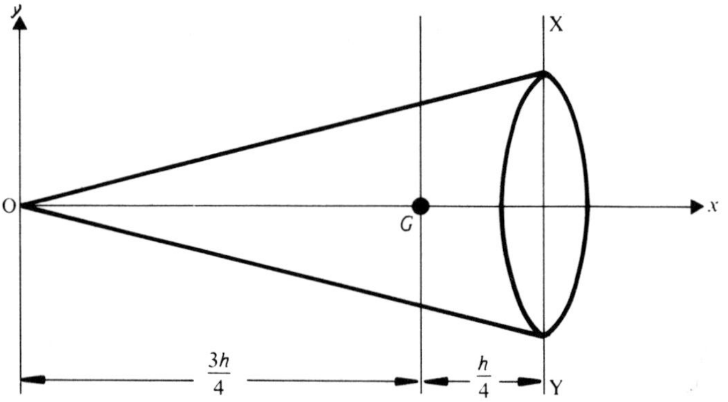

If I_G is the moment of inertia of the cone about an axis parallel to Oy through G, the centre of mass of the cone, using the parallel axis theorem:

$$\frac{3M}{20}(r^2 + 4h^2) = I_G + M\left(\frac{3h}{4}\right)^2 \qquad (1)$$

Also $$I_{XY} = I_G + M\left(\frac{h}{4}\right)^2 \qquad (2)$$

From (1) $$I_G = M\left(\frac{3r^2}{20} + \frac{3h^2}{80}\right)$$

Therefore $$I_{XY} = \frac{M}{20}(3r^2 + 2h^2)$$

MOMENTS OF INERTIA OF STANDARD BODIES

Not all problems on rotating bodies require the calculation of the appropriate moment of inertia. The table below gives the moments of inertia of some standard bodies about a specified axis. If the moment of inertia of the body is required about another axis, this can usually be calculated using the parallel axis theorem or (for plane bodies) the perpendicular axes theorem.

Standard Results

Uniform Body of Mass M	Axis	Moment of Inertia
Rod of length $2l$	Perpendicular to the rod through the centre	$\frac{Ml^2}{3}$
Rod	Parallel to the rod and distant d from it	Md^2
Rectangle of length $2l$ and width $2d$	Perpendicular to the sides of length $2l$ and passing through their midpoints	$\frac{Ml^2}{3}$
Ring of radius a	Perpendicular to the ring and through its centre	Ma^2
Disc of radius a	Perpendicular to the disc and through its centre	$\frac{Ma^2}{2}$
Solid sphere of radius a	A diameter	$\frac{2Ma^2}{5}$
Hollow sphere of radius a	A diameter	$\frac{2Ma^2}{3}$
Solid cylinder of radius a	The axis	$\frac{Ma^2}{2}$
Hollow cylinder of radius radius a	The axis	Ma^2

SUMMARY

The moment of inertia of a body about a specified axis is $\Sigma\, mr^2$ where m is the mass of a constituent particle and r is its distance from the axis.
To calculate the moment of inertia of a solid body about the specified axis: (a) divide the body into elements, each of whose moment of inertia about the axis is known or can easily be calculated using the parallel axis theorem or the perpendicular axes theorem, (b) sum the moment of inertia of all the elements about the axis over the whole body. (*Note* that if the moment of inertia of the element about the axis is independent of its position relative to the axis, the summation does not need to be done by integration.)

EXERCISE 11c

1) Find the moment of inertia of a uniform rectangle of side $2a$ and $2b$ and mass M about an axis along one edge of length $2b$.

2) Find the moment of inertia of a uniform ring of radius r and mass M about a tangent.

3) Find the moment of inertia of a uniform disc of radius a about an axis perpendicular to its plane passing through a point on its circumference.

4) A ring of radius a and mass M has six particles each of mass M fastened to it such that the particles form the vertices of a regular hexagon. Find the moment of inertia of the system about an axis passing through two adjacent particles.

5) Find the moment of inertia of a uniform rectangle of sides $2a$ and $2b$ and mass M about a perpendicular axis through the midpoint of a side of length $2a$.

6) Find the moment of inertia of a uniform cube of edge l and mass M, about an axis through the centre of the cube and perpendicular to two faces.

7) Find the moment of inertia of a uniform solid cylinder of radius r and mass M about a generator of the cylinder.

8) Find the moment of inertia of a uniform hollow cylinder of radius r, height h and mass M about a diameter of one end.

9) An equilateral triangle is formed from three equal uniform rods, each of mass M and length $2l$, rigidly jointed together. Find the moment of inertia of the triangular framework about an axis perpendicular to the plane of the triangle passing through one vertex.

10) A uniform lamina is in the form of an isosceles triangle ABC and has a mass M. AB = AC = $5l$, and BC = $6l$. Find the moment of inertia of this lamina about an axis (a) passing through A and the midpoint of BC, (b) perpendicular to the triangle through A.

MULTIPLE CHOICE EXERCISE 11

(Instructions for answering these questions are given on page xi)

TYPE I

1)

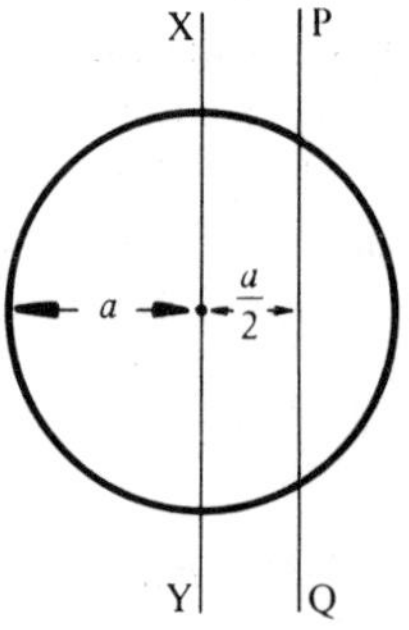

The diagram shows a uniform ring of mass M and radius a. If the moment of inertia of the ring about the axis PQ is I, the moment of inertia of the ring about the axis XY is:

(a) $I+\frac{Ma^2}{4}$ (b) $I+\frac{Ma^2}{2}$ (c) $I-\frac{Ma^2}{4}$ (d) $I-\frac{Ma^2}{2}$.

2)

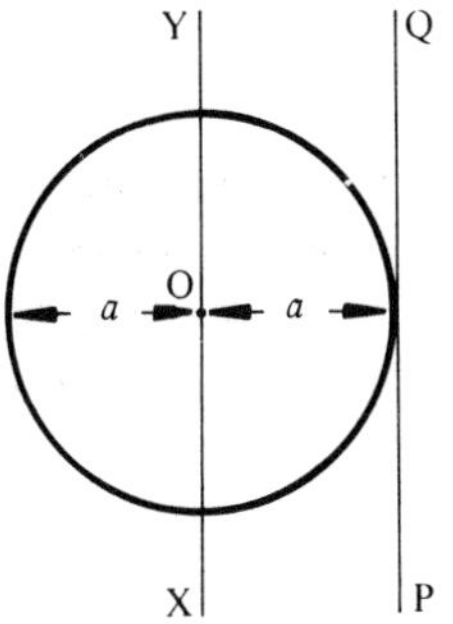

The diagram shows a non-uniform disc of radius a and mass M whose centre of mass is not at O. If the moment of inertia of the disc about XY is I, its moment of inertia about PQ is:

(a) $I - Ma^2$ (b) $I + Ma^2$ (c) Ma^2 (d) none of these.

3)

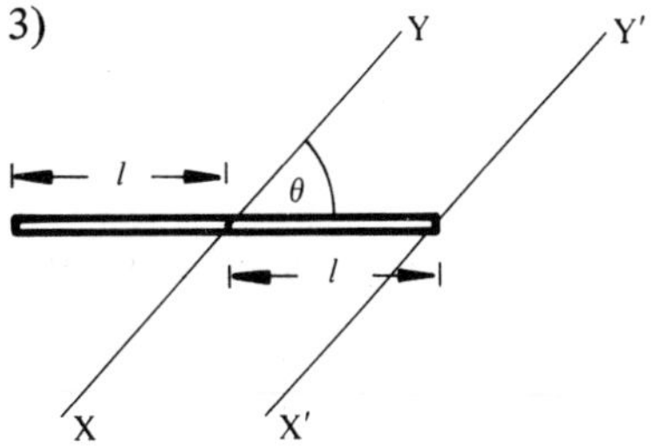

The diagram shows a uniform rod of mass M and length $2l$. The moment of inertia of the rod about XY is I. The moment of inertia of the rod about the axis X′Y′ (which is parallel to XY) is:

(a) $I + Ml^2 \sin^2\theta$ (b) $I + Ml^2$ (c) $\dfrac{Ml^2 \sin^2\theta}{3}$ (d) $I - Ml^2 \sin^2\theta$

4)

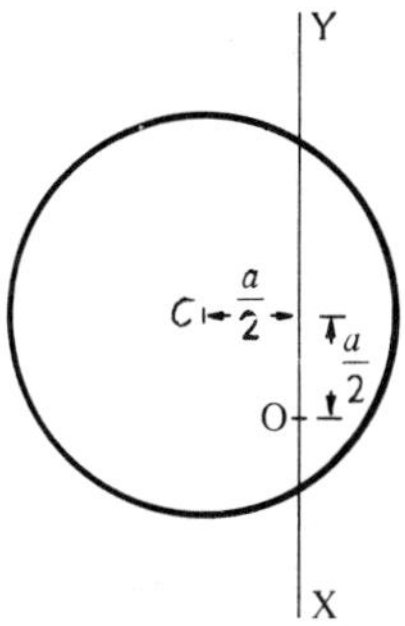

The diagram shows a uniform disc of mass M and radius a. If the moment of inertia of the disc about the axis XY is I, its moment of inertia about an axis through O perpendicular to the plane of the disc is:

(a) $I + \dfrac{Ma^2}{2}$ (b) $I + \dfrac{Ma^2}{4}$ (c) $2I$ (d) $\dfrac{I}{2}$ (e) $I - \dfrac{Ma^2}{2}$.

5)

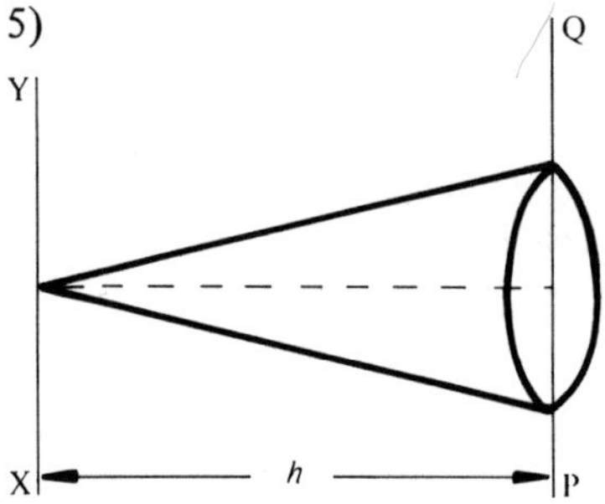

The diagram shows a uniform solid cone of height h and mass M. If the moment of inertia of the cone about XY is I, its moment of inertia about PQ is:

(a) $I + Mh^2$ (b) $Mh^2 - I$ (c) $I + \dfrac{Mh^2}{2}$ (d) $I + \dfrac{Mh^2}{4}$ (e) $I - \dfrac{Mh^2}{2}$.

TYPE II

6)

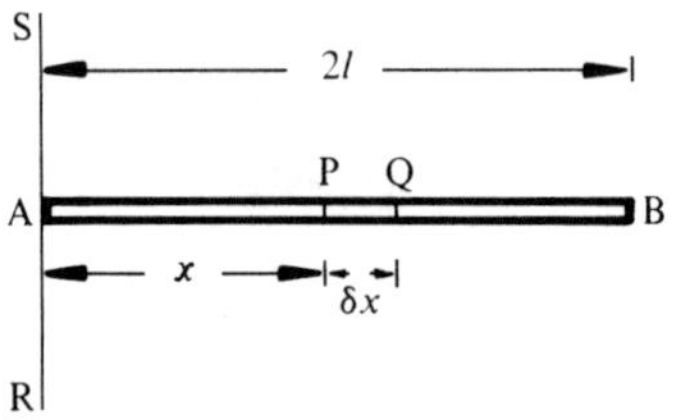

AB is a rod of length $2l$ and mass mx per unit length at a point distant x from AB.

(a) The element PQ has a mass of approximately $mx\,\delta x$.
(b) The moment of inertia of the element PQ about the axis RS is approximately $mx^3\,\delta x$.
(c) The moment of inertia of the rod about RS is given by $\int_0^l mx^3\,dx$.

7)

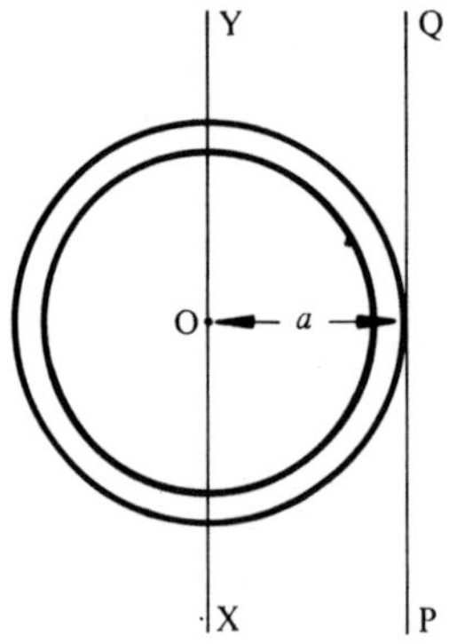

The diagram shows a non-uniform ring of radius a and mass M, whose moment of inertia about the diameter XY is I.

(a) The moment of inertia of the ring about an axis through O perpendicular to plane of the ring is Ma^2.
(b) The moment of inertia about any diameter other than XY is also I.
(c) The moment of inertia about the tangent PQ is $I + Ma^2$.

8)

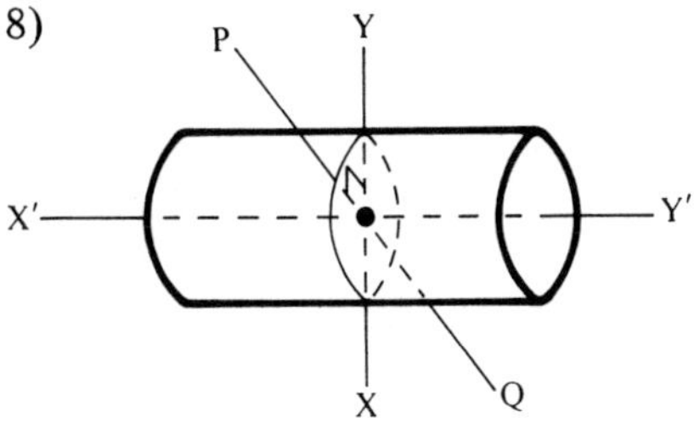

The diagram shows a uniform solid cylinder of mass M and length $2l$. XY and PQ are axes passing through the centre of the cylinder and parallel to its plane face.

(a) If the moment of inertia of the cylinder about the axis XY is I, the moment of inertia about PQ is also I.
(b) The moment of inertia of the cylinder about its axis X′Y′ is $2I$.
(c) The moment of inertia about a diameter of one end is $I + Ml^2$.

TYPE IV

9) Find the moment of inertia of a hollow cylinder about its axis.
(a) The mass of the cylinder is M.
(b) The radius of the cylinder is a.
(c) The length of the cylinder is l.

10) Find the moment of inertia of a rectangular lamina about an axis passing through its centre and the midpoints of the longer sides.
(a) The length of the rectangle is $2l$.
(b) The width of the rectangle is $2a$.
(c) The rectangle is uniform and of mass M.

11) Find the moment of inertia of a solid sphere about a diameter.
(a) The mass of the sphere is M.
(b) The radius of the sphere is r.
(c) The sphere is uniform.

12) Find the moment of inertia of a uniform triangular lamina ABC about the median passing through A.
(a) $AB = AC = 2l$.
(b) The triangle has a mass M.
(c) $BC = l$.

TYPE V

13) The moment of inertia of a hollow cylinder about its axis does not depend on whether the cylinder is uniform or not.

14) The perpendicular axis theorem can be applied to any solid body.

15) The parallel axis theorem can be applied to any solid body provided the centre of mass is known.

MISCELLANEOUS EXERCISE 11

1) A solid consists of a uniform solid cylinder of radius a and two uniform discs of radius $2a$. The discs are placed one at each end of the cylinder such that the centres of the discs are on the axis of the cylinder. If the mass of the cylinder is M and the mass of each disc is $\frac{1}{2}M$ find the moment of inertia of the complete solid about the axis of the cylinder.

2) A uniform rod is of mass M and length l. Find, by integration, the moment of inertia of the rod about an axis perpendicular to the rod and passing through one end.

3) A uniform lamina of mass M consists of two uniform discs of equal density and radii a and b, $(a < b)$. They are fixed together concentrically. Find the moment of inertia of the lamina about an axis perpendicular to its plane and passing through the centre.

4) Prove, by integration, that the moment of inertia of a uniform rectangular lamina of mass M about an axis along one side is $\dfrac{4Mb^2}{3}$ where $2b$ is the length of the sides perpendicular to the axis.

5) A uniform lamina is in the form of a semicircle of radius a and mass M. Find, by integration, its moment of inertia about an axis along its straight edge.

6) A rod AB of length l has a mass per unit length mx^2 at a point distant x from A. Find the moment of inertia of the rod about a perpendicular axis through A.

7) A thin circular wire of radius a, mass m has fastened to it three particles each of mass m at the corners of an equilateral triangle ABC. Find the moment of inertia of the system about the tangent to the circle at A.
If P is a point on the circle such that the arc AP subtends an angle θ at the centre of the circle, show that the moment of inertia of the system about the tangent at P is independent of θ. (U of L)

8) Find the radius of gyration of a uniform right circular solid cone, of height h and base radius a, about (i) the axis of the cone, (ii) a diameter of the base. Show that these two radii of gyration are equal if $2h^2 = 3a^2$. (U of L)

9) A uniform lamina of mass m is bounded by two concentric circles of radii $a_1, a_2\ (< a_1)$ respectively. Show that the moment of inertia of the lamina about an axis perpendicular to its plane through its centre of mass is $\frac{1}{2}(a_1^2 + a_2^2)$.
A solid of mass M consists of what remains of a uniform solid sphere of radius $2a$ after a circular hole of radius a, with axis along a diameter, has been drilled through it. Show that the moment of inertia of the solid about the axis of the hole is $\frac{11}{5}Ma^2$. (WJEC)

10) Show that the radius of gyration of a uniform triangular lamina PQR about the edge QR is $PS/\sqrt{6}$, where PS is an altitude of the triangle PQR.
In a kite-shaped uniform lamina ABCD of mass M the angles B and D are right angles, $AB = AD = 2a$ and $BC = DC = a$. Show that the moment of inertia of the lamina about BD is $13Ma^2/30$.
Find also the moment of inertia about the axis through the point of intersection of AC and BD perpendicular to the plane of the lamina. (U of L)

11) The radius of gyration of a uniform triangular lamina PQR about an axis through P parallel to QR is k. Show that $k^2 = \frac{1}{2}h^2$, where h is the perpendicular distance from P to QR. Find the radius of gyration about QR.
Hence, or otherwise, show that the radius of gyration of a uniform regular hexagonal lamina ABCDEF of side $2a$ about AD is $a\sqrt{(\frac{5}{6})}$. If the mass of the hexagonal lamina is M, find the moment of inertia of this lamina about an edge. (U of L)

12) The radius of gyration of a uniform thin circular disc, of radius a, about a line through its centre and perpendicular to its plane is $a/\sqrt{2}$. Deduce that the radius of gyration of a uniform solid sphere of radius r about a diameter is $r\sqrt{(\frac{2}{5})}$.
A spherical cavity of radius r is made inside a uniform solid sphere of radius R, in such a way that the two spherical surfaces touch at the point P. If G is the centre of mass of the remaining material, find (i) the distance PG, and (ii) the radius of gyration of the remaining material about a common tangent at P to the spherical surfaces. (JMB)

CHAPTER 12

ROTATION ABOUT A FIXED AXIS

CONSERVATION OF MECHANICAL ENERGY

The Principle of Conservation of Mechanical Energy can be expressed as follows:

The total mechanical energy of a system remains constant so long as no external work is done (other than by the weight of the body) and no sudden change in motion takes place.

This principle, which has already been used to advantage in solving many problems involving linear motion, applies equally well to rotating bodies.
If the axis about which a body is turning offers no resistance to the rotation, the axis is said to be smooth and is not responsible for any external work being done.
Consequently, when a body is rotating about a smooth axis and no work is done by any external force other than its own weight, the total mechanical energy of that body remains constant. In these circumstances the body is said to *rotate freely* about the axis.

EXAMPLES 12a

1) A uniform circular disc of mass m, radius r and centre O is free to turn in its own plane about a smooth horizontal axis passing through a point A on the rim of the disc. The disc is released from rest in the position in which OA is horizontal and the disc is vertical. Find the angular velocity of the disc when OA first becomes vertical.

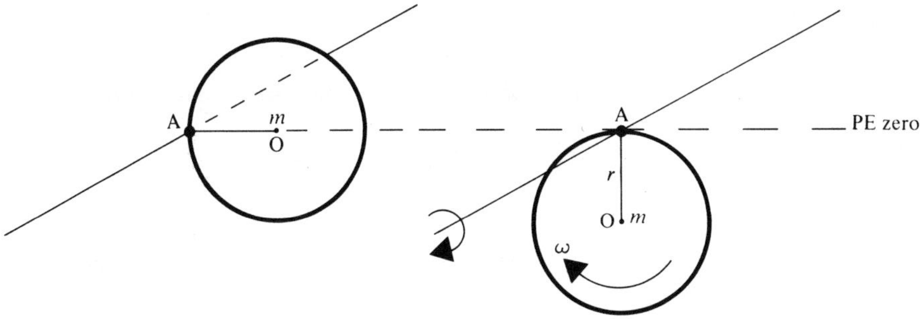

Initial position
(i)

Position with AO vertical
(ii)

(The disc will rotate in a vertical plane since the axis of rotation is horizontal.)

The moment of inertia of the disc about the axis through A perpendicular to the disc is given by

$$I = \tfrac{1}{2}mr^2 + mr^2 \quad \text{(Parallel Axis Theorem)}$$

i.e. $$I = 3mr^2/2$$

Initially (diagram (i)):

$$\text{KE} = 0$$

$$\text{PE} = 0$$

When AO is vertical (diagram (ii)) and the angular velocity is ω:

$$\text{KE} = \frac{1}{2}I\omega^2 = \frac{1}{2}\left(\frac{3mr^2}{2}\right)\omega^2$$

$$\text{PE} = -mgr$$

Using the Principle of Conservation of Mechanical Energy we have:

$$0+0 = 3mr^2\omega^2/4 - mgr$$

Hence $$\omega^2 = 4g/3r$$

i.e. $$\omega = 2\sqrt{g/3r}$$

2) A uniform rod AB of mass $3m$ and length $2l$ is free to turn in a vertical plane about a smooth horizontal axis through A. A particle of mass m is attached to the rod at B. When the rod is hanging in equilibrium, it is set moving with an angular velocity $\sqrt{kg/l}$.

(a) If $k = 2$, find the height of B above the level of A when the rod first comes to instantaneous rest.

(b) Find the range of values of k for which the particle describes complete circles about A.

About the given axis, the moment of inertia of the rod is $4ml^2$ and that of the particle is $m(2l)^2$.
Therefore the moment of inertia of the rod with attached particle is

$$4ml^2 + 4ml^2 = 8ml^2$$

Consider the rod in a general position, making an angle θ with the downward vertical.
The angular velocity of the rod in this position is $\dot{\theta}$

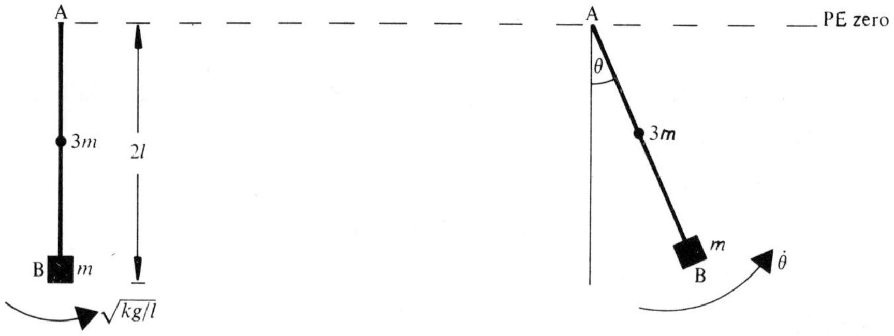

Initial position
(i)

General position
(ii)

Initially: $\text{PE} = -3mgl - mg(2l) = -5mgl$

$\text{KE} = \frac{1}{2}I(\sqrt{kg/l})^2 = 4kmgl$

In general $\text{PE} = -3mg(l\cos\theta) - mg(2l\cos\theta) = -5mgl\cos\theta$

$\text{KE} = \frac{1}{2}I\dot{\theta}^2 = 4ml^2\dot{\theta}^2$

Using Conservation of Mechanical Energy we have:

$$4kmgl - 5mgl = 4ml^2\dot{\theta}^2 - 5mgl\cos\theta$$

Hence $$4l\dot{\theta}^2 = 4kg - 5g(1 - \cos\theta) \quad (1)$$

(a) When the rod comes to instantaneous rest, $\dot{\theta} = 0$, therefore

$$4kg - 5g(1 - \cos\theta) = 0$$

But $$k = 2$$

Hence $$1 - \cos\theta = \tfrac{8}{5}$$

i.e. $$\cos\theta = -\tfrac{3}{5}$$

In this position θ is an obtuse angle so the height h of B above A is given by:

$$h = 2l\cos(\pi - \theta)$$
$$= 2l(\tfrac{3}{5})$$

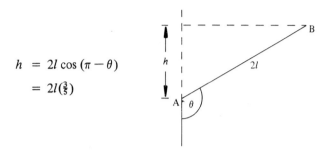

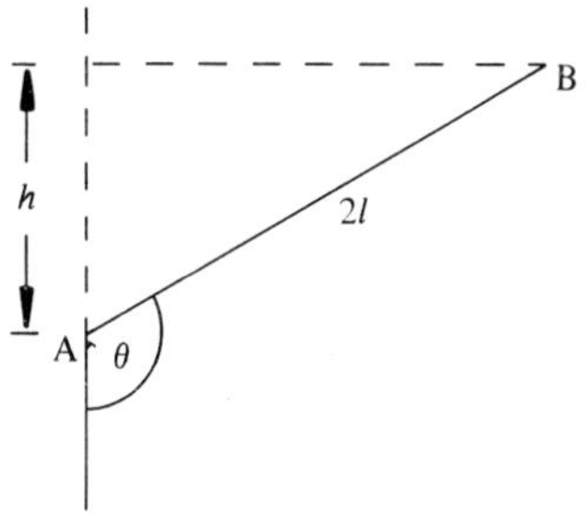

Hence B is distant $6l/5$ above the level of A.

(b) Since the particle is attached to a rigid rod, its motion is restricted to a circular path. It will therefore describe complete circles if its speed at the highest point is greater than zero.
Thus, in equation (1), $\dot{\theta}$ has a real value when $\theta = \pi$.

Hence $$4kg - 5g(1 - \cos\pi) > 0$$

The particle will therefore describe complete circles provided that

$$k > 5/2$$

3) ABC is a triangular framework of three uniform rods each of mass m and length $2l$. It is free to rotate in its own plane about a smooth horizontal axis through A which is perpendicular to ABC. If it is released from rest when AB is horizontal and C is above AB find the maximum velocity of C in the subsequent motion.

[The velocity of any point in the framework is greatest when the angular velocity is maximum, i.e. when the kinetic energy is greatest. Because the total mechanical energy is constant, the body has maximum kinetic energy when its potential energy is least, i.e. when the centre of mass is in its lowest position which is when BC is horizontal and below A.]

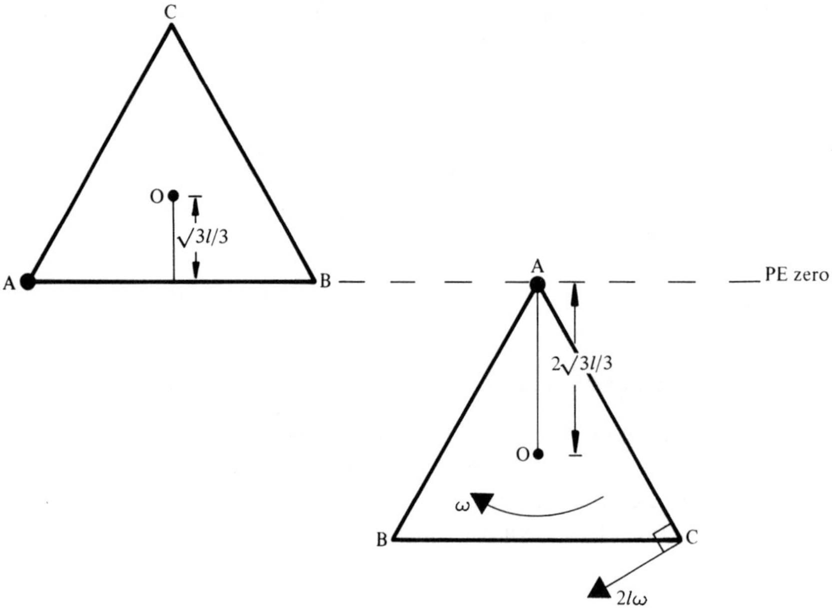

The moment of inertia about an axis through A perpendicular to the framework is:

for the rod AB $\frac{4}{3}ml^2$

for the rod AC $\frac{4}{3}ml^2$

for the rod BC $\frac{1}{3}ml^2 + m(l\sqrt{3})^2 = \frac{10}{3}ml^2$

hence for the framework $I = 6ml^2$.
If ω is the angular velocity in the second position then, using conservation of mechanical energy, we have:

$$3mg\left(\frac{l\sqrt{3}}{3}\right) = \tfrac{1}{2}(6ml^2)\,\omega^2 + 3mg\left(-\frac{2l\sqrt{3}}{3}\right)$$

Hence $$\omega^2 = g\sqrt{3}/l$$

Now the velocity of C at this instant is $2l\omega$, i.e. $2l\sqrt{\dfrac{g\sqrt{3}}{l}}$.

(*Note*: When drawing diagrams showing various positions of a rotating body, the level of the axis of rotation should be kept constant and potential energy measured from this level.)

EXERCISE 12a

1) A uniform rod of length $2l$ and mass m is free to rotate in a vertical plane about a smooth fixed horizontal axis through one end of the rod. The rod is

held in a horizontal position and is then released. Find the maximum angular velocity of the rod in the subsequent motion.

2) A uniform ring of radius a and mass m is free to rotate in a vertical plane about a fixed smooth axis which is perpendicular to the plane of the ring and passes through a point A on the ring. A particle of mass m is attached to the ring at B, where AB is a diameter. When the ring is hanging in a position of stable equilibrium the particle is struck a blow which gives it a velocity $3\sqrt{ga}$. Find the height above A to which the particle rises.

3) A uniform square lamina of side $2a$ and mass m is free to rotate about a fixed smooth horizontal axis along one edge of the lamina. When the lamina is hanging vertically below the axis it is given an angular velocity $\sqrt{kg/a}$. Find the range of values of k for which the lamina makes complete revolutions.

4) A uniform circular disc of mass m, radius r and centre O is free to rotate about a smooth, horizontal axis which is tangential to the disc at a point A. The disc is held in a vertical plane with A below O and is then slightly displaced from this position. Find the angular velocity of the disc when its plane is next vertical.

5) Four uniform rods of equal length l and mass m are rigidly joined together at their ends to form a square framework ABCD. The framework is free to rotate in a vertical plane about a fixed smooth horizontal axis passing through A. The framework is slightly displaced from its position of unstable equilibrium. Find the maximum angular velocity reached in the subsequent motion.

6) A uniform rod of length $6a$ and mass m is free to rotate in a vertical plane about a fixed smooth axis through one end A. A light elastic string of natural length a and modulus of elasticity mg connects the mid-point B of the rod to a fixed point C which is level with A, distant $3a$ from A and in the vertical plane containing the rod. The rod is released from rest with AB horizontal and when it first comes to instantaneous rest, the angle BAC $= 2\theta$. Show that

$$6 \sin 2\theta = (6 \sin \theta - 1)^2$$

Find also the angular velocity of the rod when $\theta = \pi/6$.

7) Three uniform rods, each of length $2l$ and mass m, are rigidly joined together to form a triangular framework ABC. The framework is free to rotate in a vertical plane about a fixed smooth horizontal axis through A. B is attached to one end of a light elastic string of natural length l and modulus mg, the other end of the string is attached to a point D, level with A and distant $2l$ from A such that A, B, C and D are all in the same vertical plane. The framework is held with the vertex B at D and is then released. Find the angular velocity of the framework when BC is first horizontal.

THE EFFECT OF AN EXTERNAL COUPLE

When the state of rotation of a body is changed by the action of an external couple, the total mechanical energy is no longer constant but undergoes a change equal to the amout of work done by the couple.

CALCULATION OF THE WORK DONE BY A COUPLE

Consider a body which is rotating about an axis XY, under the action of a couple comprising two equal and opposite forces of magnitude F, each distant a from XY.

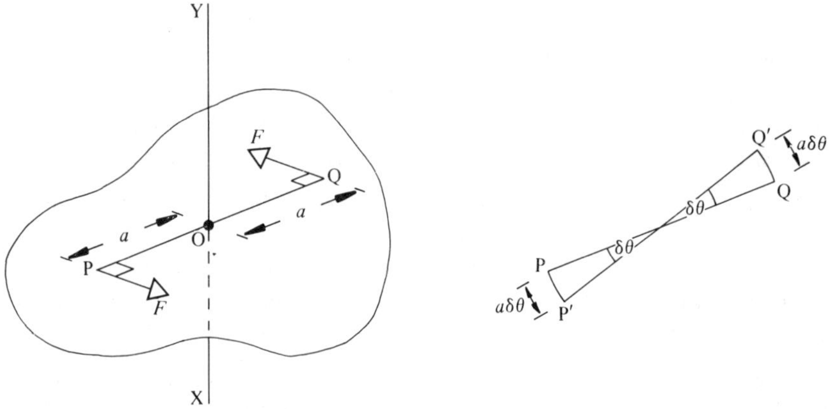

When the body rotates through a small angle $\delta\theta$, the point of application of each force moves through a distance $a\delta\theta$.
The work done by each force, therefore, is given approximately by $Fa\,\delta\theta$.
Then, if the work done by the couple is δW, we have

$$\delta W \simeq 2Fa\,\delta\theta$$

But $2Fa$ is the moment of the couple and can be represented by C where

$$C = 2Fa$$

Therefore $\qquad \delta W \simeq C\,\delta\theta$

or $\qquad \dfrac{\delta W}{\delta\theta} \simeq C$

Hence, as $\delta\theta \to 0$ we have $\qquad \dfrac{dW}{d\theta} = C$

Thus W, the work done by the couple as the body rotates through an angle α, is given by

$$W = \int_0^{\alpha} C\,d\theta$$

When C is constant, this result becomes

$$W = C\alpha$$

EXAMPLES 12b

1) A uniform rod AB of mass m and length $2l$ can rotate in a vertical plane about a horizontal axis through A. It is released from rest with AB horizontal and its angular velocity when AB first becomes vertical is $\sqrt{g/l}$. Find the magnitude of the constant frictional couple exerted by the axis.

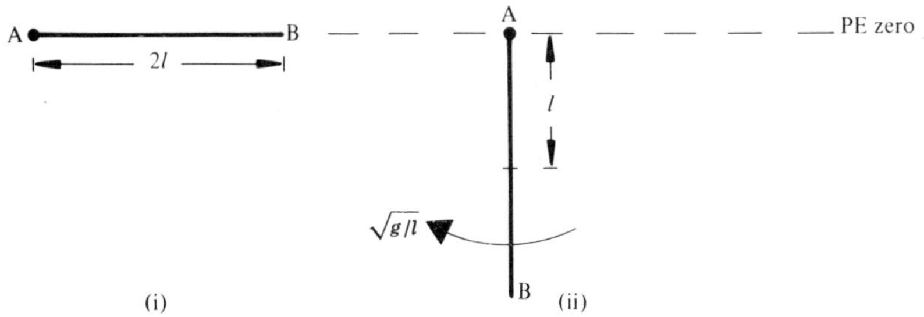

(i) (ii)

The moment of inertia of the rod about the axis through A is $\frac{4}{3}ml^2$.
The initial mechanical energy (diagram (i)) is zero
The final mechanical energy (diagram (ii)) is

$$\tfrac{1}{2}(\tfrac{4}{3}ml^2)\left(\sqrt{\frac{g}{l}}\right)^2 - mgl$$

The loss in mechanical energy during the displacement from (i) to (ii) is therefore

$$0-(\tfrac{2}{3}mgl-mgl) = \frac{mgl}{3}$$

But the loss in mechancial energy is equal to the work done by the constant frictional couple C, i.e. $C(\pi/2)$

Hence $$\tfrac{1}{2}\pi C = \frac{mgl}{3}$$

and $$C = 2mgl/3\pi$$

2) A disc of mass m and radius a can rotate in a horizontal plane about a smooth vertical axis perpendicular to the disc and passing through the centre of the disc. The disc is made to rotate from rest by a torque whose magnitude, when the disc has rotated through an angle θ, is $4m\theta$. Find the angular velocity of the disc which is uniform when it has turned through one revolution.

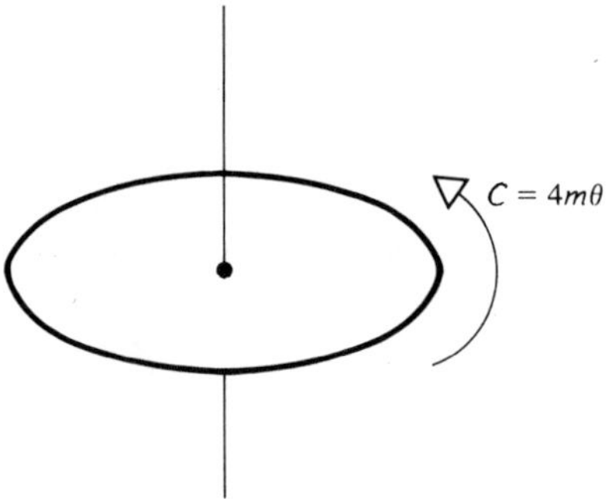

The work done by the couple is given by

$$\int_0^{2\pi} 4m\theta \, d\theta = \left[2m\theta^2\right]_0^{2\pi}$$
$$= 8m\pi^2$$

The increase in mechanical energy of the disc is equal to the work done by the couple.
Thus, if ω is the angular velocity after rotation through one revolution, we have:

$$\tfrac{1}{2}I\omega^2 = 8m\pi^2$$

Where I is the moment of inertia of the disc about the given axis

i.e. $$I = \tfrac{1}{2}ma^2$$

Hence $$\omega^2 = \frac{32\pi^2}{a^2}$$

and $$\omega = 4\pi\sqrt{2}/a$$

EXERCISE 12b

1) A uniform solid cylinder of mass m and radius a is free to rotate about its axis which is smooth and vertical. A light inextensible string is wound round the cylinder and its free end is pulled horizontally with a constant force $2mg$. Find the angular velocity of the cylinder when the free end of the string has moved through a distance $4a$. At this instant the string slips off and a retarding torque C is applied to the cylinder bringing it to rest after it has turned through two revolutions. Find the magnitude of C.

2) A uniform circular disc of radius r and mass m is free to rotate in a vertical plane about a horizontal axis through a point A on the circumference of the disc. The disc is released from rest with the diameter AB horizontal and it next comes to rest when the diameter AB has turned through an angle $\pi/3$. Find the moment of the constant frictional couple acting at the axis.

3) A uniform rod of length $2l$ and mass m is free to rotate in a vertical plane about a horizontal axis through one end A of the rod. When the other end B of the rod is hanging vertically below A it is given an angular velocity $\sqrt{\frac{5g}{l}}$. The rod next comes to rest when AB is horizontal. Find the moment of the constant frictional couple acting at the axis.

4) A uniform rod of length $4l$ and mass m is free to rotate in a vertical plane about a horizontal axis through one end of the rod. A constant frictional couple of moment mgl acts at the axis. The rod is released from rest in a horizontal position. If θ is the angle the rod turns through before it next comes to instantaneous rest show that $2 \sin \theta = \theta$.

5) A uniform solid cylinder of radius a and mass m is free to rotate about its axis which is smooth and fixed horizontally. The cylinder rotates from rest under the action of a couple whose magnitude is given by $3m\theta^2$ where θ is the angle through which the cylinder has turned. Find the angular velocity of the cylinder when it has completed one revolution.

ANGULAR ACCELERATION

Consider a body of mass m which is rotating about a smooth horizontal axis passing through a point A. The centre of mass is at a point C distant h from the axis of rotation and I is the moment of inertia of the body about that axis. The system is such that no work is done by any external torque or force (other than the weight of the body).

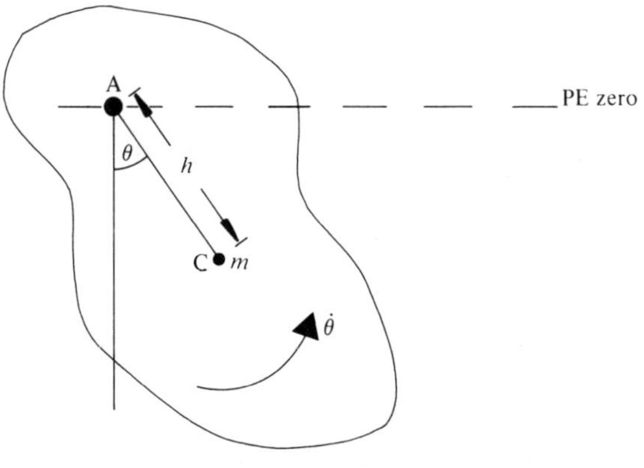

If, at any instant, the line AC is inclined at an angle θ to the downward vertical, the angular velocity of the body at this instant is $\dot{\theta}$.
Suppose that the initial mechanical energy of the body is K, then applying the principle of conservation of energy gives:

$$\tfrac{1}{2}I\dot{\theta}^2 - mgh\cos\theta = K$$

Differentiating this expression with respect to time we have:

$$I\dot{\theta}\ddot{\theta} + mgh\sin\theta\,\dot{\theta} = 0$$

Hence $$I\ddot{\theta} = -mgh\sin\theta \qquad (1)$$

In this way the general energy equation for the body can be used to derive an expression for its angular acceleration at any instant. The equation (1) above is called the equation of motion of the body.

It is not always convenient, or even possible, to use this method to find the angular acceleration of a rotating body. It would not be used, for instance, if work is done by forces other than the weight of the body. The equation of motion of a rotating body can, in such cases, be determined by applying Newton's Law $C = I\ddot{\theta}$ (see Examples 10a).

The application of Newton's Law also provides an alternative method for analysing the rotation of a system whose mechanical energy *is* constant.

For example, consider again the body rotating about a smooth horizontal axis through a point A.

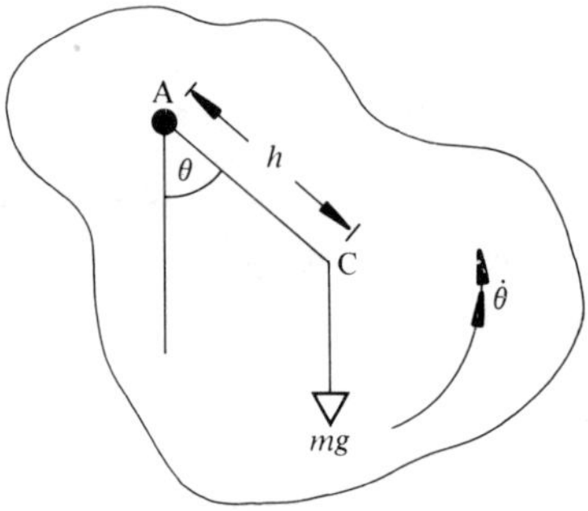

When CA is inclined at an angle θ to the downward vertical, the torque exerted on on the body by its weight is $mgh\sin\theta$ clockwise.

Hence $$mgh\sin\theta = -I\ddot{\theta} \quad \text{(see equation (1))}$$

SMALL OSCILLATIONS

When a body supported by a smooth horizontal axis swings through a very small angle on either side of its stable equilibrium position, it is said to perform *small oscillations*, and is called a *compound pendulum*. In this case the general equation of motion (equation (1) above) becomes:

$$I\ddot{\theta} \simeq -mgh\theta$$

since, for small angles, $$\sin\theta \simeq \theta$$

When θ is *very* small
$$\ddot{\theta} = -\frac{mgh}{I}\theta$$

Now this is the equation of angular simple harmonic motion about the vertical through A and the constant of proportion is $\dfrac{mgh}{I}$.

The period T of such small oscillations is therefore given by

$$T = 2\pi\sqrt{\frac{I}{mgh}}$$

Equivalent Simple Pendulum

The period of small oscillations of a simple pendulum of length l is $2\pi\sqrt{\dfrac{l}{g}}$.

Comparing this with the period of oscillation of a compound pendulum, we see that if $\dfrac{I}{mh} = l$ the periods are equal.

The quantity $\dfrac{I}{mh}$ is therefore known as *the length of the equivalent simple pendulum*.

(*Note*: When solving problems which require the equation of motion, the period of small oscillations or the length of the equivalent simple pendulum, these should be derived from first principles in each case. The *methods* demonstrated in the preceeding paragraphs can be used but the expressions derived there should not be quoted.)

EXAMPLES 12c

1) A ring of mass m and radius r has a particle of mass m attached to it at a point A. The ring can rotate about a smooth horizontal axis which is tangential to the ring at a point B diametrically opposite to A. The ring is released from rest when AB is horizontal. Find the angular velocity and the angular acceleration of the body when AB has turned through an angle $\pi/3$.

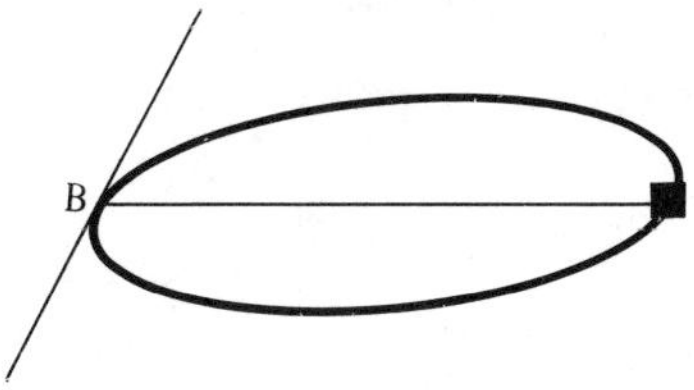

The moment of inertia about the tangent at B is:

for the ring alone $\quad \frac{3}{2}mr^2$

for the particle $\quad m(2r)^2$

Hence for the whole body $\quad I = \dfrac{11mr^2}{2}$

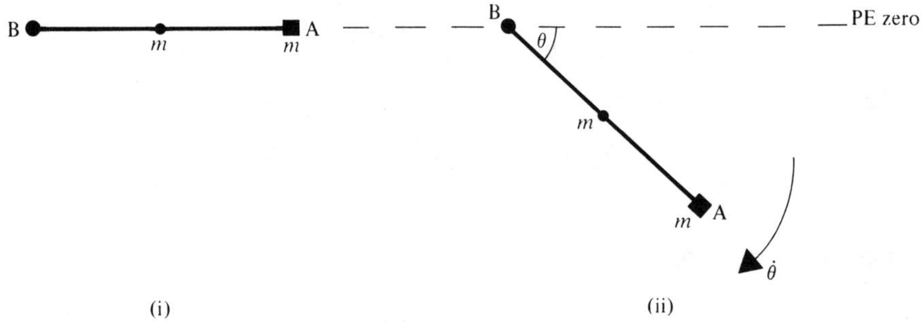

Initially (diagram (i)) the total mechanical energy is zero.
In position (ii)

$$\text{PE of ring alone} = -mgr\sin\theta$$

$$\text{PE of particle} = -mg(2r)\sin\theta$$

$$\text{PE of whole body} = -3mgr\sin\theta$$

$$\text{KE of whole body} = \frac{1}{2}\left(\frac{11mr^2}{2}\right)\dot{\theta}^2$$

Using Conservation of Mechanical Energy:

$$0 = \tfrac{11}{4}mr^2\dot{\theta}^2 - 3mgr\sin\theta$$

Hence $$\dot{\theta}^2 = \frac{12g}{11r}\sin\theta \qquad (1)$$

When $\theta = \dfrac{\pi}{3}$, $$\dot{\theta}^2 = \frac{12g}{11r}\left(\frac{\sqrt{3}}{2}\right)$$

Hence $$\dot{\theta} = \sqrt{\frac{6g\sqrt{3}}{11r}}$$

Differentiating equation (1) with respect to time:

$$2\dot{\theta}\ddot{\theta} = \frac{12g}{11r}\cos\theta\,\dot{\theta}$$

When $\theta = \dfrac{\pi}{3}$, $$\ddot{\theta} = \frac{3g}{11r}$$

Thus the angular velocity and acceleration of the body after turning through the angle $\pi/3$ are $\sqrt{\dfrac{6g\sqrt{3}}{11r}}$ and $\dfrac{3g}{11r}$ respectively.

2) A disc of mass m and radius a performs small oscillations about a smooth horizontal axis which is tangential to the disc. Find the length of the equivalent simple pendulum.

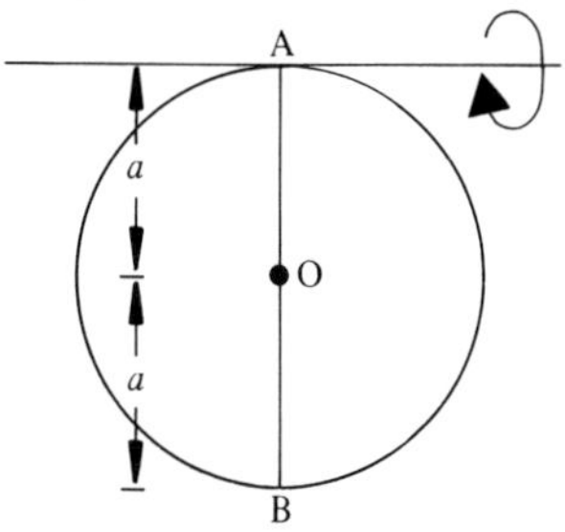

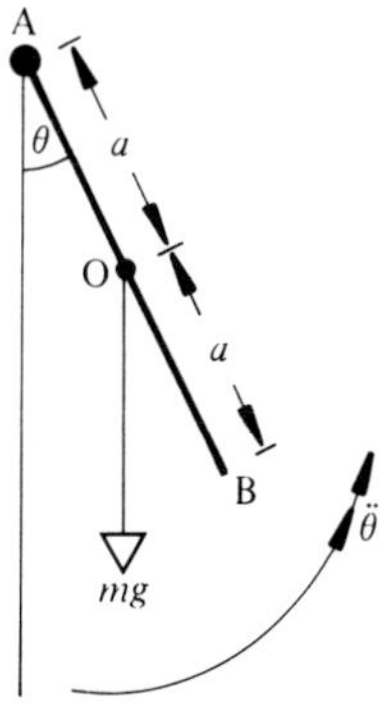

[In order to achieve a clear diagram, θ is not shown as a very small angle.]

The moment of inertia of the disc about the tangent at A is given by:

$$I = \frac{ma^2}{4} + ma^2 = \frac{5ma^2}{4}$$

Using Newton's Law for rotation we have:

$$\text{A)} \qquad mga \sin\theta = -\frac{5ma^2\ddot{\theta}}{4}$$

But for small oscillations, $\sin\theta \simeq \theta$

Hence $$\ddot{\theta} \simeq -\frac{4g}{5a}\theta$$

This is the equation for angular simple harmonic motion of period T where

$$T = 2\pi\sqrt{\frac{5a}{4g}}$$

Comparing this with the time of oscillation of a simple pendulum of length l we have

$$2\pi\sqrt{\frac{5a}{4g}} = 2\pi\sqrt{\frac{l}{g}}$$

Hence $$l = 5a/4$$

The length of the equivalent simple pendulum is therefore $5a/4$.

3) A uniform rod AB of mass m and length $4l$ is free to rotate in a vertical plane about a smooth horizontal axis through a point P distant x from the centre of

the rod. If the rod performs small oscillations about its equilibrium position find the period of oscillation and show that this is least when $x = 2l/\sqrt{3}$.

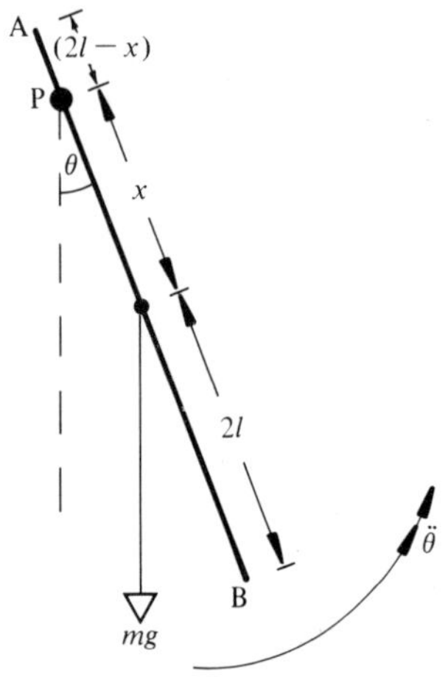

The moment of inertia of the rod about the axis through P is given by:

$$I = \tfrac{1}{3}m(2l)^2 + mx^2 \qquad \text{(Parallel Axis Theorem)}$$

Using Newton's Law for rotation we have:

$$\text{P)} \qquad mgx \sin\theta = -I\ddot{\theta}$$

i.e.

$$gx \sin\theta = -\tfrac{1}{3}(4l^2 + 3x^2)\ddot{\theta}$$

But, for small oscillations, θ is a small angle,

so

$$\sin\theta \simeq \theta$$

Hence

$$\ddot{\theta} \simeq -\frac{3gx}{(4l^2 + 3x^2)}\theta$$

This is the equation of angular simple harmonic motion about the vertical through P in which $\dfrac{3gx}{4l^2 + 3x^2}$ is the constant of proportion.

Hence T, the period of oscillations, is given by

$$T = 2\pi\sqrt{\frac{4l^2 + 3x^2}{3gx}}$$

The period will be least when $\dfrac{4l^2 + 3x^2}{x}$ is least

i.e. when

$$\frac{d}{dx}\left(\frac{4l^2}{x} + 3x\right) = 0$$

Thus
$$-\frac{4l^2}{x^2}+3 = 0$$

Taking x to be positive
$$x = 2l/\sqrt{3}$$

The minimum period therefore occurs when $x = 2l/\sqrt{3}$.

(To check that this value of x gives a minimum rather than a maximum period, note that the second differential coefficient, $+\dfrac{8l^2}{x^3}$, is positive when $x = 2l/\sqrt{3}$.)

EXERCISE 12c

1) A uniform rod of length $2l$ and mass m is free to rotate in a vertical plane about a fixed smooth horizontal axis passing through the rod at a point distant $l/2$ from one end. Find general expressions for the angular velocity and angular acceleration of the rod, if it is released from rest when in a horizontal position.

2) A uniform square lamina of side $2a$ and mass m is free to rotate about a fixed, smooth, horizontal axis along one edge of the lamina. Find an expression for the angular acceleration of the lamina when its angular displacement from the vertical plane through the axis is θ. If the lamina swings through a small angle on either side of the vertical through the axis, find the period of its oscillations.

3) A uniform rod of length $2l$ and mass m has a particle of mass $2m$ attached to one end B of the rod. The rod is free to rotate in a vertical plane about a smooth, fixed, horizontal axis through its other end A. The rod performs small oscillations about its position of stable equilibrium. Find the period of these oscillations and hence the length of the equivalent simple pendulum.

4) A uniform circular disc of radius a and mass m is free to rotate in a vertical plane about a smooth, fixed, horizontal axis passing through a point B on the disc, distant h from O, the centre of mass of the disc. A particle of mass m is attached to the circumference of the disc at the end of the diameter through B. The combined body performs small oscillations about its position of stable equilibrium. Find the period of these oscillations and the value of h for which this period is least.

5) A uniform rod AB of length $2l$ and mass m, has a particle of mass m attached at A and another particle, also of mass m, attached at its mid-point C. The rod is free to rotate in a vertical plane about a horizontal axis. If the axis passes through A, the period of small oscillations about the stable equilibrium position is T_1 and if the axis passes through C, the period of small oscillations about the stable equilibrium position is T_2. Prove that $8T_1^{\,2} = 7T_2^{\,2}$.

6) Three uniform rods, each of length $2l$, are rigidly joined together at their ends to form a triangular framework ABC. The rod CA is light, and the rods AB and BC are each of mass m. The framework is free to rotate in a vertical

plane about a fixed horizontal axis and performs small oscillations about its stable equilibrium position. Prove that the period of oscillation about an axis through B is equal to the period of oscillation about an axis through the mid-point of AC.

7) A compound pendulum consists of a uniform rod AB of length $2l$ and mass m with a particle of mass m fixed at the end B. It is free to rotate about a smooth horizontal axis through a point C of the rod where $AC = x$. Find x if the length of the equivalent simple pendulum is $3l$.

FORCE EXERTED BY THE AXIS

We saw in Chapter 8 that the acceleration of the centre of mass of a system of particles is the same as that of a single particle whose mass is the total mass of the system, acted upon by the resultant of the forces acting on the system. This principle applies to a rigid body, which is made up of a set of particles. Consider a body of mass m, rotating freely about a smooth horizontal axis. Diagram (i) shows a cross section of the body, containing the centre of mass C which is distant h from the axis through A. The general angular displacement of AC from the downward vertical is θ.

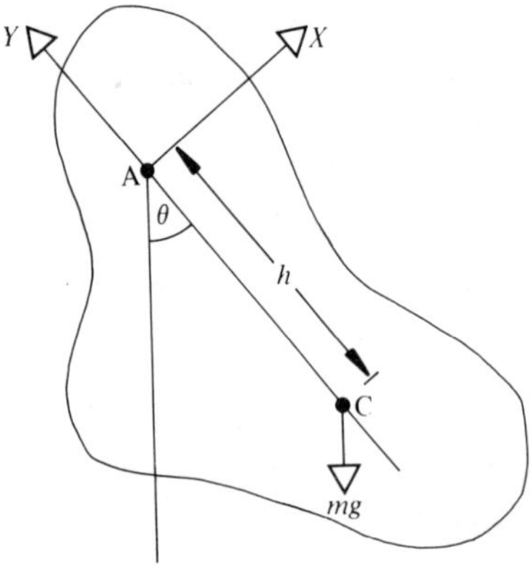

(i)

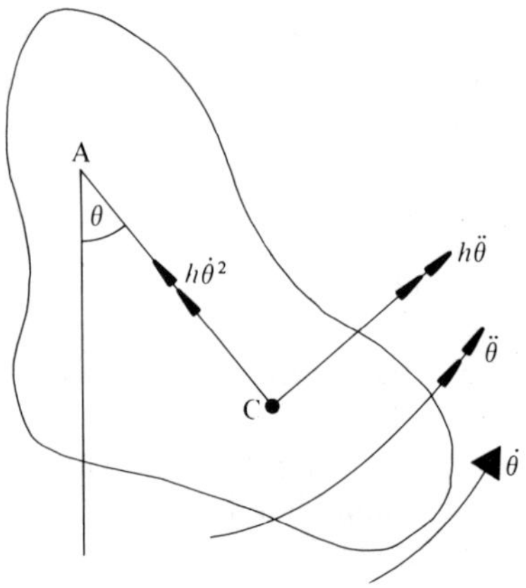

(ii)

The forces acting on the body are its weight and an unknown force exerted by the axis. The force at the axis is shown in the diagram as a pair of perpendicular components acting parallel and perpendicular to CA.
The acceleration components of C are shown in diagram (ii). These are the acceleration components which would be given to a particle of mass m placed at C and acted upon by the resultant of the forces acting on the system, see diagram (iii).

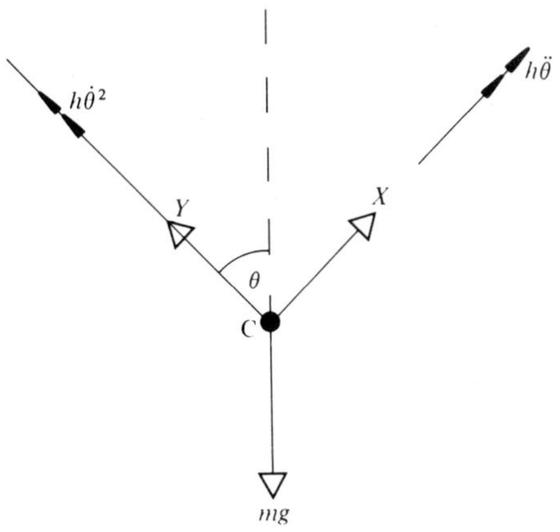

(iii)

Newton's Law gives:

$$\nwarrow \quad Y - mg\cos\theta = mh\dot{\theta}^2 \qquad (1)$$

$$\nearrow \quad X - mg\sin\theta = mh\ddot{\theta} \qquad (2)$$

In order to determine the force exerted on the body by the axis, both X and Y must be found. This can be done only if the angular velocity ($\dot{\theta}$) and the angular acceleration ($\ddot{\theta}$) of the body are known.
Now the value of $\dot{\theta}$ at any instant can be found using Conservation of Mechanical Energy. Then, by differentiating the energy equation (or by applying Newton's Law for rotation) $\ddot{\theta}$ can also be determined.
(*Note*: The angle θ need not, in every problem, be measured from the downward vertical, but $\ddot{\theta}$ and $\dot{\theta}$ must always be marked in the sense in which θ increases.)

EXAMPLE

A uniform rod AB of mass m and length $2a$ has a particle of mass m attached to the end B. The rod can rotate in a vertical plane about a smooth axis through A. If the body is slightly displaced from the position in which B is vertically above A, find the magnitude of the reaction at the axis, which is horizontal, when the rod has rotated through
(a) $\pi/3$ radians, (b) π radians.

The centre of mass of the body is at a point C midway between B and the mid-point of the rod.

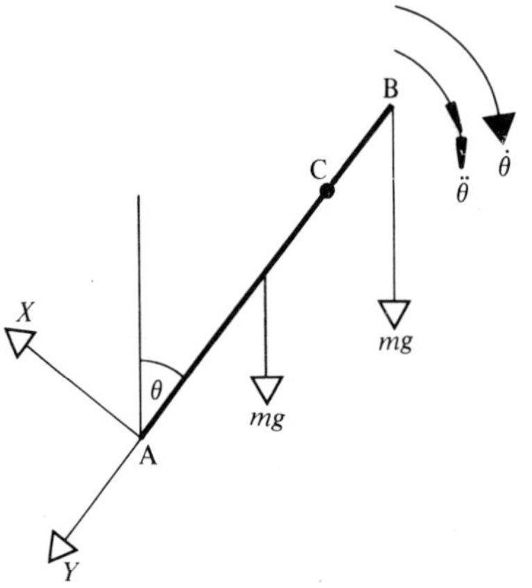

(i)

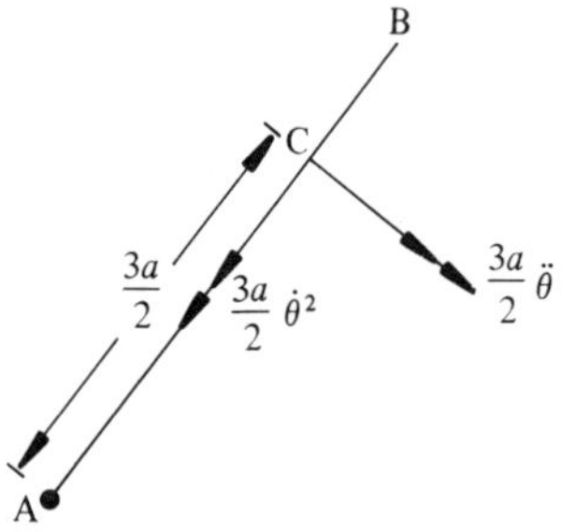

(ii)

Diagram (i) shows the forces acting on the body when it has rotated through an angle θ. Diagram (ii) shows the acceleration components of the point C in this position. Diagram (iii) shows the same forces acting on a particle of mass $2m$ (the total mass of the system) placed at C.

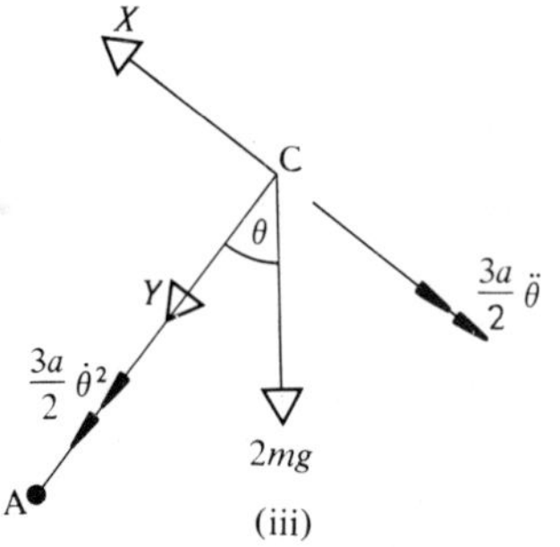

(iii)

Applying Newton's Law we have:

$$\swarrow \quad Y + 2mg\cos\theta = 2m\left(\frac{3a}{2}\dot{\theta}^2\right)$$

$$\nwarrow \quad 2mg\sin\theta - X = 2m\left(\frac{3a}{2}\ddot{\theta}\right)$$

Thus $$Y = 3ma\dot{\theta}^2 - 2mg\cos\theta \qquad (1)$$

and $$X = 2mg\sin\theta - 3ma\ddot{\theta} \qquad (2)$$

Now comparing the initial and general positions of the body:

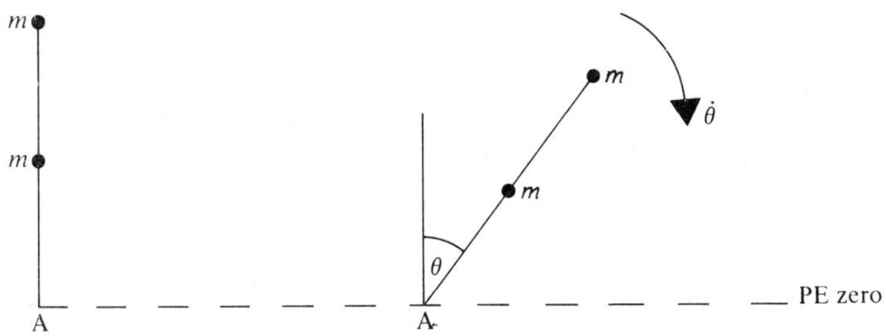

Conservation of Mechanical Energy gives:

$$mga + mg(2a) = mga\cos\theta + mg(2a\cos\theta) + \tfrac{1}{2}I\dot{\theta}^2$$

where I is the moment of inertia of the body about the axis through A.

i.e. $$I = \tfrac{4}{3}ma^2 + m(2a)^2 = \tfrac{16}{3}ma^2$$

Hence $$3mga(1-\cos\theta) = \tfrac{1}{2}(\tfrac{16}{3}ma^2)\dot{\theta}^2$$

i.e. $$\dot{\theta}^2 = \frac{9g}{8a}(1-\cos\theta)$$

Differentiating with respect to time gives:

$$2\dot{\theta}\ddot{\theta} = \frac{9g}{8a}(\sin\theta\,\dot{\theta})$$

i.e. $$\ddot{\theta} = \frac{9g}{16a}\sin\theta$$

Substituting these values into equations (1) and (2) we have:

$$Y = \frac{mg}{8}(27 - 43\cos\theta)$$

$$X = \frac{5\,mg\sin\theta}{16}$$

(a) When $\theta = \pi/3$ $Y = 11mg/16$ and $X = 5\sqrt{3}mg/32$

The magnitude of the resultant force R at the axis is therefore given by:

$$R = \sqrt{X^2 + Y^2} = \frac{mg}{32}\sqrt{75 + 484}$$

$$R = \tfrac{1}{32}mg\sqrt{559}$$

(b) When $\theta = \pi$ $Y = \tfrac{70}{8}mg$ and $X = 0$

Hence $R = \tfrac{70}{8}mg$

(*Note*: The value of $\ddot{\theta}$, and hence the value of X, is zero whenever the rod is vertical.)

EXERCISE 12d

1) A uniform circular disc with centre O, radius r and mass m is free to rotate in a vertical plane about a horizontal axis passing through a point A on its circumference. The disc is held at rest with AO horizontal and is then released. Find the component perpendicular to AO of the reaction at the axis when AO makes an angle θ with the downward vertical through A.

2) A uniform rod AB of length $2l$ and mass m is free to rotate in a vertical plane about a horizontal axis through the end A of the rod. The rod is released from rest when AB is horizontal. Find the component along AB of the reaction at the axis when AB makes an angle θ with the downward vertical through A and hence find the total reaction at the axis when AB is vertical.

3) A uniform square lamina ABCD has sides of length $2a$ and is of mass m. The lamina is free to rotate in a vertical plane about a horizontal axis through A. The lamina is slightly displaced from its position of unstable equilibrium. Find the components, parallel and perpendicular to AC, of the reaction at the axis when AC makes an angle θ with the upward vertical through A. Find the total reaction at the axis when the lamina has rotated through an angle $2\pi/3$ from rest.

4) A uniform rod AB of length $2l$ and mass $2m$ has a particle of mass m attached at B. The rod is free to rotate in a vertical plane about a horizontal axis through A. When the rod is hanging at rest, with B below A, it is given an angular velocity $\frac{7}{2}\sqrt{g/l}$. Find the reaction at the axis when the rod first becomes horizontal.

MULTIPLE CHOICE EXERCISE 12

(*Instructions for answering these questions are given on page xi*)

TYPE II

1) A body rotates freely about a smooth fixed axis. The body is slightly displaced from the position in which the centre of mass is vertically above a point A on the axis.
(a) The potential energy is least when the centre of mass is level with A.
(b) The body turns through a complete revolution.
(c) The angular acceleration is constant.
(d) The angular velocity is maximum when the centre of mass is vertically below A.

2) A body rotates freely about a smooth horizontal axis for which its moment of inertia is I. When the body has turned through an angle θ from rest in the position of unstable equilibrium,
(a) the kinetic energy is $\frac{1}{2}I\dot{\theta}^2$,
(b) the angular acceleration is $\ddot{\theta}$,
(c) the loss in potential energy is $\frac{1}{2}I\dot{\theta}^2$,
(d) the change in mechanical energy is $I\ddot{\theta}$.

3) A uniform rod AB of length $2l$ and mass m performs small oscillations in a vertical plane about a smooth horizontal axis through A.
(a) The period of oscillation is independent of m.
(b) The length of the equivalent simple pendulum is l.
(c) The moment of inertia of the rod about the axis of rotation is $\frac{1}{3}ml^2$.
(d) The total mechanical energy of the rod is constant.

4) A disc is rotating about an axis under the action only of a torque C. The angular velocity changes from ω_1 to ω_2 while the disc turns through an angle θ in time t. If I is the moment of inertia of the disc about the axis of rotation,
(a) $C\theta = \frac{1}{2}I(\omega_2{}^2 - \omega_1{}^2)$,
(b) the initial angular momentum of the disc is $I\omega_1$,
(c) the mechanical energy of the disc is constant,
(d) $2\theta = (\omega_1 + \omega_2)t$.

TYPE III

5) (a) Every point in a solid body is rotating in a horizontal plane.
(b) A solid body is rotating about a vertical axis.

6) (a) A rod rotates freely, about a fixed horizontal axis, through an angle θ on either side of its stable equilibrium position.
(b) A rod, which can rotate freely about a fixed horizontal axis, performs angular simple harmonic motion.

7) (a) A disc rotates freely in its own plane about a horizontal axis.
(b) The total mechanical energy of a rotating disc is constant.

8) (a) A compound pendulum has a periodic time $\pi\sqrt{\frac{a}{g}}$
(b) A compound pendulum has a periodic time equal to that of a simple pendulum of length $4a$.

9) A rigid body rotates about a fixed axis through its centre of mass. Apart from the friction at the axis of rotation, no external force exerts any torque on the body. If I is the moment of inertia about the axis of rotation,
(a) the angular velocity decreases from 20 rad s^{-1} to 10 rad s^{-1} in one revolution,
(b) the work done by the frictional couple at the axis in one revolution is $150I$ joules.

TYPE IV

10) A uniform ring rotates about a horizontal tangential axis. Find the work done in one complete revolution by the frictional couple acting at the axis if:
(a) the mass of the ring is m(kg),
(b) the radius of the ring is a (m),

(c) apart from forces at the axis the only force acting on the ring is its own weight,
(d) the loss in kinetic energy in one revolution is $8m$ (J).

11) A body free to rotate about a horizontal axis, performs small oscillations about its position of stable equilibrium. Find the period of the oscillations.
(a) The moment of inertia about the axis is I.
(b) The distance of the centre of mass from the axis is l.
(c) The mass of the body is m.
(d) The body swings through an angle of 4° on either side of the downward vertical through the axis of rotation.

12) A body is free to rotate about a smooth horizontal axis. When it is at rest it is given a blow which sets it rotating. Find the angle through which it turns before first coming to instantaneous rest.
(a) The body has a mass M.
(b) The body has a radius of gyration k.
(c) The distance of the centre of mass from the axis of rotation is h.
(d) The body is initially in a position of stable equilibrium.

TYPE V

13) When a flywheel, rotating about a fixed horizontal axis through its centre of mass, is brought to rest by a frictional couple acting at the axis, the mechanical energy of the flywheel remains constant.

14) A uniform rod of length $2l$ and mass m is free to rotate in a vertical plane about an axis through its centre of mass. If the rod is held horizontally and then released it will perform complete revolutions.

15) A uniform rod is free to rotate in a vertical plane about an horizontal axis through one end A. If the rod is held at an angle of 4° to the downward vertical through A and is then released, it will perform angular simple harmonic motion.

16) Whenever a torque C acts on a body causing it to rotate through an angle θ about a fixed axis, the work done by the torque is $C\theta$.

17) If the total mechanical energy of a rotating body is constant, the axis of rotation must be smooth.

MISCELLANEOUS EXERCISE 12

1) A flywheel has a horizontal axle of radius r. The system has a mass M and radius of gyration k about its axis and rotates without friction. A string is wound around the axle and carries a mass m hanging freely. If the system is released from rest, prove that the acceleration of the mass m is $g\dfrac{mr^2}{Mk^2+mr^2}$.

If the string slips off the axle after the mass m has descended a distance $8r$, find the magnitude of the constant retarding couple which is necessary to bring the flywheel to rest in n more revolutions. (AEB)

2) A uniform square lamina ABCD is of mass $6m$ and side $2a$. Particles of mass m, $2m$, $3m$, $4m$ are fixed to the lamina at the vertices A, B, C, D respectively. The lamina is rotating freely with its plane vertical about a fixed horizontal axis through its centre. Prove that the angular velocity of the lamina is a maximum when CD is horizontal and below AB.
If the maximum kinetic energy of the system is three times its minimum kinetic energy, find the angular velocity of the lamina when the side BC is horizontal. Find also at this instant the rate of change of the moment of momentum of the system about the axis of rotation. (U of L)

3) Four thin uniform rigid rods, each of mass m and length $2a$, are joined together at their ends to form a rigid square frame ABCD which can turn freely about a fixed horizontal axis through A perpendicular to the plane ABCD. Show that the moment of inertia of the frame about this axis is $40ma^2/3$.
The frame executes complete revolutions. If the greatest and least angular velocities are $(1+p)\omega$ and $(1-p)\omega$, where $0<p<1$, show that

$$10pa\omega^2 = 3\sqrt{2}g$$

(JMB)

4) A uniform solid circular cylinder of mass M can rotate freely about a horizontal axis which coincides with a diameter of an end face of the cylinder. The length and the radius of the cylinder are both equal to a. Show that its moment of inertia about the axis is $7Ma^2/12$.
The cylinder is slightly disturbed from rest when its centre of mass is vertically above the axis. Show that, when the axis of the cylinder makes an angle θ with the upward vertical,

$$7a\dot{\theta}^2 = 12g(1-\cos\theta)$$

and obtain an expression for the angular acceleration $\ddot{\theta}$. (U of L)

5) A uniform rod AB, of length $2a$ and mass M, is fixed to a smooth pivot at A. One end of a light elastic string of natural length $2a$ and modulus kMg is fastened at the end B and the other end of the string is fastened at a point P vertically above A, where $AP = 2a$. The rod is held at rest with B at P and is slightly displaced. If the rod next comes momentarily to rest when the angle PAB lies between $\frac{1}{2}\pi$ and π, find the range of possible values of k.

(U of L)

6) A uniform flywheel, which can rotate freely about its axis, is set in motion by a couple of constant moment C. The moment of inertia of the flywheel about its axis is I. Find the time taken for the angular velocity of the flywheel to reach the value ω_0 from rest, and find the angle through which it has turned in this time.

The constant couple is now replaced by an increasing couple (in the same sense) proportional to the angular velocity of the flywheel, the initial magnitude of this couple being C. Find the time taken to increase the angular velocity from ω_0 to $2\omega_0$, and show that in this time the flywheel turns through an angle $I\omega_0{}^2/C$.
Sketch the graph of the angular acceleration of the flywheel against the time.
(U of L)

7) Show that the centre of gravity of a uniform lamina in the form of a semi-circle of radius a is at a distance $4a/(3\pi)$ from the straight edge. Show also that the radius of gyration about this edge is $\frac{1}{2}a$.
This lamina, whose mass is m, is hinged so that it can rotate about its straight edge which is fixed in a horizontal position. The lamina is held at rest in a horizontal position and released. The motion of the lamina is opposed by a constant resisting torque. The lamina comes momentarily to rest for the first time when it has turned through an angle of 150°. Find the magnitude of the resisting torque and the angular speed of the lamina when it is vertical for the first time.
(AEB)

8) A uniform solid circular cylinder of radius a and mass m can turn freely about its axis, which is fixed in a vertical position. A horizontal force of magnitude mg acts on the cylinder, its line of action being at a distance a from the axis. Find the time taken to make n complete revolutions from rest.
When the angular velocity is ω radians per second the horizontal force ceases to act and a braking couple is applied. This couple is proportional in magnitude to the angular velocity of the cylinder, and in T seconds the angular velocity is reduced to $\frac{1}{2}\omega$ radians per second. Find the angular velocity after a further T seconds.
(U of L)

9) A thin uniform plate of mass m has the form of a rhombus ABCD, of side $2a$ and with $\angle$ BAD $= \frac{1}{3}\pi$. Show that the moment of inertia of the plate about the line normal to its plane and passing through its centre of mass is $\frac{2}{3}ma^2$.
The plate is suspended with its plane vertical from a fixed horizontal axis passing through B and is free to swing about this axis. Show that the period of small oscillations is the same as that of a simple pendulum of length $5a/3$. (Oxford)

10) A uniform rod of length $2a$ and mass m can swing freely in a vertical plane about a fixed horizontal axis through a point of the rod at a distance h from the mid-point of the rod. Find the period T of small oscillations of the rod and find the value of h for which T is a minimum.
If $h = a$ and the rod is released from rest when horizontal, find the force exerted by the rod on the pivot at the instant when the rod passes through the vertical position.
(U of L)

11) Obtain a formula for the period of small oscillations of a compound pendulum, stating clearly the meaning of each symbol used.

A thin uniform circular disc of mass M and diameter d is free to rotate about a fixed horizontal axis which is perpendicular to the disc and passes through its centre. A particle of mass m is attached to a point on the edge of the disc. The disc is then displaced slightly from the position of stable equilibrium. Find the period of the resulting small oscillations. If this is the same as the period of a simple pendulum of length equal to the diameter of the disc, show that $M = 2m$. (U of L)

12) A uniform thin rod has length $2a$ and mass M. Prove by integration that the moment of inertia of the rod about an axis through its centre O, perpendicular to its length, is $\frac{1}{3}Ma^2$.
A small nut of mass $\frac{1}{2}M$ can be screwed to any point of the rod. The system is free to rotate in a vertical plane about a fixed frictionless horizontal axis through O. Find the period T of a small oscillation when the nut is at a distance x from O and the value of x which makes T a minimum. (AEB)

13) Prove that the moment of inertia of a uniform rod, of length $2a$ and mass m, about an axis perpendicular to the rod at a distance x from its midpoint is

$$\tfrac{1}{3}m(a^2 + 3x^2)$$

A uniform rod, of weight W and length $6b$, is smoothly pivoted to a fixed point at a distance $2b$ from one end and is free to swing in a vertical plane. The rod is held horizontal and released. Prove that after it has rotated through an angle θ its angular velocity is

$$\sqrt{\left(\frac{g \sin \theta}{2b}\right)}$$

and that when the rod is vertical the reaction on the pivot is $3W/2$. (JMB)

14) Prove that for a compound pendulum the period of small oscillations is the same as that of a simple pendulum of length k^2/h where k is the radius of gyration about the axis of rotation and h is the distance of the centre of mass from this axis.
A compound pendulum is made from a light rod of length $\frac{1}{2}a$ and a heavy circular uniform disc of radius a, one end of the rod being fixed to the disc at its centre with the rod perpendicular to the plane of the disc. The pendulum hangs from the other end of the rod. Prove that the length of the equivalent simple pendulum is a. (Oxford)

15) A rigid body is free to rotate about a fixed horizontal axis. The radius of gyration of the body about a parallel axis through its centre of gravity G is k and the distance of G from the axis of rotation is h. Show that the period of small oscillations is

$$2\pi\sqrt{\left(\frac{k^2 + h^2}{gh}\right)}$$

A non-uniform rod AB of length l and mass M has its centre of gravity at G, where AG $= \frac{1}{3}l$. When suspended from a smooth hinge at A it performs small

oscillations of period T. The radius of gyration of the rod about an axis through G perpendicular to the rod is k. Find T in terms of l and k.
If, when the rod is similarly suspended from B, the period of small oscillations is also T, find k. (U of L)

16) Prove that the moment of inertia of a uniform solid sphere of mass m and radius a about a diameter is $\frac{2}{5}ma^2$.
A compound pendulum consists of a uniform solid sphere of mass m and radius a attached to a pivot by means of a uniform rod of mass $\frac{1}{4}m$ and length $2a$. (The centre of the sphere is at a distance $3a$ from the pivot.) Find the moment of inertia about an axis through the pivot at right angles to the rod, and show that the equivalent simple pendulum for small oscillations about a horizontal axis through the pivot is less than $3a$ by approximately 0.17 per cent. (SUJB)

17) A uniform rod AB of length $2a$ and mass $3m$ has a particle of mass m attached to it at B. The rod is free to rotate in a vertical plane about a horizontal axis perpendicular to the rod through a point X of the rod at a distance $x (< a)$ from A. Find the length of the simple equivalent pendulum when the rod is slightly displaced from its equilibrium position with B below A.
Show that the length is least when $x = \frac{1}{4}a(5 - \sqrt{7})$. (U of L)

18) A uniform circular wire of mass m and radius a is freely pivoted at a fixed point A on its circumference; a particle of mass m is fixed to the point B on the wire diametrically opposite to A. When the system is hanging freely with B below A, an initial angular velocity ω is given to the system which is just sufficient to make it perform complete revolutions in its own vertical plane. Prove that $a\omega^2 = 2g$.
If during the motion, AB makes an angle θ with the downward vertical, R is the component of the reaction of the pivot in the direction BA and S is the component in the direction perpendicular to BA in the sense of θ increasing, prove that

$$R = mg(3 + 5\cos\theta),$$

and find S. (JMB)

19) A uniform plane rectangular lamina ABCD has AB $= 2a$ and BC $= 2b$, where $a > b$. Prove that the radius of gyration of the lamina about BC is $2a/\sqrt{3}$.
Find the period of small oscillations when the lamina is suspended so that it can swing freely in its own plane, which is vertical, about a smooth pivot at the mid-point of AB.
This period is unchanged when the pivot is moved to the mid-point of BC. Show that $a/b = (3 + \sqrt{5})/2$. (U of L)

20) A uniform straight rod OA of length $2a$ and mass m can rotate freely in a vertical plane about its end O, which is fixed. The rod is released from rest when horizontal. Show that when OA makes an angle θ with the downward vertical, the component perpendicular to OA of the reaction of OA on the pivot at O is of magnitude $\frac{1}{4}mg\sin\theta$, and find the magnitude of the component along OA.

Find the greatest and least magnitudes of the reaction of OA on the pivot, and find also the angle θ at which the direction of the reaction makes an angle $\frac{1}{4}\pi$ with OA. (WJEC)

21) Show that the radius of gyration of a uniform rectangular lamina with sides of length a and b about an axis through the mid-points of the sides of length a is $a/\sqrt{12}$.
This lamina is smoothly hinged along one of its sides of length b and can perform small oscillations about this hinge which is fixed and horizontal. The period of these oscillations is T_1. When the lamina is smoothly hinged to a fixed pivot at one of its corners it can perform small oscillations in its own vertical plane with period T_2. Find T_1 and T_2 and show that, if $T_2 = 2T_1$, then $b = a\sqrt{15}$. (U of L)

22) A uniform circular hoop has radius a, mass M and centre G. It makes complete revolutions in its own plane, which is vertical, being smoothly hinged at a fixed point O of its circumference. If the least angular velocity during its motion is Ω, prove that, when OG makes an angle θ with the upward vertical at O,

$$a\dot{\theta}^2 = g(k - \cos\theta)$$

where

$$k = 1 + \frac{a\Omega^2}{g}$$

Prove also that, if R is the magnitude of the reaction at O, then

$$R^2 = \tfrac{1}{4}M^2g^2(15\cos^2\theta - 16k\cos\theta + 4k^2 + 1)$$

The hinge will break if $R > 4Mg$. Find the maximum value of Ω for which complete revolutions are possible. (Cambridge)

23) ABC is a uniform isosceles triangular lamina in which AB = AC = $5a$ and BC = $6a$. The mass of the lamina is m. Find its moment of inertia
(i) about an axis through A parallel to BC,
(ii) about an axis through A perpendicular to BC and lying in the plane of ABC.
The lamina is free to rotate about a horizontal axis through A perpendicular to the plane of the triangle. Find the period of small oscillations about the position of equilibrium in which BC is below A. (U of L)

24) A rigid body is free to swing about a horizontal axis L. Show that it can perform approximately simple harmonic oscillations of period

$$\tau = 2\pi\sqrt{\left(\frac{k^2 + h^2}{gh}\right)}$$

where k is the radius of gyration of the body about an axis through its centre of gravity G parallel to L, and h is the distance of G from L. Hence find the least value of τ attainable by variation of h.
A uniform rigid lamina, in the form of an annulus bounded by concentric circles of radii a and $b(< a)$, can be smoothly pivoted at any point of itself so as to

perform small oscillations in its own plane, which is vertical. Find the least attainable value of the period of oscillation. (JMB)

25) Two heavy particles of equal mass are fixed to the corners A and C of a light square framework ABCD of side $2a$, AC being a diagonal. The framework can rotate freely in a vertical plane about a fixed horizontal axis which bisects AD at right angles. Find the length of the equivalent simple pendulum for small oscillations about the position of stable equilibrium.
If the corner C is held vertically above the mid-point of AD and is then released from rest, show that the maximum angular velocity ω of the framework is given by

$$15a\omega^2 = (10+4\sqrt{5})g$$

(U of L)

26) A rigid body is free to swing about a horizontal axis L. Show that it can perform approximately simple harmonic oscillations of period

$$2\pi\sqrt{\left(\frac{l}{g}\right)}, \quad \text{where } l = h + \frac{k^2}{h}$$

k being the radius of gyration of the body about an axis through its centre of gravity G parallel to L, and h the distance of G from L.
A uniform lamina, bounded by concentric circles of radii a and $b (a < b)$ and with centre G is free to move about a smooth fixed horizontal axis lying in the plane of the lamina at a distance x $(a \leqslant x \leqslant b)$ from G. Find l in terms of x, a and b.
If $b = 4a$, show that $a\sqrt{17} \leqslant l \leqslant \frac{21}{4}a$.
Find the corresponding ranges for l when (i) $b = \frac{3}{2}a$, (ii) $b = 3a$. (Cambridge).

27) A rough uniform rod, of mass m and length $4a$, is held on a horizontal table perpendicularly to an edge of the table, with a length $3a$ projecting horizontally over the edge. If the rod is released from rest and allowed to turn about the edge, show, by using the principle of energy, that its angular speed after turning through an angle θ is

$$\sqrt{\left(\frac{6g\sin\theta}{7a}\right)}$$

assuming that the rod has not started to slip.
Deduce an expression, in terms of θ, for the angular acceleration, and hence determine the reaction normal to the rod. Show that the rod begins to slip when $\tan\theta = 4\mu/13$, where μ is the coefficient of friction.
(The moment of inertia of the rod about the edge of the table is $7ma^2/3$.)

(JMB)

CHAPTER 13

FURTHER PROPERTIES OF THE ROTATION OF A RIGID BODY

IMPULSE OF A TORQUE

Consider a rigid body which is free to rotate about a fixed axis. The moment of inertia of the body about the axis of rotation is I and $\ddot{\theta}$ is the angular acceleration of the body at any instant.
Now suppose that a torque C acts on the body for a time t causing the angular velocity to change from ω_1 to ω_2 in that time.
It has been established (Chapter 12) that

$$C = I\ddot{\theta}$$

Integrating this relationship with respect to time we have

$$\int C\,dt = \int I\ddot{\theta}\,dt$$

Hence, using corresponding limits:

$$\int_0^t C\,dt = \Big[I\dot{\theta}\Big]_{\omega_1}^{\omega_2}$$

$$= I\omega_2 - I\omega_1$$

When C is constant, this result becomes:

$$Ct = I\omega_2 - I\omega_1$$

But $I\omega$ is the angular momentum or *moment of momentum* of the body and $\int C\,dt$ is called the *impulse of the torque C.*
Hence *Impulse of Torque = Increase in Angular Momentum*
In some cases the torque C results from the action of a single force F applied at a distance r from the axis of rotation, i.e. $C = Fr$

Then $$\int C\,dt = \int Fr\,dt = r\int F\,dt$$

But $\int F\,dt$ is the impulse of the force F
So $r\int F\,dt$ is the moment of that impulse.
Hence Impulse of Torque $\equiv$ Moment of Impulse of Force.

INSTANTANEOUS IMPULSE OF TORQUE

If a torque is suddenly applied to a rotating body the angular momentum of the body undergoes a sudden change.
Such a torque exerts an instantaneous impulse whose magnitude can be determined only from the change in angular momentum produced, (since $\int C\,dt$ cannot be evaluated when the time of application is infinitesimal).

EXAMPLES 13a

1) A uniform disc of mass 20 kg and radius 0.5 m can turn about a smooth axis through its centre and perpendicular to the disc. A constant torque is applied to the disc for three seconds from rest and the angular velocity at the end of that time is $240/\pi$ revolutions per minute. Find the magnitude of the torque. If the torque is then removed and the disc is brought to rest in t seconds by a constant force of 10 N applied tangentially at a point on the rim of the disc, find t.

The moment of inertia of the disc about the axis of rotation is given by:

$$I = \tfrac{1}{2}ma^2 = \tfrac{20}{2}(0.5)^2 = \tfrac{5}{2}\text{ kg m}^2$$

The angular velocity after three seconds is given by

$$\omega = \frac{240 \times 2\pi}{\pi \times 60} = 8\text{ rad s}^{-1}$$

(i) While the torque C is applied

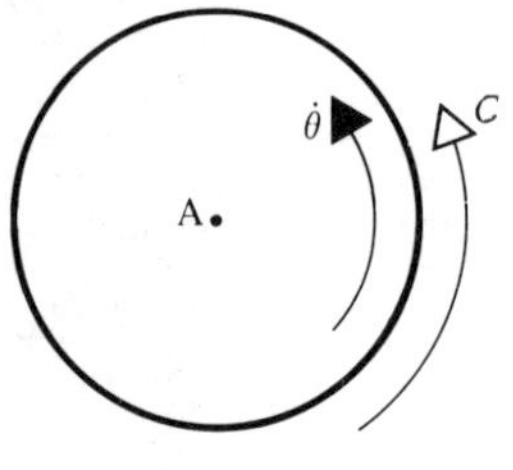

using Impulse of Torque = Increase in Angular Momentum we have:

A↻ $$\int_0^3 C\,dt = I\omega - 0$$

Hence $$3C = (\tfrac{5}{2})(8)$$

i.e. $$C = \tfrac{20}{3}\ \text{Nm}$$

(ii) While the brake is applied

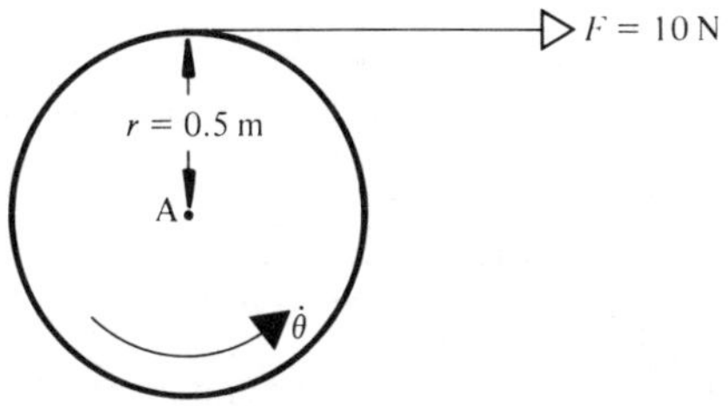

using Moment of Impulse = Increase in Angular Momentum we have

A↻ $$-r\int_0^t F\,dt = 0 - I\omega$$

Hence $$rFt = I\omega$$

or $$\tfrac{1}{2}(10)t = (\tfrac{5}{2})(8)$$

Therefore $$t = 4\ \text{s}$$

2) A flywheel whose moment of inertia about its axis of rotation is 20 kg m^2 is turning freely with an angular velocity of 3 rad s^{-1} when a braking torque is applied. The magnitude of the torque, whose initial value 32 Nm, decreases uniformly with time in such a way that after 8 seconds the magnitude would be zero. Find the time which elapses before the flywheel is brought to rest.

The magnitude C of the braking torque at time t is given by:

$$C = 32 - kt$$

But $$C = 0 \text{ when } t = 8 \quad \text{so } k = 4$$

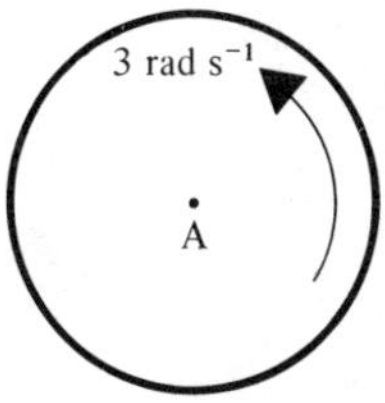

When $t = 0$
(i)

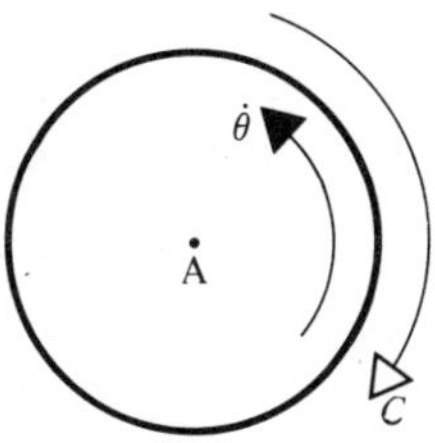

At time t
(ii)

Using Impulse of Torque = Increase in Angular Momentum

A↻ $$-\int_0^t C\,dt = I\dot{\theta} - I(3)$$

Hence $$\int_0^t (32 - 4t)\,dt = 20(3 - \dot{\theta})$$

$$32t - 2t^2 = 20(3 - \dot{\theta})$$

When the disc comes to rest, $\dot{\theta} = 0$ and $32t - 2t^2 = 60$

The smaller root of this equation is $t = 8 - \sqrt{34}$ (at this time the braking force ceases to act since rotation has stopped.)
Hence the flywheel is brought to rest in $8 - \sqrt{34}$ seconds.

3) A uniform rod AB of mass $3m$ and length $4l$, which is free to turn in a vertical plane about a smooth horizontal axis through A, is released from rest when horizontal. When the rod first becomes vertical a point C of the rod, where AC = $3l$, strikes a fixed peg. Find the impulse exerted by the peg on the rod if:
(a) the rod is brought to rest by the peg,
(b) the rod rebounds and next comes to instantaneous rest inclined to the downward vertical at an angle $\pi/3$ radians.

The moment of inertia of the rod about the axis through A is given by

$$I = \tfrac{4}{3}(3m)(2l)^2 = 16ml^2$$

Let ω_1 be the angular velocity of the rod when it first becomes vertical

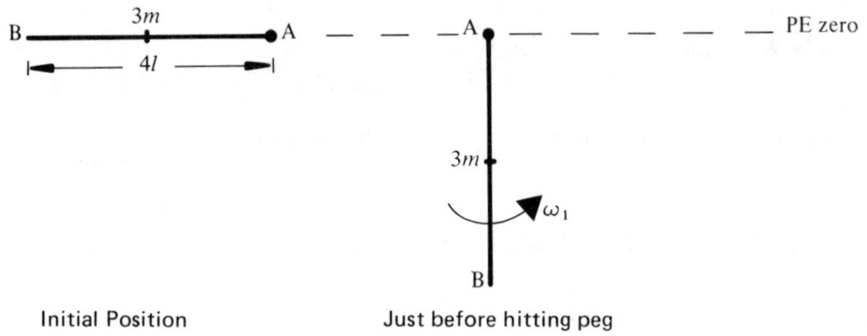

Using Conservation of Mechanical Energy we have:

$$0 = \tfrac{1}{2}I\omega_1^{\,2} - 3mg(2l)$$

Hence $$\omega_1^{\,2} = 3g/4l$$

(a) Let J_1 be the impulse exerted by the peg in bringing the rod to rest.

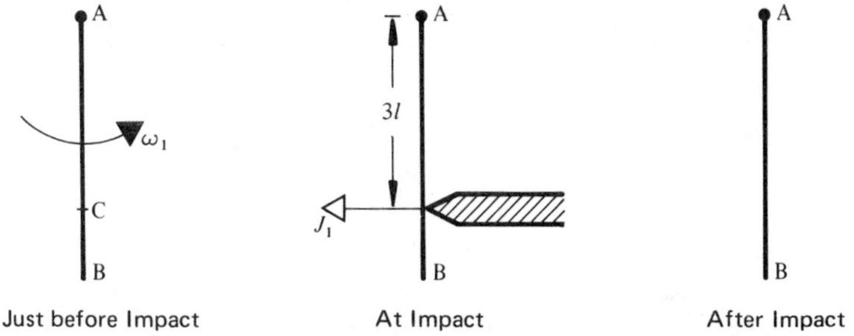

Using Moment of Impulse = Increase in Angular Momentum

A↻ $$-J_1(3l) = 0 - I\omega_1$$

Hence $$3J_1 l = 16ml^2\sqrt{\frac{3g}{4l}}$$

$$J_1 = \tfrac{8}{3}m\sqrt{3gl}$$

(b) Let J_2 be the impulse exerted by the peg in causing the rod to rebound, and let ω_2 be the initial angular velocity of rebound.

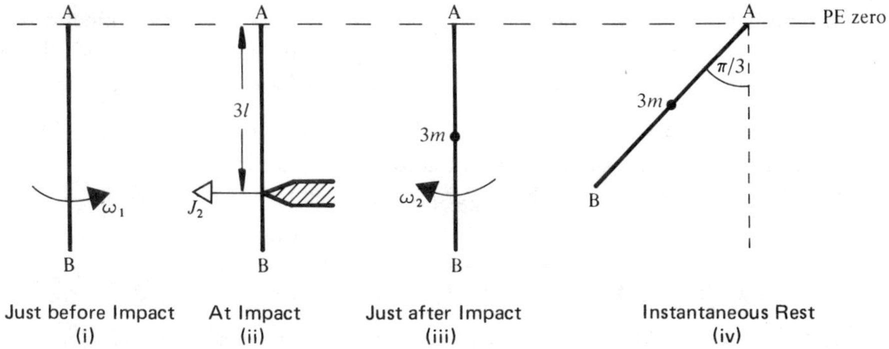

Using Moment of Impulse = Increase in Angular Momentum

A↻ $$J_2(3l) = I\omega_2 - (-I\omega_1)$$

Hence $$3lJ_2 = 16ml^2(\omega_1 + \omega_2)$$

But ω_2 can be found by using Conservation of Mechanical Energy from positions (iii) to (iv)

$$\tfrac{1}{2}I\omega_2{}^2 - 3mg(2l) = -3mg(2l\cos\pi/3)$$

Hence $$\omega_2{}^2 = 3g/8l$$

Then $$3lJ_2 = 16ml^2\left(\sqrt{\frac{3g}{4l}}+\sqrt{\frac{3g}{8l}}\right)$$

$$= 16ml^2\sqrt{\frac{3g}{8l}}\,(\sqrt{2}+1)$$

$$J_2 = \frac{4m}{3}\sqrt{6gl}\,(\sqrt{2}+1)$$

(*Note*: Because the form of the expressions derived for J_1 and J_2 is unfamiliar, a check on the dimensions of these results is advisable.

e.g. $\frac{8}{3}m\sqrt{3gl}$ has dimensions $M\sqrt{\frac{L}{T^2}\times L} = MLT^{-1}$

Now impulse (force × time) has dimensions $\frac{ML}{T^2}\times T = MLT^{-1}$

Hence the results obtained for J_1 and J_2 are dimensionally correct.)

EXERCISE 13a

1) A uniform solid cylinder of mass m and radius a rotates freely about its axis of symmetry under the action of a constant torque $4mga$. Find the angular velocity of the cylinder after 4 seconds if it started from rest.

2) Find the magnitude of the constant torque required to increase the kinetic energy of a uniform solid sphere, rotating freely about a diameter, from 400 J to 9025 J in 6 seconds if the mass of the sphere is 20 kg and its diameter is 0.5 m.

3) A flywheel whose moment of inertia about its axis of rotation is 16 kg m^2 is rotating freely in its own plane about a smooth axis through its centre. Its angular velocity is 9 rad s^{-1} when a torque is applied to bring it to rest in T seconds. Find T if:
(a) the torque is constant and of magnitude 4 Nm,
(b) the magnitude of the torque after t seconds is given by kt.

4) A uniform rod AB of length $2l$ and mass m is rotating in a horizontal plane about a vertical axis through A, with angular velocity Ω, when the mid-point of the rod strikes a fixed rail and is brought immediately to rest. Find the impulse exerted by the rail.

5) A uniform square trap-door ABCD is free to rotate about an axis along the side AB which is horizontal. When it is hanging at rest, the mid-point of CD is struck a blow which causes the trap-door to rotate through an angle of 60° before next coming to instantaneous rest. Find the impulse of the blow if the trap-door is 1 m square and its mass is 40 kg. (Take $g = 10$ ms^{-2}.)

6) ABC is a triangular framework of three uniform rods each of mass $3m$ and length $2l$. The framework is free to rotate about a smooth horizontal axis

along AB. Initially the framework is in a position of unstable equilibrium and is slightly disturbed from rest. When it is first in a horizontal plane the point C collides with a fixed peg. Find the impulse exerted by the peg if:
(a) the framework is brought to rest on impact,
(b) the framework rebounds so that C rises to a height $l\sqrt{3}/2$ above the level of AB before coming to instantaneous rest.

CONSERVATION OF ANGULAR MOMENTUM (MOMENT OF MOMENTUM)

If a torque acts on a rotating object it causes a change in angular momentum equal in magnitude to the impulse of the torque.
When two objects collide they exert equal and opposite instantaneous impulses upon each other.
About any specified axis, therefore, the moments of these two instantaneous impulses are equal and opposite.
If neither object is fixed the change in the angular momentum of each is equal to the moment of the impulse exerted on it, so the two objects undergo equal and opposite changes in angular momentum.
Thus the total angular momentum of the system is unchanged.
This is the Principle of Conservation of Angular Momentum and it can be stated in the following general form.
Unless some external force exerts a torque on a system the total angular momentum (moment of momentum) of that system is constant.

EXAMPLES 13b

1) A uniform circular disc of mass m and radius a is rotating with constant angular velocity ω in a horizontal plane about a vertical axis through its centre A. A particle P of mass $2m$ is placed gently on the disc at a point distant $a/2$ from A. If the particle does not slip on the disc find the new angular velocity of the rotating system.

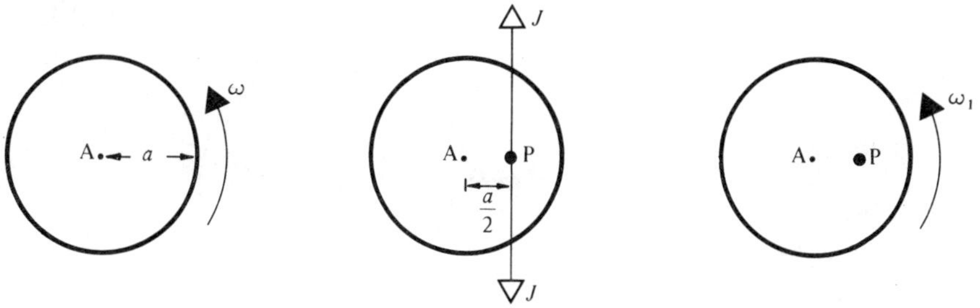

Rotation of disc alone (i)

Instantaneous addition of particle (ii)

Rotation of disc and particle (iii)

The moment of inertia of the disc about the axis through A is $\frac{ma^2}{2}$.

Hence the angular momentum of the rotating disc is $\left(\frac{ma^2}{2}\right)(\omega)$.

[At the instant when the particle is placed on the disc a pair of equal and opposite frictional impulses act, one on the disc and one on the particle, causing equal and opposite changes in moment of momentum about the axis through A.]
The particle stays in contact with a fixed point on the disc so it has the same angular velocity as the disc.
The moment of inertia of the disc and particle about the same axis is

$$\frac{ma^2}{2} + 2m\left(\frac{a}{2}\right)^2.$$

Hence the angular momentum of the rotating system is $(ma^2)(\omega_1)$
Using the Principle of Conservation of Angular Momentum we have:

$$\frac{ma^2}{2}\omega = ma^2\omega_1$$

Hence $$\omega_1 = \frac{\omega}{2}$$

Thus the system rotates with an angular velocity $\omega/2$.

2) A uniform rod AB of length $2l$ and mass m is turning freely in a horizontal plane about a vertical axis through A, with angular velocity $3v/l$. A particle P, also of mass m, is moving with constant velocity v in the same horizontal plane. The particle and the rod are moving towards each other and when AB is perpendicular to the path of the particle, P collides with the point B of the rod. If the coefficient of restitution between the rod and the particle is $\frac{1}{2}$ find the speed of the particle after impact.

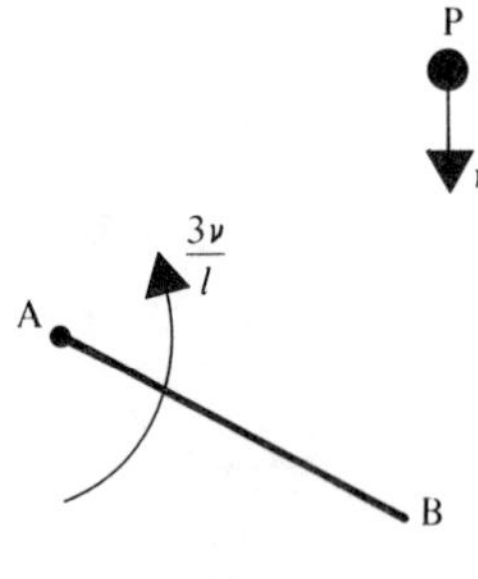

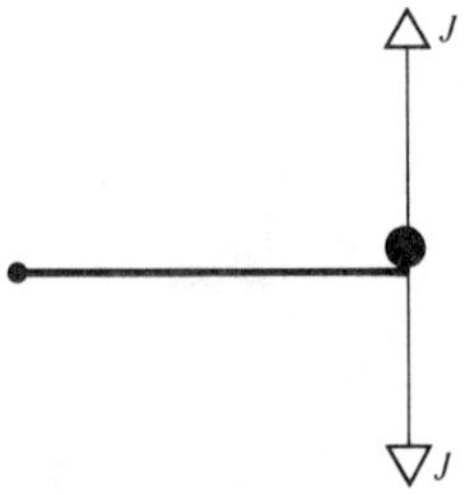

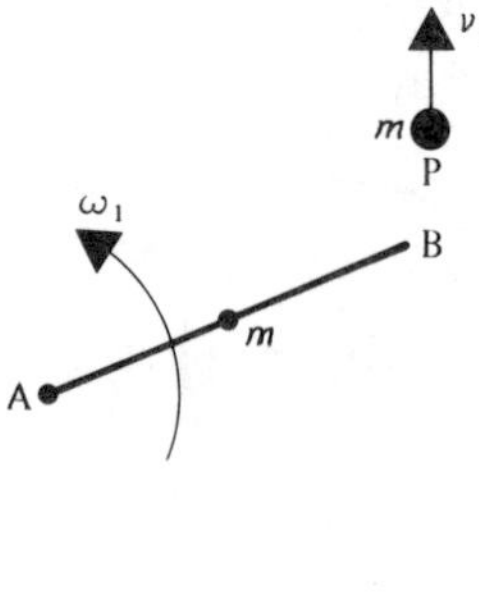

[This problem involves a particle moving with linear momentum as well as a rotating rod. In this context the use of the term moment of momentum is more logical than reference to angular momentum.]
At the moment of collision, equal and opposite impulses of magnitude J act on the rod and the particle, so moment of momentum is conserved.
Just before collision the moment of momentum about the axis through A (↻) is:

for the rod	$\frac{4}{3}ml^2\left(\frac{3v}{l}\right)$
for the particle	$-(mv)(2l)$
hence for the system	$4mlv - 2mlv$

Just after impact the moment of momentum about the axis through A (↻) is:

for the rod	$\frac{4}{3}ml^2\omega_1$
for the particle	$(mv_1)(2l)$
hence for the system	$\frac{4}{3}ml^2\omega_1 + 2mlv_1$

Using Conservation of Moment of Momentum we have:

$$2mlv = \tfrac{4}{3}ml^2\omega_1 + 2mlv_1$$

i.e.

$$3v = 2l\omega_1 + 3v_1 \qquad (1)$$

Now the Law of Restitution can be applied to the relative speeds of the particle P and the *point* B with which it collides.

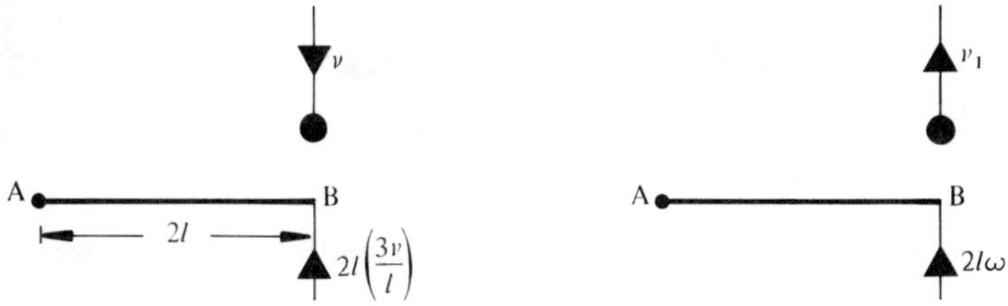

Speed of approach is	$6v + v$
Speed of separation is	$v_1 - 2l\omega_1$

Hence the Law of Restitution gives:

$$\tfrac{1}{2}(7v) = v_1 - 2l\omega_1 \qquad (2)$$

Then solving equations (1) and (2) we have:

$$\omega_1 = -\frac{15v}{16l}$$

$$v_1 = \frac{13v}{8}$$

Thus, after impact the rod begins to rotate with angular velocity $15v/16l$ in the opposite sense and the particle begins to move in the reverse direction with speed $13v/8$.

3) A pulley in the form of a uniform disc of mass $2m$ and radius r, is free to rotate in a vertical plane about a fixed horizontal axis through its centre. A light inextensible string has one end fastened to a point on the rim of the pulley and is wrapped several times round the rim. The portion of string not wrapped round the pulley is of length $8r$ and carries a particle of mass m at its free end. The particle is held close to the rim of the pulley and level with its centre. If the particle is released from this position find the initial angular velocity of the pulley and the impulse of the sudden tension in the string when it becomes taut.

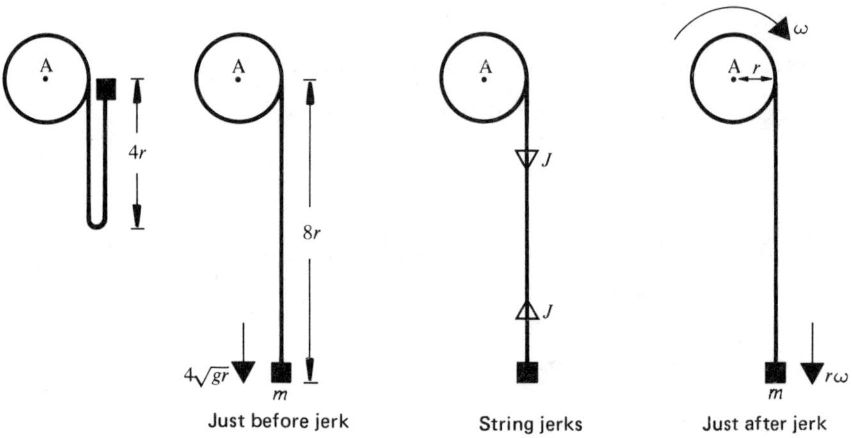

The particle falls freely through a distance $8r$ before the string becomes taut.

By then its velocity is $\sqrt{2g(8r)} = 4\sqrt{gr}$.

At this instant the sudden tension in the string exerts equal and opposite impulses J on the particle and the pulley.
If the pulley then begins to move with an angular velocity ω the linear velocity of the particle becomes $r\omega$.
Therefore applying Conservation of Moment of Momentum we have:

$$(mv)r = (mr\omega)r + I\omega$$

where I is the moment of inertia of the pulley about the axis of rotation,

i.e. $$I = \tfrac{1}{2}(2m)r^2.$$

Hence
$$(4m\sqrt{gr})r = (mr\omega)r + \tfrac{1}{2}(2m)r^2\omega$$
$$4\sqrt{gr} = 2r\omega$$

The pulley therefore begins to move with angular velocity $2\sqrt{g/r}$.
Considering the sudden change in the linear momentum of the particle caused by the impulsive tension J we have:

$$J = -mr\omega - (-4m\sqrt{gr})$$

i.e.
$$J = 2m\sqrt{gr}$$

EXERCISE 13b

1) A uniform rod AB of mass $3m$ and length $2l$ is lying at rest on a smooth horizontal table with a smooth vertical axis through the end A. A particle of mass $2m$ moves with speed $2u$ across the table and strikes the rod at its mid-point C. If the impact is perfectly elastic find the speed of the particle after impact if:
(a) it strikes the rod normally,
(b) its path before impact was inclined at $60°$ to AC.

2) A light string hangs over a pulley which is a uniform disc of mass $4m$ and radius a, and carries a scale pan of mass m at each end. If a particle of mass m is dropped into one of the scale pans from a height $10a$ above it, find the initial angular velocity of the pulley assuming that the string does not slip on the pulley and that the particle does not rebound from the scale pan.

3) A uniform rod AB, of mass m and length $4a$, is smoothly pivoted at a point O of its length, where $AO = a$, and hangs at rest with A uppermost. The rod receives a horizontal impulse of magnitude J at its centre of mass. Find the initial angular velocity of the rod.
If the rod describes complete revolutions in the subsequent motion, find an inequality for J in terms of a, m and g.

4) A uniform circular disc is rotating in its own plane about an axis through its centre. Its mass is M, its radius is a and it is rotating with angular velocity ω. A lamina of mass $6M$ in the shape of an annulus (the area between two concentric circles) with inner radius a and outer radius $2a$ is held with its centre vertically over the centre of the rotating disc and is then dropped. When the annulus circumscribes the disc there is no slipping at the circle of contact. Find the angular velocity of the compound disc.

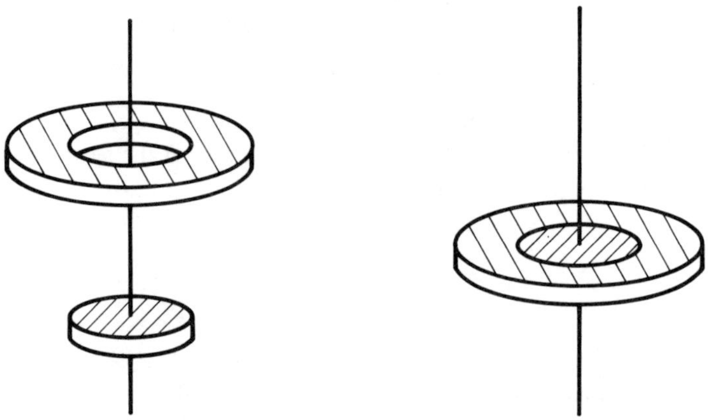

5) A uniform circular disc of mass M and radius a is pivoted at a point O on its circumference so that it can rotate about the tangent at O, which is horizontal, the centre of the disc describing a vertical circle of centre O in a plane perpendicular to the tangent. The point diametrically opposite to O is A and the disc is just displaced from rest when A is vertically above O. Find the angular velocity of the disc when A is vertically below O.
At this instant a particle of mass M travelling with velocity u in the direction of motion of the centre of the disc hits the disc at its centre, and adheres to it. Find the angular velocity of the system immediately after the impact.
If the disc just reaches its initial position show that

$$u = (3\sqrt{2} - \sqrt{5})\sqrt{(ag)}$$

ROTATION ABOUT AN AXIS MOVING IN A STRAIGHT LINE

Motion of a Lamina in its own Plane

Consider a lamina which is moving under the action of a set of coplanar forces so that a point A of the lamina moves in a straight line while the lamina rotates about an axis through A.
If initially A is at a point O on the plane and Ox is the line along which A moves, then the linear velocity and acceleration of A at any instant are $\dot{x}$ and $\ddot{x}$. Let P be any other point on the lamina where AP $= r$ and θ is the inclination of AP to Ox at any instant.

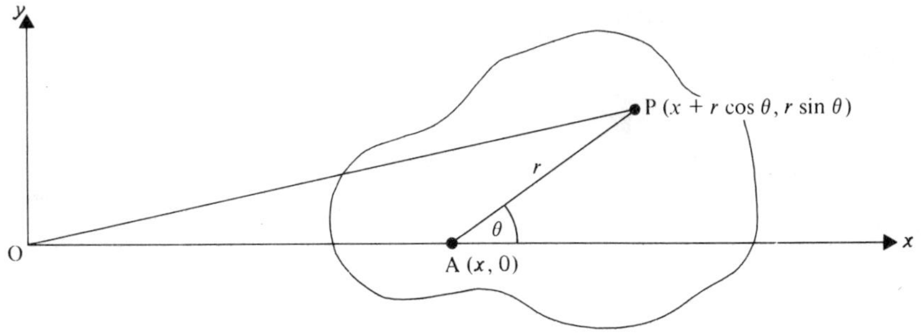

The co-ordinates (X, Y) of P at any instant are given by

$$X = x + r\cos\theta$$

$$Y = r\sin\theta$$

Now r is constant since A and P are points fixed in the lamina so, differentiating with respect to time, we have

$$\dot{X} = \dot{x} - r\sin\theta\,\dot{\theta}$$

$$\dot{Y} = r\cos\theta\,\dot{\theta}$$

So the velocity components of P can be used in the form illustrated below.

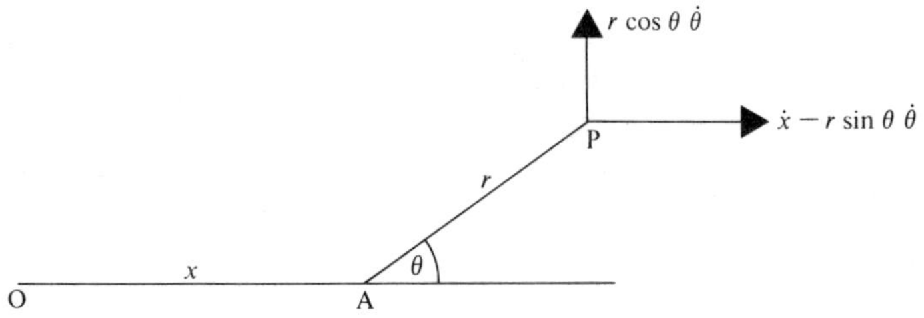

Alternatively, since $\overrightarrow{OP} = \overrightarrow{OA} + \overrightarrow{AP}$, the motion of P is a combination of the motion of A and the motion of P relative to A.
Now relative to A, P is a point with polar co-ordinates (r, θ) where r is constant. The radial and transverse components of the velocity and acceleration of a point with general polar co-ordinates (r, θ) were derived in Chapter 3.
For the motion of P relative to A, r is constant, so these general results become:

	Radial Component	Transverse Component
Velocity	0	$r\dot{\theta}$
Acceleration	$-r\dot{\theta}^2$	$r\ddot{\theta}$

Thus the motion of P relative to O is made up of the components shown in the diagrams below.

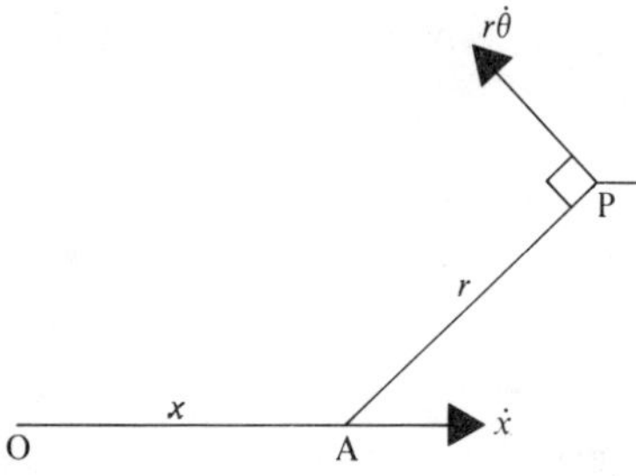

Velocity components

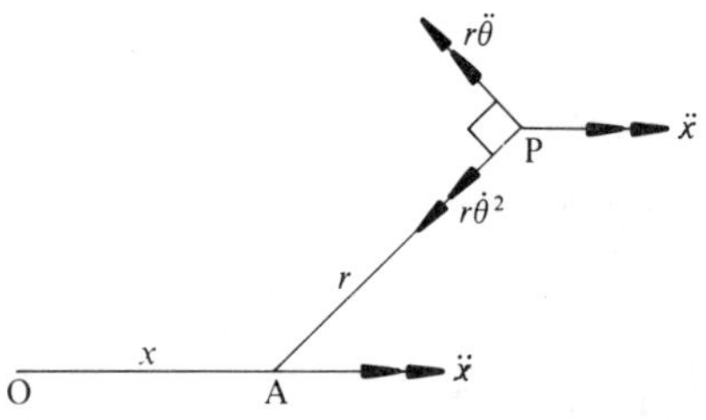

Acceleration Components

EXAMPLE

A lamina is moving in its own plane so that a point A of the lamina moves with constant acceleration a in a straight line Ox, and the lamina rotates about an axis through A with constant angular velocity ω. Find:

(i) the equation relative to A of the locus of the set of points with acceleration of magnitude $a\omega^2$,

(ii) in terms of t the equation relative to O of the set of points with speed $2a\omega$, given that the initial speed of A is u.

The angular velocity of the lamina is constant ($\dot{\theta} = \omega$) so the angular acceleration $\ddot{\theta}$ is zero.

(i) Thus the acceleration components of a point P distant r from A are as shown in the diagram below.

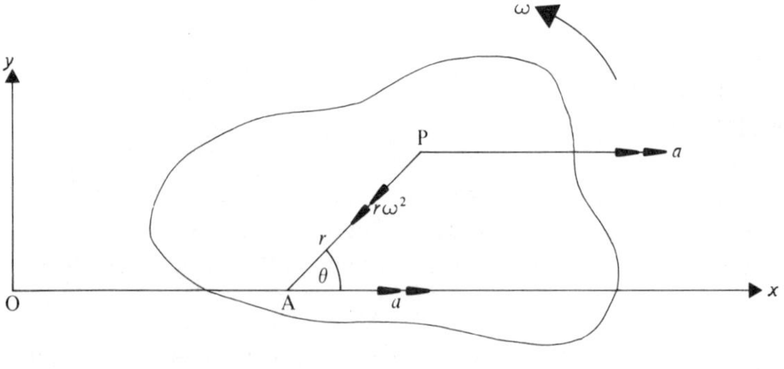

The magnitude of the resultant acceleration of P is $a\omega^2$

Hence $\qquad (a - r\omega^2 \cos\theta)^2 + (r\omega^2 \sin\theta)^2 = (a\omega^2)^2 \qquad (1)$

Then if the co-ordinates of P relative to A are (X, Y) we have

$$X = r\cos\theta \quad \text{and} \quad Y = r\sin\theta$$

Hence equation (1) becomes $\quad (a - X\omega^2)^2 + (Y\omega^2)^2 = (a\omega^2)^2$

or
$$\left(X - \frac{a}{\omega^2}\right)^2 + Y^2 = a^2$$

This equation gives the relationship between X and Y and is therefore the equation relative to A of the locus of points with acceleration $a\omega^2$.

This locus can be identified as a circle with centre $\left(\frac{a}{\omega^2}, 0\right)$ and radius a.

(ii) The velocity of A at time t is $u + at$ so the components of the velocity of P at time t are as shown in the following diagram:

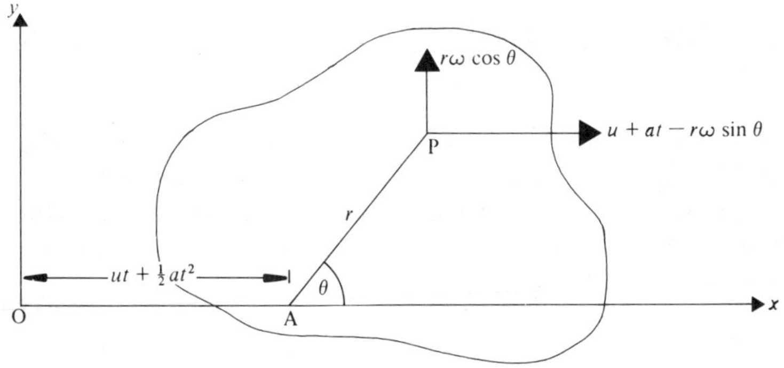

Points with speed $2a\omega$ are such that

$$(u + at - r\omega \sin\theta)^2 + (r\omega\cos\theta)^2 = (2a\omega)^2.$$

Now OA $= ut + \frac{1}{2}at^2$, so the co-ordinates (x, y) of P relative to O are:

$$x = ut + \tfrac{1}{2}at^2 + r\cos\theta$$

$$y = r\sin\theta$$

Hence $\quad (u + at - \omega y)^2 + (x - ut - \frac{1}{2}at^2)^2\omega^2 = (2a\omega)^2$

i.e. $\quad \left(y - \frac{u + at}{\omega}\right)^2 + \left(x - \frac{2ut + at^2}{2}\right)^2 = 4a^2$

This is the equation, relative to O, of the locus of points with speed $2a\omega$ at time t. The locus is seen to be a circle with centre $\left(\frac{u + at}{\omega}, \frac{2ut + at^2}{2}\right)$ and radius $2a$.

GENERAL MOTION OF A LAMINA MOVING UNDER THE ACTION OF A SET OF COPLANAR FORCES

Consider a lamina of mass M, whose centre of mass is at the point G, moving under the action of a set of forces whose resultant is $\mathbf{R}$. It has been established (in Chapter 8) that the acceleration of the point G is the same as that of a particle of mass M moving under the action of **R**.

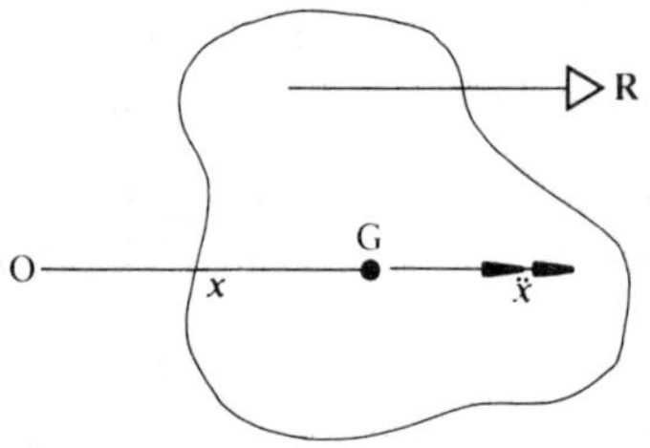

If the direction of **R** is constant and initially G was either at rest or moving in the direction of **R**, then G subsequently moves in a straight line Ox parallel to **R**.

Thus the equation of motion of G is

$$R = M\ddot{x}$$

The acceleration components of any point P have already been established and are shown in diagram (i) below. If, at P, there is a constituent particle of the lamina, of mass m, the force acting on this particle is shown in diagram (ii) as a pair of perpendicular components parallel and perpendicular to GP.

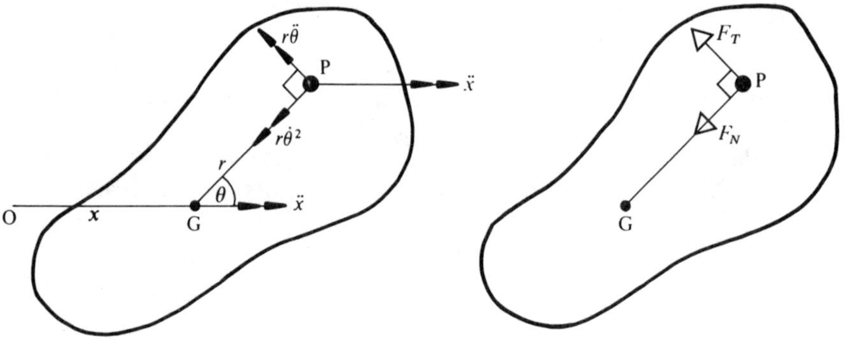

(i) (ii)

The rotational behaviour of the lamina will now be investigated.
Applying Newton's Law to the motion of the particle at P we have, in the direction of F_T

$$F_T = m(r\ddot{\theta} - \ddot{x}\sin\theta)$$

Hence

$$rF_T = mr(r\ddot{\theta} - \ddot{x}\sin\theta)$$

For the whole body therefore:

$$\sum rF_T = \ddot{\theta}\sum mr^2 - \ddot{x}\sum mr\sin\theta \qquad (1)$$

Now $\Sigma\, rF_T = C$, the resultant torque about G, $\Sigma\, mr^2 = I_G$ and $\Sigma\, mr \sin\theta = 0$ as G is the centre of mass. Therefore equation (1) becomes

$$C = I\ddot{\theta}$$

Hence the lamina rotates as through it were moving under the action of the resultant torque about a fixed axis through G.
The motion of a lamina rotating and translating in its own plane can therefore be analysed by considering independently:
(1) the linear motion of the centre of mass,
(2) rotation about an axis through the centre of mass.
These results are confirmed if we consider that the given force **R** is equivalent to an equal force **R** acting through G, together with a couple of moment C where $C = Ra$.

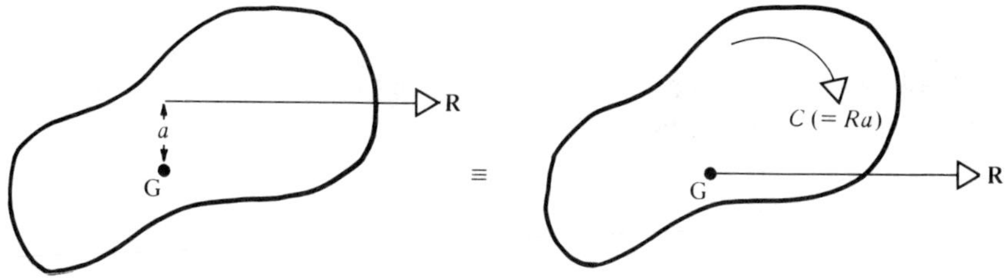

The equivalent system clearly indicates that G can be expected to move in a straight line and that the lamina will rotate about a perpendicular axis through G. If the resultant **R**, of the given set of forces, itself acts through G then $C = 0$. In this case the lamina will move in the same way as a particle of mass M at G. (This property justifies the method used in Volume One to solve a number of problems involving the motion of rigid bodies, in which the body was treated as a particle.)

EXAMPLE

One end of a light inextensible string is fastened to a point on the rim of a uniform circular disc of mass m and radius a. The string is wrapped several times round the circumference and the other end is then attached to a fixed point A. The disc is held in the vertical plane containing A so that the portion of string which is not in contact with the disc is taut and vertical. If the disc is then released from rest find the acceleration of its centre and the tension in the string.

[As the disc descends, its centre O moves vertically downward since the only forces acting on the disc are vertical. The disc also rotates as it falls because the resultant force does not act through O.]
Let the linear velocity and acceleration of O be $\dot{x}$ and $\ddot{x}$ and let the angular velocity and acceleration of the disc be $\dot{\theta}$ and $\ddot{\theta}$.

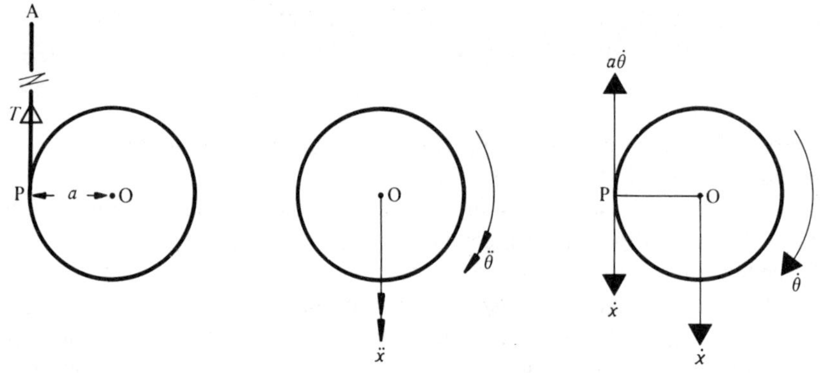

(i) (ii) (iii)

The point P where the string loses contact with the disc has a velocity $\dot{x} - a\dot{\theta}$ vertically downward (diagram (iii)). But this point is momentarily at rest (in contact with a stationary point on the string).

i.e. $$\dot{x} = a\dot{\theta}$$

Hence $$\ddot{x} = a\ddot{\theta}$$

For the linear motion of the centre of mass, Newton's Law gives:

$$mg - T = m\ddot{x} \qquad (1)$$

For the rotation of the disc about an axis through O, using $C = I\ddot{\theta}$ gives:

$$Ta = \left(\frac{ma^2}{2}\right)\ddot{\theta}$$

But $$\ddot{x} = a\ddot{\theta}$$

Hence $$Ta = \frac{ma\ddot{x}}{2} \qquad (2)$$

From equations (1) and (2) we see that the tension in the string is $\frac{mg}{3}$ and that the centre of the disc descends vertically with acceleration $\frac{2g}{3}$.

ROTATION AND TRANSLATION OF A RIGID BODY

A rigid body can be regarded as being made up of a series of parallel laminate sections.

Consider a body which is moving so that:

(1) its centre of mass is moving in a straight line,

(2) each of the constituent laminas is rotating in its own plane,

(3) the axis of rotation passes through the centre of mass of each section and hence through the centre of mass of the body.

Then each lamina moves as though it were rotating, under the action of its own resultant torque, about the common axis through the centres of mass. Also, because the body is rigid, every lamina has the same angular acceleration.

Thus the whole body moves as though it were rotating under the action of the total resultant torque about an axis through its centre of mass.

The motion of such a body of mass M can therefore be analysed in the same way as the motion of a lamina moving in its own plane, i.e. by investigating separately:

(1) the linear motion of a particle of mass M at the centre of mass,

(2) the rotation of the body about an axis through the centre of mass.

Kinetic Energy

Consider again a lamina, whose centre of mass is G, moving in its own plane under the action of a set of forces whose resultant has a constant direction. The velocity components of a constituent particle P of the body are shown below:

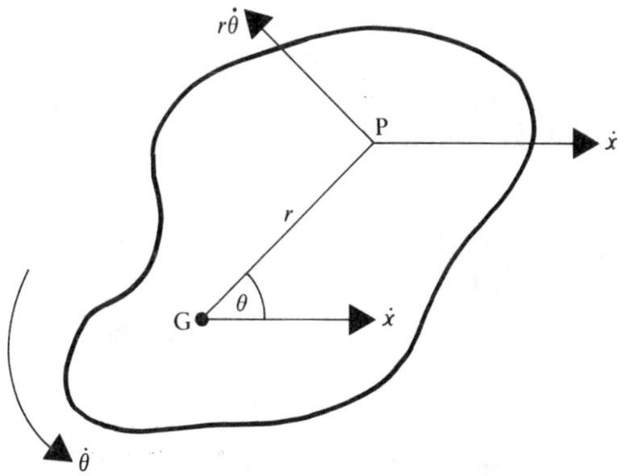

If the mass of the particle is m, its kinetic energy is given by:

$$\text{KE}_\text{P} = \tfrac{1}{2}m[(\dot{x} - r\dot{\theta}\sin\theta)^2 + (r\dot{\theta}\cos\theta)^2]$$

Summing for the whole lamina we have:

$$\text{KE} = \sum \tfrac{1}{2}m[\dot{x}^2 - 2\dot{x}r\dot{\theta}\sin\theta + r^2\dot{\theta}^2]$$

$$= \tfrac{1}{2}\dot{x}^2 \sum m - \dot{x}\dot{\theta}\sum mr\sin\theta + \tfrac{1}{2}\dot{\theta}^2 \sum mr^2$$

But $\quad \sum m = M$, the total mass of the lamina

$\sum mr^2 = I$, the moment of inertia of the lamina about the axis through G.

$\sum mr \sin\theta = 0$, (since G is the centre of mass.)

The kinetic energy of the lamina is therefore given by:

$$\text{KE} = \tfrac{1}{2}M\dot{x}^2 + \tfrac{1}{2}I\dot{\theta}^2$$

Thus it can be seen that, as far as kinetic energy is concerned, the motion of the lamina can again be regarded as being made up of:
(1) the linear motion of a particle of mass M at the centre of mass ($\text{KE} = \tfrac{1}{2}M\dot{x}^2$),
(2) the rotation of the lamina about an axis through the centre of mass ($\text{KE} = \tfrac{1}{2}I\dot{\theta}^2$).

By considering, as we did before, a rigid body to comprise a set of parallel laminas, we see that the application of the results derived above can be extended to deal with the motion of a rigid body.

EXAMPLES 13c

1) A uniform solid sphere of radius a and mass M is rolling without slipping on a horizontal surface so that its centre of mass moves in a straight line with constant speed v. Find the kinetic energy of the sphere.

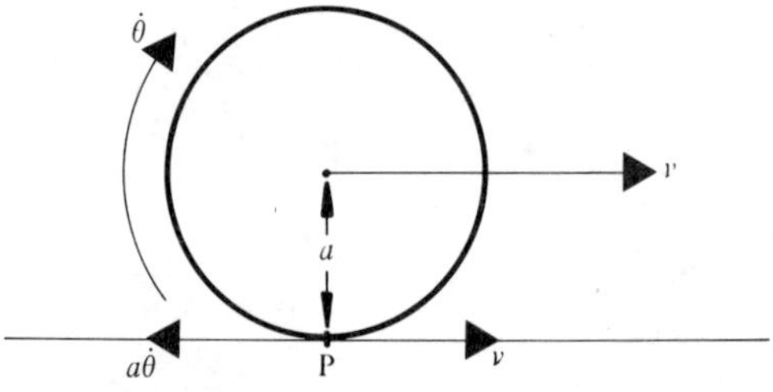

The velocity components of the point P, which is momentarily in contact with the plane, are shown in the diagram.
But the disc does not slip on the plane so P is instantaneously at rest.

Hence $\quad v - a\dot{\theta} = 0$

i.e. $\quad \dot{\theta} = v/a$

Now the kinetic energy of the sphere is given by:

$$\text{KE} = \tfrac{1}{2}Mv^2 + \tfrac{1}{2}I\dot{\theta}^2$$

$$= \tfrac{1}{2}Mv^2 + \tfrac{1}{2}(\tfrac{2}{5}Ma^2)\left(\frac{v}{a}\right)^2$$

$$= \tfrac{7}{10}Mv^2$$

2) A uniform solid cylinder of mass $2m$ and radius r rolls without slipping down a plane inclined at 30° to the horizontal. Find its angular acceleration and the least possible value of the coefficient of friction between the cylinder and the plane.

[A circular object cannot roll down an inclined plane unless the plane is rough because it is the moment of the frictional force which causes rotation about an axis through the centre of mass. If the object rolls without slipping, the point of application of the frictional force is the point of contact between the object and the plane. As this point is always momentarily at rest no work is done by the frictional force.]

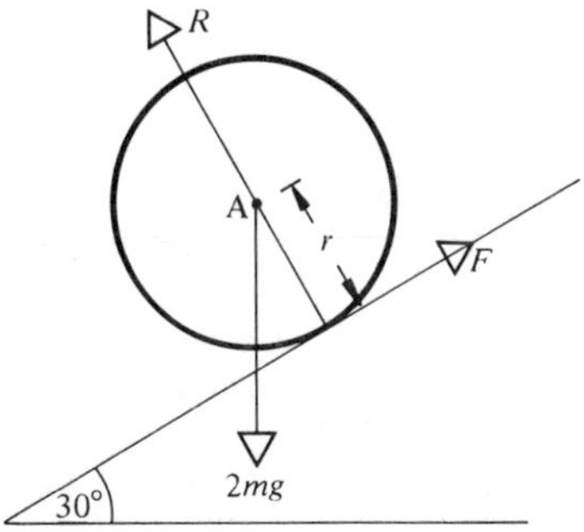

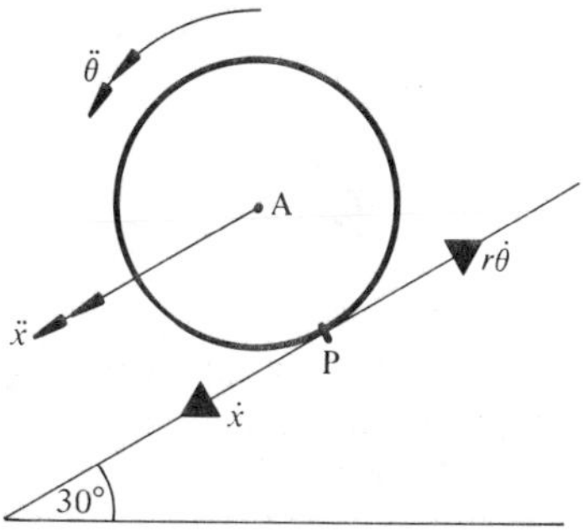

The velocity components of the point of contact P are shown in the diagram. But P is momentarily at rest so $\dot{x} - r\dot{\theta} = 0$.

Therefore $\ddot{x} - r\ddot{\theta} = 0$.

For the linear motion of a particle of mass $2m$ at A we have:

$$\swarrow \quad 2mg\sin 30° - F = 2m\ddot{x} \qquad (1)$$

$$\nwarrow \quad R - 2mg\cos 30° = 0 \qquad (2)$$

Considering rotation about an axis through A we have:

$$\text{A}\circlearrowright \quad Fr = I\ddot{\theta}$$

But $\quad I = 2m\left(\dfrac{r^2}{2}\right) \quad$ and $\quad r\ddot{\theta} = \ddot{x}$

So $\quad Fr = mr^2\left(\dfrac{\ddot{x}}{r}\right)$

i.e. $\quad F = m\ddot{x} \qquad (3)$

Solving equations (1) and (3) gives:

$$F = \tfrac{1}{3}mg$$

$$\ddot{x} = \tfrac{1}{3}g$$

Hence the angular acceleration of the disc is $g/3r$.

Now $\quad F = \tfrac{1}{3}mg \quad$ and $\quad R = mg\sqrt{3}$

But $\quad F \leqslant \mu R$

i.e. $\quad \tfrac{1}{3}mg \leqslant \mu mg\sqrt{3}$

Hence $\quad \mu \geqslant \dfrac{\sqrt{3}}{9}$

3) A ring of mass M and radius a rolls from rest without slipping down a plane inclined at an angle α to the horizontal. Find the angular velocity of the ring when it has travelled a distance d down the plane.

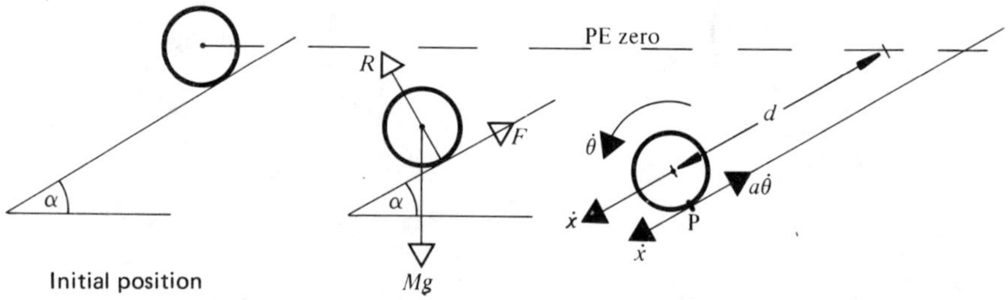

Considering the point P, as before, we have $\dot{x} = a\dot{\theta}$.
No work is done by R or F since neither force moves its point of application, so there is no change in the total mechanical energy.
The initial mechanical energy is zero.

The final potential energy $= -Mgd \sin \alpha$

The final kinetic energy $= \tfrac{1}{2}M\dot{x}^2 + \tfrac{1}{2}I\dot{\theta}^2$

$$= \tfrac{1}{2}M(a\dot{\theta})^2 + \tfrac{1}{2}(Ma^2)\dot{\theta}^2$$

Hence $\quad 0 = -Mgd \sin \alpha + Ma^2\dot{\theta}^2$

i.e. $\quad \dot{\theta}^2 = \dfrac{gd \sin \alpha}{a^2}$

Thus the angular velocity of the ring is $\dfrac{1}{a}\sqrt{gd \sin \alpha}$.

EXERCISE 13c

1) Find the kinetic energy of a body rolling without slipping on a horizontal plane if:
(i) the body is a ring of mass M and radius a and its angular velocity is ω,
(ii) the body is a uniform hollow sphere of mass $5M$ and radius a and the centre is moving with velocity v.

2) A uniform solid cylinder of mass 8 kg and diameter 1 m is placed with its axis horizontal on a rough horizontal plane. When a horizontal force of 20 N is applied to the mid-point of, and perpendicular to, the highest generator, the cylinder begins to roll without slipping on the plane. Find the angular acceleration of the cylinder and the range of values of the coefficient of friction between the cylinder and the plane for which this motion is possible. (Take $g = 10\ \text{ms}^{-2}$)

3) A body rolls without slipping down a plane inclined at an angle α to the horizontal. Find the angular acceleration and the least value of the co-efficient of friction if the body is:
(i) a uniform solid sphere of radius a,
(ii) a uniform hollow cylinder of radius a,
(iii) a uniform hollow sphere of radius a.

4) A uniform solid sphere of radius 0.25 m rolls from rest, without slipping, on a plane inclined at 30° to the horizontal. Find the velocity of the centre of the sphere:
(i) after five seconds,
(ii) after moving a distance 1 m down the plane.

5) A body with constant circular cross-section of radius a, has a radius of gyration k about its axis of symmetry. If it rolls without slipping down a plane inclined at an angle α to the horizontal, show that

$$\tan\alpha \leqslant \mu\left(1 + \frac{a^2}{k^2}\right)$$

6) A yoyo consists of two uniform circular discs each of mass m and radius $4a$ between which is a small uniform solid cylinder of mass $3m$ and radius a. The line through the axis of the cylinder passes through the centre of each disc (as shown in the diagram below). A light inextensible string has one end fastened to a point on the cylinder and is wrapped round the cylinder. The yoyo is projected vertically downward with speed $4\sqrt{ag}$ and the string, as it unwinds, becomes vertical. Find the speed of the centre when a length $126a$ of string has unwound.

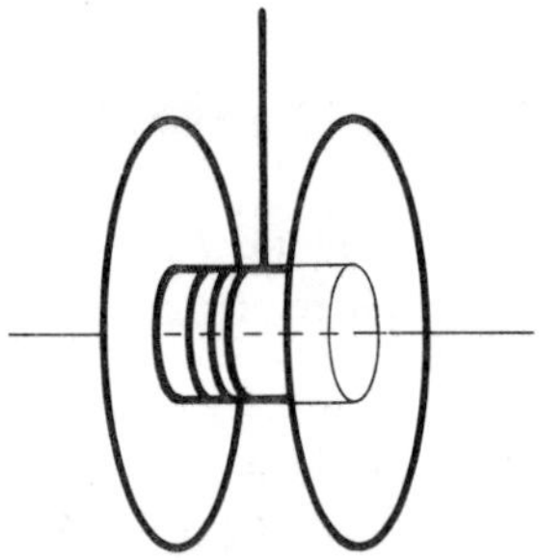

MULTIPLE CHOICE EXERCISE 13

(Instructions for answering these questions are given on page xi)

TYPE II

1) A constant torque C acts for a time t on a lamina rotating in a horizontal plane about a smooth vertical axis. If the moment of inertia of the lamina about that axis is I and the angular velocity changes from ω_1 to ω_2 in the time t,
(a) the lamina has a constant angular acceleration,
(b) the work done by the torque is equal to $I(\omega_2{}^2 - \omega_1{}^2)$,
(c) the impulse of the torque is Ct,
(d) $Ct = I(\omega_2 - \omega_1)$.

2) A rod AB is free to rotate in a vertical plane about a horizontal axis through A. It is slightly disturbed from rest in its position of unstable equilibrium and when it is next vertical the end B collides with a fixed peg and rebounds. If the rod comes to instantaneous rest when AB is horizontal,
(a) the total mechanical energy of the rod is constant throughout,
(b) the coefficient of restitution between the rod and the peg is $\frac{1}{2}$,
(c) the angular momentum of the rod is constant except for a sudden change at the instant of impact with the peg,
(d) the sudden change in the angular momentum of the rod at the instant of impact is equal to the moment of the impulse at the peg.

3) A uniform solid cylinder rolls without slipping through a distance d down an inclined plane.
(a) The angular acceleration is constant.
(b) The work done by the frictional force F is Fd.
(c) Friction is limiting.
(d) The mechanical energy of the cylinder is constant.

4) A uniform rod AB of mass m and length $2l$ is rotating freely, on a smooth horizontal table, about a vertical axis through A. A particle P of mass m collides normally with the mid-point of the rod, when moving with speed v, and as a result of the impact is brought to rest. Therefore:

(a) The rod is also brought to rest.
(b) The rod exerts an impulse of magnitude mv on the particle.
(c) The collision is inelastic.
(d) The total moment of momentum is not changed by the impact.

TYPE III

5) (a) A sphere is rolling without slipping on a horizontal plane.
(b) A sphere is moving on a rough horizontal plane.

6) (a) The impulse of a torque C is Ct.
(b) A constant torque C acts for a time t.

7) (a) There is no sudden change in the total moment of momentum of a rotating rod and a moving particle when they collide.
(b) There is a perfectly elastic collision between a rotating rod and a moving particle.

TYPE IV

8) A torque C acts on a uniform circular disc which can rotate about a diameter. If the disc starts from rest find its angular velocity after t seconds:
(a) the radius of the disc is r,
(b) the mass of the disc is m,
(c) no other force exerts a torque on the disc,
(d) the torque is constant.

9) The end B of a rod AB, hanging vertically from a fixed horizontal axis through A, is struck a blow. If the rod first comes to instantaneous rest after turning through an angle $2\pi/3$ find the impulse of the blow given that:
(a) the length of the rod is $2l$,
(b) the rod is uniform,
(c) the axis is smooth,
(d) the mass of the rod is m.

10) A ring rolls down an inclined plane. Find the least value of the coefficient of friction between the ring and the plane if:
(a) the plane is inclined at 30° to the horizontal,
(b) the mass of the ring is m,
(c) the ring does not slip on the plane,
(d) friction is limiting.

11) A particle is placed gently on a rotating turntable and does not slip. Find the angular velocity of the turntable immediately afterwards.
(a) The mass of the particle is m.
(b) The turntable is a uniform disc of mass $4m$ and radius r.

(c) The angular velocity of the turntable just before the addition of the particle is ω.
(d) Friction between the particle and the turntable is limiting.

TYPE V

12) The impulse of a torque C acting on a body for a time t is given by $\int C\,dt$ only if no other torque acts on the body.

13) A body cannot roll on a smooth surface.

14) If, throughout a specified time interval, the moment of momentum of a system is constant, the total mechanical energy of the system must also be constant during the same interval.

15) A sphere of mass $5m$ and radius a is moving, on a horizontal plane, with angular velocity ω about a horizontal diameter, and its centre of mass has a linear speed v. The kinetic energy of the sphere is $\frac{5}{2}mv^2 + ma^2\omega^2$.

MISCELLANEOUS EXERCISE 13

1) Show that the moment of inertia of a uniform circular disc of radius a and mass m about an axis through its centre perpendicular to the disc is $\frac{1}{2}ma^2$. The disc can rotate freely in a horizontal plane about a fixed vertical axis through a point in its circumference. Find the moment of the constant couple required to turn the disc through three complete revolutions in t seconds from rest, and find the moment of momentum of the disc about the axis at the end of this time. (U of L)

2) A uniform circular disc of mass $4m$ and radius $2a$ is rotating with angular velocity ω about a vertical axis through its centre perpendicular to its plane, which is horizontal. Write down expressions for its kinetic energy and its angular momentum about the axis.
The disc is brought to rest by a tangential force of magnitude $4mg$ applied at its rim. Find the time taken and the angle the disc turns through in that time.
When this disc is rotating as before with constant angular velocity ω, a small uniform disc of mass m and radius a mounted on the same axis is allowed to fall gently upon it from rest. Find their common angular velocity when slipping has finished. (U of L)

3) A flywheel, of radius a and moment of inertia $4Ma^2$ about its axis, is free to rotate about its axis which is horizontal and fixed. A light inextensible string has one end attached to the curved surface of the wheel and is wrapped round this surface and carries a mass M at its free end. This mass is released from rest and falls freely a distance h until the string becomes taut. The mass then continues in the same line and the wheel starts to rotate. The rotation of the

flywheel is opposed by a constant resisting torque of magnitude G and the system comes to rest after the mass M has fallen a further distance h. Find the value of G and the impulsive tension in the string when it becomes taut. (AEB)

4) A uniform rod AB of mass m and length $2a$ is suspended from a smooth fixed pivot at its end A. The rod is hanging at rest when a particle of mass m, moving horizontally with speed u, collides with it at a point at a distance x below A. The particle adheres to the rod.
Prove that immediately after the impact the rod starts to move with angular velocity $3ux/(3x^2 + 4a^2)$. Find the value of x for which the angular velocity is a maximum and find, with this value for x, the value of u for which the rod just becomes horizontal in the subsequent motion. (JMB)

5) Prove that the moment of inertia of a uniform solid sphere of radius a and mass M about a diameter is $\frac{2}{5}Ma^2$.
When the sphere is rotating freely about a fixed horizontal diameter with angular speed ω a stationary particle of mass m adheres to the lowest point of the sphere. Find the angular speed of the sphere immediately after picking up the particle. If the sphere subsequently comes to instantaneous rest, find the angle that the radius to the particle makes with the horizontal at this instant. Hence show that the sphere will make complete revolutions about its axis if

$$\omega^2 > \frac{5m(2M + 5m)g}{aM^2}$$

(U of L)

6) A compound pendulum of mass M oscillates about a smooth fixed horizontal axis. Its moment of inertia about an axis through its centre of gravity G, parallel to the fixed axis, is Mk^2. The centre of gravity moves in a vertical plane which cuts the fixed axis at O, and OG $= a$. Show that for small oscillations the period of the pendulum is the same as that of a simple pendulum of length l, and find l in terms of k and a.
If the pendulum is held with OG horizontal and is then released, find its angular velocity when it reaches the position where OG is vertical.
At this instant a stationary particle of mass M adheres to the pendulum at G. Find the angular velocity of the combined body immediately afterwards. Subsequently it first comes to rest with OG making an angle ϕ with the vertical. Show that

$$\cos\phi = \frac{k^2 + 3a^2}{2k^2 + 4a^2}$$

(Cambridge)

7) A uniform square lamina has mass M and side $2a$. Prove that the moment of inertia about a side is $4Ma^2/3$.
The lamina is smoothly hinged along a horizontal side and hangs vertically; a bullet of mass m, moving horizontally with speed v, strikes the lamina perpendicularly at the mid-point of the lower horizontal side and adheres to the lamina.

Show that the whole starts to rotate about the hinge with an angular speed ω given by

$$a\omega = \frac{3mv}{2(M+3m)}$$

Find, in terms of M, m, g and a, the least value of v^2 which would cause the whole to make complete revolutions. (JMB)

8) A uniform rod AB, of mass m and length $2a$, is freely pivoted to a fixed point at A and is initially hanging in equilibrium. A particle of mass m, moving horizontally with speed u, strikes the rod at its middle point and rebounds from it. If there is no loss of energy at the impact, show that immediately afterwards the speed of the particle is $u/7$, and find the angular velocity with which the rod begins to rotate. Deduce that, if $u^2 < 49ag/12$, the rod will come to rest at an inclination θ to the downward vertical at A, where

$$\cos\theta = 1 - \frac{24}{49}\frac{u^2}{ag}$$

(JMB)

9) A composite body consists of a sphere of mass m and radius a together with a rod OP of mass m and length $4a$. The end P of the rod is fixed to a point on the circumference of the sphere in such a way that OP produced would pass through the centre of the sphere. Calculate the moment of inertia of the body about an axis through O perpendicular to OP.
The body is free to rotate about a smooth horizontal axis through O. It is held with OP horizontal and is then released. Show that its angular velocity ω when OP makes an angle θ with the vertical is given by

$$\omega^2 = \frac{210g}{461a}\cos\theta$$

At the instant when OP has reached a vertical position the rod strikes an inelastic stop at B, where $OB = 3a$, and the body is brought to rest. Find the impulse on the stop. (Cambridge)

10) A body consists of two uniform discs, one of mass m and radius a, the second of mass $2m$ and radius a, the centres of which are fixed to the ends A and B respectively of a uniform rod of mass $3m$ and length $5a$. The discs and the rod are coplanar. Show that the moment of inertia of the body about an axis at right angles to the plane of the discs through a point O on the rod such that $AO = 2a$ is $\frac{61}{2}ma^2$.
The body is pivoted on a smooth horizontal axis which is perpendicular to the plane of the discs and which passes through O. Prove that the period of small oscillations of the body about O is the same as that of a simple pendulum of length $\frac{61}{11}a$.

Initially the body is resting in equilibrium with B vertically below O. When struck by a horizontal impulse J through B in the plane of the discs it just reaches the position where AB is horizontal. Find J. (Cambridge)

11) Define angular momentum for a rigid body rotating about a fixed axis.
A uniform rod, of mass m and length $2a$, is freely hinged at a fixed point at one of its ends and hangs freely in its position of stable equilibrium. The rod is suddenly made to rotate with initial angular velocity $\sqrt{(3g/a)}$ as a result of a horizontal impulse being applied at its free end.
(i) Find the magnitude of the impulse.
(ii) Show that the rod just reaches its position of unstable equilibrium.
(iii) Find the reaction at the hinge when the rod is horizontal. (AEB)

12) State the principle of conservation of angular momentum.

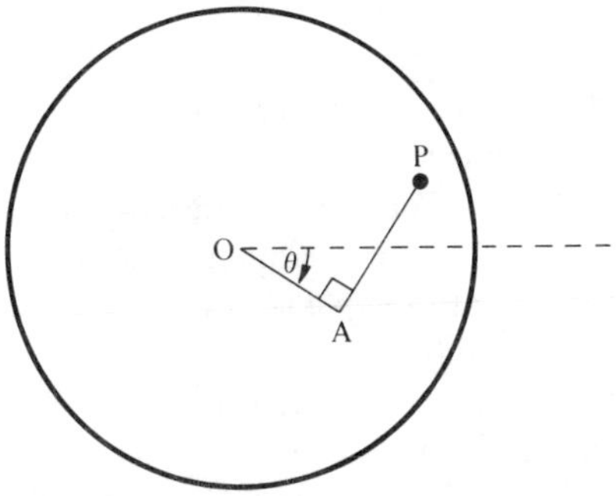

A horizontal turntable can rotate freely about a fixed vertical axis through a point O of its plane about which its moment of inertia is $8ma^2$. An insect P of mass m is at rest on the stationary turntable at a point A, where $OA = a$. The insect then crawls along the line in the turntable perpendicular to OA maintaining a uniform speed V relative to the turntable.
The diagram shows the turntable when it has turned through an angle θ after time t from the start of the motion. Prove that

$$\dot{\theta} = \frac{Va}{9a^2 + V^2t^2}$$

Deduce that the turntable never turns through an angle greater than $\frac{1}{6}\pi$ from its initial position.
By considering the kinetic energy of the turntable and insect, show that the rate of working of the insect at time t is

$$\frac{ma^2V^4t}{(9a^2 + V^2t^2)^2}$$

(Cambridge)

13) A rod of length l is rotating with angular velocity ω on a smooth horizontal table about one end O which is fixed. It hits a particle of mass m which is at rest on the table at a distance x from O. Prove that, if the moment of inertia of the rod about the vertical axis through O is I, if the coefficient of restitution is e and if the velocity imparted to the particle is v, then

$$I\omega(1+e)/v = I/x + mx$$

Hence show that, if $I < ml^2$, the greatest value of v for varying values of x is $\frac{1}{2}\omega(1+e)(I/m)^{1/2}$. (Oxford)

14) A uniform rod AB of mass m and length $2a$ has a particle of mass m attached at the end B. It is free to rotate about a fixed smooth horizontal axis through A. Find the period of small oscillations about the axis.
When the rod hangs vertically at rest a horizontal impulse J at right angles to the axis is applied at B. Show that the rod makes a complete rotation if $J > 4m\sqrt{(ga)}$. (U of L)

15) A compound pendulum consists of a uniform circular disc of radius a and mass $4M$, the centre of which is fixed to the end B of a uniform rod AB of mass $3M$ and length $4l$ $(2l > a)$. The rod and the disc may be considered to be coplanar. The axis of rotation of the pendulum is horizontal and passes through A in a direction perpendicular to the plane of the disc and rod. Find the moment of inertia of the pendulum about this axis.
The pendulum is held in a position such that B is vertically above A and is then slightly displaced from rest. Show that ω, the angular velocity of the pendulum just before the mid-point of AB hits a stop S situated at the same horizontal level as A, is given by

$$\omega^2 = 22gl/(a^2 + 40l^2)$$

If the elasticity of the stop S is such that the pendulum rebounds with an angular velocity $\frac{1}{2}\omega$, find the impulse on the stop in terms of M, g, a and l.
Show that the angle turned through by the pendulum between hitting the stop and next coming to rest is $\arcsin \frac{1}{4}$. (Cambridge)

16) A compound pendulum consists of two circular discs each of mass m and radius a, the centres of which are fixed to the ends A and B of a rod of mass m and length $5a$. The discs and the rod may be taken to be in one plane. Calculate the moment of inertia of the pendulum about an axis at right angles to the discs through a point O on the rod such that $AO = 2a$.
The pendulum is pivoted on a smooth horizontal axis which is perpendicular to the plane of the discs and passes through O. The pendulum is initially at rest with O vertically below A. A particle of mass m is now projected in the plane of the discs in such a manner that it strikes the rod at a point P below O, where $OP = a$. At the instant of its impact with the rod the particle is moving horizontally with speed v and after the impact it adheres to the rod. Show that the initial angular velocity of the pendulum is $3v/(52a)$. (Cambridge)

17) A uniform rod of mass M and length $2l$ is free to rotate in a vertical plane about a fixed horizontal axis through one end of the rod. When hanging in equilibrium, the rod receives an impulse of magnitude $2M\sqrt{(gl/3)}$ applied to its lower end in the plane of rotation and in a direction making an angle $\pi/3$ with the horizontal. Prove that when next at rest the rod makes an angle $\pi/3$ with the vertical. Prove also that, at this instant, the component perpendicular to the rod of the reaction at the axis is $\sqrt{3}Mg/8$. (U of L)

18) Two circular discs, each of radius $2r$, are attached to the ends of an axle of radius r so that the axis of the axle is perpendicular to the discs and passes through their centres. The discs stand on a horizontal table. A light inextensible string has one end attached to the curved surface of the axle and is wound several times round the axle. The string leaves the axle in a direction perpendicular to the axis of the axle and parallel to the table, passes over a smooth peg and carries a mass $2M$ at its free end. The mass of the combined discs and axle is M and their moment of inertia about the axis is Mr^2. On being released the discs roll without slipping on the table. Find the acceleration of the falling mass, the tension in the string and the least possible value of the coefficient of friction between the discs and the table.

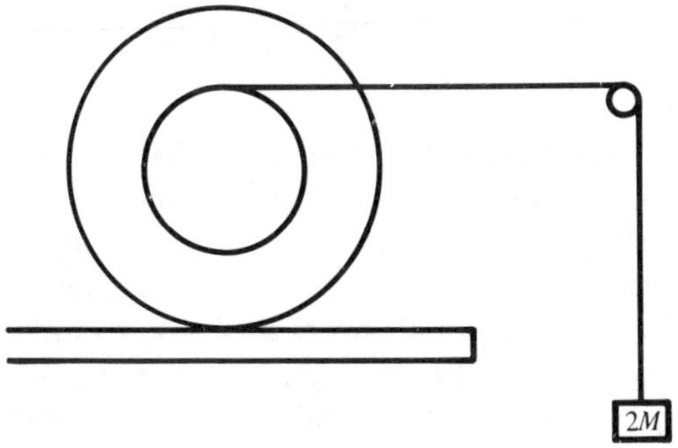

(AEB)

19) A composite body consists of three parts: (i) a uniform rod OA of mass m and length $2a$, (ii) a uniform rod AB of mass $2m$ and length $2a$, fixed to OA so that OAB is a straight line (iii) a particle of mass $3m$ attached at B. Find the moment of inertia of the body about an axis through O perpendicular to OAB. When the body is smoothly pivoted on a horizontal axis through O and is hanging in equilibrium it is struck at B with a horizontal impulse P at right angles to the plane containing OAB and the axis. Show that the angular velocity with which the body starts to move is $P/(17ma)$.
Show that when OAB makes an angle θ with the downward vertical through O, the angular velocity ω of the body is given by

$$\omega^2 = \frac{P^2}{289m^2a^2} - \frac{19g}{34a}(1-\cos\theta)$$

Check the correctness of the dimensions of this equation.
Find the smallest value of P if the body is to perform complete revolutions.

(Cambridge)

20) Two uniform circular cylinders, of masses m_1, m_2 and radii a_1, a_2 respectively, are free to rotate about their axes, which are fixed, horizontal and parallel. A light inextensible string hangs in a loop between the cylinders; each cylinder has an end portion of the string wrapped round it, the end of the string being fastened to the cylinder. The loop passes under a smooth light pulley from which hangs a particle of mass m. The whole of the string is in a vertical plane perpendicular to the axes of the cylinders, and the straight lengths between the pulley and the cylinders are vertical. Show that when the angular velocities of the cylinders are ω_1 and ω_2 respectively, the positive sense of rotation in each case being that which causes more string to be wound on to the cylinder, the upward velocity of the particle is

$$\tfrac{1}{2}(a_1\omega_1 + a_2\omega_2)$$

(i) Find the tension in the string when the system is moving freely.
(ii) Show that if the particle is suddenly fixed when $a_1\omega_1 + a_2\omega_2 > 0$, the pulley being left free to rotate, the resulting impulsive tension in the string is

$$\frac{1}{2}\frac{m_1 m_2}{m_1 + m_2}(a_1\omega_1 + a_2\omega_2)$$

(JMB)

21) A particle P of mass m rests on a rough horizontal table, the coefficient of friction being μ, and is connected to a particle Q of mass M by a light inextensible string which passes over a circular pulley of radius a and moment of inertia I about its axis. The string from P to the pulley is horizontal, and from the pulley to Q is vertical. The pulley can rotate freely about its axis which is fixed and horizontal. The system is released from rest. Assuming that the system moves and that the string does not slip on the pulley, prove that the magnitude of the acceleration of each particle is

$$\frac{(M - \mu m)g}{M + m + (I/a^2)}$$

and find the tension in the vertical portion of the string.
When the speed of the particles is V the particle P is seized and held fixed, causing Q and the pulley to stop also. Find the impulse which must be applied to P to achieve this. (JMB)

22) A uniform solid cylinder rolls without slipping down a rough plane inclined at an angle α to the horizontal. Find the acceleration of the centre of gravity of the cylinder and the least possible value of the coefficient of friction between the cylinder and the plane.
A uniform solid sphere rolls without slipping down a rough plane inclined at an angle β to the horizontal. The acceleration of the centre of the sphere is the same

as the acceleration of the centre of gravity of the cylinder. Show that $14 \sin \alpha = 15 \sin \beta$. (AEB)

23) A uniform circular disc of mass m and radius a can rotate freely about a fixed horizontal axis which passes through a point on its circumference and is perpendicular to its plane. The disc is hanging in equilibrium when it is struck by an impulse P acting in a horizontal line in the plane of the disc at a distance x below its centre. Prove that the disc starts to rotate with angular velocity $\omega = 2(a+x)P/(3ma^2)$. Find in terms of a, x and P the horizontal component, R, of the impulsive reaction of the axis on the disc and deduce the value of ω when x is chosen so that $R = 0$.
Prove that with this value of x and with $P = \frac{2}{3}m\sqrt{(2ga)}$ the diameter of the disc which is initially vertical turns through an angle arccos $(\frac{1}{3})$ before coming to instantaneous rest. (JMB)

24) A rod of mass M can swing freely in a vertical plane about one end A. The distance of the centre of gravity of the rod (not necessarily uniform) from A is h, and the radius of gyration about A is K.
While the rod is hanging in equilibrium, it is struck by a horizontal impulse of magnitude P at a point X at a depth x below A. (i) Find the angular velocity with which the rod begins to move. (ii) Prove that the kinetic energy imparted to the rod is $\frac{1}{2}Pv$, where v is the velocity with which the point X begins to move. (iii) If the rod just reaches the upward vertical, prove that $Px = 2MK\sqrt{(gh)}$. (JMB)

25) A uniform sphere is released from rest on a rough plane inclined at an angle α to the horizontal. The coefficient of friction is μ. Prove that, if $\mu > \frac{2}{7} \tan \alpha$, the sphere will roll down the plane and its centre will have an acceleration $\frac{5}{7}g \sin \alpha$ down the plane.
Prove also that, if $\mu < \frac{2}{7} \tan \alpha$, the acceleration of the centre of the sphere down the plane is greater than $\frac{5}{7}g \sin \alpha$. (Oxford)

26) A closed container in the form of a right circular cylinder, of radius a and height $4a$, is made from thin uniform sheet metal. If the mass of the container is M, show that its moment of inertia about its axis is $\frac{9}{10}Ma^2$.
This container rolls without slipping down a plane inclined at an angle α to the horizontal. Calculate the angular speed of the container when its axis has travelled, from rest, a distance $10a$ down the plane. (AEB)

27) Show that the moment of inertia of a uniform solid cylinder of mass M and radius r about its axis is $\frac{1}{2}Mr^2$.
State the moment of inertia of a thin cylindrical ring, also of mass M and radius r, about its axis.
The solid cylinder and the ring roll from rest, without slipping, down lines of greatest slope of a plane inclined at α to the horizontal. Show that the solid cylinder covers a distance $\frac{1}{12}gt^2 \sin \alpha$ further than the ring in time t.

Find the minimum value of the coefficient of friction between the plane and the rolling bodies. (AEB)

28) A thin hollow uniform spherical shell has radius a. Prove that its moment of inertia about a diameter is $\frac{2}{3}Ma^2$, where M is the mass of the shell.
The shell rolls from rest, without slipping, down a rough plane inclined at an angle arcsin $\frac{1}{12}$ to the horizontal. Find its angular velocity when its centre has travelled through a distance $5a$, and the time taken to travel this distance. (U of L)

29) Prove that the moment of inertia of a uniform solid sphere, of mass M and radius r, about a diameter is $(2Mr^2)/5$.
The sphere is placed in contact with a rough plane inclined at 30° to the horizontal. On being released the sphere rolls, without slipping, down a line of greatest slope of the plane. Show that the centre of the sphere has constant acceleration and that it acquires a speed of 7 metres per second after rolling 7 metres down the plane. What is the smallest possible value of the coefficient of friction between the sphere and the plane? (AEB)

30) A uniform sphere of radius a is standing at rest on the rough upper surface of a flat board which is lying on a horizontal table. The board is now moved along the table with a constant acceleration f. Prove that, if the coefficient of friction μ between the sphere and the board is greater than $\frac{2}{7}(f/g)$, the sphere rolls on the board with angular acceleration $\frac{5}{7}(f/a)$.
Find the angular acceleration of the sphere if $\mu < \frac{2}{7}(f/g)$. (U of L)

31) A cylinder of mass M and radius a rolls, without slipping and with its axis horizontal, down a plane inclined at an angle α to the horizontal. In two successive seconds it is observed to make two complete revolutions and three complete revolutions. Find:
(i) the angular acceleration of the cylinder,
(ii) the moment of inertia of the cylinder about its axis,
(iii) the minimum value of the coefficient of friction between the cylinder and the plane. (AEB)

32) A uniform rod AB of length l and mass m hangs from a fixed point A at which it is freely hinged. From the end B hangs a uniform disc of radius a and mass M, the rod being freely hinged to the disc at a point B of its circular edge. When the system hangs in equilibrium an impulse I, horizontal and in the same vertical plane as the disc, is given to the rod at a point a distance x below A. Find the angular velocities imparted to the rod and the disc. Prove that, if the impulsive reaction at the fixed end A is zero, then $x = 2l(M+m)/(2M+3m)$. (Oxford)

CHAPTER 14

SHEARING FORCE AND BENDING MOMENT

INTERNAL STRESSES IN A RIGID BODY

Hitherto in our study of the equilibrium of rigid bodies we have confined ourselves to the effect of external forces. We must now give some consideration to the internal forces which act between adjacent parts of a rigid body. A beam, or girder, is a particularly important body to study in this way because of its many uses in the construction industry.

In general the forces acting on a beam give rise to stresses within the beam which can cause it to stretch, break, bend or twist. To deal fully with these effects requires complex analysis beyond the scope of this book. Consequently we will confine ourselves to the study of a horizontal girder to which coplanar vertical forces only are applied and which is rigid enough for the distortion that those forces produce to be ignored.

SHEARING FORCE AND BENDING MOMENT

Consider first a uniform light rigid girder AB which rests horizontally on two supports, one at each end. A concentrated load of weight $2W$ is carried at the mid-point C.

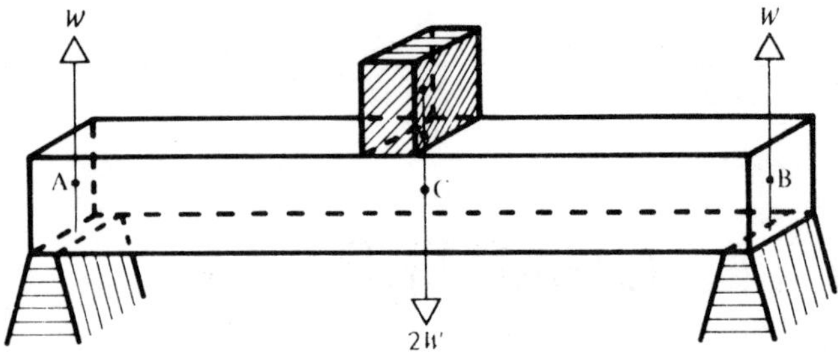

If we were to cut vertically through the girder at a point P and remove the part PB, then the section AP would collapse.
To prevent this from happening an upward force S and an anticlockwise torque $\mathfrak{M}$ have to be applied at P so that the equilibrium of the section AP is restored.

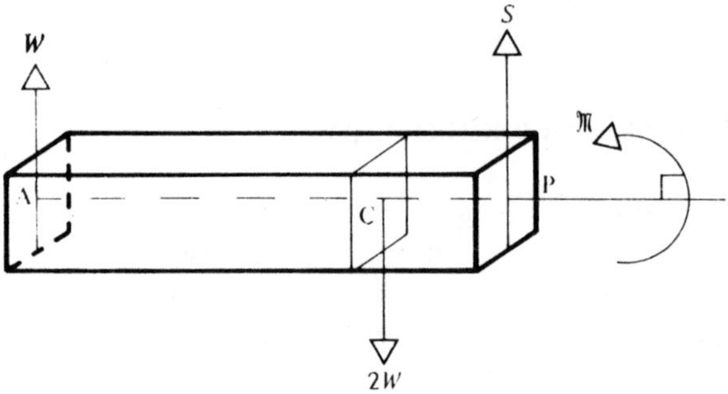

We therefore conclude that, because the girder was in equilibrium before it was divided at P, the section PB must have been responsible for exerting on the section AP a force equal to S and a torque equal to $\mathfrak{M}$.
But internal forces act in equal and opposite pairs so the section AP must also have exerted on the section PB a force equal and opposite to S and a torque equal and opposite to $\mathfrak{M}$. (See diagram.)

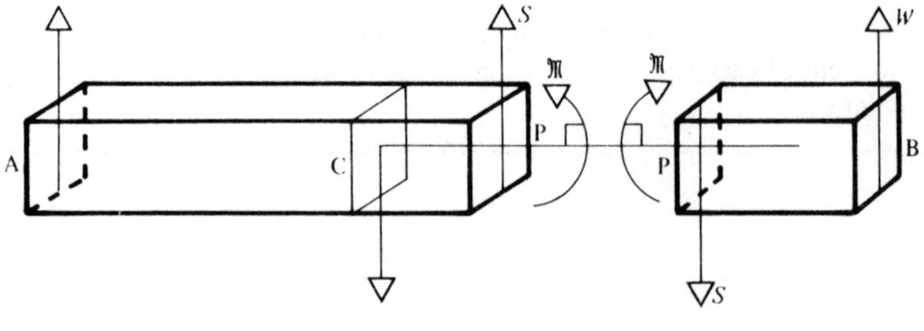

The force S is called the *shearing force* at P. The pair of equal and opposite shearing forces at that point tends to make the beam break across its vertical cross-section at P (i.e. to *shear*).

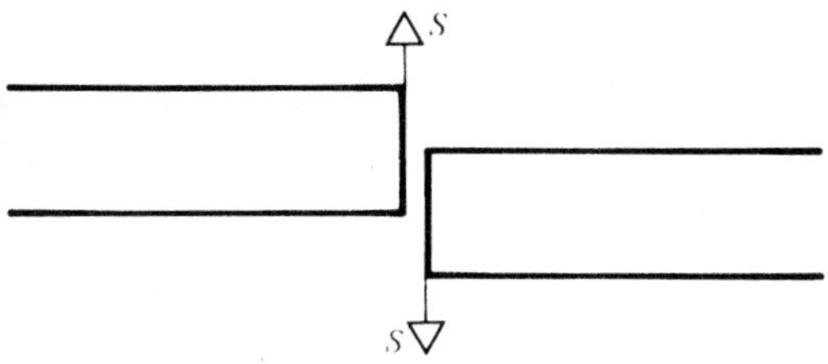

The torque $\mathfrak{M}$ is called the *bending moment* at P. Its effect on the beam is to produce a tendency to bend in a vertical plane.

SIGN CONVENTION

The sense in which S and $\mathfrak{M}$ are taken to be positive is a purely arbitrary choice. In this book, when considering the left-hand portion of the beam the upward direction is taken as positive for the shearing force while the positive sense for bending moment is anticlockwise.

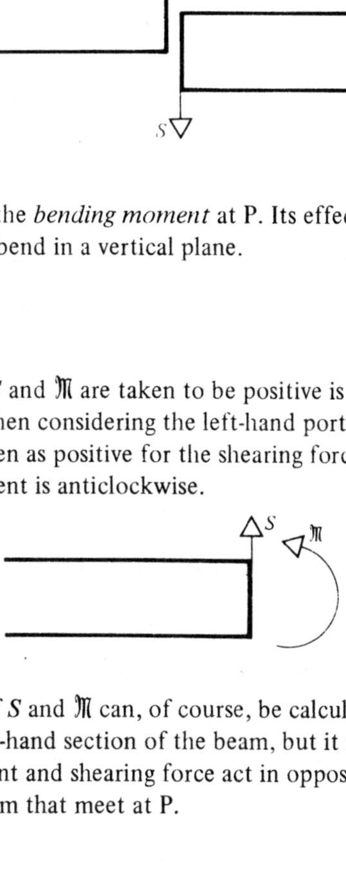

The numerical values of S and $\mathfrak{M}$ can, of course, be calculated by considering the equilibrium of the right-hand section of the beam, but it must be remembered that the bending moment and shearing force act in opposite directions on the two portions of the beam that meet at P.

CONCENTRATED AND DISTRIBUTED FORCES

The effect of some of the forces which act on beams is concentrated at a point (or on an area which is so small compared with the area of the beam that it can be regarded as a point).

Such a force is called a *concentrated* or *point force*. An example of a force of this type is the reaction exerted on a beam by a knife-edge support.

Other forces are such that their effect is spread out over an appreciable length of the beam; these are called *distributed* forces. The commonest example of this type is the weight of a uniform heavy beam. Although for overall analysis the total weight of the beam can be taken to act through its mid-point, when studying the internal stresses within the beam it must be appreciated that the weight of each element or section of the beam acts through the centre of gravity of that portion.

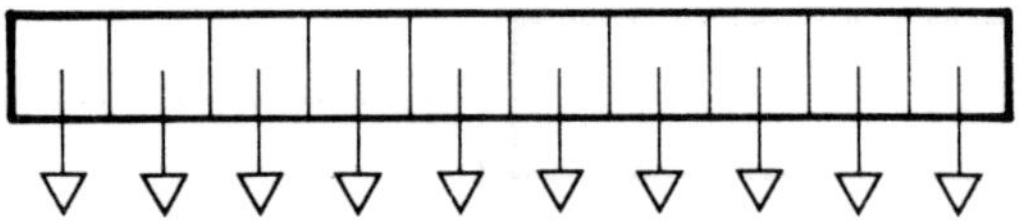

In the diagrams illustrating the examples that follow, concentrated forces will be marked in the usual way but distributed forces will be indicated by shading along the beam and by a broken line marking the overall force, e.g. a uniform heavy beam of weight W will be shown as:

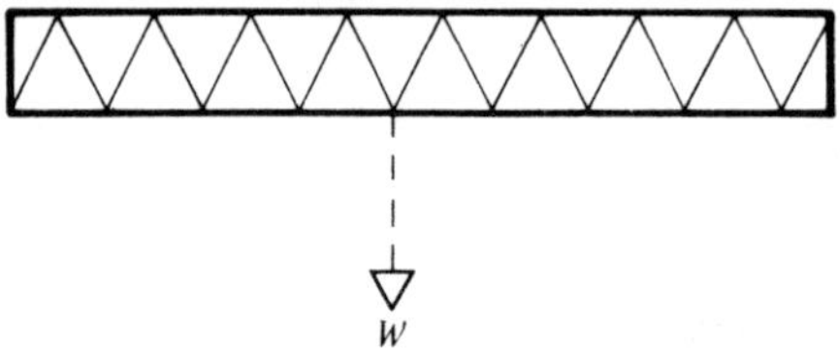

The internal stresses caused by concentrated forces differ somewhat from those caused by distributed forces so we shall consider these two cases separately.

CALCULATION OF SHEARING FORCE AND BENDING MOMENT

CASE 1. LIGHT BEAM WITH CONCENTRATED LOAD

Consider a light beam AB of length $2a$, resting horizontally on two end supports and carrying a concentrated load $2W$ at the mid-point C.
From symmetry we see that each end support exerts an upward force W on the beam.

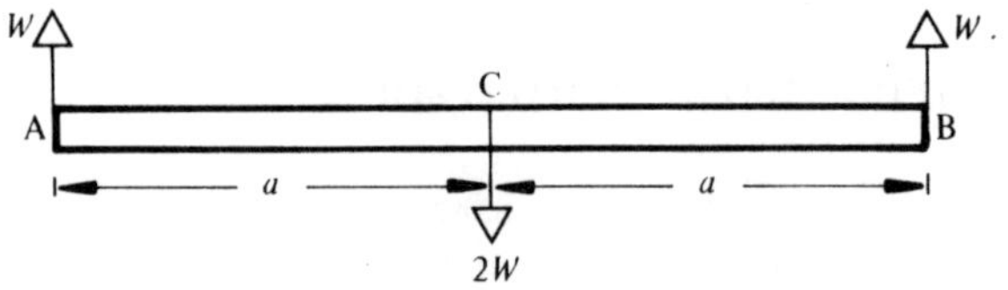

Let P be any point on the beam such that AP $= x$. Then, if S and $\mathfrak{M}$ are the shearing force and bending moment at P, we have:

(i) $0 < x < a$

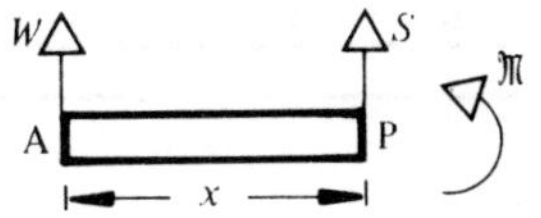

Resolving vertically and taking moments about a horizontal axis through P:

$$\uparrow \qquad W + S = 0$$

$$\text{P}\circlearrowright \qquad \mathfrak{M} - Wx = 0$$

Hence $S = -W$ and $\mathfrak{M} = Wx$

(ii) $a < x < 2a$

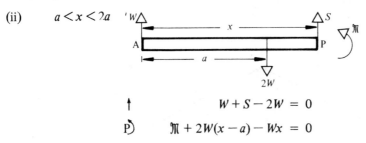

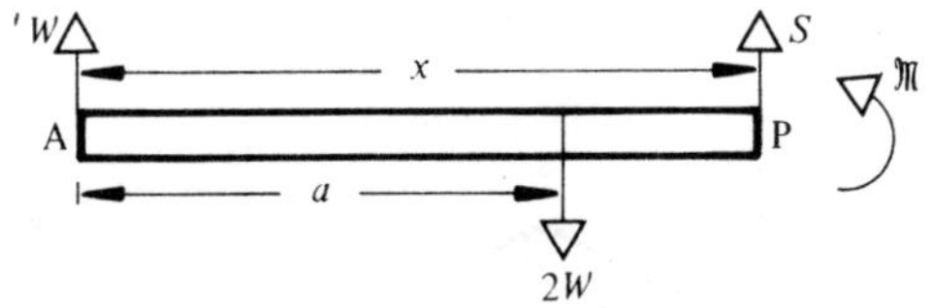

$$\uparrow \qquad W + S - 2W = 0$$

$$\text{P}\circlearrowright \qquad \mathfrak{M} + 2W(x-a) - Wx = 0$$

Hence $S = W$ and $\mathfrak{M} = W(2a - x)$
From these results we see that the expressions for S and $\mathfrak{M}$ are *different* for AC and CB.

Between A and C $\begin{cases} S \text{ is constant} \quad (S = -W) \\ \mathfrak{M} \text{ increases uniformly with } x \end{cases}$

Between C and B $\begin{cases} S \text{ is constant} \quad (S = +W) \\ \mathfrak{M} \text{ decreases uniformly with } x \end{cases}$

At C there is a sudden change in the value of S from $-W$ to W. The magnitude of this sudden change (discontinuity) is equal to the magnitude of the concentrated load at C.
At A and B, $\mathfrak{M} = 0$.
Both for $0 < x < a$ and $a < x < 2a$ (but *not* when $x = a$ where there is a discontinuity) $\dfrac{d\mathfrak{M}}{dx} = -S$.
At C there is a sudden change in the gradient of $\mathfrak{M}$.
These properties can be illustrated graphically as follows:

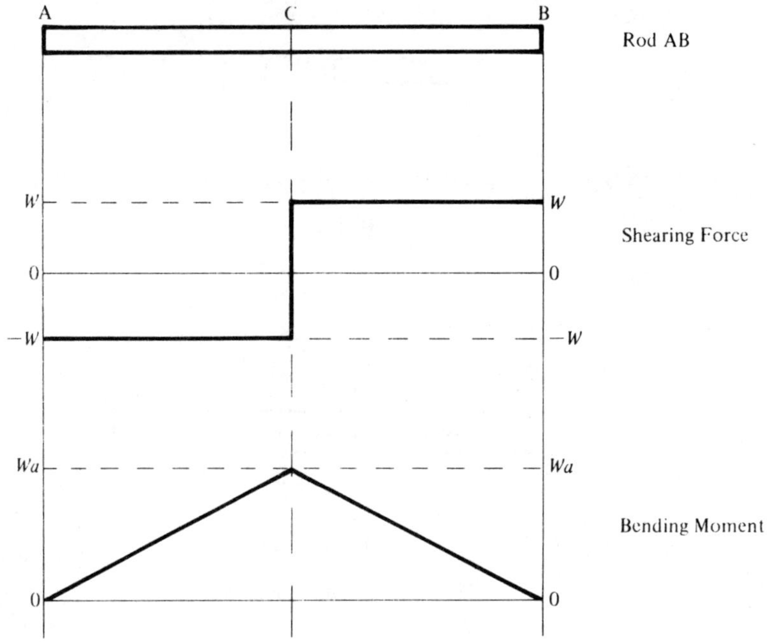

CASE 2. UNIFORM HEAVY BEAM

Again we shall consider a beam AB of length $2a$ resting horizontally on two end supports but this time the weight of the beam, $2W$, is distributed uniformly along the length of the beam. The weight per unit length of the beam is therefore $\frac{W}{a}$.

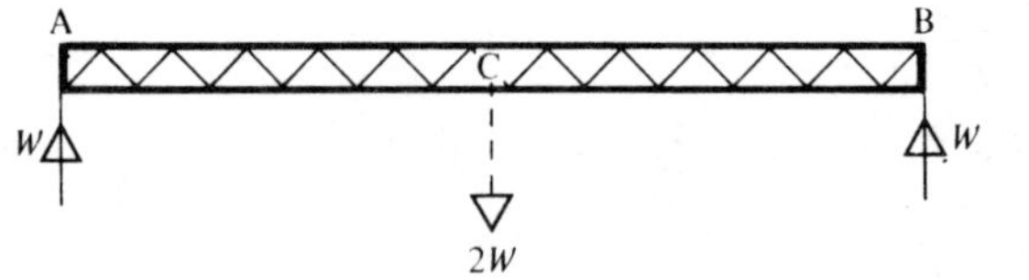

If P is any point such that AP $= x$ then, considering the equilibrium of the section AP, we have:

(i) $0 < x < a$

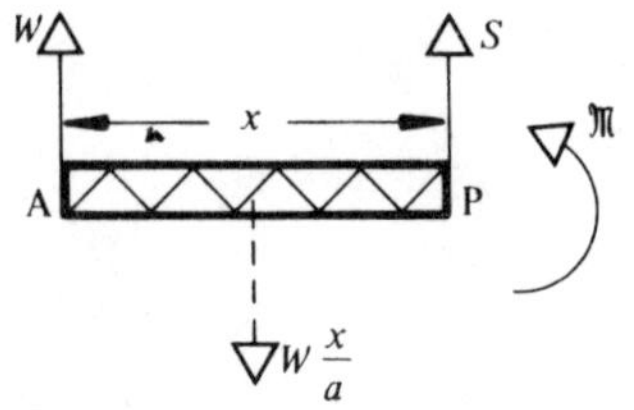

$$\uparrow \qquad W + S - W\frac{x}{a} = 0$$

$$\overset{\curvearrowright}{P} \qquad \mathfrak{M} + W\frac{x}{a}\left(\frac{x}{2}\right) - Wx = 0$$

Hence $\quad S = \dfrac{W}{a}(x-a) \quad$ and $\quad \mathfrak{M} = \dfrac{Wx}{2a}(2a-x)$

(ii) $a < x < 2a$

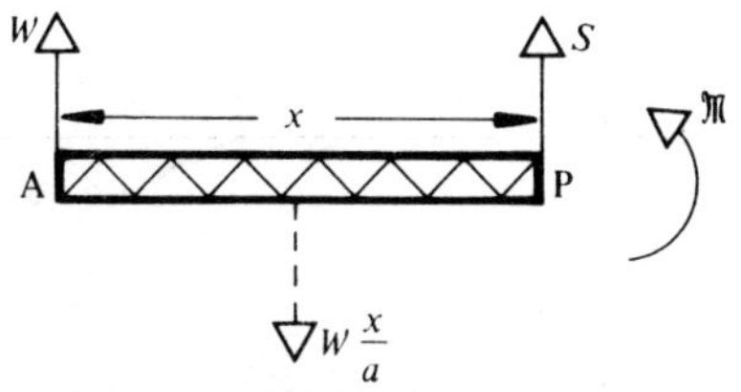

$$\uparrow \qquad W + S - W\frac{x}{a} = 0$$

$$\overset{\curvearrowright}{P} \qquad \mathfrak{M} + W\frac{x}{a}\left(\frac{x}{2}\right) - Wx = 0$$

So again $\quad S = \dfrac{W}{a}(x-a) \quad$ and $\quad \mathfrak{M} = \dfrac{Wx}{2a}(2a-x)$

This time the characteristic properties are:

The expressions for S and $\mathfrak{M}$ are *the same* for AC and CB so it was not necessary to consider two separate sections.
S increases uniformly with x throughout.
$\mathfrak{M}$ is a quadratic function of x throughout.
At A, $S = -W$ and $\mathfrak{M} = 0$.
At B, $S = W$ and $\mathfrak{M} = 0$.
At C, $S = 0$.
For all values of x, $\dfrac{d\mathfrak{M}}{dx} = -S$.
There is no discontinuity in the value of S or the gradient of $\mathfrak{M}$.
The graphical illustrations of these properties are shown below.

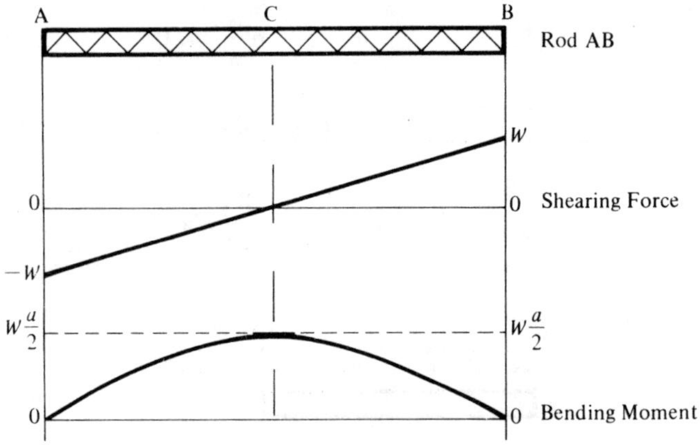

RELATIONSHIP BETWEEN BENDING MOMENT AND SHEARING FORCE

In the two particular cases we have so far studied, observation of the functions derived for $\mathfrak{M}$ and S shows that $\dfrac{d\mathfrak{M}}{dx} = -S$.
This, in fact, is a property of all uniform beams and it can be proved by considering a small section of a heavy beam whose weight per unit length is w.
If, at a point P where AP $= x$, the shearing force and bending moment are S and $\mathfrak{M}$, then at an adjacent point Q, where AQ $= x + \delta x$, the shearing force and bending moment are $S + \delta S$ and $\mathfrak{M} + \delta\mathfrak{M}$. Provided that no concentrated forces act between P and Q the forces which act on the section PQ are as follows:

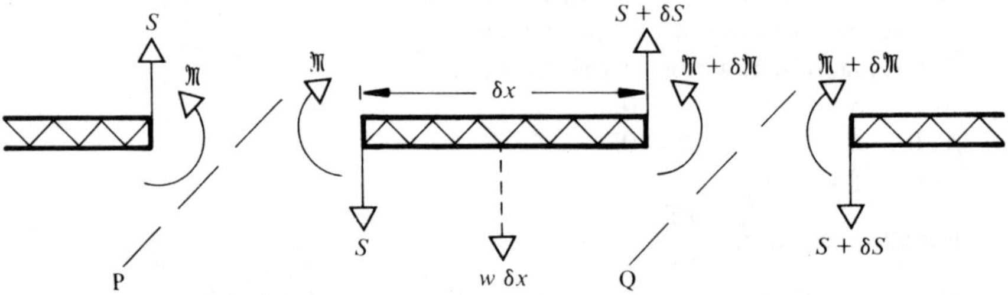

Taking moments about a horizontal axis through Q we have:

$$(\mathfrak{M} + \delta\mathfrak{M}) + S\,\delta x + w\,\delta x\left(\frac{\delta x}{2}\right) - \mathfrak{M} = 0 \qquad (1)$$

i.e.
$$\delta\mathfrak{M} = -S\,\delta x - \frac{w}{2}(\delta x)^2$$

Hence $$\frac{\delta \mathfrak{M}}{\delta x} = -S - \frac{w}{2}\delta x$$

As $\delta x \to 0$ $$\frac{d\mathfrak{M}}{dx} = -S$$

If we now consider a light beam in the same way, equation (1) becomes

$$(\mathfrak{M} + \delta\mathfrak{M}) + S\,\delta x - \mathfrak{M} = 0$$

i.e. $$\delta \mathfrak{M} = -S\,\delta x$$

So again $$\frac{d\mathfrak{M}}{dx} = -S$$

Thus, for all uniform beams, $$\frac{d\mathfrak{M}}{dx} = -S$$

[*Note.* If a different convention is used for the positive direction of $\mathfrak{M}$ and S, this relationship may appear in the form $\frac{d\mathfrak{M}}{dx} = +S$]

Now consider a section of a uniform beam bounded by $x = a$ and $x = b$.

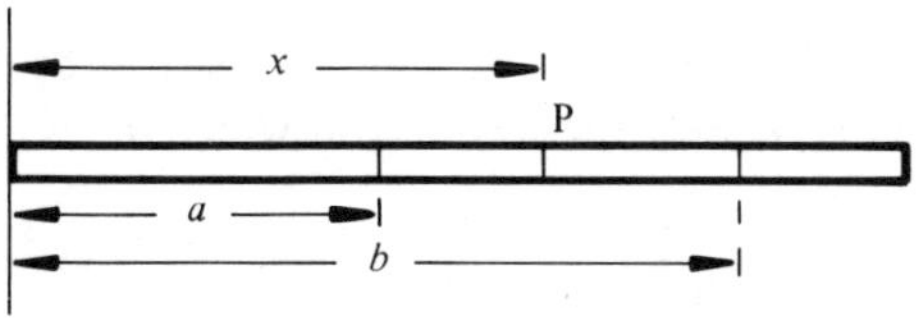

If no concentrated forces are applied within this section then at a general point P,

$$\frac{d\mathfrak{M}}{dx} = -S$$

Hence $$\int_{x=a}^{x=b} d\mathfrak{M} = -\int_{x=a}^{x=b} S\,dx$$

i.e. $$\mathfrak{M}_{x=a} - \mathfrak{M}_{x=b} = \int_a^b S\,dx$$

But $\int_a^b S\,dx$ is the area under the shearing force graph between $x = a$ and $x = b$. *Thus the change in bending moment over a uniform section is equal in magnitude to the corresponding area under the shearing force diagram.*

EXAMPLES 14a

1) A light girder ABCD (AB = BC = CD = a) rests horizontally on supports at A and C and carries concentrated loads $2W$ and W at B and D respectively. Draw

diagrams showing the variation in shearing force and bending moment along the girder and calculate the greatest bending moment.

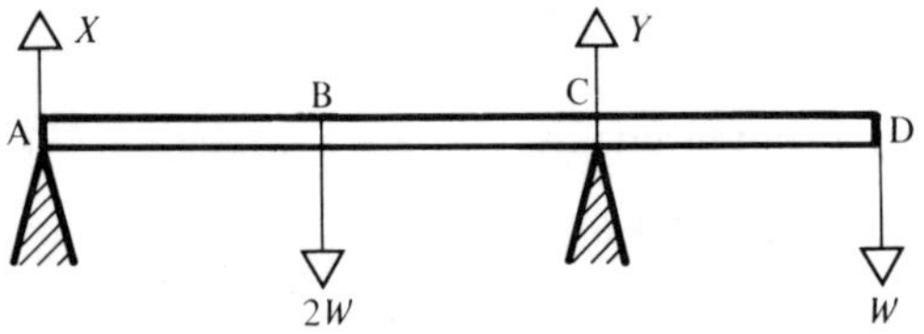

[This girder is not symmetrically supported or loaded so we must first calculate the values of the supporting forces X and Y.]
Considering the equilibrium of the girder under the action of the external forces only, and taking moments about horizontal axes through A and C we have:

$$\text{A)} \qquad 2Wa + 3Wa - 2Ya = 0$$

$$\text{C)} \qquad Wa + 2Xa - 2Wa = 0$$

Hence $\qquad X = W/2 \quad \text{and} \quad Y = 5W/2$

Now consider the shearing force S and the bending moment $\mathfrak{M}$ at a point P on the beam, where $\text{AP} = x$.
[There are concentrated forces (and hence discontinuities) at B and C so it is necessary to consider separately the sections for which $0 < x < a$, $a < x < 2a$ and $2a < x < 3a$.]

(i) $0 < x < a$

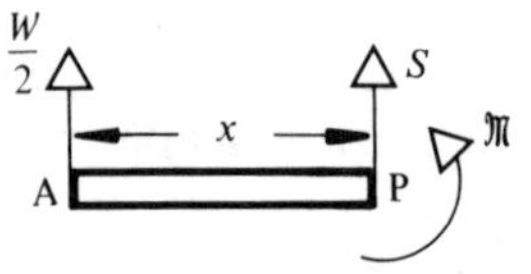

$$\uparrow \qquad S + W/2 = 0$$

$$\text{P)} \qquad \mathfrak{M} - Wx/2 = 0$$

Hence $\quad S = -W/2 \quad \text{and} \quad \mathfrak{M} = Wx/2$

(ii) $a < x < 2a$

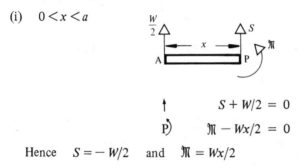

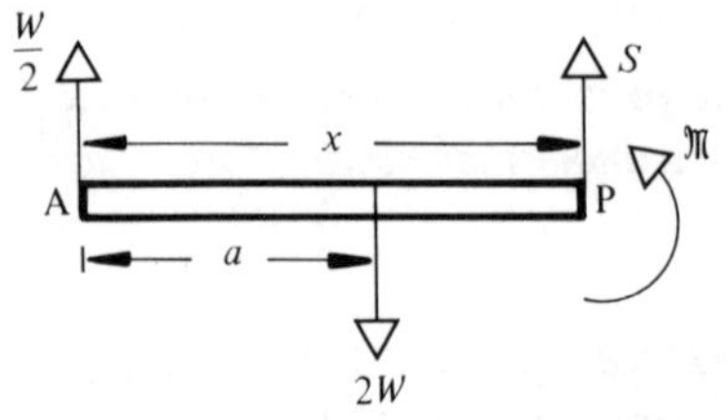

$$\uparrow \qquad S + W/2 - 2W = 0$$

$$\overset{\curvearrowright}{\text{P}} \qquad \mathfrak{M} - Wx/2 + 2W(x-a) = 0$$

Hence $S = 3W/2$ and $\mathfrak{M} = W(4a - 3x)/2$

(iii) $2a < x < 3a$

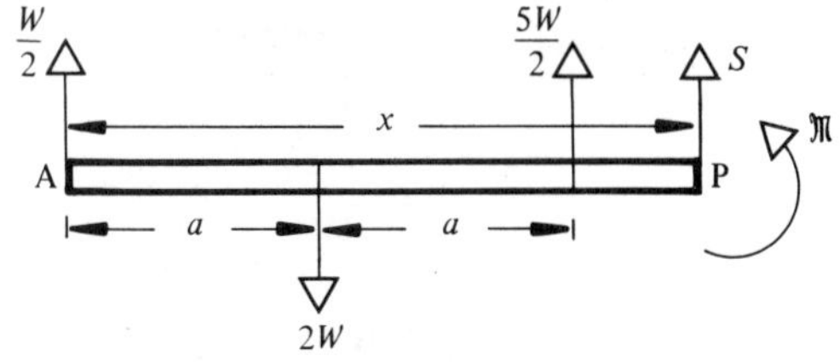

$$\uparrow \qquad W/2 + 5W/2 - 2W + S = 0$$

$$\overset{\curvearrowright}{\text{A}} \qquad \mathfrak{M} + Sx + 5Wa - 2Wa = 0$$

Hence $S = -W$ and $\mathfrak{M} = W(x - 3a)$

The values of $\mathfrak{M}$ at A, B, C and D are 0, $Wa/2$, $-Wa$ and 0 respectively. The shearing force and bending moment diagrams can now be drawn.

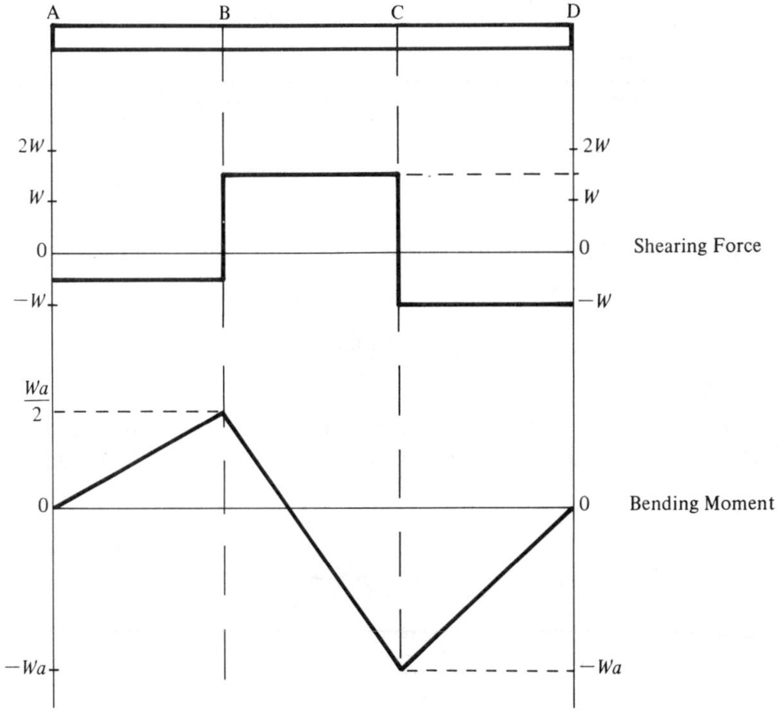

The greatest bending moment (i.e. the greatest *magnitude* of $\mathfrak{M}$ regardless of sign) occurs at C and is of magnitude Wa.

Note: The expressions for S and $\mathfrak{M}$ in the section CD of the beam can, if preferred, be derived as follows:

(iii) $2a < x < 3a$

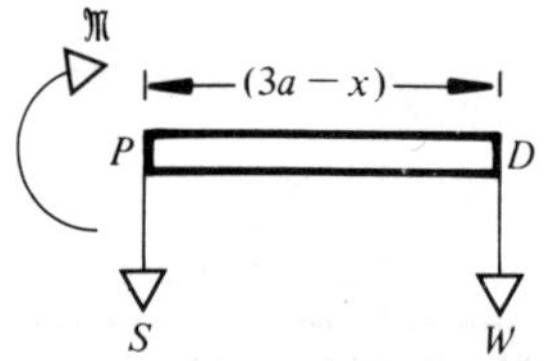

$$\downarrow \qquad S + W = 0$$

$$P\circlearrowright \qquad \mathfrak{M} + W(3a - x) = 0$$

Thus $S = -W$ and $\mathfrak{M} = W(x - 3a)$ as before.

2) A light beam AB of length $2a$ is supported horizontally at A and B. A load of weight $4W$ is uniformly distributed between A and C where AC $= a$. Find expressions for the bending moment and shearing force of the beam, indicating the point at which bending is most likely to occur.

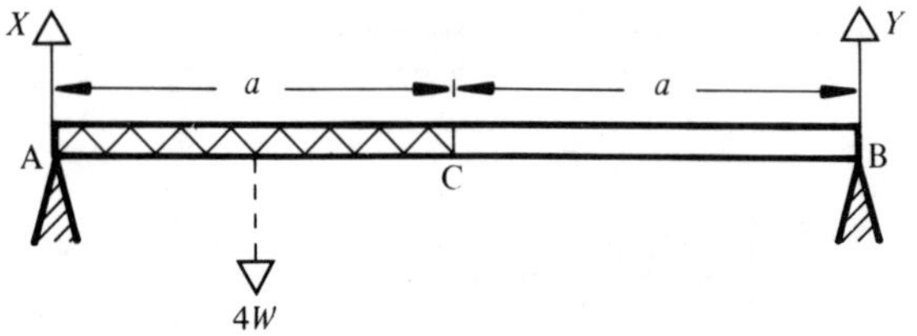

Considering external forces only:

$$A\circlearrowleft \qquad 2aY - 4W(a/2) = 0$$

$$B\circlearrowright \qquad 2aX - 4W(3a/2) = 0$$

Hence $X = 3W$ and $Y = W$

Now using $4W/a$ as the weight per unit length along AC and considering the shearing force S and bending moment $\mathfrak{M}$ at a point P where AP $= x$, we have:

(i) $0 < x < a$

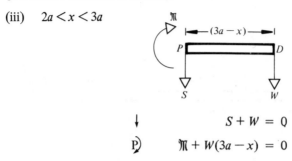

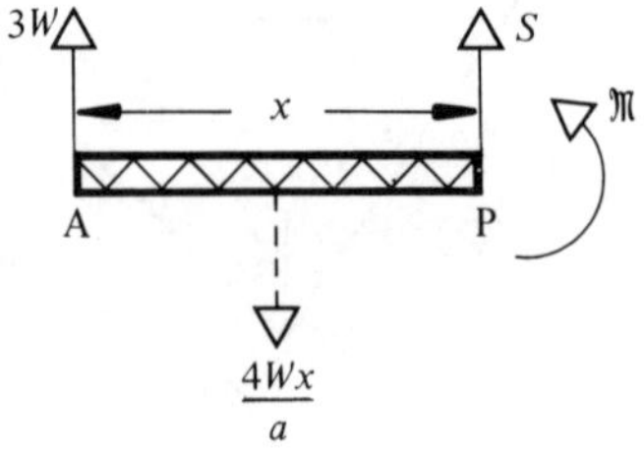

$$\uparrow \qquad 3W + S - 4Wx/a = 0$$

$$\text{P}\circlearrowright \qquad \mathfrak{M} + 4W\frac{x}{a}\left(\frac{x}{2}\right) - 3Wx = 0$$

Hence $S = W(4x - 3a)/a$

and $\mathfrak{M} = Wx(3a - 2x)/a$

When $x = 0$ $\quad S = -3W$ and $\mathfrak{M} = 0$

When $x = a$ $\quad S = W$ and $\mathfrak{M} = Wa$

(ii) $a < x < 2a$

(using the right-hand section)

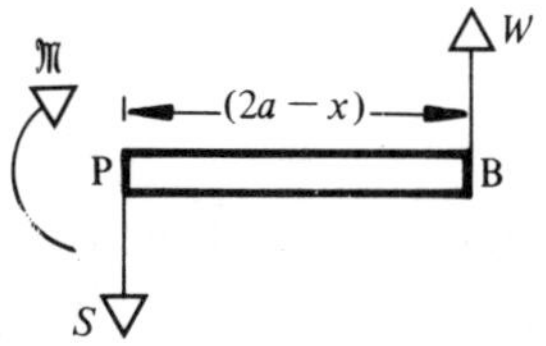

$$\uparrow \qquad W - S = 0$$

$$\text{P}\circlearrowright \qquad \mathfrak{M} - W(2a - x) = 0$$

Hence $S = W$ and $\mathfrak{M} = W(2a - x)$

When $x = 2a$ $\quad \mathfrak{M} = 0$

Now for $0 < x < a$, $\mathfrak{M}$ is a quadratic function of x so its graph is a parabola whose vertex is at a point where $\dfrac{d\mathfrak{M}}{dx} = 0$.

But $\dfrac{d\mathfrak{M}}{dx} = -S$, so the vertex of the bending moment parabola is at the point where $x = \dfrac{3a}{4}$. For this value of x, $\mathfrak{M} = 9Wa/8$ and this is the greatest value of $\mathfrak{M}$ between A and C.

For $a < x < 2a$, $\mathfrak{M}$ is a linear function of x with extreme values Wa and 0.

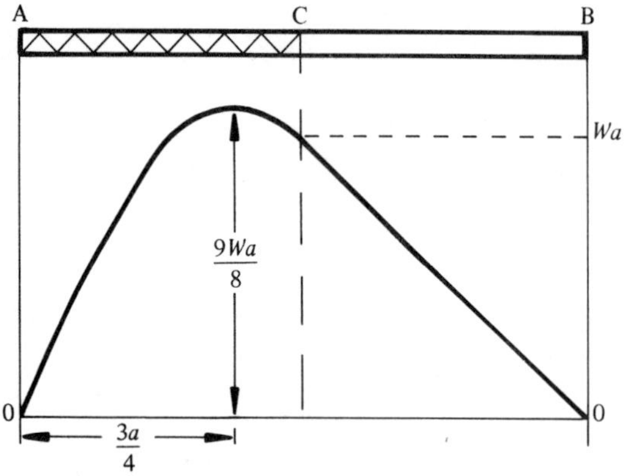

Bending Moment

So the beam is most likely to bend at a point distant $3a/4$ from A.

3) A uniform heavy beam AB is held in a horizontal position by a clamp at the end A. Sketch the bending moment and shearing force diagrams.

A beam which is fixed by a clamp at one end is called a *cantilever*.
In order to support the beam, the clamp must exert on it both an upward force and a torque, whose magnitudes must be evaluated before the internal stress analysis can begin.
Let the weight and length of the rod be W and l respectively, and let F and C be the force and couple exerted by the clamp.

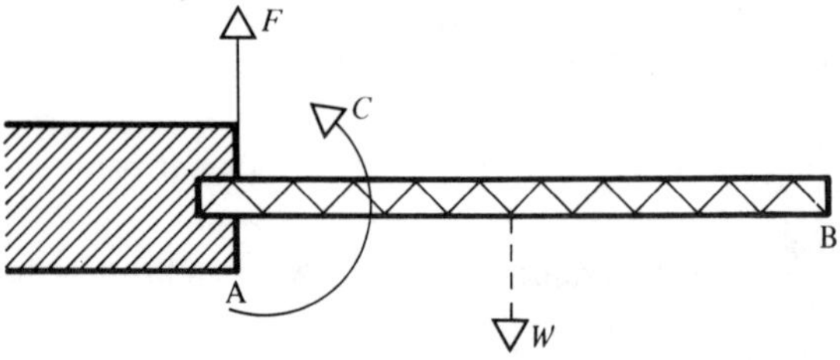

Considering the equilibrium of the beam

$$\uparrow \qquad F - W = 0$$

$$\text{A}\circlearrowleft \qquad C - Wl/2 = 0$$

Now consider the shearing force and bending moment at a point P where AP $= x$.
As there is no concentrated force acting between A and B, the expressions for S and $\mathfrak{M}$ will not change at any intermediate point.

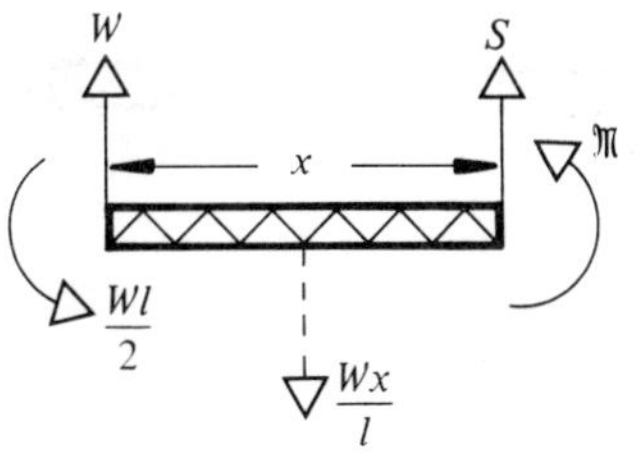

$$\uparrow \qquad S + W - Wx/l = 0$$

$$\circlearrowright P \qquad W\frac{l}{2} + \mathfrak{M} - Wx + W\frac{x^2}{2l} = 0$$

Hence $\quad S = \dfrac{W}{l}(x - l) \quad$ and $\quad \mathfrak{M} = \dfrac{W}{2l}(2lx - x^2 - l^2)$

i.e. $\quad S = -\dfrac{W}{l}(l - x) \quad$ and $\quad \mathfrak{M} = -\dfrac{W}{2l}(l - x)^2$

The shearing force and bending moment diagrams therefore become:

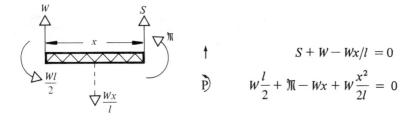

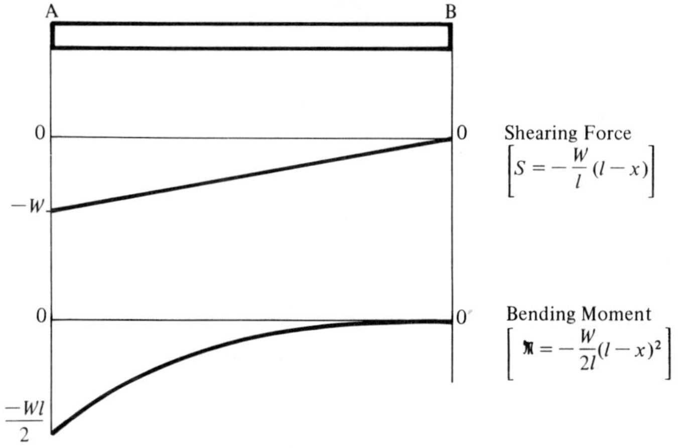

4) A rigid beam of length 10 m rests horizontally on two supports at points distant 2 m from each end, and carries a uniformly distributed load of 1000 N per m along its complete length. Draw shearing force and bending moment diagrams for this beam.
If the supports are to remain equidistant from the ends of the beam, calculate their position so that the magnitudes of the greatest positive and negative bending moments shall be the same.

[The reactions at the two supports are equal since the beam is mechancially symmetrical.]

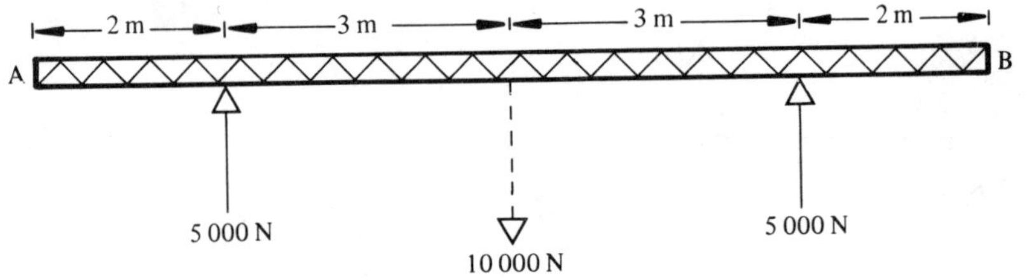

Consider the shearing force and bending moment at a point P distant x from A. Since the beam is symmetrical we need deal with only half of it.
We therefore consider two separate ranges of values of x;
$0 < x < 2$ and $2 < x < 5$.

(i) $0 < x < 2$

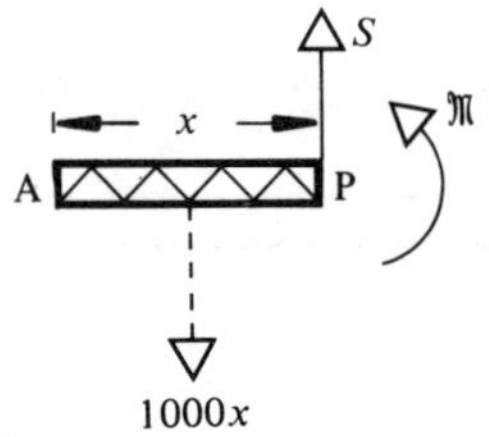

$$\uparrow \qquad S - 1000x = 0$$

$$\overset{\curvearrowright}{P} \qquad \mathfrak{M} + 1000x\left(\frac{x}{2}\right) = 0$$

Hence $\mathfrak{M} = -500x^2$ (a parabola with vertex where $x = 0$)

When $x = 0$ $S = 0$ and $\mathfrak{M} = 0$

When $x = 2$ $S = 2000$ N and $\mathfrak{M} = -2000$ Nm

(ii) $2 < x < 5$

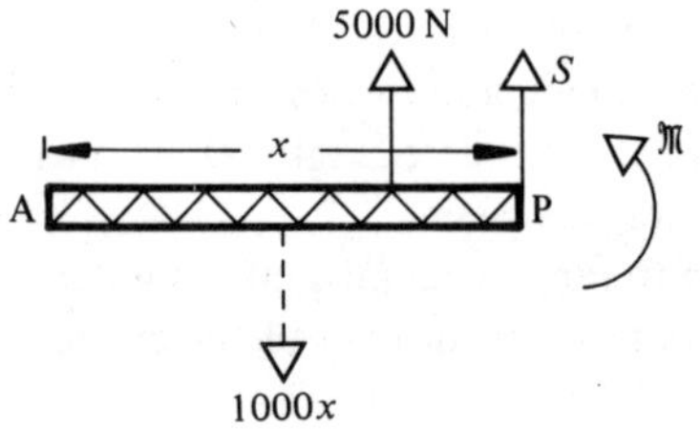

$$\uparrow \qquad S - 1000x + 5000 = 0$$

$$\overset{\curvearrowright}{P} \qquad \mathfrak{M} - 5000(x-2) + 500x^2 = 0$$

(The vertex of this bending moment parabola is at the point where $\frac{d\mathfrak{M}}{dx} = -S = 0$; i.e. when $x = 5$ m.)

When $x = 2$ $\quad S = -3000$ N and $\mathfrak{M} = -2000$ Nm
When $x = 5$ $\quad S = 0$ and $\mathfrak{M} = 2500$ Nm
So the bending moment and shearing force diagrams are:

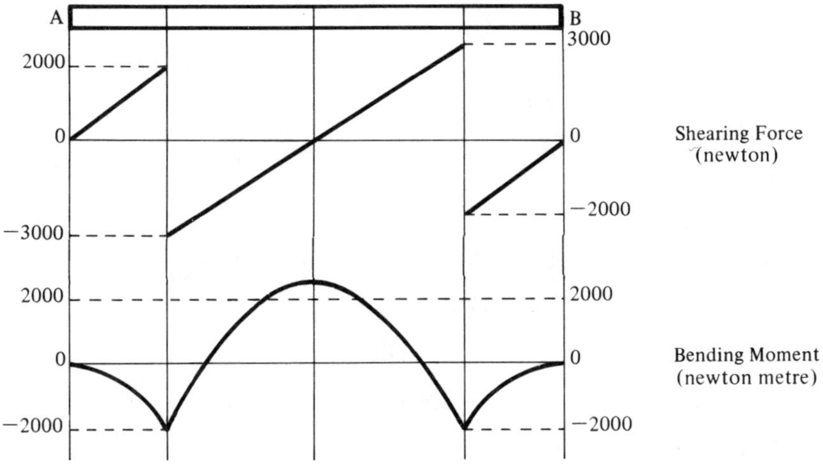

Now let the supports be moved to points distant d from each end so that the magnitudes of the greatest positive and the greatest negative bending moments are equal.
The expressions for S and $\mathfrak{M}$, and the corresponding graphs become:

$0 < x < d \qquad S = 1000x$

$\qquad\qquad\quad \mathfrak{M} = -500x^2$

$d < x < 5 \qquad S = 1000x - 5000$

$\qquad\qquad\quad \mathfrak{M} = 5000(x - d) - 500x^2$

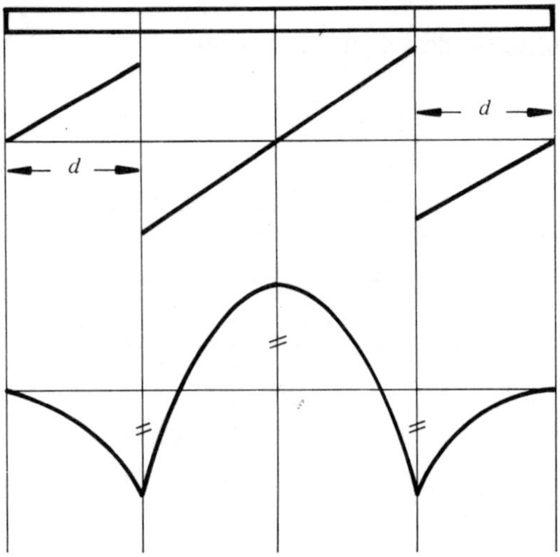

The greatest positive value of $\mathfrak{M}$ occurs at the midpoint.

When $x = 5$ $\qquad \mathfrak{M} = 5000(5-d) - 12\,500$

$$= 12\,500 - 5000d$$

The greatest negative value of $\mathfrak{M}$ occurs at the supports.

When $x = d$ $\qquad \mathfrak{M} = -500d^2$

If the magnitudes of these bending moments are equal,

$$12\,500 - 5000d = 500d^2$$

i.e. $\qquad d^2 + 10d - 25 = 0$

Hence $\qquad d = 5(\sqrt{2} - 1)$

Thus the supports should be placed at a distance $5(\sqrt{2}-1)$ m from each end.

SUMMARY

For a uniform horizontal beam under the action of coplanar vertical forces only:

1. At the junction P of two sections of the beam, each section exerts on the other a force S and a torque $\mathfrak{M}$.
   S is vertical and is called the *shearing force* at P.
   $\mathfrak{M}$ is in a vertical plane parallel to the axis of the beam and is called the *bending moment* at P.

2. The sign convention used in this book is: considering the left-hand section of the beam, S is positive upwards and $\mathfrak{M}$ is positive in the anticlockwise sense.

3. Over a uniform section of the beam:
   (a) if the beam is light, S is constant and $\mathfrak{M}$ is a linear function of x (the distance of P from the left-hand end of the beam),
   (b) if the beam is heavy, S is a linear function and $\mathfrak{M}$ a quadratic function of x.
4. At a point of application of a concentrated load there is a discontinuity in the value of S and in the gradient of $\mathfrak{M}$.
5. At any point other than the point of application of a concentrated load,
   $$\frac{d\mathfrak{M}}{dx} = -S.$$
6. The change in the value of $\mathfrak{M}$ over a uniform section of the beam is equal in magnitude to the area under the corresponding section of the shearing force graph.

EXERCISE 14a

In each of the following examples calculate (where necessary) the forces and/or torque supporting the beam. Find expressions for the shearing force S and the bending moment $\mathfrak{M}$ at any point on the beam and illustrate your results graphically.

1)
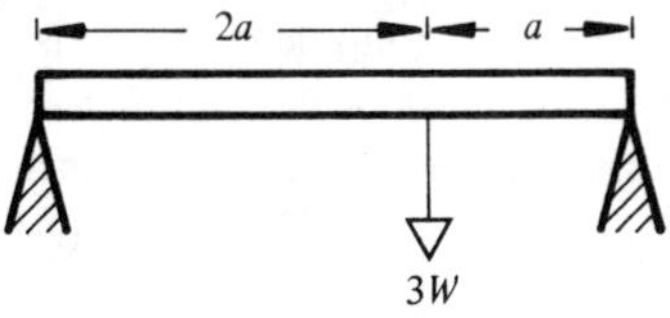

2)
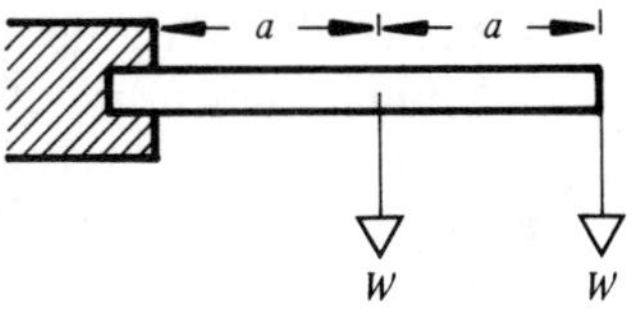

3)
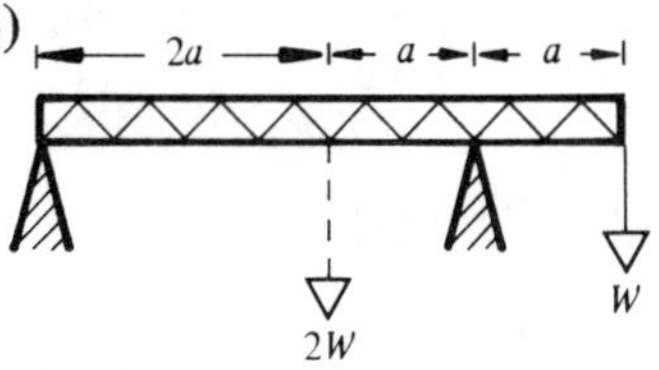

4)
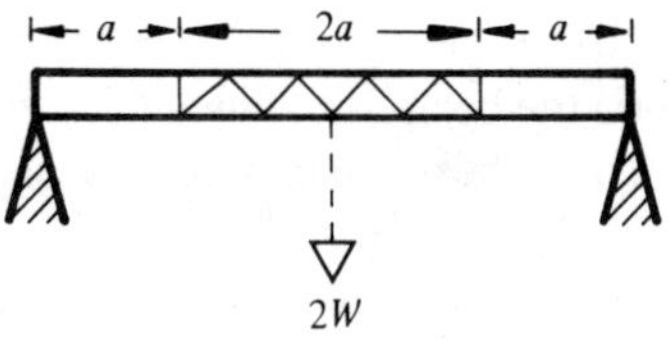

5)
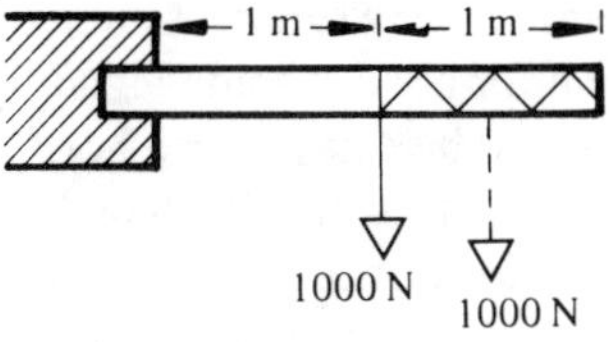

6)
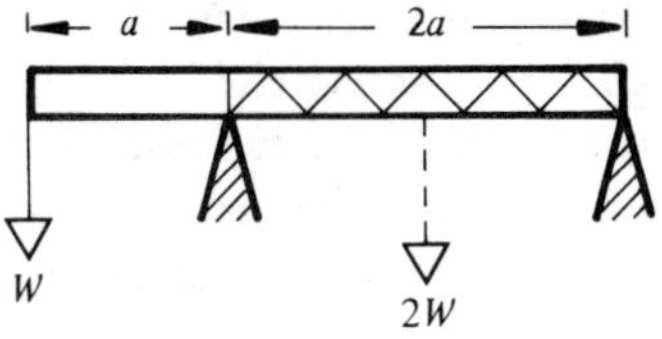

7)

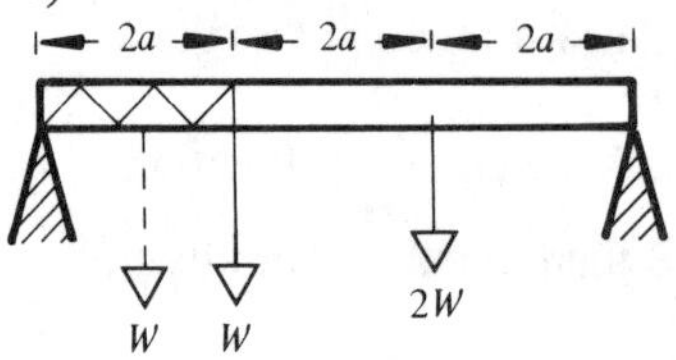

8)

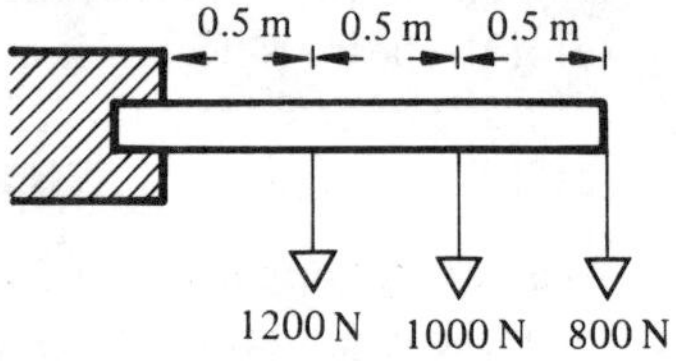

MISCELLANEOUS EXERCISE 14

1) A uniform beam ACDB, of weight 800 newtons and length 5 metres, rests horizontally on two supports at A and D and carries a load of 400 newtons at C, where AC = DB = 1 metre. Sketch the shearing force and the bending moment diagrams and find the greatest bending moment.
When an additional load is attached at B the bending moment at C is zero. Calculate the magnitude of this additional load and the new bending moment at D. (AEB)

2) A light rigid horizontal beam ACDB, in which AC = CD = DB = d m, is simply supported at A and D. It carries a point load W N at C and a uniformly distributed load from C to B.
(i) If the load uniformly distributed from C to B is $2W$ N, find the reactions at A and D and sketch the shearing force and the bending moment diagrams.
(ii) If the uniformly distributed load is q N/m and the value of the bending moment is zero at the centre of the beam, find q and the reactions in terms of W. (AEB)

3) A straight rigid beam ACDB, 4 metres long, rests horizontally on two supports at A and D and carries a load of 20 newtons at C where AC = DB = 1 metre. Sketch the shearing force and bending moment diagrams when:
(i) the beam is of negligible weight,
(ii) the beam weighs 10 newtons per metre, and the weight is uniformly distributed.
In each case, calculate the forces at the supports and find the maximum bending moment. (AEB)

4) On a light rigid beam ACDB of length $4l$ the points C and D are such that AC = DB = l. A weight $2W$ is attached at C and a uniformly distributed load of $4W$ is spread over the beam from C to B. The beam is held at rest in a horizontal position by means of two vertical strings attached at A and D. Draw the shearing force and bending moment diagrams for this loading of the beam and state the maximum bending moment to which the beam is subjected and where this occurs.
A weight is now attached at B. Find its magnitude if it is just sufficient to reduce the tension in one string to zero. Draw the shearing force and bending moment diagrams in this case. (AEB)

5) A load W is uniformly distributed over a length l of a light horizontal beam which is of length $2l$ and is simply supported at its ends. Sketch the shearing force and bending moment diagrams when:
(i) the centre of the distributed load is at a point of trisection of the beam,
(ii) the centre of this load is at the mid-point M of the beam and an additional point load W is attached at M.
In each case calculate the magnitude and position of the maximum bending moment. (AEB)

6) A uniform horizontal beam ABC is of length $4a$ and weight $4W$. It is supported in a horizontal position by vertical supports at B and C where $AB = a$. A particle of weight $2W$ is suspended from the mid-point of the rod.
Show that the reactions at the supports at B and C are $4W$ and $2W$ in magnitude, respectively.
Find the shearing force and bending moment for each point of the rod and sketch graphs to display your results. (Accurate graphs are *not* required.)
Find the points of the rod which have greatest and least bending moment. (SUJB)

7) A uniform beam is supported in a horizontal position. Derive the relationships between bending moment, shearing force and the weight of the beam.
A uniform beam ABC, of weight 160 N and length 16 m, rests horizontally on two supports, one at the end A and the other at B which is 6 m from the other end C. A load of 40 N is attached at the end C. Sketch the shearing force and bending moment diagrams, showing the position and the magnitude of the maximum bending moment. (AEB)

8) A uniform beam AB of length $5a$ and weight w per unit length is supported in a horizontal position by vertical forces at C and D, where $AC = DB = a$, and a load of $3wa$ is suspended from the mid-point E of the beam. Copy and complete the following table which shows the shearing force S and the bending moment $\mathfrak{M}$ for different distances x from A:

Distance from A	S	$\mathfrak{M}$
$0 \leqslant x < a$	wx	
$a < x < 5a/2$		
$5a/2 < x < 4a$		$-w(x^2 - 2ax - 7a^2)/2$
$4a < x \leqslant 5a$		

Sketch the shearing force and bending moment diagrams separately, showing each stage clearly. (AEB)

9) Prove that the change in bending moment between two points on a span is equal to the area of the shearing force diagram over that portion of the span. A rigid beam of length 6 m is supported at two points 1 m from each end and carries a uniformly distributed load of 400 N per m over its whole length.

Draw the shearing force and bending moment diagrams for this beam.
If the supports are to remain equidistant from the ends of the beam calculate their position so that the maximum positive and negative bending moments shall be the same.

10) A uniform rigid beam AB of length 10 m and of weight 500 N per m rests horizontally on two supports P and Q, where AP = 1 m and QB = 3 m. A weight of 800 N is suspended from the mid-point of AB. Draw the shearing force and bending moment diagrams, and find the distances from A of all sections of the beam where the bending moment is zero.

11) A light horizontal beam ABCDE of length 18 m rests on two supports at B, D, where AB = 5 m, DE = 3 m. The mid-point C of the span BD carries a load of 7×10^4 N and there are loads of 5×10^4 N at A and 3×10^4 N at E. Find the magnitude of the bending moment at C and sketch the shearing force and bending moment diagrams.
An additional load is now uniformly distributed from B to D causing the bending moment at C to increase numerically to 13×10^4 Nm. Find the additional loading in newtons per metre and show how the diagrams should be modified.

12) A rigid light horizontal beam ACDB of length 12 m is simply supported at C and D, where AC = DB = 3 m, and carries a uniformly distributed load of 250 N per m over its whole length AB. Draw the shearing force and bending moment diagrams.
An extra load added at E, where CE = 5 m, has the effect of making the support at D carry twice the load of that at C. Draw the shearing force and bending moment diagrams for this new loading and find the greatest values of the shearing force and the bending moment.

13) A light rigid beam ACDB, in which AC = CD = DB = 3 m, is simply supported, in a horizontal position, at the end B and at C. Between A and C the beam carries a uniformly distributed load of 180 N per m. Draw the shearing force and bending moment diagrams for the beam in each of the following cases:
(i) When an additional point load of 400 N is attached at D.
(ii) When this additional load of 400 N is spread uniformly from the mid-point of CD to the mid-point of DB.

14) A uniform bridge span, of length 200 m and weight 6×10^6 N, is freely supported at its ends. Draw shearing force and bending moment diagrams.
A uniform train, of length 50 m and weight 3×10^6 N, runs on to the span and stops just on it. Draw, on the same figure as before, the new bending moment and shearing force diagrams. Calculate:
(a) the point at which the shearing force is unchanged,
(b) the ratio of the new bending moment to the old at the centre point of the span.

APPENDIX

ROTATION ABOUT A FIXED AXIS

Moment of Inertia	$\sum mr^2$
Equation of Motion	$C = I\ddot{\theta}$
Kinetic Energy	$\frac{1}{2}I\omega^2$
Angular Momentum	$I\omega$

STANDARD MOMENTS OF INERTIA

Uniform Body of Mass M	*Axis*	I
Rod of length $2l$	Perpendicular to the rod through the centre	$\frac{Ml^2}{3}$
Rod	Parallel to the rod and distant d from it	Md^2
Rectangle of length $2l$	Perpendicular to the sides of length $2l$ and passing through their midpoints	$\frac{Ml^2}{3}$
Ring of radius a	Perpendicular to the ring through its centre	Ma^2
Disc of radius a	Perpendicular to the disc through its centre	$\frac{Ma^2}{2}$

Uniform Body of Mass M	*Axis*	*I*
Solid sphere of radius a	A diameter	$\frac{2Ma^2}{5}$
Hollow sphere of radius a	A diameter	$\frac{2Ma^2}{3}$
Solid cylinder of radius a	The axis	$\frac{Ma^2}{2}$
Hollow cylinder of radius a	The axis	Ma^2

RADIAL AND TRANSVERSE COMPONENTS OF ACCELERATION

Radial: $\ddot{r} - r\dot{\theta}^2$ Transverse: $r\ddot{\theta} + 2\dot{r}\dot{\theta}$

NEWTON'S LAW OF MOTION

$\mathbf{F} = m\frac{d\mathbf{v}}{dt} + \mathbf{u}\frac{dm}{dt}$ where $\mathbf{u}$ is the relative velocity of the mass increment.

DAMPED HARMONIC MOTION

$$\ddot{x} + 2k\dot{x} + n^2x = 0$$

If $k^2 < n^2$,

$$x = Ce^{-kt}\cos[t\sqrt{(n^2 - k^2)} + \epsilon]$$

$$T = 2\pi/\sqrt{(n^2 - k^2)}$$

ANSWERS

Exercise 1a

1) (i) $3\mathbf{i} + 6\mathbf{j} + 4\mathbf{k}$ (ii) $\mathbf{i} - 2\mathbf{j} - 7\mathbf{k}$
(iii) $\mathbf{i} - 3\mathbf{k}$

2) (i) $(5, -7, 2)$ (ii) $(1, 4, 0)$
(iii) $(0, 1, -1)$

3) (i) $6; \frac{1}{3}, -\frac{2}{3}, \frac{2}{3}$ (ii) $7; \frac{6}{7}, \frac{2}{7}, -\frac{3}{7}$
(iii) $\sqrt{206}; \dfrac{11}{\sqrt{206}}, \dfrac{-7}{\sqrt{206}}, \dfrac{-6}{\sqrt{206}}$

4) (i) $4\mathbf{j} + 5\mathbf{k}$ (ii) $16\mathbf{i} - 16\mathbf{j} - 8\mathbf{k}$
(iii) $\pm 4\sqrt{2}\mathbf{i} + 4\mathbf{j} + 4\mathbf{k}$
(iv) $\pm (\frac{8}{3}\mathbf{i} + \frac{1}{3}\mathbf{j} + \frac{4}{3}\mathbf{k})$

5) (i) $5\sqrt{2}$; 64°54′, 55°34′, 45°
(ii) $\sqrt{3}$; 125°16′, 54°44′, 125°16′

6) (i) $\dfrac{1}{\sqrt{3}}\mathbf{i} - \dfrac{1}{\sqrt{3}}\mathbf{j} + \dfrac{1}{\sqrt{3}}\mathbf{k}$
(ii) $\frac{3}{7}\mathbf{i} + \frac{2}{7}\mathbf{j} - \frac{6}{7}\mathbf{k}$

7) $2\sqrt{3}(\mathbf{i} + \mathbf{j} + \mathbf{k})$

8) (i) $12\mathbf{i} - 4\mathbf{j} + 6\mathbf{k}$ (ii) $\pm(12\mathbf{i} + 3\mathbf{j} + 24\mathbf{k})$

Exercise 1b

1) $\sqrt{14}; \dfrac{2}{\sqrt{14}} : \dfrac{-3}{\sqrt{14}} : \dfrac{1}{\sqrt{14}};$
$4\sqrt{2}; \dfrac{-1}{\sqrt{2}} : \dfrac{-1}{\sqrt{2}} : 0;$
$4\sqrt{2}; \dfrac{-1}{\sqrt{2}} : \dfrac{-1}{\sqrt{2}} : 0$

2) (i) $3\mathbf{i} + \frac{4}{3}\mathbf{j} + \frac{1}{3}\mathbf{k}$ (ii) $2\mathbf{i} - \frac{1}{2}\mathbf{j} + \frac{1}{2}\mathbf{k}$
(iii) $-6\mathbf{i} - \frac{7}{2}\mathbf{j} - \frac{1}{2}\mathbf{k}$

3) (a) No (b) Yes (c) Yes

4) (i) No (ii) Yes (iii) Yes (iv) No

5) $4\mathbf{i} + \frac{3}{2}\mathbf{j} + \frac{1}{2}\mathbf{k}$; $5\mathbf{i} + \mathbf{k}$

6) (i) $\pm \frac{16}{3}(\mathbf{i} + 2\mathbf{j} - 2\mathbf{k})$ (ii) $-6\mathbf{i} - 18\mathbf{j} - 9\mathbf{k}$

7) $12\sqrt{3}; 0, \dfrac{\pi}{2}, \dfrac{\pi}{2}$ or $4\sqrt{55};$
$\arccos\sqrt{\dfrac{3}{55}}, \arccos\dfrac{4}{\sqrt{55}}, \arccos\dfrac{6}{\sqrt{55}}$

Exercise 1c

In all questions requiring the vector or Cartesian equations of a line, the answer given is only one of several forms in which the equation can be written.

1) (i) $x - 2 = y - 3 = z + 1$
(ii) $\dfrac{x}{3} = \dfrac{z}{5}$ and $y = 4$
(iii) $\dfrac{x}{2} = \dfrac{y}{3} = \dfrac{z}{4}$

2) (i) $\mathbf{r} = 3\mathbf{i} + \mathbf{j} + 7\mathbf{k} + \lambda(4\mathbf{i} + 2\mathbf{j} + 6\mathbf{k})$
(ii) $\mathbf{r} = 2\mathbf{i} - 5\mathbf{j} + \mathbf{k} + \lambda(3\mathbf{i} + \mathbf{j} + 4\mathbf{k})$
(iii) $\mathbf{r} = \mathbf{i} + \lambda(-3\mathbf{i} + 5\mathbf{j} + \mathbf{k})$

3) (i) $\mathbf{r} = \mathbf{i} - 3\mathbf{j} + 2\mathbf{k} + \lambda(5\mathbf{i} + 4\mathbf{j} - \mathbf{k})$;
$\dfrac{x-1}{5} = \dfrac{y+3}{4} = \dfrac{z-2}{-1}$
(ii) $\mathbf{r} = 2\mathbf{i} + \mathbf{j} + \lambda(3\mathbf{j} - \mathbf{k})$;
$x = 2, \dfrac{y-1}{3} = \dfrac{z}{-1}$
(iii) $\mathbf{r} = \lambda(\mathbf{i} - \mathbf{j} - \mathbf{k})$; $x = -y = -z$

4) (i) No (ii) Yes (iii) Yes (iv) Yes (v) No

5) $\mathbf{r} = 4\mathbf{i} + 5\mathbf{j} + 10\mathbf{k} + \lambda(-2\mathbf{i} - 2\mathbf{j} - 6\mathbf{k})$;
$\dfrac{4-x}{2} = \dfrac{5-y}{2} = \dfrac{z-10}{-6}$
$\mathbf{r} = 2\mathbf{i} + 3\mathbf{j} + 4\mathbf{k} + \lambda(-\mathbf{i} - \mathbf{j} - 5\mathbf{k})$;
$\dfrac{2-x}{1} = \dfrac{3-y}{1} = \dfrac{4-z}{5}$
$3\mathbf{i} + 4\mathbf{j} + 5\mathbf{k}$

6) (i) $\mathbf{r} = 3\mathbf{i} + \mathbf{j} - 4\mathbf{k} + \lambda(2\mathbf{i} - 6\mathbf{j} - 12\mathbf{k})$
$(\frac{7}{3}, 3, 0); (0, 10, 14); (\frac{10}{3}, 0, -6)$

(ii) $r = i + j + 7k + \lambda(2i + 3j - 6k)$;
$(\frac{10}{3}, \frac{9}{2}, 0); (0, -\frac{1}{2}, 10); (\frac{1}{3}, 0, 9)$
7) $r = 5i - 2j - 4k + \lambda(3i + 4j + 5k)$;
$(\frac{13}{2}, 0, -\frac{3}{2})$

Exercise 1d

1) (a) $\arccos \dfrac{19}{21}$ (b) $\arccos \dfrac{2}{3}$

(c) $\arccos \dfrac{8\sqrt{3}}{15}$

2) (a) parallel   (b) intersecting; $r = i + 2j$
(c) skew

3) $r = 3i + 2j + k + \lambda(16i + 13j + 2k)$
4) $a = -3; r = -i + 3j + 4k$
5) $r = 4i + j + 5k + \lambda(3i + 14j + 12k)$
6) $r = i + 2j + 3k + \lambda(3j)$;
$r = -2i + 5k + \mu(3i + 3j - 2k)$;
$r = 4i + 7j + k + \nu(-3i - 4j + 2k)$;
$(1, 3, 3)$

Multiple Choice Exercise 1

Type 1
1) b   2) c   3) a   4) e   5) b
Type 2
1) d   2) a, b, d   3) a, b   4) d
Type 3
1) D   2) D   3) D   4) D
Type 4
1) b, d   2) d and *one* of a, b, c
3) I   4) c
Type 5
1) T   2) F   3) F   4) F

Miscellaneous Exercise 1

1) (a) $\sqrt{6}, \sqrt{10}$ (b) $\dfrac{9\sqrt{35}}{70}$ (c) $\dfrac{\sqrt{59}}{2}$

2) 9; $\arccos \frac{4}{9}$, $\arccos (-\frac{1}{9})$, $\arccos \frac{8}{9}$
(a) $r = \lambda(-2i - 10j + 5k)$

(b) $\dfrac{x}{-2} = \dfrac{y}{-10} = \dfrac{z}{5}$

3) (a) $r = 5i + 3j - 13k + \lambda(2i + j - 5k)$
(b) $r = \mu(i - 2j)$ (c) $\frac{1}{5}(-i + 2j)$
4) $(6, 9, 10); 19/21$
5) (a) $r = 3j + 12k + \lambda(-i + 6j + 4k)$;
$r = -3i + 3j + 6k + \mu(5i - 6j + 4k)$
(b) $r = \frac{3}{4}i - \frac{3}{2}j + 9k$
7) $n = 3; 5i + \frac{4}{3}j - \frac{2}{3}k; \frac{160}{3}$ J, $\frac{109}{3}$ J
8) $2\sqrt{6}/5$
9) -4; $r = 3i + 2j - 5k + \lambda(3i - 2j - k)$
11) $a = -8, b = 1, c = -2$ or $a = 16$,
$b = -7, c = 10$

12) $\sqrt{41}$; $\arccos \dfrac{6}{\sqrt{41}}$, $\arccos \dfrac{-1}{\sqrt{41}}$,

$\arccos \dfrac{-2}{\sqrt{41}}$

13) $r = sb + (1 - s)c$;
$r = \frac{1}{2}ta + \frac{1}{2}(1 - t)(b + c)$;
$r = \frac{1}{4}(a + b + c)$
14) 13 newtons at $\arctan \frac{5}{12}$ to Ox;
$r = \frac{21}{5}i + \frac{29}{15}j + \lambda(12i + 5j)$
15) $m = -5; n = 6$
16) (a) $7\sqrt{3}$ (b) $2i - j + 3k$
(c) $r = (\lambda + 2)i + (\lambda - 1)j + (\lambda + 3)k$
17) $\frac{1}{2}(a + b)$, $\frac{1}{3}(b + 2c)$, $2d - c$.
18) $i + 10j + 3k$; $\arccos \dfrac{1}{\sqrt{140}}$; $\sqrt{110}$

Exercise 2a

1) $r = i - 5j + 2k + t(2i + 7j - 6k)$
2) $r = 2tv(2i + 2j - k)$
3) $4i - 13j - 3k$
4) a) No   b) Yes at $i - 2j + k$ c) No
5) a) $(89t^2 - 190t + 150)^{1/2}$; $5\sqrt{173/89}$
b) $(5t^2 - 18t + 18)^{1/2}$; $3\sqrt{5/5}$
6) $4\sqrt{5}$ miles
7) A and C; 3 seconds later; $i + 2j + 3k$
8) $35i$ and $32i + 4j$

Exercise 2b

1) (i) $2i + 3j - 6k; 7$ (ii) $4i - 3j; 5$
(iii) $2j + 2k; 2\sqrt{2}$
2) $4\sqrt{2}i - (18 + 4\sqrt{2})j$
3) $4i - 5j + 7k$; $-2i - 3j + 3k$
4) $5\sqrt{3}i - 15j$
6) (i) $i - 2j - 6k$ (ii) 20   (iii) $\frac{1}{2}\sqrt{205}$

Exercise 2c

1) $30; 0; -1$
2) $7, \sqrt{7/3}$; $14, \sqrt{7/19}$
3) 4
4) $-4; 15; 11; -12; 29$
5) $\sqrt{7/34}$; $\pi - \arccos \sqrt{7/10}$

Exercise 2d

1) 23
2) 176, 174, 278

3) Magnitude of components $\dfrac{1}{\sqrt{2}}, \dfrac{5}{\sqrt{2}}$;

$\sqrt{5}, 2\sqrt{5}$; $\frac{7}{5}, \frac{1}{5}$

4) Magnitude of components $\dfrac{10}{\sqrt{17}}, \dfrac{11}{\sqrt{17}}$

5) $\dfrac{\mathbf{i}+9\mathbf{j}-5\mathbf{k}}{\sqrt{107}}$; $\dfrac{-\mathbf{i}-2\mathbf{j}+\mathbf{k}}{\sqrt{6}}$
6) $2\sqrt{2}/3$
7) $3\sqrt{3}$
8) (i) 76 J (ii) $\mathbf{i}-\mathbf{j}+2\mathbf{k}$ (iii) 112 J
9) 12 J, $2\ \mathrm{ms}^{-1}$

Miscellaneous Exercise 2

1) $\sqrt{2}:\sqrt{3}$; $\sqrt{5}/2\sqrt{2}$
2) $v_1=\sqrt{3}u/3$, $v_2=2\sqrt{3}u/3$
3) $2\mathbf{i}-\mathbf{j}+2\mathbf{k}$; $6\mathbf{i}+2\mathbf{j}+4\mathbf{k}$;
$\sqrt{11}$, $\mathbf{i}+\mathbf{j}+3\mathbf{k}$
4) $20\ \mathrm{ms}^{-1}$ (a) 1.35 m (b) 2.24 m
5) $(3-2T)\mathbf{i}+(4-5T)\mathbf{j}+(T-2)\mathbf{k}$;
$\sqrt{\frac{43}{15}}$; $\frac{17}{15}\mathbf{i}-\frac{2}{3}\mathbf{j}-\frac{16}{15}\mathbf{k}$
6) 13 units; $14\mathbf{i}+19\mathbf{j}$; $9\mathbf{i}+7\mathbf{j}$
$-\mathbf{i}+4\mathbf{j}$, $\frac{1}{65}(-191\mathbf{i}+188\mathbf{j})$
7) $3\mathbf{i}-2\mathbf{j}-3\mathbf{k}$;
$(a-7+3t)\mathbf{i}+(5-2t)\mathbf{j}-(2+3t)\mathbf{k}$;
$[(a-7+3t)^2+(5-2t)^2+(2+3t)^2]^{\frac{1}{2}}$;
$a=-19/3$
8) $-6\mathbf{i}-25\mathbf{j}-16\mathbf{k}$; $13\mathbf{i}+53\mathbf{j}+34\mathbf{k}$
9) $7\mathbf{i}+2\mathbf{j}$; 12.30; $\mathbf{r}=\frac{9}{2}\mathbf{i}+\mathbf{j}$
10) $\sqrt{45}$, 1; $2\mathbf{i}+3\mathbf{j}+3\mathbf{k}$, 3
11) 3; $\sqrt{134}$ metres
12) $8\mathbf{i}-7\mathbf{j}+6\mathbf{k}$; $6\mathbf{i}-6\mathbf{j}+2\mathbf{k}$;
$16\mathbf{i}+12\mathbf{j}+8\mathbf{k}$
13) $n=3$; $5\mathbf{i}+\frac{4}{3}\mathbf{j}-\frac{2}{3}\mathbf{k}$; $\frac{160}{3}$ joules, $\frac{109}{3}$ joules
14) $\mathbf{r}=(\mathbf{i}+\mathbf{j})+t(\mathbf{i}+\mathbf{k})$; $1/\sqrt{3}$
16) $\frac{11}{5}$, $\frac{2}{5}$; $\frac{11}{5}$, $\frac{23}{5}$; $\frac{22}{25}(4\mathbf{i}+3\mathbf{j})$
17) 33 units; 90°
18) $\dfrac{1}{\sqrt{2}}\mathbf{v}_1+\dfrac{1}{\sqrt{3}}\mathbf{v}_2+\dfrac{1}{\sqrt{6}}\mathbf{v}_3$
19) 92 J; $2\sqrt{23}\ \mathrm{ms}^{-1}$
20) $2\mathbf{i}-5\mathbf{j}+\mathbf{k}$; $F\sqrt{84}$

Exercise 3a

3) a) $x^2+y^2=4$
b) $x^2+y^2-4x-8y+19=0$
c) $x^2-y^2=4$
d) $y^2-8y+4ax+16-24a=0$
4) a) $x^2+z^2=a^2$, $y=0$
b) $y^2=z$, $x=0$
c) $\dfrac{x^2}{4}+\dfrac{z^2}{9}=1$, $y=0$
d) $x^2+y^2=1$, $z=1$
5) a) $\mathbf{r}=\pm(\mathbf{i}+2\mathbf{j})/\sqrt{5}$
b) $\mathbf{r}=\left(\dfrac{\sqrt{37}-1}{2}\right)\mathbf{i}\pm\left(\dfrac{\sqrt{37}-1}{2}\right)^{\frac{1}{2}}\mathbf{j}$
c) $\mathbf{r}=3\mathbf{j}+\mathbf{k}$
6) $\mathbf{r}=(1+a\cos\theta)\mathbf{i}+(b\sin\theta-2)\mathbf{j}$
7) $\mathbf{r}=(1+at^2)\mathbf{i}+(2at-2)\mathbf{j}$
8) $\mathbf{r}=3\cos\theta\,\mathbf{i}+3\sin\theta\,\mathbf{j}+\dfrac{2\theta}{\pi}\mathbf{k}$
9) $\mathbf{r}=5\cos(\theta-53°8')\mathbf{i}$
$+5\sin(\theta-53°8')\mathbf{j}+\dfrac{3\theta}{2\pi}\mathbf{k}$
10) $\mathbf{r}=(3\cos\theta+2)\mathbf{i}+\mathbf{j}+3\sin\theta\,\mathbf{k}$
11) $\mathbf{r}=-\frac{21}{4}\mathbf{i}+\lambda\mathbf{j}$

Exercise 3b

1) $3at^2\mathbf{i}+a\mathbf{j}+2at\mathbf{k}$, $6at\mathbf{i}+2a\mathbf{k}$
2) $\frac{1}{4}(\mathbf{i}+4\mathbf{j})$, $-\frac{1}{16}\mathbf{i}$
3) $\frac{1}{2}p(\mathbf{i}-\sqrt{3}\,\mathbf{j})$, $-\frac{1}{2}p^2(\sqrt{3}\,\mathbf{i}+\mathbf{j})$
4) $(e^8+4)^{1/2}$
5) $\mathbf{v}=(t^2+1)\mathbf{i}+(t+2)\mathbf{j}$,
$\mathbf{r}=(\frac{1}{3}t^3+t)\mathbf{i}+(\frac{1}{2}t^2+2t)\mathbf{j}$
6) $9\mathbf{i}+5\mathbf{j}$
7) $\mathbf{r}=\left(\dfrac{1}{2}-\sin t\right)\mathbf{i}+\left(\dfrac{\sqrt{3}}{2}-\cos t\right)\mathbf{j}$
8) $\mathbf{r}=\dfrac{1}{\omega}(\cos\omega t\,\mathbf{i}-\sin\omega t\,\mathbf{j})$,
$\mathbf{a}=-\omega(\cos\omega t\,\mathbf{i}-\sin\omega t\,\mathbf{j})$
9) $\mathbf{v}=t(2\mathbf{i}-\mathbf{j}+\mathbf{k})/3$,
$\mathbf{r}=\left(\dfrac{t^2}{3}+1\right)\mathbf{i}-\left(\dfrac{t^2}{6}+2\right)\mathbf{j}+\dfrac{t^2}{6}\mathbf{k}$
10) $\mathbf{r}=\dfrac{t^2}{2m}(4\mathbf{i}+\mathbf{j}-\mathbf{k})$
11) $2m\sqrt{(p^4+1)}$
12) $\mathbf{r}=\cos vt\,\mathbf{i}-\sin vt\,\mathbf{j}+\dfrac{vt}{2\pi}\mathbf{k}$,
$\mathbf{v}=-v\sin vt\,\mathbf{i}-v\cos vt\,\mathbf{j}+\dfrac{v}{2\pi}\mathbf{k}$
$\mathbf{a}=-v^2\cos vt\,\mathbf{i}+v^2\sin vt\,\mathbf{j}$
13) $2\mathbf{i}+2\mathbf{j}$, $\mathbf{i}-2\mathbf{j}$
14) $\mathbf{r}=vt\cos\alpha\,\mathbf{i}+(vt\sin\alpha-\frac{1}{2}gt^2)\mathbf{j}$
15) $x^2-4y+2x=0$

Exercise 3c

1) a) $-a\theta\omega^2$, $2a\omega^2$
b) $-a(1+\theta)\omega^2$, $2a\omega^2$
c) $-a\omega^2(1+2\sin\theta)$, $2a\omega^2\cos\theta$
d) $-2a\omega^2(\cos\theta+\sin\theta)$,
$2a\omega^2(\cos\theta-\sin\theta)$
2) a) increases with θ
b) increases with θ
c) $\left(2a, (4n+1)\dfrac{\pi}{2}\right)$
d) acceleration constant.
3) $-2r\omega^2$, $-2\omega^2\sqrt{(a^2-r^2)}$
4) $r=2+\cos\theta$

5) $\dfrac{a\omega \sin\theta}{(1+\cos\theta)^2}$, $\dfrac{a\omega}{1+\cos\theta}$
6) $r = e^{\theta}$

Multiple Choice Exercise 3

1) c	2) b	3) d	4) c	5) a
6) a	7) b, c	8) b	9) a, b, c	10) *A*
11) *B*	12) *C*	13) *I*	14) a, b	15) *A*
16) *A*	17) c	18) *F*	19) *T*	20) *F*

Miscellaneous Exercise 3

1) $\mathbf{r} = \left(\dfrac{t^2}{2} + vt\right)\mathbf{i} + \dfrac{1}{a}\left(t - \dfrac{1}{a}\sin at\right)\mathbf{j}$
2) $2 - \sqrt{5} < a < 2 + \sqrt{5}$
3) $-2(5\cos t\,\mathbf{i} + 3\sin t\,\mathbf{j})$, 17 units
4) $5(3\mathbf{i} + \mathbf{j} - 4\mathbf{k})$, 16.25 units
5) $128\mathbf{i} + 16\mathbf{j} - 112\mathbf{k}$, $3264m$ J
6) $\arccos(2\cos\omega t - \sin\omega t)/\sqrt{5}$, $\dfrac{1}{\omega}(n\pi + \arctan 2)$
7) $\dfrac{x^2}{9} + \dfrac{y^2}{16} = 1,\ z = 1;\ y^2 = 4x,\ z = 1;$ $z = 1;\ \mathbf{r}.\mathbf{k} = 1;\ \dfrac{n\pi}{\omega}$
8) $k = 2/c^2$
9) $\pm(12\mathbf{i} + 16\mathbf{j})/5$
11) $0, \dfrac{3}{2}\mathbf{i} + \dfrac{\sqrt{3}}{2}\mathbf{j},\ f = \dfrac{9u^2\sqrt{3}}{2\pi^2}$
12) $a\mathbf{i} \pm a\sqrt{3}\mathbf{j}$, $(2n+1)\pi/2\omega$
13) $t = 0,\ \mathbf{r} = a\mathbf{j},\ \mathbf{v} = a\mathbf{i} + a\mathbf{k};$ $t = \dfrac{\pi}{2},\ \mathbf{r} = a\mathbf{i} + \dfrac{a\pi}{2}\mathbf{k},\ \mathbf{v} = -a\mathbf{j} + a\mathbf{k};$ $\dfrac{\pi}{3};\ ma^2;\ ma;\ \dfrac{\pi a^2 m}{4}$
14) $-\omega^2\cos\omega t\,\mathbf{i}$, $\mathbf{r} = \cos\omega t\,\mathbf{i}$
15) $-a\omega^2\cos\omega t\,\mathbf{i}$
17) $r = a(1 + 2\sin\theta)$, max. acceleration when $r = 3a$
18) $r = ae^{\theta};\ r = a(1+\theta);$ $r = a(\cos\theta + \sin\theta)$
19) $2ma\omega^2\sqrt{5}$ at $\arctan(-\frac{1}{2})$ to the initial line.

Exercise 4a

1) i) $\mathbf{a}\times\mathbf{b}$ ii) $\mathbf{0}$ iii) $2\mathbf{a}\times\mathbf{b}$ iv) $\mathbf{a}\times\mathbf{c}.\mathbf{b}$ v) $\mathbf{0}$ vi) $\mathbf{0}$
2) i) $-3\mathbf{j} - 3\mathbf{k}$ ii) $-3\mathbf{j} - 3\mathbf{k}$
3) $(3\mathbf{i} - \mathbf{j} + \mathbf{k})/\sqrt{11}$, $\sqrt{(11/12)}$
4) $-\mathbf{i} - 3\mathbf{j} + 2\mathbf{k}$, $\mathbf{i} - \mathbf{j}$
5) $\sqrt{3/5}$, $\pm(\mathbf{i} - \mathbf{j} - 2\mathbf{k})/\sqrt{6}$
6) $\pm(4\mathbf{i} - 5\mathbf{j} + 7\mathbf{k})/3\sqrt{10}$
8) $k\mathbf{a} = 3\mathbf{b} - 2\mathbf{c}$
10) $-2\mathbf{k} + t(\mathbf{i} - 3\mathbf{k})$

Exercise 4b

1) $\frac{1}{2}\sqrt{35}$
2) $\frac{1}{2}\sqrt{5}$
4) 1
5) $\sqrt{265}$
6) 3
7) $\frac{1}{3}$
8) $\frac{11}{6}$
9) $\frac{1}{6}|\mathbf{b}\times\mathbf{c}.\mathbf{a} + \mathbf{b}\times\mathbf{a}.\mathbf{d} + \mathbf{c}\times\mathbf{b}.\mathbf{d} + \mathbf{a}\times\mathbf{c}.\mathbf{d}|$
10) a) $\sqrt{2}$ b) $\sqrt{2}/2$ c) $\dfrac{|\mathbf{c}\times\mathbf{b}|}{|\mathbf{c}|}$ d) $\dfrac{5|\mathbf{p}\times\mathbf{r}|}{|\mathbf{r} + 2\mathbf{p}|}$
11) $3\sqrt{2}/2$
12) $8/\sqrt{29}$
13) a) no b) yes

Exercise 4c

1) a) $3\mathbf{k}$ b) $-5\mathbf{j}$ c) $3\mathbf{i} + 3\mathbf{j}$ d) $\mathbf{i} + 7\mathbf{j} + 4\mathbf{k}$
2) a) $-3\mathbf{k}$ b) $3\mathbf{i} + 3\mathbf{j} + 6\mathbf{k}$ c) $2\mathbf{i} - \mathbf{j} - \mathbf{k}$
3) $(\mathbf{a} - \mathbf{b})\times\mathbf{F}$
4) $-2\mathbf{i} + \mathbf{j} - 3\mathbf{k}$, $2\mathbf{i} - \mathbf{j} - 4\mathbf{k}$, $-7\mathbf{k}$
5) a) 5 b) $3\sqrt{3}$
6) a) $\mathbf{r} = \lambda\mathbf{k}$ b) $\mathbf{r} = 2\mathbf{j} - \mathbf{k} + \lambda\mathbf{i}$ c) $\mathbf{r} = -\mathbf{i} + 2\mathbf{j} - \mathbf{k} + \lambda(-\mathbf{i} + 4\mathbf{j} + 2\mathbf{k})$
7) a) $\mathbf{r} = -4\mathbf{j} + \lambda(\mathbf{i} + \mathbf{j})$ b) $\mathbf{r} = \mathbf{k} + \lambda(2\mathbf{i} - \mathbf{j})$ c) $\mathbf{r} = -\mathbf{j} - 2\mathbf{k} + \lambda(\mathbf{i} + \mathbf{j} - \mathbf{k})$
8) $4\mathbf{i} + \mathbf{j} - \mathbf{k}$, $\sqrt{2}$
9) $\sqrt{14}$, $3\mathbf{i} + 2\mathbf{j} - \mathbf{k}$

Exercise 4d

2) $-\mathbf{i} + 4\mathbf{j} + 2\mathbf{k}$
3) $\mathbf{r} = 2\mathbf{i} - \mathbf{j} + \lambda\mathbf{i}$
4) 1
5) a) a single force $4\mathbf{i} + \mathbf{j}$ along the line $\mathbf{r} = \frac{7}{2}\mathbf{j} + \lambda(4\mathbf{i} + \mathbf{j})$
b) a couple of vector moment $\frac{1}{2}(\mathbf{a}\times\mathbf{b} + \mathbf{b}\times\mathbf{c} + \mathbf{c}\times\mathbf{a})$
c) a single force $\mathbf{i} + 3\mathbf{j}$ along the line $\mathbf{r} = -3\mathbf{j} + \lambda(\mathbf{i} + 3\mathbf{j})$
d) a single force $-\mathbf{i} + \mathbf{k}$ through 0
6) $\mathbf{F}_4 = -3\mathbf{i} + 4\mathbf{j} - \mathbf{k}$, $\mathbf{r} = 3\mathbf{i} - \mathbf{k} + \lambda(3\mathbf{i} - 4\mathbf{j} + \mathbf{k})$
7) $-\mathbf{i} + 3\mathbf{j} + 2\mathbf{k}$
8) $\mathbf{r} = -2\mathbf{j} + \mathbf{k} + \lambda(\mathbf{i} + \mathbf{j} - \mathbf{k})$

Multiple Choice Exercise 4

1) d	2) b	3) c
4) a	5) b	6) b
7) a, c	8) a, b, c	9) b, c
10) a, b	11) *E*	12) *A*
13) *A*	14) *C*	15) *D*
16) b	17) *A*	18) *I*
19) b	20) *F*	21) *T*
22) *T*	23) *F*	

Miscellaneous Exercise 4

1) 2, $4\mathbf{i}+\mathbf{j}-2\mathbf{k}$, 2/5, $\frac{1}{2}\sqrt{21}$
2) $\Delta OPQ + \Delta OQR + \Delta ORP$,
$\Delta OPQ - \Delta OQR - \Delta ORP$
3) $\frac{5}{3}(-2\mathbf{i}+\mathbf{j})$
4) $\mathbf{u}=(\lambda+\frac{1}{2})\mathbf{i}+(\lambda+\frac{1}{2})\mathbf{j}+2\lambda\mathbf{k}$
5) b) $\pm\frac{1}{\sqrt{3}}, \mp\frac{1}{\sqrt{3}}, \pm\frac{1}{\sqrt{3}}$
6) b) $-\frac{1}{4}, \frac{1}{4}$
7) $3\sqrt{2}$
9) i) $\mathbf{r}=\lambda\mathbf{i}+(\lambda-2)\mathbf{j}+\mathbf{k}$, ii) $2n\pi+\frac{\pi}{4}$
10) $(3\cos t+3\sin t)\mathbf{i}+(2\sin t+\cos t)\mathbf{j}$,
$-4(\cos t+\sin t)$,
$\mathbf{r}=\lambda(3\cos t+3\sin t)\mathbf{i}$
$+[\frac{4}{3}+\lambda(2\sin t+\cos t)]\mathbf{j}$
11) ii) a) $3F$ in the direction AD along a line $5a/3$ or a beyond AD.
b) $10aF$ in the sense ADC.
13) $a=1, b=-2, c=1$, $(18\mathbf{i}+19\mathbf{j})/13$,
$4\sqrt{10}$ at arctan $\frac{1}{3}$ to Ox,
$\mathbf{r}=-\frac{11}{4}\mathbf{j}+\lambda(3\mathbf{i}+\mathbf{j})$
14) a) $y+x\sqrt{3}-2a\sqrt{3}=0$,
b) $7\sqrt{7}$ at arctan $\sqrt{3}/2$ to Ox, 21
15) $-\mathbf{i}+33\mathbf{j}+21\mathbf{k}$
16) $2\sqrt{6}/5$
17) $18\sqrt{26}$, $-9\mathbf{i}+18\mathbf{j}-18\mathbf{k}$,
$(4\mathbf{i}-\mathbf{j}-3\mathbf{k})/\sqrt{26}$
18) $3\sqrt{14}$, $\mathbf{r}=2\mathbf{i}+5\mathbf{j}+5\mathbf{k}+\lambda(3\mathbf{i}+\mathbf{j}+2\mathbf{k})$,
$\frac{x-2}{3}=\frac{y-5}{1}=\frac{z-5}{2}$, $\frac{5}{2\sqrt{39}}$

Exercise 5a

1) a) $\mathbf{r}=\mathbf{i}+2\mathbf{j}-\mathbf{k}+\lambda(\mathbf{j}+3\mathbf{k})+\mu(\mathbf{i}-2\mathbf{k})$,
$\mathbf{r}.(2\mathbf{i}-3\mathbf{j}+\mathbf{k})=-5$
b) $\mathbf{r}=\mathbf{i}+\mathbf{j}-2\mathbf{k}+\lambda(\mathbf{j}-3\mathbf{k})+\mu(3\mathbf{i}+\mathbf{k})$,
$\mathbf{r}.(-\mathbf{i}+9\mathbf{j}+3\mathbf{k})=2$
2) a) $\mathbf{r}.(5\mathbf{i}-2\mathbf{j}-3\mathbf{k})=7$
b) $\mathbf{r}.(-\mathbf{j}+2\mathbf{k})=3$
c) $\mathbf{r}.(2\mathbf{i}+\mathbf{k})=5$
3) a) $x+y-z=2$ b) $2x+3y-4z=1$
c) $2x-5y-z=-15$
4) $\mathbf{r}.(\mathbf{i}-\mathbf{j}-\mathbf{k})=0$
5) $\mathbf{r}=-3\mathbf{i}-2\mathbf{j}+\lambda(\mathbf{i}-2\mathbf{j}+\mathbf{k})$
$+\mu(2\mathbf{i}-\mathbf{j}+2\mathbf{k})$
6) $\mathbf{r}=\mathbf{i}+\mathbf{j}+\lambda(\mathbf{i}+\mathbf{k})+\mu(\mathbf{i}-\mathbf{j}+\mathbf{k})$,
second line is contained in the plane.
7) $\mathbf{r}.(7\mathbf{i}+2\mathbf{j}-3\mathbf{k})=3$, $3/\sqrt{62}$
8) $3/\sqrt{5}$
9) 2
10) $\mathbf{r}=\lambda(\mathbf{i}-2\mathbf{j}+\mathbf{k})$
11) $\mathbf{r}=2\mathbf{i}+\mathbf{j}+\mathbf{k}+\lambda(\mathbf{i}+2\mathbf{j}-3\mathbf{k})$
12) $\mathbf{r}.(\mathbf{i}-2\mathbf{j})=-2$
13) $\mathbf{r}.(\mathbf{j}-\mathbf{k})=0$

Exercise 5b

1) a) $3/\sqrt{11}$	b) $1/\sqrt{1102}$	c) $20/\sqrt{442}$
2) a) $\sqrt{3}/9$	b) $\sqrt{15}/5$	c) 8/21
3) a) 2	b) 22/9	c) $9/\sqrt{5}$

4) a) $\mathbf{r}=-\frac{1}{3}(10\mathbf{j}+11\mathbf{k})+\lambda(3\mathbf{i}+5\mathbf{j}+7\mathbf{k})$
b) $\mathbf{r}=\mathbf{j}+4\mathbf{k}+\lambda(\mathbf{i}-\mathbf{j}-3\mathbf{k})$

Exercise 5c

1) 1
2) $\frac{1}{5}(13\mathbf{i}+10\mathbf{j}-7\mathbf{k})$
3) $\frac{4}{\sqrt{259}}$
4) (4, 4, 2), $\mathbf{r}.(\mathbf{i}+4\mathbf{j}-\mathbf{k})=18$
5) (0, 0, 0), (2, 0, 0), (0, −4, 0),
(0, 0, 4), $\frac{16}{3}$
6) $\frac{16}{3}$, $\sqrt{50}/6$, $200\pi/27$
7) 4, 16π
8) $\mathbf{r}=2\mathbf{i}+\mathbf{k}+\lambda(\mathbf{i}-2\mathbf{j}+2\mathbf{k})$,
$(\frac{11}{3},-\frac{10}{3},\frac{13}{3})$, $(-\frac{4}{3},-\frac{25}{3},-\frac{17}{3})$, $\frac{625}{3}\pi$
9) $\frac{1}{\sqrt{2}}(\mathbf{j}+\mathbf{k})$, $\mathbf{r}=(3\mathbf{i}+2\mathbf{j})+\lambda(\mathbf{j}+\mathbf{k})$
$(3, 2+3\sqrt{2}, 3\sqrt{2})$
10) $\mathbf{r}=\mathbf{i}+\mathbf{j}+\lambda(3\mathbf{i}-\mathbf{j}+5\mathbf{k})+\mu(\mathbf{i}-2\mathbf{j})$
11) $\mathbf{r}=\lambda(2\mathbf{i}+\mathbf{j}-5\mathbf{k})$,
$\mathbf{r}=\mathbf{i}+2\mathbf{j}-\mathbf{k}+\mu(2\mathbf{i}+\mathbf{j}-5\mathbf{k})$
$\mathbf{r}.(2\mathbf{i}+\mathbf{j}-5\mathbf{k})=0$, $\sqrt{30}/5$

Miscellaneous Exercise 5

9) a) arccos $(\sqrt{2}/2)$ b) arccos $(\sqrt{6}/6)$
10) $\mathbf{r}.(2\mathbf{i}+\mathbf{k})=0$, $\sqrt{5}$
12) a) $\frac{x}{4}=\frac{y-1}{-6}=\frac{z+1}{4}$
b) $2x-3y+2z+5=0$, (1, 3, 1)
13) 12
14) (3, 4, 1), $3x-2y-2z+1=0$
18) $\frac{1}{6}(3\mathbf{a}+2\mathbf{b}+\mathbf{c})$, $\frac{1}{6}(\mathbf{a}+3\mathbf{b}+2\mathbf{c})$,
$\frac{1}{6}(2\mathbf{a}+\mathbf{b}+3\mathbf{c})$
19) $\frac{1}{5}\mathbf{c}$, −3
21) i) $(15\mathbf{a}+\mathbf{b})/\sqrt{226}$ ii) $(15\mathbf{a}+\mathbf{b})/11$

22) a − c
23) i) $\lambda(\mathbf{a}+\mathbf{b})$ ii) $\mathbf{b}+\mu(\frac{1}{2}\mathbf{a}-\mathbf{b})$

Exercise 6a

1) a) $3\sqrt{13}/4\ \text{ms}^{-1}$ at arctan $\sqrt{3}/6$ to the wall
b) $3\ \text{ms}^{-1}$ at 30° to the wall
c) $3\sqrt{3}/2\ \text{ms}^{-1}$ along the wall.
2) $\frac{1}{3}$
3) a) $-2\mathbf{i}+2\mathbf{j}, \mathbf{j}$ b) $-3\mathbf{i}+2\mathbf{j}, \mathbf{i}+\mathbf{j}$
c) $-\mathbf{i}+2\mathbf{j}, -\mathbf{i}+\mathbf{j}$
4) $\frac{1}{2}, \mathbf{i}+3\mathbf{j}$
5) 1, 45° to AB
6) Along AC, arccos $\frac{1}{8}$ to AC
7) $4\mathbf{j}, 2\mathbf{i}+\mathbf{j}, 5:\sqrt{13}, 5$
9) $\frac{1}{6}, v=4u, u\sqrt{77}/14$
10) $(u\sqrt{58})/5,\ (u\sqrt{61})/5$
11) $e=\frac{1}{5}$, $\tan\alpha=\sqrt{5}/5$
13) 15×10^{-5} Nm at arctan $\frac{4}{3}$ below Ox, $\frac{1}{10}(33\mathbf{i}-7\mathbf{j}), \frac{1}{20}(31\mathbf{i}+11\mathbf{j}), \frac{9}{40}$
15) $\mathbf{v}_1=\frac{9}{5}\mathbf{n}+\frac{12}{5}\mathbf{t}, \mathbf{v}_2=\frac{7}{5}\mathbf{n}+\frac{1}{5}\mathbf{t}, \frac{3}{2}\mathbf{n}+\frac{12}{5}\mathbf{t}, \frac{17}{10}\mathbf{n}+\frac{1}{5}\mathbf{t}, \frac{1}{50}(141\mathbf{i}-12\mathbf{j}), \frac{1}{50}(59\mathbf{i}+62\mathbf{j})$
17) $9u^2/320g$

Exercise 6b

1) a) 1.79, 43.9 m b) 6.69 s, 164 m
2) a) 667 m b) 2000 m
3) OA = 56.2 m, $2\frac{2}{3}$ s
4) a) 11.25 m b) 54.5 m
5) $\mathbf{r}=(50t+5)\mathbf{i}+(30t-\frac{1}{2}gt^2)\mathbf{j}+(50t-3)\mathbf{k}$,
$\mathbf{v}=50\mathbf{i}+(30-gt)\mathbf{j}+50\mathbf{k}, 60/g$,
$\mathbf{r}\cdot(\mathbf{i}-\mathbf{k})=8$
6) 2.29 s, 44.3 m
7) a) 50 m b) 10 m c) 30 m
8) a) up and along the plane
b) down the plane at arctan $\frac{37}{9}$ to the horizontal
9) arcsin $\frac{1}{3}$

Multiple Choice Exercise 6

1) d	2) d	3) a	4) c
5) d	6) b, d	7) a, b, c, d	8) a, c
9) b	10) *D*	11) *E*	12) *C*
13) c	14) *I*	15) *A*	16) *I*
17) *A*	18) *T*	19) *T*	20) *F*

Exercise 7a

1) $m\ddot{x}+kx+R=0$
2) $m\ddot{x}+k\dot{x}+mg=0$
3) $|x|>a,\ a\ddot{x}+ka\dot{x}+gx+ga=0$;
$|x|<a,\ \ddot{x}+k\dot{x}=0$
4) $M\ddot{x}=\dfrac{H}{\dot{x}}-Mk\dot{x}^2$
5) $(M-mt)\ddot{x}=mu-(M-mt)g$
9) $1-\frac{4}{9}\ln\frac{13}{4}$
10) $\ddot{x}=5(1-\dot{x}^2),\ \dfrac{e^2-1}{e^2+1}$
11) $100\dfrac{dv}{dt}=\dfrac{2500}{v}-v$, 247 m.
12) $g+u\ln\frac{3}{2},\ \dfrac{g}{2}-u[1+\ln\frac{4}{9}]$

Exercise 7b

1) a) yes, $4\pi/\sqrt{7}$, b) no, c) yes, $\pi\sqrt{\dfrac{2l}{g}}$
2) $\dfrac{3\sqrt{2}}{2}e^{-2}\sin(2\sqrt{2})$
3) $e^{-2\pi/3}$
4) $\ddot{x}+\dot{x}\sqrt{\dfrac{g}{2a}}+\dfrac{2g}{a}x=0$, yes
5) $\dfrac{3a}{4}e^{-\pi/\sqrt{15}}$

Miscellaneous Exercise 7

2) $x=4e^{-3t/2}(\cos t+3\sin t)$
3) $x=\dfrac{1}{2}ft^2+\dfrac{f}{n^2}\cos nt-\dfrac{f}{n^2}$
4) $x=\dfrac{V}{2n}e^{-nt}\sin 2nt$
7) $\mu=43.2,\ m_1=1353,\ a_{max}=79.9,\ a_{min}=10.8$
9) $\omega/n(1-e^{-k\pi/n})$

Exercise 8a

1) $\frac{3}{2}\mathbf{i}+\frac{3}{4}\mathbf{j}+\frac{2}{3}\mathbf{k}$
2) $\frac{3}{2}\mathbf{a}+\frac{5}{6}\mathbf{b}$
3) $m=2;\ p=\frac{3}{2};\ q=\frac{11}{2}$
4) (2, −1, −1)
5) $\dfrac{\mathbf{a}}{2}+\dfrac{7\mathbf{c}}{10}$
6) $a=1; b=-8; c=1; (\frac{7}{3}, 1, -3)$

Exercise 8b

1) (i) $\frac{3}{5}\mathbf{i}+\mathbf{j}+\frac{7}{10}\mathbf{k}$ (ii) $(2\mathbf{i}+\mathbf{j}-\mathbf{k})/m$
2) (i) $\frac{2}{3}\mathbf{i}+\frac{4}{3}\mathbf{j}-\frac{5}{3}\mathbf{k}$ (ii) $\frac{7}{3}\mathbf{i}-\frac{2}{3}\mathbf{j}+\frac{7}{3}\mathbf{k}$
(iii) $14\mathbf{i}-4\mathbf{j}+14\mathbf{k}$
3) $F/2m$ at arctan $\frac{4}{3}$ to AB
(a) increases it in ratio 5 : 3
(b) none
4) $1\ \text{ms}^{-1}$ at 60° to PQ; 10 Ns; 50 J; not at all.

5) $-\frac{1}{2}\mathbf{i} + 3\mathbf{j}$; 324 J

6) $(\frac{8}{3}, 2)$; $-7\mathbf{i} - 5\mathbf{j}$ where i and j are in the directions $\overrightarrow{OA}$ and $\overrightarrow{OC}$ respectively.

7) $3\mathbf{i} + 5\mathbf{j}$; $\frac{8}{5}\mathbf{i} + \frac{17}{5}\mathbf{j}$, $994\frac{1}{8}$ J

Exercise 8c

2) $\pi/3$ or 0

3) AC vertical, unstable; $A\hat{C}B = \frac{\pi}{3}$, stable;

A at C, unstable (if ring stops at pulley).

5) AB horizontal; No

6) (i) Symmetrical, unstable; BC at

$\arccos \frac{a}{2b}$ to the horizontal, stable.

(ii) Symmetrical only, stable.

7) PQ vertically downward, unstable;
PQ vertically upward, unstable;

PQ at $\frac{\pi}{6}$ to the horizontal (Q above P),

stable.

Miscellaneous Exercise 8

1) $\frac{5}{3}\mathbf{i} + \mathbf{j}$; $-7\mathbf{i} - 4\mathbf{j}$

2) $\frac{a}{8}(2\mathbf{i} + \mathbf{j})$; $3\mathbf{i} + 4\mathbf{j}$

3) $\frac{m_1(v_2 - v_1)}{m_1 + m_2}$

4) $\frac{11}{7}\mathbf{i} + \mathbf{j}$; $\frac{5\sqrt{3}}{7}$;
$[(3 + 4\sqrt{2})\mathbf{i} + (4 - 3\sqrt{2})\mathbf{j}](15/\sqrt{154})$

5) a, $2a$; $\sqrt{\frac{ga}{3}}$, $2\sqrt{\frac{ga}{3}}$

6) $3(\mathbf{i} + \mathbf{j} + \mathbf{k})$; $6(\mathbf{i} + 2\mathbf{j} + 2\mathbf{k})$; 54 units

7) $\mathbf{i} + 3\mathbf{j}$; $5\sqrt{21}$, $3\mathbf{j}$; $5\sqrt{5}$

8) $128\mathbf{i} + 16\mathbf{j} - 112\mathbf{k}$; $3264m$ joules

9) (i) $\frac{1}{10}(-8\mathbf{i} - 30\mathbf{j} + 39\mathbf{k})$ (ii) 15, 9, 25

10) $Wa[\cos\theta + \frac{1}{4}(1 + 8\sin^2\theta)^{1/2}]$
+ any constant; unstable

11) (i) stable (ii) unstable (iii) unstable

12) OA vertically upward, stable;
OA vertically downward, unstable

13) stable

14) $0 < k < 2$; $k = \sqrt{2}$

15) (i) $\theta = 0°$, stable; $\theta = 180°$, unstable
(ii) $\theta = 0°$, unstable; $\theta = 60°$, stable;
$\theta = 180°$, unstable
(iii) $\theta = 0°$, unstable; $\theta = 180°$, stable

16) $2Wa \sin\theta + \frac{Wa\sqrt{2}}{2}(-1 + 6\cos\theta - 2\cos^2\theta)^{1/2}$ relative to 0.

17) $\theta = 0$, unstable; $\theta = \pm\frac{\pi}{3}$, stable

Exercise 9a

1) ML^2T^{-2}, MLT^{-1}, ML^2T^{-2}, T^{-1}, T^{-2}, L^2, L^3, MLT^{-2}, MLT^{-1}, ML^2T^{-3}, MLT^{-2}

2) $T = kml\omega^2$

5), 6), 7), 8), 10), 11), 12) are wrong.

Exercise 10a

1) 400 J

2) 32 kg m^2

3) 2.59×10^{29} J

4) 18 kg m^2

5) 128π Nm

6) 80π Nm

7) $g/5r$

8) $\frac{1}{8}$ rad s^{-2}; $\frac{25}{16}$ rad

9) $\frac{(m_1 \sim m_2)gI}{I + (m_1 + m_2)r^2}$; $\left[\frac{I(m_1 + m_2) + 4m_1m_2r^2}{I + (m_1 + m_2)r^2}\right]g$

10) $g/6r$

Miscellaneous Exercise 10

1) 40; 3 minutes

2) $\sqrt{8\pi r/g}$

3) a) $\frac{\pi(n_1 - n_2)}{450}$ rad s^{-2} b) $\frac{(3n_2 - n_1)^2}{8(n_1 - n_2)}$

4) $\frac{(M - m)g}{(M + m + n)a}$

5) $8\pi/\omega$; $16\pi/\omega$; 8

6) $3\pi L/75$

7) $\frac{Wr^2(3g - 14l)}{2lg}$; $\frac{2Wr(6l - g)}{g}$

8) $\frac{2(m_1 - m_2)g}{M + 2m_1 + 2m_2}$; $m = \frac{M}{4}$

9) (i) $gt^2/8$ (ii) $2mga$

10) a) $(\omega_2 - \omega_1)I/N$ b) $(\omega_2^2 - \omega_1^2)I/2N$;
$10\pi/9$; $\pi/18$; $45/8$

11) $\frac{g}{2}(1 - \mu\cos\alpha - \sin\alpha)$; $2Ma^2$

12) $M = 2m$; $G = \frac{1}{2}mga$; $7g/10$

14) $Mg/3$

Exercise 11a

1) $144a^2$ b) $20a^2$ c) $128a^2/5$ d) $16a^2$

2) $Ma^2/2$

3) $Ma^2/2$

4) $2Ma^2/5$, $2M'a^2/5$

5) $4Mb^2/3$

6) $Ml^2/6$
7) a) $4Ml^2/3$, b) $\frac{4}{3}Ml^2 \sin^2\theta$
8) $Ml^2/2$
9) $M = 8\pi a^3 m$, $8Ma^2/3$
10) $5Ma^2/2$

Exercise 11b

1) $5Ml^2/6$
2) $3Ma^2$
3) $\frac{Ma^2}{10}\left(\frac{15l + 8a}{3l + 2a}\right)$
4) $5Ml^2/6$
5) $\frac{1}{3}\left(\frac{2 + 3\pi}{2 + \pi}\right)Ma^2$
6) a) $2Ml^2$ b) $\frac{1}{3}Ml^2$
7) $\frac{4}{9}Ma^2$

Exercise 11c

1) $4Ma^2/3$
2) $3Mr^2/2$
3) $3Ma^2/2$
4) $35Ma^2/4$
5) $\frac{1}{3}M(a^2 + 4b^2)$
6) $Ml^2/6$
7) $3Mr^2/2$
8) $M\left(\frac{r^2}{2} + \frac{h^2}{3}\right)$
9) $6Ml^2$
10) a) $3Ml^2/2$ b) $19Ml^2/2$

Multiple Choice Exercise 11

1) c 2) d 3) a 4) c 5) e 6) a, b
7) a 8) a, c 9) c 10) b 11) A
12) A 13) T 14) F 15) T

Miscellaneous Exercise 11

1) $5Ml^2/2$
2) $Ml^2/3$
3) $\frac{M(a^4 + b^4)}{2(a^2 + b^2)}$
5) $Ma^2/4$
6) $ml^5/5$
7) $6ma^2$
8) i) $\sqrt{(3a^2/10)}$ ii) $\sqrt{[(3a^2 + 2h^2)/20]}$
10) $17Ma^2/30$
11) $h\sqrt{6}$, $23Ma^2/6$
12) i) $\frac{(R + r)(R^2 + r^2)}{R^2 + rR + r^2}$, ii) $[\frac{7}{5}(R^2 + r^2)]^{1/2}$

Exercise 12a

1) $\sqrt{3g/2l}$
2) $2a$
3) $k > 3$
4) $4\sqrt{g/5r}$
5) $\left(\frac{12g\sqrt{2}}{5l}\right)^{1/2}$
7) $\sqrt{\frac{(2\sqrt{3} - 1)g}{3l}}$ if C is initially below AB;
$\sqrt{\frac{(10\sqrt{3} - 13)g}{6l}}$ if C is initially above AB

Exercise 12b

1) $4\sqrt{2g/a}$; $2mga/\pi$
2) $3\sqrt{3}\, mgr/2\pi$
3) $14mgl/3\pi$
5) $\frac{4\sqrt{2\pi^3}}{a}$

Exercise 12c

1) $\sqrt{12g \sin\theta/7l}$; $6g\cos\theta/7l$
2) $4\pi\sqrt{a/3g}$
3) $4\pi\sqrt{7l/15g}$; $28l/15$
4) $2\pi\sqrt{\frac{3a^2 + 4ah + 4h^2}{2g(a + 2h)}}$; $h = \frac{a}{2}(\sqrt{2} - 1)$
7) $l\sqrt{\frac{11}{6}}$

Exercise 12d

1) $\frac{1}{3}mg \sin\theta$
2) $\frac{5}{2}mg\cos\theta$; $\frac{5}{2}mg$
3) $\frac{1}{4}mg\sin\theta$; $\frac{1}{2}mg(5\cos\theta - 3)$;
$\frac{1}{8}mg\sqrt{487}$ at $\arctan\sqrt{3}/22$ to CA.
4) $mg\sqrt{1954}$

Multiple Choice Exercise 12

Type II

1) b, d 2) a, b, c 3) a, d 4) b

Type III

5) C 6) B 7) A 8) C 9) A

Type IV

10) A 11) d 12) I

Type V

13) F 14) F 15) T 16) F 17) T

Miscellaneous Exercise 12

1) $\left(\frac{4mgr}{\pi n}\right)\left(\frac{Mk^2}{Mk^2 + mr^2}\right)$
2) $\sqrt{\frac{2g}{3a}}$; $4mga$
4) $6g\sin\theta/7a$
5) $2 < k < 3 + 2\sqrt{2}$
6) $I\omega_0/C$; $I\omega_0^2/2C$; $(I\omega_0/C)\ln 2$
7) $4mga/5\pi^2$; $\sqrt{112g/15\pi a}$
8) $\sqrt{2\pi na/g}$; $\frac{1}{4}\omega$
10) $2\pi\left(\frac{a^2 + 3h^2}{3gh}\right)^{1/2}$; $\frac{a}{\sqrt{3}}$; $\frac{5mg}{2}$
11) $2\pi\left\{\frac{(M + 2m)d}{4mg}\right\}^{1/2}$
12) $2\pi\sqrt{\frac{2a^2 + 3x^2}{6gx}}$; $a\sqrt{\frac{2}{3}}$

15) $2\pi\sqrt{\dfrac{9k^2+l^2}{3gl}}$; $\frac{1}{3}l\sqrt{2}$
16) $146ma^2/15$
17) $2(2x^2-5ax+4a^2)/(5a-4x)$
19) $2\pi\left(\dfrac{a^2+4b^2}{3gb}\right)^{1/2}$
20) $\frac{5}{2}mg\cos\theta$; $\frac{5}{2}mg$, $\frac{1}{4}mg$; arctan 10
21) $2\pi\left(\dfrac{2a}{3g}\right)^{1/2}$; $2\pi\left\{\dfrac{2(a^2+b^2)^{1/2}}{3g}\right\}^{1/2}$
22) $\sqrt{g/a}$
23) (i) $8ma^2$ (ii) $\frac{3}{2}ma^2$; $\dfrac{\pi}{2}\sqrt{\dfrac{57a}{g}}$
24) $2\pi\sqrt{\dfrac{2k}{g}}$; $2\pi\left(\dfrac{\sqrt{2(a^2+b^2)}}{g}\right)^{1/2}$
25) $3a$
26) $x+\dfrac{a^2+b^2}{4x}$; (i) $\frac{29}{16}a \leqslant l \leqslant \frac{49}{24}a$
(ii) $a\sqrt{10} \leqslant l \leqslant \frac{23}{9}a$
27) $\dfrac{3g\cos\theta}{7a}$; $\frac{4}{7}mg\cos\theta$

Exercise 13a

1) $32g/a$
2) 12.5 Nm
3) a) 36 b) $12\sqrt{2/k}$
4) $\frac{4}{3}ml\Omega$
5) $40\sqrt{5/3}$ Ns
6) a) $6m\sqrt{gl/\sqrt{3}}$ b) $6m(\sqrt{2}+1)\sqrt{gl\sqrt{3}/6}$

Exercise 13b

1) a) $2u/3$ b) $2u/\sqrt{3}$
2) $2\sqrt{g/5a}$
3) $3J/7ma$; $J > 2m\sqrt{7ga/3}$
4) $\omega/31$
5) $4\sqrt{g/5a}$; $\dfrac{4}{9a}(u+\sqrt{5ga})$

Exercise 13c

1) (i) $Ma^2\omega^2$ (ii) $\dfrac{25Mv^2}{6}$
2) $\frac{20}{3}$ rad s^{-2}; $\mu \geqslant \dfrac{5}{6g}$
3) (i) $\dfrac{5g\sin\alpha}{7a}$; $\mu \geqslant \frac{2}{7}\tan\alpha$
(ii) $\dfrac{g\sin\alpha}{2a}$; $\mu \geqslant \frac{1}{2}\tan\alpha$
(iii) $\dfrac{3g\sin\alpha}{5a}$; $\mu \geqslant \frac{2}{5}\tan\alpha$
4) (i) $25g/14$ (ii) $\sqrt{5g/7}$
6) $6\sqrt{2ag}$

Multiple Choice Exercise 13

Type II

1) a, c, d 2) d 3) a, d 4) b, d

Type III

5) A 6) C 7) B

Type IV

8) A 9) A 10) b, d 11) I

Type V

12) F 13) T 14) F 15) T

Miscellaneous Exercise 13

1) $18\pi ma^2/t^2$; $18\pi ma^2/t$
2) $4ma^2\omega^2$; $8ma^2\omega$; $a\omega/g$; $a\omega^2/2g$; $16\omega/17$
3) $\frac{6}{5}Mga$; $\frac{4}{5}M\sqrt{2gh}$
4) $2a/\sqrt{3}$; $[4ga(\sqrt{3}+2)/\sqrt{3}]^{\frac{1}{2}}$
5) $2M\omega/(2M+5m)$; arcsin $\left[\dfrac{2M^2a\omega^2}{5mg(2M+5m)}-1\right]$
6) $\dfrac{k^2+a^2}{a}$; $\sqrt{\dfrac{2ga}{k^2+a^2}}$; $\dfrac{\sqrt{2ga(k^2+a^2)}}{k^2+2a^2}$
7) $v^2 > 4ag(M+2m)(M+3m)/3m^2$
9) $\dfrac{m}{3}\sqrt{\dfrac{461\times 14ga}{15}}$
10) $\dfrac{m}{3}\sqrt{305ga}$
11) (i) $2m\sqrt{ga/3}$
(iii) Components $\frac{3}{2}mg$, $\frac{7}{4}mg$
14) $\dfrac{8\pi}{3}\sqrt{\dfrac{a}{g}}$
15) $2M(a^2+40l^2)$; $\dfrac{3M}{2}\left[22g\,\dfrac{(a^2+40l^2)}{l}\right]^{1/2}$
16) $\dfrac{49}{3}ma^2$
18) $18g/23$; $10Mg/23$; $2/23$
19) $68ma^2$; $P = m\sqrt{323ag}$
20) (i) $\dfrac{mm_1m_2g}{2m_1m_2+m(m_1+m_2)}$
21) $V(M+m+I/a^2)$
22) $\frac{2}{3}g\sin\alpha$; $\frac{1}{3}\tan\alpha$
23) $P(a-2x)/3a$; $a/2$
25) (i) Px/MK^2
26) $(200g\sin\alpha/19a)^{1/2}$
27) $\frac{1}{2}\tan\alpha$
28) $\sqrt{g/2a}$; $10\sqrt{2a/g}$
29) $2\sqrt{3}/21$
30) $5\mu g/21$
31) 2π rad s^{-2}; $\dfrac{ma}{2\pi}(g\sin\alpha-2\pi a)$;

$\tan\alpha - \dfrac{2\pi a}{g\cos\alpha}$

32) $3xI/(M+m)l^2$; $-2xI/(M+m)al$

Throughout Chapter 14, x is measured from the left hand end of the beam.
In answers to questions where diagrams are required, the bending moment and shearing force functions for each section are given.

Exercise 14a

1) $0 < x < 2a$ $S = -W$; $\mathfrak{M} = Wx$
$2a < x < 3a$ $S = 2W$; $\mathfrak{M} = 2W(3a - x)$

2) $0 < x < a$ $S = -2W$;
$\mathfrak{M} = W(2x - 3a)$
$a < x < 2a$ $S = -W$;
$\mathfrak{M} = W(x - 2a)$

3) $0 < x < 3a$ $S = W\left(\dfrac{x}{2a} - \dfrac{1}{3}\right)$;
$\mathfrak{M} = Wx\left(\dfrac{1}{3} - \dfrac{x}{4a}\right)$
$3a < x < 4a$ $S = W\left(\dfrac{x}{2a} - 3\right)$;
$\mathfrak{M} = W\left(3x - 8a - \dfrac{x^2}{4a}\right)$

4) $0 < x < a$ $S = -W$; $\mathfrak{M} = Wx$
$a < x < 3a$ $S = W\left(\dfrac{x}{a} - 2\right)$;
$\mathfrak{M} = W\left(2x - \dfrac{a}{2} - \dfrac{x^2}{2a}\right)$
$3a < x < 4a$ $S = W$; $\mathfrak{M} = W(4a - x)$

5) $0 < x < 1$ $S = -2000$ N;
$\mathfrak{M} = (2000x - 2500)$ Nm
$1 < x < 2$ $S = 1000(x -)$ N;
$\mathfrak{M} = 500(4x - x^2 - 4)$ Nm

6) $0 < x < a$ $S = W$; $\mathfrak{M} = -Wx$
$a < x < 3a$ $S = W\left(\dfrac{x}{a} - \dfrac{5}{2}\right)$;
$\mathfrak{M} = \dfrac{W}{2}\left(5x - 6a - \dfrac{x^2}{a}\right)$

7) $0 < x < 2a$ $S = \dfrac{W}{6}\left(\dfrac{3x}{a} - 13\right)$;
$\mathfrak{M} = Wx\left(\dfrac{13}{6} - \dfrac{x}{4a}\right)$
$2a < x < 4a$ $S = -\dfrac{W}{6}$;
$\mathfrak{M} = \dfrac{W}{6}(x + 18a)$
$4a < x < 6a$ $S = \dfrac{11W}{6}$;
$\mathfrak{M} = \dfrac{11W}{6}(6a - x)$

8) $0 < x < 0.5$ $S = -3000$ N;
$\mathfrak{M} = (3000x - 2800)$ Nm
$0.5 < x < 1.0$ $S = -1800$ N;
$\mathfrak{M} = (1800x - 2200)$ Nm
$1.0 < x < 1.5$ $S = -800$ N;
$\mathfrak{M} = (800x - 1200)$ Nm

Miscellaneous Exercise 14

1) $0 < x < 1$ $S = 160x - 600$;
$\mathfrak{M} = 600x - 80x^2$
$1 < x < 4$ $S = 160x - 200$;
$\mathfrak{M} = 400 + 200x - 80x^2$
$4 < x < 5$ $S = 160x - 800$;
$\mathfrak{M} = 800x - 2000 - 80x^2$
525 Nm; 2080 N; 2160 Nm

2) (i) $W/2$, $5W/2$;
$0 < x < d$ $S = -\dfrac{W}{2}$; $\mathfrak{M} = \dfrac{Wx}{2}$
$d < x < 2d$ $S = \dfrac{W}{2d}(2x - d)$;
$\mathfrak{M} = \dfrac{W}{2}\left(d + x - \dfrac{x^2}{d}\right)$
$2d < x < 3d$ $S = \dfrac{W}{d}(3d - x)$;
$\mathfrak{M} = -\dfrac{W}{2d}(3d - x)^2$
(ii) $\dfrac{2W}{d}$; $\dfrac{9W}{2}$, $\dfrac{W}{2}$

3) (i) $0 < x < 1$ $S = -\frac{40}{3}$; $\mathfrak{M} = \dfrac{40x}{3}$
$1 < x < 3$ $S = \frac{20}{3}$; $\mathfrak{M} = 20 - \dfrac{20x}{3}$
$3 < x < 4$ $S = 0$; $\mathfrak{M} = 0$
$\frac{20}{3}$ N, $\frac{40}{3}$ N; $\frac{40}{3}$ Nm
(ii) $0 < x < 1$ $S = 10x - \frac{80}{3}$;
$\mathfrak{M} = \dfrac{80x}{3} - 5x^2$
$1 < x < 3$ $S = 10x - \frac{20}{3}$;
$\mathfrak{M} = 20 + \dfrac{20x}{3} - 5x^2$
$3 < x < 4$ $S = 10x - 40$;
$\mathfrak{M} = 40x - 5x^2 - 80$
$\frac{100}{3}$ N, $\frac{80}{3}$ N; $\frac{65}{3}$ Nm

4) $0 < x < l$ $S = -2W$; $\mathfrak{M} = 2Wx$
$l < x < 3l$ $S = \dfrac{4W}{3l}(x - l)$;
$\mathfrak{M} = 2Wl - \dfrac{2W}{3l}(x - l)^2$

$3l < x < 4l \quad S = \frac{4W}{3l}(x-l) - 4W;$

$\mathfrak{M} = 4Wx - 10Wl - \frac{2W}{3l}(x-l)^2$

$2Wl$ at C.

$0 < x < l \quad S = 0; \ \mathfrak{M} = 0$

$l < x < 3l \quad S = 2W + \frac{4W}{3l}(x-l);$

$\mathfrak{M} = \frac{2W}{3l}(l-x)(2l+x)$

$3l < x < 4l \quad S = \frac{4W}{3l}(x-l) - 10W;$

$\mathfrak{M} = \frac{2W}{3l}(x-4l)(13l-x)$

$6W$

5) (i) $0 < x < \frac{l}{6} \quad S = -\frac{2W}{3}; \ \mathfrak{M} = \frac{2Wx}{3}$

$\frac{l}{6} < x < \frac{7l}{6} \quad S = W\left(\frac{x}{l} - \frac{5}{6}\right);$

$\mathfrak{M} = \frac{2Wx}{3} - \frac{W}{2l}\left(x - \frac{l}{6}\right)^2$

$\frac{7l}{6} < x < 2l \quad S = \frac{W}{3}; \ \mathfrak{M} = \frac{W}{3}(2l-x)$

$\frac{Wl}{3}$ when $x = \frac{5l}{6}$

(ii) $0 < x < \frac{l}{2} \quad S = -W; \ \mathfrak{M} = Wx$

$\frac{l}{2} < x < l \quad S = W\left(\frac{x}{l} - \frac{3}{2}\right);$

$\mathfrak{M} = Wx - \frac{W}{2l}\left(x - \frac{l}{2}\right)^2$

$l < x < \frac{3l}{2} \quad S = W\left(\frac{1}{2} - \frac{x}{l}\right);$

$\mathfrak{M} = Wl - \frac{W}{2l}\left(x - \frac{l}{2}\right)^2$

$\frac{3l}{2} < x < 2l \quad S = W; \ \mathfrak{M} = W(2l-x)$

$7Wl/8$ at M

6) $0 < x < a \quad S = \frac{Wx}{a}; \ \mathfrak{M} = -\frac{Wx^2}{2a}$

$a < x < 2a \quad S = \frac{Wx}{a} - 4W;$

$\mathfrak{M} = 4W(x-a) - \frac{Wx^2}{2a}$

$2a < x < 4a \quad S = \frac{Wx}{a} - 2W;$

$\mathfrak{M} = 2Wx - \frac{Wx^2}{2a}$

The magnitude of the bending moment is least when zero at $x = 0, 1.17a, 4a$.
The magnitude of the bending moment is greatest when $2Wa$ at the midpoint.

7) $0 < x < 10 \quad S = 10x - 8;$

$\mathfrak{M} = 8x - 5x^2$

$10 < x < 16 \quad S = 10x - 200;$

$\mathfrak{M} = 200x - 1920 - 5x^2$

Greatest bending moment is 420 Nm at B

8) $0 \leqslant x < a \quad \mathfrak{M} = -\frac{Wx^2}{2}$

$a < x < \frac{5a}{2} \quad S = W(x-4a);$

$\mathfrak{M} = 4Wa(x-a) - \frac{Wx^2}{2}$

$\frac{5a}{2} < x < 4a \quad S = W(x-a)$

$4a < x \leqslant 5a \quad S = W(x-5a);$

$\mathfrak{M} = -\frac{W}{2}(x-5a)^2$

9) $0 < x < 1 \quad S = 400x \quad \mathfrak{M} = -200x^2$

$1 < x < 5 \quad S = 400(x-3)$

$\mathfrak{M} = 1200(x-1) - 200x^2$

$5 < x < 6 \quad S = 400(x-6)$

$\mathfrak{M} = 2400(x-3) - 200x^2$

Distant $3(\sqrt{2}-1)$ m from each end

10) $0 < x < 1 \quad S = 500x; \ \mathfrak{M} = -250x^2$

$1 < x < 5 \quad S = 500x - \frac{5800}{3};$

$\mathfrak{M} = \frac{5800}{3}(x-1) - 250x^2$

$5 < x < 7 \quad S = 500x - \frac{3400}{3};$

$\mathfrak{M} = 6200 + \frac{3400x}{3} - 250x^2$

$7 < x < 10 \quad S = 500x - 5000;$

$\mathfrak{M} = 5000x - 250x^2 - 2500$

$\mathfrak{M} = 0$ at $x = 0, 1.18, 5.03, 10$

11) At C, $\mathfrak{M} = 5 \times 10^3$ N

The following quantities are measured in kN or kNm

$2 < x < 5 \quad S = 50; \quad \mathfrak{M} = -50x$

$5 < x < 10 \quad S = -51;$

$\mathfrak{M} = 51x - 505$

$10 < x < 15 \quad S = 19;$

$\mathfrak{M} = -19x + 195$

$15 < x < 18 \quad S = -30;$

$\mathfrak{M} = 30x - 540$

Extra loading 10 kN per m

12) $0 < x < 3 \quad S = 250x; \quad \mathfrak{M} = -125x^2$

$3 < x < 9 \quad S = 250x - 1500;$

$\mathfrak{M} = 1500(x-3) - 125x^2$

$9 < x < 12 \quad S = 250x - 3000;$

$\mathfrak{M} = 3000x - 18000 - 125x^2$

$0 < x < 3 \quad S = 250x; \ \mathfrak{M} = -125x^2$
$3 < x < 8 \quad S = 250(x-8);$
$\mathfrak{M} = -125(x^2 - 16x + 48)$
$8 < x < 9 \quad S = 250(x+4);$
$\mathfrak{M} = -125(x^2 + 8x - 144)$
$9 < x < 12 \quad S = 250(x-12);$
$\mathfrak{M} = -125(x^2 - 24x + 144)$
750 N; 1125 Nm

13) (i) $0 < x < 3 \quad S = 180x; \ \mathfrak{M} = -90x^2$
$3 < x < 6 \quad S = -335;$
$\mathfrak{M} = 335x - 1815$
$6 < x < 9 \quad S = 65;$
$\mathfrak{M} = 585 - 65x$
(ii) $0 < x < 3 \quad S = 180x;$
$\mathfrak{M} = -90x^2$
$3 < x < 4.5 \quad S = -335;$
$\mathfrak{M} = 335x - 1815$
$4.5 < x < 7.5 \quad S = \frac{400x}{3} - 935;$
$\mathfrak{M} = -\frac{200x^2}{3} + 935x - 3165$
$7.5 < x < 9 \quad S = 65;$
$\mathfrak{M} = 585 - 65x$

14) Bridge only. $S = 10^4(3x - 300)$
$\mathfrak{M} = 10^4\left(-\frac{3x^2}{2} + 300x\right)$

With train.
$0 < x < 50 \quad S = 10^4(9x - 562.5)$
$\mathfrak{M} = \frac{10^4 x}{2}(1125 - 9x)$
$50 < x < 200$
$S = 10^4(3x - 262.5)$
$\mathfrak{M} = 10^4\left(\frac{3x^2}{2} + 262.5x + 7500\right)$
(a) 43.75 m from end carrying train
(b) 13:4

INDEX